D1751437

*Edited by*
*Felix Tretter,*
*Peter J. Gebicke-Haerter,*
*Eduardo R. Mendoza, and*
*Georg Winterer*

**Systems Biology in Psychiatric Research**

## Related Titles

Soreq, H., Friedman, A., Kaufer, D. (eds.)

**Stress - From Molecules to Behavior**

A Comprehensive Analysis of the Neurobiology of Stress Responses

2010

ISBN: 978-3-527-32374-6

Klipp, E., Liebermeister, W., Wierling, C., Kowald, A., Lehrach, H., Herwig, R.

**Systems Biology**

A Textbook

2009

ISBN: 978-3-527-31874-2

Klipp, E., Herwig, R., Kowald, A., Wierling, C., Lehrach, H.

**Systems Biology in Practice**

Concepts, Implementation and Application

2005

ISBN: 978-3-527-31078-4

Nothwang, H., Pfeiffer, E (eds.)

**Proteomics of the Nervous System**

2008

ISBN: 978-3-527-31716-5

Müller, H. W. (ed.)

**Neural Degeneration and Repair**

Gene Expression Profiling, Proteomics and Systems Biology

2008

ISBN: 978-3-527-31707-3

*Edited by*
*Felix Tretter, Peter J. Gebicke-Haerter,*
*Eduardo R. Mendoza, and Georg Winterer*

# Systems Biology in Psychiatric Research

From High-Throughput Data to Mathematical Modeling

WILEY-BLACKWELL

WILEY-VCH Verlag GmbH & Co. KGaA

**The Editors**

**Prof. Dr. Felix Tretter**
Isar-Amper-Klinikum GmbH
Klinikum München Ost
Vockestr. 72
85540 Haar
Germany

**Prof. Dr. Peter J. Gebicke-Haerter**
Department for Psychopharmacology
Central Institute of Mental Health
J5
68159 Mannheim
Germany

**Dr. Eduardo R. Mendoza**
Ludwigs Maximilian University Munich
Department of Physics
Geschwister-Scholl-Platz 1
80539 Munich
Germany

**Prof. Dr. Georg Winterer**
University of Düsseldorf
Psychiatric Hospital
Bergische Landstr. 2
40629 Düsseldorf
Germany

All books published by **Wiley-VCH** are carefully produced. Nevertheless, authors, editors, and publisher do not warrant the information contained in these books, including this book, to be free of errors. Readers are advised to keep in mind that statements, data, illustrations, procedural details or other items may inadvertently be inaccurate.

**Library of Congress Card No.:** applied for

**British Library Cataloguing-in-Publication Data**
A catalogue record for this book is available from the British Library.

**Bibliographic information published by the Deutsche Nationalbibliothek**
The Deutsche Nationalbibliothek lists this publication in the Deutsche Nationalbibliografie; detailed bibliographic data are available on the Internet at http://dnb.d-nb.de.

© 2010 WILEY-VCH Verlag GmbH & Co. KGaA, Weinheim

All rights reserved (including those of translation into other languages). No part of this book may be reproduced in any form – by photoprinting, microfilm, or any other means – nor transmitted or translated into a machine language without written permission from the publishers. Registered names, trademarks, etc. used in this book, even when not specifically marked as such, are not to be considered unprotected by law.

**Typesetting**   Thomson Digital, Noida, India
**Printing and Binding**   Strauss GmbH, Mörlenbach
**Cover Design**   Adam-Design, Weinheim

Printed in the Federal Republic of Germany
Printed on acid-free paper

**ISBN:** 978-3-527-32503-0

# Foreword

Our mental activities are obviously results of very complex and dynamic systems. Systems biology has been mostly concerned with understanding cellular systems as well as disease states of human beings. Insufficient attention has been paid to linking our understanding of molecular systems and psychology. This book embarks on challenges of paving the ground for systems biology of psychiatry. While effects of molecular level perturbations on neuronal activities are actively investigated, mostly in the context of molecular neuroscience, framing such research in the context of psychiatry is a novel enterprise. At the cellular level, challenges of systems biology are not only on computing, but also on obtaining well coordinated and quantitative data so that precision of the model can be improved and verified. Quantitatively describing and measuring psychiatric observations in a proper manner is a true challenge. At the same time, introduction of systems biology into psychiatry may require a new level of understanding, as modeling efforts may force psychiatrists to describe the subject more quantitatively and with broader coverage. Dynamic models have to be calibrated using data supposed to exist. Systems biology of psychiatry entails a whole new set of problems that are even more difficult to solve. This book marks the beginning of a long journey.

Bon voyage  *Hiroaki Kitano*
Tokyo, February 2010

## Contents

Foreword  *V*
Preface  *XV*
List of Contributors  *XVII*

**Part One  Introduction**  *1*

**1  Philosophical Aspects of Neuropsychiatry**  *3*
*Felix Tretter*
1.1  Development of Research Paradigms and Strategies in Psychiatry  *4*
1.2  The Mind–Body Problem – Philosophy of Mind  *5*
1.2.1  Monism and Dualism  *8*
1.2.2  Correlation  *9*
1.2.3  Identity Theory and its Problems  *9*
1.2.4  Causation  *11*
1.2.5  Supervenience  *11*
1.3  The Conditions of Scientific Knowledge – Philosophy of Science  *11*
1.4  Experimental Research – From Observation to Theory  *12*
1.4.1  Hypotheses and Theory  *14*
1.4.2  The "Epistemic Cycle"  *14*
1.4.3  Top-Down Analysis – Reductionism?  *15*
1.4.4  Bottom-Up Explanations – Holism?  *16*
1.5  Theoretical (Neuro)psychiatry  *17*
1.6  Systems Thinking  *19*
1.7  Perspectives – Towards a "Neurophilosophy"  *19*
References  *22*

**2  Neuropsychiatry – Subject, Concepts, Methods, and Computational Models**  *27*
*Felix Tretter and Peter J. Gebicke-Haerter*
2.1  Introduction  *27*
2.2  Psychiatric Fundamentals of Neuropsychiatry  *27*
2.2.1  General Psychiatry  *27*

*Systems Biology in Psychiatric Research.*
Edited by F. Tretter, P.J. Gebicke-Haerter, E.R. Mendoza, and G. Winterer,
Copyright © 2010 WILEY-VCH Verlag GmbH & Co. KGaA, Weinheim
ISBN: 978-3-527-32503-0

| | | |
|---|---|---|
| 2.2.2 | Psychopathology | 28 |
| 2.2.2.1 | Quantitative Psychopathology | 28 |
| 2.2.2.2 | Theoretical Psychopathology | 29 |
| 2.2.3 | Psychiatric Diagnoses | 30 |
| 2.2.3.1 | Diagnostic Criteria | 30 |
| 2.2.4 | Theoretical Psychiatry | 30 |
| 2.2.4.1 | "Computational Neuropsychiatry" | 31 |
| 2.2.4.2 | "Systems Neuropsychiatry" | 32 |
| 2.3 | Neurobiological Fundamentals of Neuropsychiatry | 33 |
| 2.3.1 | Basic Findings of (Neuro)biological Psychiatry | 33 |
| 2.3.1.1 | Neuropsychopathology | 33 |
| 2.3.1.2 | Neurobiological Methods | 34 |
| 2.3.1.3 | Experimental Paradigms | 35 |
| 2.3.1.4 | Structure and Function of the Brain | 38 |
| 2.3.1.5 | Global Circuits and their Connectivities | 40 |
| 2.3.1.6 | Local Networks of Neurons | 43 |
| 2.3.1.7 | Prefrontal Network in Schizophrenia | 46 |
| 2.3.2 | Neuron | 50 |
| 2.3.2.1 | Electrical Signaling of the Neuron | 52 |
| 2.3.3 | Synapse | 54 |
| 2.3.3.1 | Receptors | 56 |
| 2.3.4 | The Cell as a System of Interacting Molecules | 63 |
| 2.3.4.1 | Intracellular Signal Cascades – From Receptor to Genome | 63 |
| 2.3.4.2 | Modeling Signal Transduction Networks Relevant in Schizophrenia | 64 |
| 2.3.4.3 | Genomics and Proteomics | 66 |
| 2.3.4.4 | Gene Regulation – Circular Signaling Pathways | 67 |
| 2.3.4.5 | Systems Biology of the Neuron | 69 |
| 2.3.5 | The Brain as a Neurochemical Oscillator | 70 |
| 2.3.5.1 | Neurochemical Interaction Matrix | 71 |
| 2.3.5.2 | "Neurochemical Mobile" | 72 |
| 2.4 | Conclusions and Perspectives | 73 |
| | References | 76 |
| | | |
| **3** | **Introduction to Systems Biology** | **81** |
| | *Marvin Schulz and Edda Klipp* | |
| 3.1 | Introduction | 81 |
| 3.1.1 | What is Systems Biology? | 81 |
| 3.1.2 | Purpose of Modeling | 81 |
| 3.1.3 | Levels of Modeling | 82 |
| 3.2 | Data Analysis | 83 |
| 3.2.1 | Types of Data | 83 |
| 3.2.1.1 | Purification | 83 |
| 3.2.1.2 | Detection | 83 |
| 3.2.1.3 | Large-Scale Analyses | 84 |
| 3.2.1.4 | Identification of Components | 84 |

| | | |
|---|---|---|
| 3.2.2 | Working with Data | 84 |
| 3.2.2.1 | Different Clustering Approaches | 85 |
| 3.2.2.2 | Principal Component Analysis | 85 |
| 3.3 | ODE Modeling | 86 |
| 3.3.1 | Differential Equations | 86 |
| 3.3.2 | Stoichiometric Matrix | 86 |
| 3.3.3 | Reaction Kinetics | 87 |
| 3.3.4 | Steady States | 88 |
| 3.3.5 | Metabolic Control Analysis | 88 |
| 3.3.6 | Simulating Models | 89 |
| 3.3.7 | Parameter Estimation | 90 |
| 3.4 | Results Gained from Systems Biology | 91 |
| 3.4.1 | Just-in-Time Transcription | 91 |
| 3.5 | Standard Formats, Databases, and Tools | 91 |
| 3.5.1 | XML-Based Formats for ODE Models | 91 |
| 3.5.2 | Databases | 92 |
| 3.5.3 | Tools for the Construction, Simulation, and Analysis of ODE Models | 93 |
| 3.6 | Future Directions in Systems Biology | 93 |
| | References | 94 |
| | | |
| 4 | **Mind Over Molecule: Systems Biology for Neuroscience and Psychiatry** | 97 |
| | *Denis Noble* | |
| 4.1 | Introduction: Mind and Molecule Meet | 97 |
| 4.2 | First Steps: Modeling Excitable Cells | 98 |
| 4.3 | Higher-Level Simulation | 101 |
| 4.4 | Genetic Programs? | 104 |
| 4.5 | Programs in the Brain? | 105 |
| 4.6 | Conclusions | 107 |
| | References | 108 |
| | | |
| Part Two | **Basics** | 111 |
| | | |
| 5 | **Neuropsychiatry, Psychopathology, and Nosology – Symptoms, Syndromes, and Endophenotypes** | 113 |
| | *Wolfram Kawohl and Paul Hoff* | |
| 5.1 | Introduction | 113 |
| 5.2 | Conceptual and Historical Introduction | 113 |
| 5.3 | Finding the "Atomic Unit" in Psychopathology: Endophenotypes | 116 |
| 5.3.1 | Susceptibility Genes | 117 |
| 5.3.2 | Requirements for Endophenotypes | 118 |
| 5.3.3 | Identified and Possible Endophenotypes | 118 |
| 5.3.4 | Endophenotypes and the Role of Psychopathology | 119 |
| 5.4 | Basic Methodological Problem: Time/Spatial Resolution | 119 |

| 5.4.1 | An Example: Libet's Experiment   *120* |
| 5.4.2 | The First Problem: The Estimation of W   *121* |
| 5.4.3 | The Second Problem: The Explanatory Power in the Light of a Questionable Time Resolution   *121* |
| 5.5 | Future New Diagnostic Schedules and Research   *122* |
| 5.6 | On the Future Role of Psychopathology   *123* |
| | References   *125* |

| **6** | **System Properties of Populations of Neurons in Cerebral Cortex**   *129* |
| | *Walter J. Freeman* |
| 6.1 | Introduction   *129* |
| 6.2 | Spatial Structure of Brain Waves   *131* |
| 6.3 | Temporal Structure of the EEG/ECoG   *133* |
| 6.4 | Behavioral Correlates in Spatio-Temporal Patterns of the EEG   *136* |
| 6.5 | Synthesis of Two Levels of Function in the Cortical System   *140* |
| 6.6 | Conclusions and Applications   *141* |
| | References   *143* |

| **7** | **Dopamine and the Electrophysiology of Prefrontal Cortical Networks**   *145* |
| | *Patricio O'Donnell* |
| 7.1 | Introduction   *145* |
| 7.2 | Electrophysiological Actions of DA in Prefrontal Cortical Circuits   *146* |
| 7.3 | Changes in DA Modulation of Pyramidal Neurons during Adolescence   *148* |
| 7.4 | Changes in DA Modulation of GABA Interneurons during Adolescence   *149* |
| 7.5 | Abnormal Periadolescent Maturation of DA Actions in Developmental Animal Models of Schizophrenia   *151* |
| 7.6 | Implications for Schizophrenia Pathophysiology and Novel Treatments   *153* |
| | References   *154* |

**Part Three  Research in Molecular Psychiatry**   *159*

| **8** | **Nicotinic Cholinergic Signaling in the Human Brain – Systems Perspective**   *161* |
| | *Arian Mobascher and Georg Winterer* |
| 8.1 | Introduction   *161* |
| 8.2 | Epidemiological Relevance of the Nicotinic Cholinergic System   *162* |
| 8.3 | nAChrRs and the Cellular Effects of Nicotine   *163* |
| 8.4 | Nicotine and Cognition   *169* |
| 8.5 | Nicotine and Reward   *171* |

| | | |
|---|---|---|
| 8.6 | Nicotine and Stress Response | *173* |
| 8.7 | Variation in nAChR Genes and Smoking | *174* |
| 8.8 | Future Perspectives | *178* |
| 8.9 | Role of Systems Neuroscience | *179* |
| | References | *181* |

| | | |
|---|---|---|
| **9** | **Progress in Psychopharmacotherapy though Molecular Imaging** | *189* |
| | *Ingo Vernaleken, Gerhard Gruender, and Paul Cumming* | |
| 9.1 | Optimizing Psychopharmacotherapy through Molecular Imaging | *189* |
| 9.1.1 | Techniques for Molecular Imaging in Living Brain | *189* |
| 9.1.1.1 | Methodologic Background | *189* |
| 9.2 | Characterization of Neurotransmitter Systems with Molecular Imaging | *194* |
| 9.3 | Focus Schizophrenia | *196* |
| 9.4 | Action of Psychopharmaceuticals in Schizophrenia | *199* |
| | References | *203* |

| | | |
|---|---|---|
| **10** | **The Marriage of Phenomics and Genetical Genomics: A Systems Approach to Complex Trait Analysis** | *207* |
| | *Laura M. Saba, Paula L. Hoffman, Lawrence E. Hunter, and Boris Tabakoff* | |
| 10.1 | Introduction and Brief History of Genetical Genomics | *207* |
| 10.1.1 | Characteristics of eQTLs | *209* |
| 10.1.2 | Recent Developments | *210* |
| 10.1.3 | Extension of Genetical Genomics to Include a Systems Approach to Phenomics | *213* |
| 10.2 | Potential Pitfalls in Phenotype Selection and Current Technology | *214* |
| 10.3 | General Strategy for Identifying Candidate Pathways | *218* |
| 10.4 | Conclusions: Contributions of Genetical Genomic Phenomics to Systems Biology and Medicine | *222* |
| | References | *224* |

| | | |
|---|---|---|
| **Part Four** | **Data Mining and Modeling** | *229* |

| | | |
|---|---|---|
| **11** | **From Communicational to Computational: Systems Modeling Approaches for Psychiatric Research** | *231* |
| | *Eduardo R. Mendoza* | |
| 11.1 | Introduction | *231* |
| 11.2 | Steps of the Modeling Process | *233* |
| 11.3 | From Diagrams to Qualitative Models through Petri Nets | *235* |
| 11.4 | Petri Net Modeling of Apoptosis in Leukemic Cells and Neurons | *237* |
| 11.5 | From Stoichiometric (Qualitative) to Kinetic (Quantitative) Genome-Scale Models | *239* |

| | | |
|---|---|---|
| 11.6 | From Diagrams Directly to Quantitative Canonical Models: The Concept Map Method | 239 |
| 11.7 | Summary and Outlook | 241 |
| | References | 242 |

## 12 Network Dynamics as an Interface between Modeling and Experiment in Systems Biology 243
*James Smith and Marc-Thorsten Hütt*

| | | |
|---|---|---|
| 12.1 | Introduction | 243 |
| 12.2 | Aspects of Graph Theory | 245 |
| 12.3 | A Network Perspective on Systems Biology | 251 |
| 12.3.1 | Large-Scale Systems | 251 |
| 12.3.2 | TRNs | 253 |
| 12.3.3 | Metabolic Networks | 255 |
| 12.4 | Network Dynamics | 260 |
| 12.4.1 | General Aspects | 260 |
| 12.4.2 | Binary Dynamics and Cellular Automata on Graphs | 261 |
| 12.5 | Applicability in Psychiatry | 267 |
| | References | 270 |

## 13 Some Useful Mathematical Tools to Transform Microarray Data into Interactive Molecular Networks 277
*Franziska Matthäus, V. Anne Smith, and Peter J. Gebicke-Haerter*

| | | |
|---|---|---|
| 13.1 | Introduction | 277 |
| 13.2 | Microarray Data | 278 |
| 13.3 | Dimensionality Reduction | 279 |
| 13.3.1 | Principal Components Analysis (PCA) | 279 |
| 13.3.2 | Clustering Methods | 281 |
| 13.3.2.1 | Partitional Clustering: $k$-Means | 282 |
| 13.3.2.2 | Hierarchical Clustering: Bottom-Up and Top-Down Approaches | 282 |
| 13.4 | Statistical Tests: ANOVA and the Naïve Bayes Classifier | 284 |
| 13.5 | Bayesian Networks | 286 |
| 13.5.1 | Definition of a Bayesian Network | 286 |
| 13.5.2 | Learning Bayesian Networks from Data | 288 |
| 13.5.3 | Bayesian Networks and Microarrays | 291 |
| 13.6 | Final Considerations | 294 |
| | References | 295 |

## 14 Biochemical Networks in Psychiatric Disease 301
*Maria Lindskog, Geir Halnes, Rodrigo F. Oliveira, Jeanette Hellgren Kotaleski, and Kim T. Blackwell*

| | | |
|---|---|---|
| 14.1 | Introduction | 301 |
| 14.2 | The Example of Schizophrenia | 301 |
| 14.3 | Looking for Nodes of Interaction | 302 |
| 14.4 | Dopamine Signaling and DARPP-32 | 303 |

| | | |
|---|---|---|
| 14.5 | Physiological Role of DARPP-32 | 305 |
| 14.6 | DARPP-32 in Psychiatric Disease | 306 |
| 14.7 | Modeling Signaling Pathways with a Deterministic Model | 308 |
| 14.8 | Biological Conclusions from the DARPP-32 Model | 309 |
| 14.9 | Stochastic Models | 310 |
| 14.10 | PKA Activation: A Case Study | 311 |
| 14.10.1 | Conceptual Model | 311 |
| 14.10.2 | Empirical Estimates of Rate Constants | 312 |
| 14.10.3 | Model Validation against Steady-State Data | 313 |
| 14.10.4 | Model Validation against Dynamic Data | 314 |
| 14.10.5 | Deterministic versus Stochastic Algorithms | 314 |
| 14.11 | Conclusions | 316 |
| | References | 316 |
| | | |
| **15** | **Local Cortical Dynamics Related to Mental Illnesses** | **321** |
| | *Marco Loh, Edmund T. Rolls, and Gustavo Deco* | |
| 15.1 | Introduction | 321 |
| 15.2 | A Recurrent Neural Network | 321 |
| 15.3 | Concept of an Attractor Network | 324 |
| 15.4 | Noise and Stability | 325 |
| 15.5 | Mental Illnesses | 328 |
| 15.5.1 | Schizophrenia | 328 |
| 15.5.2 | OCD | 332 |
| 15.6 | Outlook | 334 |
| | References | 335 |
| | | |
| **16** | **Epilogue** | **341** |
| | *Peter J. Gebicke-Haerter* | |
| 16.1 | Some General Remarks | 341 |
| 16.2 | Pharmaceutical Discourse | 342 |
| 16.3 | From Molecular to Cellular Networks | 342 |
| 16.4 | Modeling Strategies | 343 |
| | References | 344 |

**Index** 347

# Preface

Mentally ill individuals frequently report abnormal experiences such as hearing voices of absent persons. Although this is taken as a subjective report on a dysfunctional mental state, modern psychiatry increasingly conceives mental disorders as brain disorders. This view is strengthened by the increasing influence of the biological sciences on psychiatry. However, the reduction of mental states to brain states has not been successful until now. Therefore, from a clinical perspective, it is important to keep these two points of view apart. This issue of philosophy of mind is covered in the introductory section of this book.

Nevertheless, biological psychiatry has provided so much new information about brain processes related to mental illnesses that explanatory concepts have to be developed to put these pieces of data together into a conceptual framework of brain function and dysfunction. These attempts to construct systemic relations between biological processes in brain areas or between molecular processes within neurons usually end up in graphical diagrams with boxes and arrows that might be called qualitative models.

In the wake of cybernetics and systems theory, analogies between brain and electrical circuits have already been drawn and very early on in the 1950s, computer-based models were constructed to insert neurobiological data into the framework of artificial neuronal networks. To achieve these tasks, a great deal of mathematics had to be applied – a field disliked by many physicians and biologists. Those network models were elaborated on the basis of electrical signals, spikes from single neurons, and global brain currents, which are recorded by the electroencephalogram. Despite great progress made in the development of mathematical tools of signal analysis, the main question remained focused on the detection of local distribution patterns of synchronization or desynchronization of various frequencies in brain currents.

The beginning of the twenty-first century marked the advent of a new strategy in molecular biology and genetics termed "systems biology". Systems biology tries to tie together experimentation and building of theoretical models in a dialectic mode. This new approach fits very well with the present focus of biological psychiatry – the analysis of the genome and proteome. There is also an urgent demand in

*Systems Biology in Psychiatric Research.*
Edited by F. Tretter, P.J. Gebicke-Haerter, E.R. Mendoza, and G. Winterer
Copyright © 2010 WILEY-VCH Verlag GmbH & Co. KGaA, Weinheim
ISBN: 978-3-527-32503-0

pharmaceutical companies to better understand mental disorders on the molecular biological level for the development of new and more efficient substances for psychiatric treatment. Many drugs used in clinical practice do not fulfill modern standards of specificity and low unwanted side-effects. It seems to be timely to join the rather new field of "molecular psychiatry" with systems biology. There are presently not many conferences, publications, student courses, or other academic activities that try to connect molecular psychiatry and neurobiology with systems biology. To this end, we have organized several interdisciplinary workshops devoted to the advancement of "computational neuropsychiatry" and "systems neuropsychiatry."

In this book, we try to put together some recent developments in specialized fields of systems biology in psychiatry. We hope that one or more of the chapters will raise the reader's interest and curiosity to read further in other publications. We also would like to direct the reader's attention to some basic concepts and findings of psychiatry, and to methodological approaches from computational sciences that permit us to build computational models of disorders of brain functions. This may lead to the design of new types of experiments that explicitly take into account hypotheses on a systems level.

Munich, Mannheim, and Düsseldorf             *Felix Tretter, Peter J. Gebicke-Haerter*
February 2010                                 *Eduardo Mendoza, and Georg Winterer*

# List of Contributors

**Kim T. Blackwell**
George Mason University
School of Computational Sciences and
Krasnow Institute for Advanced Studies
4400 University Drive
Fairfax, VA 22030-444
USA

**Paul Cumming**
Ludwig Maximilians University Munich
Department of Nuclear Medicine
Marchioninistrasse 15
81377 Munich
Germany

**Gustavo Deco**
Universitas Pompeu Fabra
Department of Technology
Computational Neuroscience
Passeig de Circumvallació 8
08003 Barcelona
Spain

and

Institució Catalana de Recerca i Estudis
Avançats (ICREA)
Passeig Lluís Companys 23
08010 Barcelona
Spain

**Walter J. Freeman**
University of California at Berkeley
Department of Molecular & Cell Biology
Donner 101
Berkeley, CA 94720-3206
USA

**Peter J. Gebicke-Haerter**
Central Institute of Mental Health
Department of Psychopharmacology
J5
68159 Mannheim
Germany

**Gerhard Gruender**
RWTH Aachen University
Department of Psychiatry
Pauwelsstrasse 30
52074 Aachen
Germany

**Geir Halnes**
Royal Institute of Technology
School of Computer Science and
Communication
Lindstedtsvägen 3
100 44 Stockholm
Sweden

*Systems Biology in Psychiatric Research.*
Edited by F. Tretter, P.J. Gebicke-Haerter, E.R. Mendoza, and G. Winterer,
Copyright © 2010 WILEY-VCH Verlag GmbH & Co. KGaA, Weinheim
ISBN: 978-3-527-32503-0

**Jeanette Hellgren Kotaleski**
Karolinska Institutet
Department of Neuroscience
Retziusväg 8
171 77 Stockholm
Sweden

and

Royal Institute of Technology
School of Computer Science and
Communication
Lindstedtsvägen 3
100 44 Stockholm
Sweden

**Paul Hoff**
University of Zurich
Clinic for General
and Social Psychiatry
Lenggstrasse 31
8032 Zurich
Switzerland

**Paula L. Hoffman**
University of Colorado Denver
Department of Pharmacology
12801 East 17th Avenue
Aurora, CO 80045
USA

**Lawrence E. Hunter**
University of Colorado Denver
Department of Pharmacology
12801 East 17th Avenue
Aurora, CO 80045
USA

**Marc-Thorsten Hütt**
Jacobs University Bremen
School of Engineering and Science
Campus Ring 1
28759 Bremen
Germany

**Wolfram Kawohl**
University of Zurich
Clinic for General
and Social Psychiatry
Lenggstrasse 31
8032 Zurich
Switzerland

**Edda Klipp**
Humboldt University Berlin
Theoretical Biophysics
Invalidenstrasse 42
10115 Berlin
Germany

and

Royal Institute of Technology
School of Computer Science and
Communication
Lindstedtsvägen 3
100 44 Stockholm
Sweden

**Maria Lindskog**
Karolinska Institutet
Department of Neuroscience
Retziusväg 8
171 77 Stockholm
Sweden

**Marco Loh**
Oxford University
Centre for Computational Neuroscience
South Parks Road
Oxford OX1 3UD
UK

*Franziska Matthäus*
University of Heidelberg
Center for Modeling and Simulation in
the Biosciences
Im Neuenheimer Feld 368
69120 Heidelberg
Germany

*Eduardo R. Mendoza*
Ludwig Maximilians University Munich
Department of Physics
Geschwister-Scholl-Platz 1
80539 Munich
Germany

and

University of the Philippines Diliman
Institute of Mathematics
C.P. Garcia Street
1101 Quezon City
Philippines

*Arian Mobascher*
University of Düsseldorf
Psychiatric Hospital
Bergische Landstrasse 2
40629 Düsseldorf
Germany

*Denis Noble*
University of Oxford
Department of Physiology
Anatomy and Genetics
Parks Road
Oxford OX1 3PT
UK

*Patricio O'Donnell*
University of Maryland School of
Medicine
Department of Anatomy &
Neurobiology
20 Penn Street
Baltimore, MD 21201
USA

*Rodrigo F. Oliveira*
George Mason University
School of Computational Sciences and
Krasnow Institute for Advanced Studies
4400 University Drive
Fairfax, VA 22030-444
USA

*Edmund T. Rolls*
Oxford University
Centre for Computational Neuroscience
South Parks Road
Oxford OX1 3UD
UK

*Laura M. Saba*
University of Colorado Denver
Department of Pharmacology
12801 East 17th Avenue
Aurora, CO 80045
USA

*Marvin Schulz*
Humboldt University Berlin
Theoretical Biophysics
Invalidenstrasse 42
10115 Berlin
Germany

*James Smith*
Jacobs University Bremen
School of Engineering and Science
Campus Ring 1
28759 Bremen
Germany

**V. Anne Smith**
University of St. Andrews
School of Biology
Westburn Lane
St. Andrews, Fife KY16 9TH
UK

**Boris Tabakoff**
University of Colorado Denver
Department of Pharmacology
12801 East 17th Avenue
Aurora, CO 80045
USA

**Felix Tretter**
Isar-Amper-Klinikum GmbH
Klinikum München Ost
Vockestrasse 72
85540 Haar
Germany

**Ingo Vernaleken**
RWTH Aachen University
Department of Psychiatry
Pauwelsstrasse 30
52074 Aachen
Germany

**Georg Winterer**
University of Düsseldorf
Psychiatric Hospital
Bergische Landstrasse 2
40629 Düsseldorf
Germany

# Part One
# Introduction

Part I presents an introduction to the basic issues of psychiatry. Chapter 1 on philosophical aspects by Felix Tretter delineates some aspects of the historical development of science and philosophy. Focus is on philosophy of mind, philosophy of science, and "neurophilosophy" – special fields of fundamental importance for reflection of psychiatry, its concepts, its methods, and its theories. This chapter also describes the rise of philosophy in biology and systems biology.

The second basic chapter, Chapter 2 by Felix Tretter and Peter Gebicke-Haerter, gives an overview on general psychiatry, beginning with an introduction into the methods, diagnostics, and therapy of mental disorders. The main part is an introduction to the neurobiological basis of modern psychiatry. Brain anatomy, cellular physiology, and molecular mechanisms are briefly outlined, and examples are given by referring to findings in schizophrenic patients. Finally, first attempts at computational modeling are presented, mainly with respect to working memory functions in schizophrenia.

Subsequently, in Chapter 3, Marvin Schulz and Edda Klipp provide a brief but comprehensive outline of systems biology from a biochemical point of view. The reader will find methods used in the "wet" laboratory of biochemists and also methods from the "dry" laboratory of computational scientists.

Chapter 4 provides an overview by one of the pioneers of systems biology, Denis Noble. From his seminal work on the heart, he extends the view to systems biology of the brain. He also includes a brief outlook on the philosophical dimension of his research project.

# 1
# Philosophical Aspects of Neuropsychiatry
*Felix Tretter*

> Mental disorders are brain disorders.
>
> [after Griesinger, 1882, 1845]

Psychiatry as the science of mental disorders must integrate biological, psychological-clinical and social data and aspects. This implies several philosophical problems that are usually overlooked. First, biological psychiatry aims to relate mental phenomena to brain phenomena. This is a fruitful effort, but it might end up in a vision of total reduction and substitution of mental phenomena to brain mechanisms. Regarding this tendency in research, several philosophical restrictions have to be considered:

- *Philosophy of mind* presents several limitations of identifying the mind with the brain that might relate to clinical psychiatry. One basic limitation is related to the reductionistic aim to substitute the subjective experience by categories of brain research. It is also not sufficient to reduce consciousness to the physicochemical properties of neurons (the "emergence problem").
- *Philosophy of science* presents results of the analysis of the history of concepts and methods of physics that should also be considered. In that respect it is to be determined if brain correlates "explain" mental disorders. In a philosophical sense, "explanation" means the application of general laws to specific cases. This is more than description by observational data because explanatory propositions imply logical operations. In addition, the part–whole problem tackles the consistent understanding of the brain by detailed knowledge of the behavior of molecules. This is important if one considers that systems biology aims to create a computer-based model of the cell. For this project mathematics plays a crucial role. Taking into account that psychiatry depends on the methods and results of numerous academic disciplines it seems to be interesting to establish the new field of "neurophilosophy." Such a platform seems to be very important when studying the effects of molecules on mental states.

## 1.1
### Development of Research Paradigms and Strategies in Psychiatry

At the early times of scientific psychiatry, about 100 years ago, clinical practice dominated the knowledge of psychiatry. Psychiatrists could only observe human behavior disorders and describe them verbally. Case studies were used to characterize the different disorders. At that time the explanations of the causes of mental disorders were very speculative. One approach was to explain mental disorders as a consequence of sins. Only a few therapeutic tools were available, and it is well known that therapeutic chairs with restraints of body movements and shock treatments were very usual from that time up until the 1970s. In the 1950s psychoanalysis also became influential and therefore psychological mechanisms were claimed to be the causes of mental disorders.

Little by little, psychiatry was established within medicine as a natural science of mental disorders. This took quite a lot of time (see Chapter 5 by Kawohl and Hoff). Emil Kraepelin was maybe the first to establish psychiatry as a science concerned with the quantification of psychic functions and states (Kraepelin, 1902). He was interested in measuring the cognitive performance of psychiatric patients compared to normal subjects, as was proposed by Wilhelm Wundt when starting experimental psychology (Wundt, 1896). In this situation, Kraepelin was also trying to distinguish the different forms of madness by objective criteria. As a consequence, Kraepelin identified different disease entities such as "dementia praecox" and "depression" (Kraepelin, 1902). Dementia praecox was subsequently named "schizophrenia" by Eugen Bleuler (Bleuler, 1911). Later, diseases such as Alzheimer's disease, anxiety disorders and addictions came into the catalogue of psychiatric disorders (Sadock and Sadock, 2007). Details of the history of psychiatry are presented in Chapter 5 by Kawohl and Hoff.

Increasingly, the tools of natural sciences as they were established in medicine were also applied in psychiatry from the 1930s. As a consequence, neuropathology and genetics were already developed before World War II. After the war, the diagnostic tools were improved, and rating scales and operational definitions were established. The severity of a disease could be "measured" by objective and/or subjective rating scales. In the 1960s, animal models for mental disorders were additionally developed, and more data were obtained by using neurobiological experiments (e.g., from the brains of socially deprived animals modeling depression).

After several decades of psychological psychiatry from the 1960s to the 1990s, mental disorders were related more and more to their biological roots. The main reasons were several observations:

- The induction of psychosis by brain disorders such as infections and by drug consumption.
- The clustering of mental disorders in some families that indicated heredity.
- The treatment effects of some pharmaceuticals.

At present, the dominating research paradigm in psychiatry is the research strategy of neuropsychiatry and biological psychiatry (Andreasen, 2004).

The aim of neuropsychiatry or biological psychiatry is to relate mental phenomena to brain phenomena. This approach was already initiated by Wilhelm Griesinger

(Griesinger, 1882, 1845) who stated that "insanity is merely a symptom complex of various anomalous states of the brain." He also coined the phrase "Mental disorders are brain disorders" used as the opening quotation in this chapter. Methodologically, experimental and clinical brain research focused on imaging, electrophysiology, histology and molecular biology studies. Today, from brain downstream, over neuronal circuits and local networks, the neurons and their molecular structures are studied in order to identify pathologies. By the development of electrophysiological methods such as electroencephalography (EEG) and various imaging methods, brain correlates of mental disorders were identified that suggested that the brain is the organ that can "cause" or "produce" mental disorders. For instance, progress in molecular biological methods has helped to identify genes that could be candidates for the causation of schizophrenia. However, this approach has not succeeded in "explaining" schizophrenia. It also has to be assumed that only the symptoms, not the time-course, of schizophrenia can be "explained" by molecular biology. Additionally, rather than molecules alone, both cells and cellular networks must be identified to explain the symptoms. Presumably, only processes at circuits of local cell assemblies can be the basis of understanding symptoms of mental disorders.

We are now in the situation where the classification of mental disorders established by international classification systems such as the *International Statistical Classification of Diseases and Related Health Problems*, 10th revision (WHO, 1992) or the *Diagnostic and Statistical Manual of Mental Disorders*, text revision, 4th edn (American Psychiatric Association, 2000) are being criticized again – some authors suggest that the symptoms, the signs, or the so-called "endophenotypes" that are related to neuroscience should be the focus of reference (Hyman, 2007). The new classification system, the 11th revision of the *International Statistical Classification of Diseases and Related Health Problems*, will have to integrate several new radical points of view that are partially based on biological data that were obtained by the study of mental illness (see Chapter 5 by Kawohl and Hoff).

Additionally, some neuroscientists suggest that phenomenological terms should be avoided in the scientific context and should be substituted by neurobiological terms (Crick, 1994). In this view, psychiatric examination can be described as the "behavioral examination of the brain" (Taylor and Vaidya, 2009, p. 56). However, it is not proven that contents of experience such as hallucinations can be completely represented by behavioral observations (Kim, 1998). Regarding this philosophical issue, this behavioristic position is related to the origin of reductionistic "neurophilosophy" – a term that has been used by Patricia and Paul Churchland, for instance, since the 1980s (Churchland, 1986, 2007).

## 1.2
### The Mind–Body Problem – Philosophy of Mind

Everyday experience provides evidence that we are awake and consciously living organisms, subjects, persons that can initiate and inhibit motor behavior by thoughts. This (self-)experience suggests that the so-called "mind" can control the body. Most

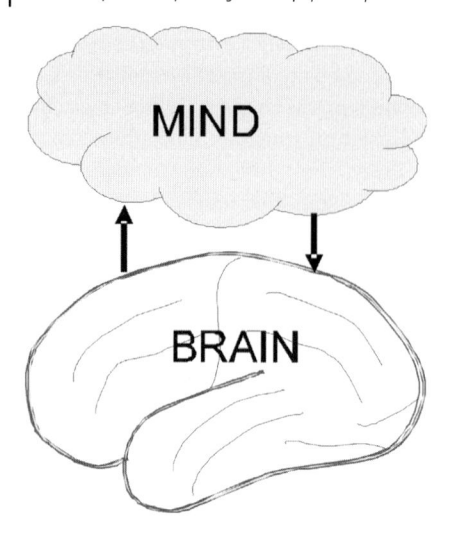

**Figure 1.1** The basic problem of brain and mind – who controls whom?

individuals have also experienced that alcohol can change the mind. Therefore, it is evident that a drug that is in the body can influence the mind. Here, "mind" means mental states and processes such as conscious experience, thinking, feeling, planning, imagination, desires, and so on. These experiences have been known since the ancient Greek philosophers, and therefore the "mind–body problem" has a very long tradition in the history of our philosophical and psychological concepts (Figure 1.1). It was resolved in a dualistic conception in the sense of Rene Descartes until recent years, when neurobiology showed much progress in studying mental processes. These findings started a wave of monistic conceptions that claim that the mind is just a function and a state of the brain, and that there is no special entity that can be called the "mind," "soul," and so on (Place 1956; Block 1980; Churchland 1984; Chalmers 1995).

This discussion is important for neuropsychiatry and therefore some basic aspects are mentioned here. Recommended interesting textbooks on the philosophy of mind are Heil (2004), and on neurophilosophy are Bennett and Hacker (2003) and Northoff (2000, 2004).

In principle, only a few main positions can be distinguished in the brain–mind debate (Figure 1.2):

i) The brain controls/produces the mind (*materialism, physicalism, epiphenomenalism, supervenience*; e.g., Churchland, 1981, 1984; Kim, 2002). This concept has growing influence at present and it is preferred by most neuroscientists.
ii) The mind controls/influences the brain (*mentalism, idealism*). This most traditional position is supported by the everyday experience that I can move my hand if I intend to do it. Traditional philosophical idealists think that the world and also the brain are the result of the action of a distinct mental entity. This position is hard to combine with views of natural science.

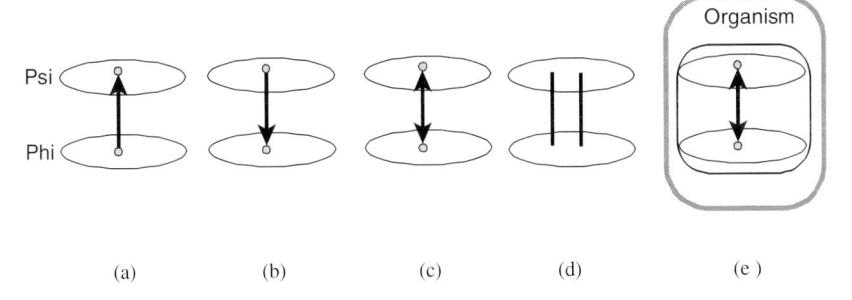

**Figure 1.2** Five main concepts of mind brain relations. (a) The brain produces and/or controls the mind. (b) The mind controls the brain. (c) There are interactions between the brain and mind. (d) The mind is identical with the brain. (e) The brain/mind is an organ/function of the person/organism.

iii) The mind and the brain interact and influence each other (*dualism*; e.g., Popper and Eccles, 1977; Libet, 2005). This position is quite common, with difficulties of explanation of downward causation (Walter and Heckmann, 2003). In scientific psychology and psychiatry, interactive dualism has reached a wide acceptance; at present only aspect dualism or property dualism are proposed based on the difference of methods of studying the brain.
iv) The mind is the brain and the brain is the mind (*identity concept, materialism, monism*; Churchland, 1981; Davidson, 1970). Owing to reasons of logic, this concept was preferred in the professional debate since the activities of the Vienna Circle (Stadler, 2001). The psychological terms should be eliminated and substituted by neurological terms (Carnap, 1928, 1932). However, this position also has logical difficulties.
v) The brain is the organ of the person, of the subject (*phenomenology*; McGinn, 1989). In the traditional concept of phenomenology, the experiencing subject is the frame of reference so that the brain is only an organ of the whole.

The most interesting question in the brain–mind debate is (Chalmers, 1995; Jackson, 1982): what is the mind? The presently preferred answer is: it is a (dispositional) property of the brain. However, what is the brain? This question is not trivial, in such a way that the question of the nature of matter is interesting: if matter is mass then matter is, according to Einstein's famous equation, the ratio of energy divided through the squared speed of light. In this view, a trivial understanding of matter is not sufficient (Levine, 1983).

At present, many neuroscientists claim that the mind is only an epiphenomenon of the body (respectively, of the brain). In this view, the mind is similar to the piping of the steam locomotive – it is the product of the brain, but it cannot influence the producer (Crick, 1994; Edelman and Tononi, 2000)! Additionally, many experts in the field of the mind–body debate claim that there is no ego and no self (Bennett and Hacker, 2003; Metzinger, 2009). Also, consciousness is supposed to be a "mirror" that can only "represent" some actions of the brain. But what is the mirror?

## 1.2.1
### Monism and Dualism

The mind, in a preliminary definition, is approximately equivalent to subjective "experience," and is a phenomenon that can be expressed directly only by living organisms and that – with regard to consciousness – can be ascribed to "subjects" (Davidson, 1970; Jackson, 1982). From a scientific point of view, mental functions and activities must be expressed in functional terms that are characterized usually by "if–then" relations and that represent the typical input–output relation or black box perspective (Block, 1980; Putnam, 1965). This is obviously not completely possible if a system exerts "spontaneous" (i.e., intrinsically conditioned) behavior.

In his historical paper "What is it like to be a bat," Thomas Nagel has shown that it is nearly impossible to identify or substitute observations that are made by a subject by observations that are made by a brain researcher (Nagel, 1974). This is known as the basic problem of the complete substitution of the "subjective" first-person perspective by the ("objective") third-person perspective of science (Levine, 1983; Shoemaker, 1996).

In contrast, the monistic position is closely related to identity theory (Section 1.2.3) and denies the functional relevance of mental events. This position says that there is only a brain that is relevant for mental processes that are epiphenomena. Some authors state that mental states are illusions of the subject or of the brain (Crick, 1994; Dennett, 2006). Only a few famous neuroscientists such as Benjamin Libet are at least methodological dualists (Libet, 2005).

The experimental bases for monistic positions are seen in the experiments testing the "free will" that were conducted by Benjamin Libet (Libet *et al.*, 1983). By recording EEG signals, these experiments gave evidence that decisions made by a subject occur about 300 ms prior to their conscious intentions to act. However, the subjects have to be trained to participate in these experiments so that they only execute a trained reaction to a special experimental situation and do not exert a free will that is related to a personal important event such as a marriage, buying a car, and so on. For this reason, from a methodological point of view, it is not conclusive to propose that the mind cannot influence motor actions and to substitute the terms for mental phenomena by terms for brain phenomena.

This controversy between monists and dualists is very complicated as not only the concept "mind" cannot be defined easily and directly, but also the concept "brain" is not as clear as it seems. This is important, because a precise definition must be presented in order to enable a precise discussion. It must also be kept in mind that the brain is necessary for mental states and processes but it is not a sufficient condition for them. The mind cannot be expressed in kilograms or cubic centimeters as properties of the brain, and also localizations of functions are very limited – we have about 40 areas for visual functions and many areas such as the striatum have many functions. Also, the cerebellum is not involved in conscious processes, so the "brain" is a different category too global. Mental states, on the other hand, are cognitions, emotions, memories, drives, and so on. These states and processes differ quite markedly. Finally, there is no strong correlation between intelligence and brain size or brain weight.

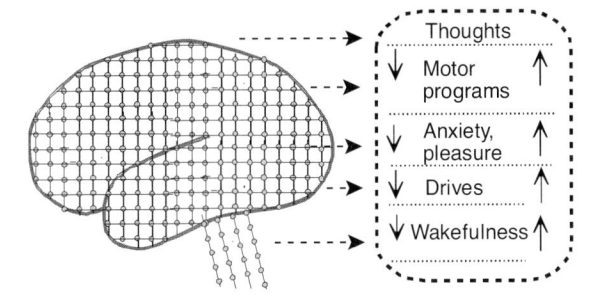

**Figure 1.3** The brain as a network of neurons and some inter-related mental functions.

Therefore, a simple brain–mind relation obviously cannot hold. For the reasons mentioned above, regarding complexity and connectedness, the perspective that conceives the brain as a network and the mental to be a system seems to be more appropriate (Figure 1.3). These aspects have not yet been discussed sufficiently and, therefore, we do not have a "solution" of the brain–mind problem at present.

## 1.2.2
### Correlation

Psychological measures and physiological measures, in a first step, can only be correlated. However, the identification of "experiences" with electrical brain signals is problematic. For instance, many brain signals are analog signals, whereas mental events such as a thought, a perception, a recognition, a memory, or a decision are sudden mental events that can be interpreted as digital signals (Figure 1.4). This can be seen in the discussion of the experiments of Libet mentioned above that were performed to determine the temporal structure of volitional processes (Libet et al., 1983). The punctual decision was experienced later than the deviation of the averaged EEG signal from the reference line. In this way, discrete phenomena of single units (action potentials of neurons) can induce qualitative new properties if they are seen as phenomena of a collective of units (field potentials) and as population phenomena: the degree of coherence or coincidence of discharge of single units can determine the coding properties of the activity of these neuronal elements (Freeman, 2000, also see Chapter 6 by Freeman).

## 1.2.3
### Identity Theory and its Problems

One of the leading concepts in the brain–mind debate is the concept of the identity of the brain and mind (Lewis, 1966). From the point of view of some philosophers, the "same" properties of the mind and the brain justify the assumption of an identity. However, a light sensation is not the same as an electrical brain event, just as not every electrical brain event is related to a light sensation. Obviously, this position has its limitations. However, there is a striking advantage – by considering identity theory,

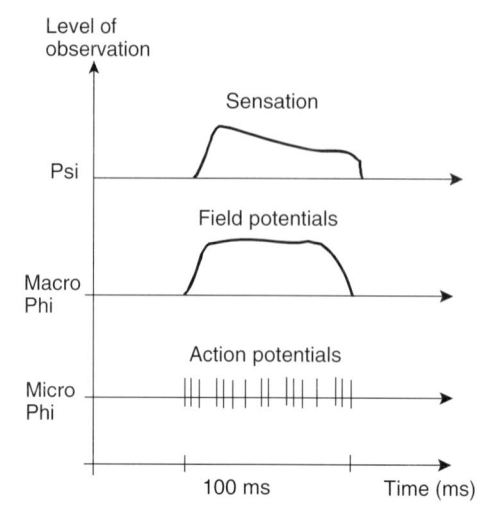

**Figure 1.4** Relations of psychophysical observations – sensations as continuous processes (analog signals) are related to discrete events of neurons. Even if the time–activity curve is similar, how are these signals correlated and is there an "identity?" This problem is already given when micro events (action potentials) are related to macro states of a neuronal network (field potentials).

the question regarding the ontological quality of brain and mind and the nature of the relation between brain and mind can be resolved in principle. See Figure 1.5.

However, regarding the identity concept, it becomes clear that not only the mind is loosely defined, but also the term "brain": Northoff (2000) shows that it is important to distinguish between the structural brain, the functional brain, and the mental brain. The structural brain, for example, also refers to the cerebellum that usually is not seen as an important structure for conscious processes as it seems to be concerned mainly with motor coordination. Therefore, it should be declared what kind of function the cerebellum has. The functional brain is that part of the brain that has well-defined functions such as automatic habitual behavior, which is organized in subcortical structures or conscious experience that requires highly interconnected neuronal networks. The mental brain is that subsystem of the total brain that is related to mental functions.

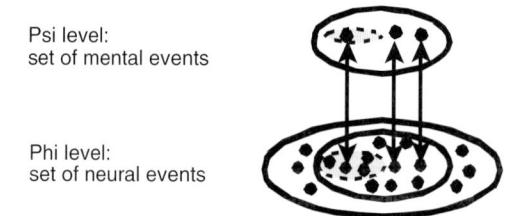

**Figure 1.5** The problem of identity relations between mental events (Psi level) and neural events (Phi level).

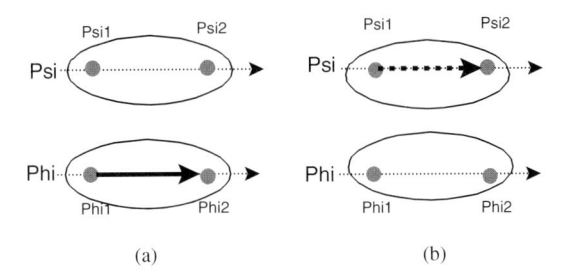

**Figure 1.6** Parallel causality in the physical level (Phi) and in the mental level (Psi) and in between.

### 1.2.4
### Causation

The everyday experience that I can move my limb if I want raises the concept of a mental power that can control the physical. An extreme position in philosophy of mind assumes that the mind controls the brain that controls the body. However, in this concept the mind is understood as an immaterial entity. This implies the problem that it cannot be explained how an immaterial system can cause processes in a physical system. As a result of this difficulty, most neuroscientists reject mentalism and prefer a monism that identifies the mind and the brain. However, it is not possible to eliminate the methodological difference between subjective experience (first-person perspective) and objective scientific observations (third-person perspective), so that total reductions of mind onto brain are not conclusive. See Figure 1.6.

### 1.2.5
### Supervenience

The supervenience concept is a very sophisticated concept of duality that says that the mind "supervenes" the brain, meaning that there is a special entity that is influenced by brain processes, but that has no influence on brain processes (Kim, 1998). It says that several brain processes can influence a mental event, but no two mental events can be caused by the same neural event. Supervenience is not saying that "the mental is the physical," but it assumes that the mental is caused or produced by the physical brain (Kim, 2002). Noting that different physical conditions can cause identical mental events ("multiple realization"), several critical arguments against the identity thesis are answered (Figure 1.7).

## 1.3
## The Conditions of Scientific Knowledge – Philosophy of Science

Some philosophers have tried to analyze the structure of science. This specialized approach is usually called the "philosophy of science" or "metascience" (Bunge, 1998) and first attempts at a philosophy of science of psychiatry have already been made

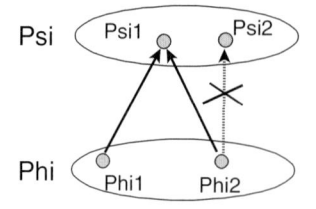

**Figure 1.7** Scheme of the "supervenience" concept – multiple possibility of the physical causation of psychological events. A psychological event Psi1 can be generated by two or several physical events Phi1, Phi2, and so on; however, a Phi event (e.g., Phi2) cannot generate two Psi events; or a Psi event can be generated by different Phi events, but a Phi event can only generate a specific Psi event.

(Cooper, 2007). The main subject of these studies was physics. Mainly in the 1920s, theoretical physicists such as Albert Einstein, Niels Bohr, Werner Heisenberg, Erwin Schrödinger, and others, generated a discussion that was devoted to metaproblems of theoretical physics: Is light matter or a wave? What does measurement mean? Is measurement independent of the observer? Is quantum physics the basis that the universe is developing in an undetermined way? Is the world a set of stochastic events?

One controversy was between researchers that focused on experimental research (empiricists) and others that focused on theoretical research: experimentalists, for instance, claim that we need more data in order to understand the functions and dysfunctions of the subjects that are studied. Theoreticians, on the other hand, criticize that we do not see the wood because of looking at the trees: observation without theory seems not to be possible.

One main center of this metascientific discussion was the Vienna Circle of the 1920s with Rudolf Carnap, Karl Popper, and others (Stadler, 2001). For instance, this group discussed the difference of observations and theory. This discussion is still ongoing as observations imply a theory of observation and as theories need empirical terms in order to be tested. In the meanwhile, many meta-analyses were performed regarding these problems. Some of the philosophical results can be used as guidelines for a metatheory of biology and also of biological psychiatry. In recent years, biology and philosophy have cooperated increasingly (Sober, 2000; Boogerd et al., 2007). This will be presented briefly later.

## 1.4
### Experimental Research – From Observation to Theory

For experimentalists, the field of theories is disliked. In discussions it is very often said that we do not have enough data to build theories and, in some cases, theory is called "speculation." Such a position, from a view point of the philosophy of science, is not appropriate, because how do we know at which time we do "know enough?" Who knows why and when we should start to build theories? Is observation possible without theory? Furthermore, what do "understanding" and "explaining" a phenomenon like mental disorders actually mean?

Many experimentalists think that (experimental) research means to put questions to "nature" and obtain answers. Often, answers raise more questions that require more data to understand the functions of the system under study (e.g., the brain). Usually experimentalists understand their work as data collecting and hypothesis testing – they have observations, then they formulate a hypothesis, and then they set up a new experiment in order to test the hypothesis. A hypothesis is a proposition that states that an observation is not a random deviation from a reference value, but is a systematic variation (difference hypothesis). Another type of hypothesis should clarify if the observable is the result of the action of another conditional variable or not (causal hypothesis). The hypotheses are tested by experimental set-ups and by statistical analysis of the observed values of the measured variables. By variation of conditions the results of experimentation allow us to make the hypothesis more precise. In early stages of research only qualitative measurements and hypotheses are possible, whereas in stages of more sophisticated experimental research large quantitative databases are obtained. In principle, the cycle of empirical research (i.e., observation, hypothesis, experiment, observation, new hypothesis, etc.) seems to iterate in an endless cycle, suggesting that there is no need for theory (Figure 1.8). However, any observation implies theory, for instance regarding criteria whether the observation is "true" or an artifact (e.g., theory of measurement and theory of errors). Experimentalists often state also that they do not need "theory." In this context, it should be kept in mind that, for instance, even simple hypotheses are based on theoretical assumptions that are very often implicitly integrated into the hypothesis: any observation is based on a theory of measurement. When measuring some

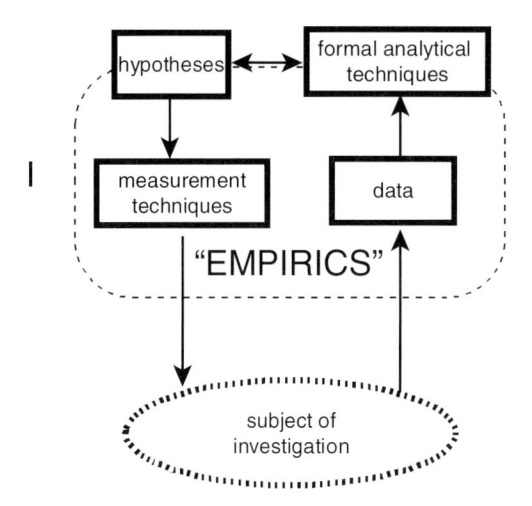

**Figure 1.8** Cycle of empirical research (I, "empirics"; Tretter, 2005; Kell and Knowles, 2006). Observations and databases induce the formation of hypotheses that can be tested by specific experimental set-ups and measurement techniques.

phenomenon, it is assumed that the respective measure is an indicator for the respective phenomenon, such as a clinical symptom.

This antitheoretical position is similar to that of the logic empiricism as it was designed, for instance, by Rudolf Carnap (1928, 1932). In Carnap's view, psychology should be reduced to physics. At present, we have to state that this program has not succeeded.

### 1.4.1
### Hypotheses and Theory

In this context it has to be mentioned that hypotheses are propositions that relate different observations to each other. Therefore, a set of hypotheses can be classified as a theory, especially if "explanatory" hypotheses are included. Explanatory hypotheses can be recognized if they are constituted by the word "because." In this case it is referred to a general law (Hempel and Oppenheim, 1948; Hempel, 1965). Also, the distinction between empirical propositions and theoretical propositions is not clear-cut. Therefore, even hypotheses are theoretical propositions that can be tested empirically. Theories can also be understood as (complex) hypotheses. In this view, several connected hypotheses can be understood as a "theoretical framework" (Bunge, 1998). See Figure 1.9.

Science is not only based on experimental (or at least observational) data, but also has a field that can be called "theoretical research." Theoretical research is not "speculation," but it has its own methodology of reasoning and theory building, as will be demonstrated later.

Regarding these aspects, we can note the gestalt psychologist Kurt Lewin: "There is nothing more practical than a good theory" (Lewin, 1952, p. 169).

### 1.4.2
### The "Epistemic Cycle"

There is a reciprocal inter-relation between theory and experimentation. A theoretical proposition must be tested empirically again by experiments. Then, additional propositions can be made that could be integrated into the theoretical framework of the respective research field. For this reason experimental work does not only encompass observing and measuring alone, but also implies thinking and theory

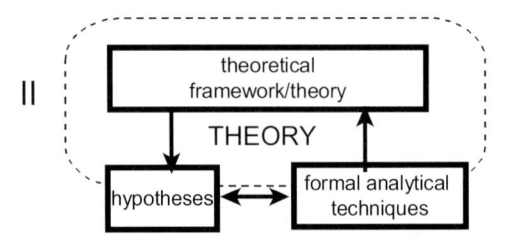

**Figure 1.9** Theory, analytical tools and theory-derived hypotheses.

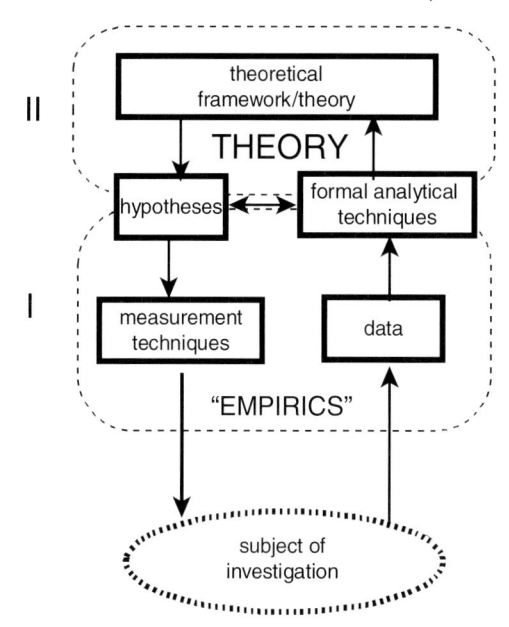

**Figure 1.10** The cycle of scientific knowledge with two subcycles between empirical research (I, "empirics") and theoretical research (II, "theory"; Bunge, 1998; Tretter, 2005; Kell and Knowles, 2006). Empirical data have to be related to a theoretical framework (model, theory) by complex data analysis that can be tested by computer experiments and that again can induce gathering of new data.

formation. This interplay between experimental research and theoretical research can be demonstrated by examples in the history of physics (Newton, 2007).

Several terms have been coined for this process, like the "epistemic cycle of research" or "research cycle" or "knowledge cycle" (Bunge, 1998; Tretter, 2005; Kell and Knowles, 2006). See Figure 1.10.

In the context of this view it also has to be mentioned that stages of qualitative research are followed by stages of quantitative research, both in empirical research as well as in theoretical research. In early stages of research, only qualitative observations are available. Later, when more sophisticated measurement technologies are developed, quantitative data are retrievable. In parallel, initially only qualitative hypotheses and theories can be set-up; however, when quantitative data can be used, quantitative theoretical models and hypotheses can also be constructed.

### 1.4.3
### Top-Down Analysis – Reductionism?

Empirical science aims to isolate a single factor that generates the phenomenon that should be explained – light is composed of photons that could be isolated theoretically and by experiments, and the constitution of water could be reduced to one atom of oxygen and two atoms of hydrogen. In biology and medicine, the "reductionistic

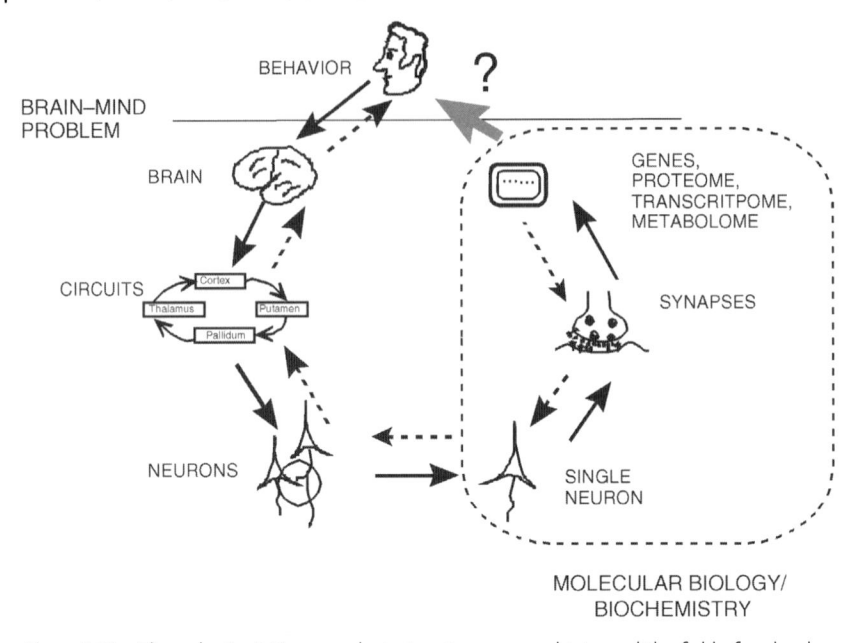

**Figure 1.11** The reductionistic research strategy in neuropsychiatry and the field of molecular biological investigation. This implies the micro–macro problem as one aspect of the brain–mind problem – how can behavior be "explained" by genes?

paradigm" is also driven by the aim and hope to find the "master molecule" or the "master gene." It is questionable whether or not it can be expected that the molecular basis of mental disorders can be identified by this reductionistic research strategy (Figure 1.11).

This reductionistic research strategy is generally a successful way to understand the structure of nature. However, detailed information on structure does not necessarily allow us to understand the global function of the system – knowledge of the parts of a clock does not imply the understanding of the mechanics of the whole clock. At present, the methodology of systems thinking might be helpful in the interplay with new multiple-unit measurement techniques in experimental molecular biology.

### 1.4.4
**Bottom-Up Explanations – Holism?**

From a philosophical point of view this top-down approach entails the problem of bottom-up explanation of the behavioral and clinical macro-phenomena by micro-phenomena such as molecular peculiarities. There are explanatory gaps that cannot be explained by physical and chemical categories, but that need bridge concepts that explain new emergent phenomena (Bedau and Humphreys, 2007) such as self-organized synchronization of activity of neurons and so on (Singer, 1999). For example, the interactions of molecules of a system evoke the temperature of the respective system; however, temperature is not a property of single molecules. In this way, the explanation of spike activity of a neuron is based on the fact that activation of several ion channels allows the rapid exchange of ions between the intracellular and extracellular space, as demonstrated by Hodgkin and Huxley (1952). Edelman and Tononi tried to "explain" consciousness as one property of re-entrant activity of highly connected neuronal circuits only by neurophysiological terms (Edelman and Tononi, 2000). Regarding psychiatry, there is the need for a neurobiological explanation of phenomena such as "craving" in addiction, "hallucinations" in schizophrenia, or "suicidality" in depression.

One research strategy that opens new perspectives for understanding complex biosystems is called "systems biology" (Kitano, 2002a, 2002b; Klipp *et al.*, 2005, 2009; Palsson, 2006). In principle, systems biology develops strategies to design experimental research in a systemic view, and to use computer-based modeling and experimental simulations in order to understand the network processes in cells, tissues, organs, and the whole organism (Noble, 2006, 2008). Promising approaches of systems biology start at the level of genomics, transcriptomics, proteomics, metabolomics, and other fields of "-omics" (Palsson, 2006). Methodologically, it is based on mathematics, informatics, and experimental molecular biology.

In the context of this new way of thinking in biology, the term "scientific knowledge" needs to be clarified.

## 1.5
## Theoretical (Neuro)psychiatry

Regarding the "epistemic cycle" (Section 1.4.2), it has to be seen that there is no explicit field that can be called "theoretical psychiatry." Instead, nearly all progress in psychiatry is thought to be based on experimental biological research. Only a few attempts can be seen that aim to construct a theoretical perspective.

Therefore, theoretical psychiatry could be defined as the field of hypotheses, models, and theories that describe and explain the mechanisms of mental disorders. In the case of neuropsychiatry, this description and explanation is provided by neurobiology. Biological research in the context of psychiatric issues has produced a rapidly expanding amount of data on the functions and dysfunctions of the brain. At present, neuropsychiatry is making much progress by using new molecular biological technologies such as microarrays. By this method, the molecular biological analysis of the genome, the transcriptome, the proteome, and the metabolome, in relation to psychiatric diseases, is proceeding very rapidly (Gebicke-Haerter, 2008). The data that are collected by high-throughput technologies demand more sophisticated instruments for data analysis in order to detect the coincidence of activation or inhibition of thousands of genes (Bhave *et al.*, 2007; Gebicke-Haerter, 2008; also see Chapter 13).

Basically, for theoretical neuropsychiatry, several methodological problems of theoretical brain research have to noted (*cf.* Tretter, Müller, and Carlsson, 2006; Tretter and Albus, 2007; Tretter, Gallinat, and Müller, 2008):

- The *complexity of neurobiological data* has to be reduced and the technically caused heterogeneity of data has to be controlled, In order to explain mental processes and their disorders, data that are obtained by different procedures such as imaging methods, electrophysiology, histology, biochemistry, pharmacology, molecular biology, and other techniques need integrative theoretical constructs that can bridge these gaps. Integrating distinct time or spatial domains is one of these challenges – electrophysiology has a good time resolution, but a bad spatial resolution as it is hard to measure the spatial meso-level (local networks) even with multiple electrode arrays. On the contrary, imaging methods have good spatial resolution, but need some seconds to generate the signals. Therefore, there always seems a methodological gap between the measurements.
- The *reconstruction of the whole brain* (or of functionally significant modules) on the basis of experimental and clinical data has to bridge some explanatory gaps regarding the difficulty in explaining single phenomena such as gestalt perception or emotions. This analytical challenge also includes the problem of integrating various timescales ranging from nanoseconds to weeks, months, or years. These aspects of the brain–mind problem is tackling neuropsychiatry and neuropsychology(Dennett, 2006; Searle, 1997, 2005; Tretter, 2007). Therefore, it is not sufficient to substitute psychological terms by neurobiological terms.
- The *structural and functional complexity* of the brain cannot be understood completely. For the brain, there are estimated to be about $10^{11}$ neurons each with $10^3$ connections. This implies that in a neuronal pathway a signal or information "re-enters" the pathway after about four neurons (or synapses). Therefore, feedback loops are essential for the functional structure of the brain. If $10^2$ spikes per second can occur, for $10^{14}$ synapses $10^{16}$ impulses have to be recorded and computed for modeling. This is beyond computability ("transcomputability"; von Foerster, 2002) and implies a challenge for modeling. In this view, the brain has to be conceptualized as a complex dynamic network (Edelman and Tononi, 2000; Tretter, 2005;Tretter, Gallinat, and Müller, 2008). As a consequence, systems science seems to be the appropriate approach in order to build theoretical models of the brain and its disorders, and to understand mental disorders.

The enormous complexity of data implies that mathematical tools such as multivariate statistics or graph theory have to be used for data analysis (see also Marin-Sangiuno and Mendoza, 2008). In a next step, for theory building in psychiatry, the integration of different and complex sets of data is necessary in order to represent this information in the respective model. For this reason, the model complexity exceeds the options of imagination alone. Therefore, computer-based data analysis and modeling becomes an increasingly important tool to study the operations of the various molecular networks. In a next important step, the results of these *in silico* experiments are transferred to the "wet" laboratory of biochemical and molecular biological experimentation. This procedure of integrating experimental research and theoretical research is cultivated in the context of systems biology (Savageau, 1976). Therefore, it seems to be promising to integrate systems biology into the field of psychiatry (Tretter and Albus, 2008; Tretter, Gallinat, and Müller, 2008). As the term "systems" is very general, it seems to be useful to characterize one possible academic reference discipline that is concerned with systems thinking, namely "systems science."

## 1.6
Systems Thinking

Systems thinking as a science could be a new direction and way of understanding organisms (Tretter, 1989; Ahn *et al.*, 2006a, 2006b). It is routed in the work of Norbert Wiener (1948) and Ludwig von Bertalanffy (1968) and it aims to understand systems on the basis of their signaling networks. Dynamics and complexity are central issues (Watts and Strogatz, 1998; Strogatz, 2001; Periwal *et al.*, 2006; Quarteroni, Formaggia, and Veneziani, 2006; Stelling *et al.*, 2006). The academic development of the systemic methodology is sometimes named "systems science," although it is hard to find university courses that teach these skills.

Systems thinking understands a system as being composed of elements and their relations. It also assumes that we are only able to understand the reality by constructing maps and models. In this view, modeling is a procedure that starts with qualitative concepts and ends up in mathematical models that can be tested by computer experiments (*in silico* experiments). Additionally, these models should help to explore the real systems (e.g., neurons) in experimental set-ups. With new data, the models are modified again and new computer simulations can be performed. Thus, a "viable" concept of the functional structure of the respective system is generated by iterative development of the models (Figure 1.12). Models are collected in order to provide a freeware set for other researchers (Finney *et al.*, 2006).

Some theories, such as catastrophe theory, chaos theory, or complexity theory, provide concepts that can be used to characterize the observed type of dynamics of the system under study. It should be mentioned here that cybernetics and systems science was substituted by informatics using the concept of "artificial neuronal networks" (Arbib and Grethe, 2001; Arbib, 2002) and now "computational (neuro) science" is the dominating field of modeling complex dynamical neuronal systems (Dayan and Abbott, 2005; Shiflet and Shiflet, 2006; Fishwick, 2007). "Complexity science" is also now booming (Boccara, 2004; Erdi, 2008).

We think that systems biology, together with ecology, will be the fields where the most fruitful developments in the theory of biosystems are going to be developed.

## 1.7
Perspectives – Towards a "Neurophilosophy"

The incredible amount of data that is generated by neurobiology was not noticed in the philosophy of mind up until the end of the 1970s. At that time, Karl Popper and Sir John Eccles tried to established a new philosophy of the brain and mind (Popper and Eccles, 1977). They proposed a dualistic concept of mind and brain, and even a third dimension that they called the world of cultural objects. In the 1980s, Patricia and Paul Churchland criticized the classical brain–mind debate and proposed a "neurophilosophy" based not only on neurobiology, but also on neuroinformatics, neurocybernetics, and artificial intelligence (Churchland, 1984, 1986). Several other authors followed this new approach (e.g., Northoff, 2000). See Figure 1.13.

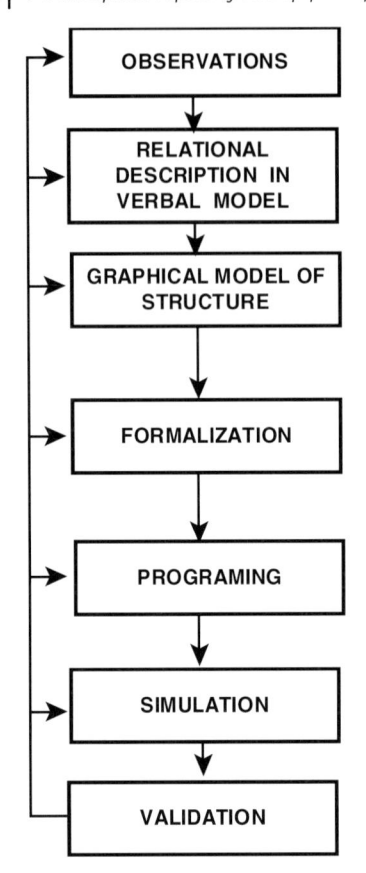

**Figure 1.12** Steps of systemic modeling (Tretter, 2005): critical steps are the formalization and computerization of the model.

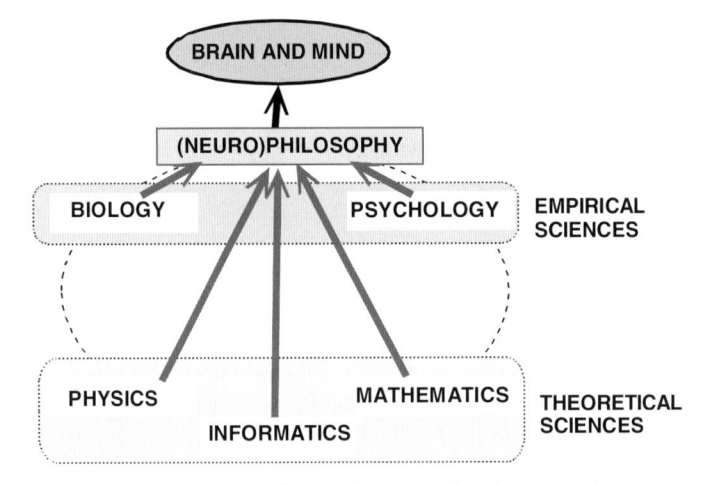

**Figure 1.13** The structure of a multidisciplinary-based "neurophilosophy."

**Table 1.1** Problem survey to the current brain–mind discourse as a questionnaire with examples (Tretter and Gruenhut, 2010).

1. Questions to the essence of brain and mind
    1.1 Materialism versus idealism
    If there are two different qualities, is there a mind as a special entity? If there is only one entity, which "entity" explains the other one?
    1.2 Dualism versus monism
    Does duality exist in a quality dualism?
    Is a methodological dualism justifiable?
    Can a monism be established now conclusively?
2. Questions focusing on the relationship between brain and mind
    2.1 Correlation and causality
    Methodologically considered: are only "correlations" between biological and psychological variables possible?
    Are statements regarding causality relations more than (pure) hypotheses?
    2.2 Stochastics and determinism
    Is there already a deterministic theory of brain functions or is the statement about the determination of brain processes a hypothesis?
3. Questions regarding the methodology of analysis of brain and mind
    3.1 The internal viewpoint versus external viewpoint (first-person perspective/third-person perspective)
    Is a complete substitution of the internal subjective viewpoint by the external objective viewpoint possible (external is internal)?
    Is the subjective viewpoint the precondition for the brain–mind debate and also the problem?
    Is a priority of the external scientific viewpoint only possible in the case of elimination of the subjective perspective and does it lead to monistic materialism?
    3.2 Plurality of neurobiological methods and generalizations
    How can the different pictures of the brain be integrated that arise through neurophysiology, histology, radiology, and so forth?
    How can single discoveries be used for explicit generalizations?
    3.3 Relation of structure and function: "neuropsychological uncertainty relation"
    Can the following relation be formulated? "The more precisely the neurobiological area is determined in the brain, the less the psychological function becomes accurately definable" (multifunctionality of brain areas, unspecificity of ion channels) and "The more precisely the function is determined, the more inaccurate becomes the identification of the brain area of this function" (multilocality of functions).
    The multifunctionality can be derived reasonably from brain areas: there is an unspecificity of ion channels; the striatum is an area related to compulsive behavior, schizophrenia, addiction, and so on.
    The multilocality can be demonstrated easily: about 40 areas in the cortex are related to vision.
    3.4 Micro–macro problem
    Can conscious functions be explained by activity of a circumscribed number of neurons?
4. Problems of the participating disciplines
    4.1 Representation of the professional competence in the discussion
    Empirics, theory, and metatheory.
    Psychology as the science of experience and behavior is rarely integrated in the discussion: it is not very convincing if, for example, philosophers themselves develop theories without psychology.
    Brain research still has hardly any systematized theories and represents only research in an ensemble of theorems (theory elements).

*(Continued)*

**Table 1.1** (Continued)

---

There exists a lack of cooperation of theory-competent disciplines such as physics, mathematics, and, above all, systems theory.
4.2 Interdisciplinary language problems
For interdisciplinary communication elaborated ordinary language is better than expert languages: is there an essential loss of precision?
Propositions very often show little systematization and therefore also logic consistency is poor
Dichotomic nominal category pairs ("determinism/indeterminism") should be avoided by "scaling" in "strong," "middle," and "weak".
4.3 Deficits of a theory of the brain
There are only few comprehensive theories for brain functions.
Concepts of brain theories lack systemic formulation that seems to be appropriate to the network character of the brain.

---

Here, we finally propose a concept of neurophilosophy that is based by multidisciplinarity. In this concept, philosophy should play a crucial role, and should integrate the various positions and arguments. It seems to be crucial for the mind–brain discussion that not only philosophers or only neurobiologists, but also other disciplines should discuss the empirical evidence and the theoretical and metatheoeretical aspects of the brain–mind relation. Some of these aspects are summarized in Table 1.1.

Although there is a wide debate in this field, there is still not enough institutionalization of such an interdisciplinary discourse that should be constituted by the integration of philosophy, psychology, neurobiology, and theoretical disciplines, such as informatics, systems research, and physics.

### Acknowledgment

I am very grateful to Peter Gebicke-Haerter for valuable criticism, corrections, and discussions of the manuscript.

### References

Ahn, A.C., Tewari, M., Poon, C.S., and Phillips, R.S. (2006a) The limits of reductionism in medicine: could systems biology offer an alternative? *PLoS Medicine*, **3**, e208.

Ahn, A.C., Tewari, M., Poon, C.S., and Phillips, R.S. (2006b) The clinical applications of a systems approach. *PLoS Medicine*, **3**, e209.

Andreasen, N. (2004) *Brave New Brain – Conquering Mental Illness in the Era of The Genome*, Oxford University Press, New York.

American Psychiatric Association (2000) *Diagnostic and Statistical Manual of Mental Disorders, text revision*, 4th edn, American Psychiatric Association, Washington, DC.

Arbib, M.A. (ed.) (2002) *The Handbook of Brain Theory and Neural Networks*, 2nd edn, MIT Press, Cambridge, MA.

Arbib, M.A. and Grethe, J.S. (2001) *Computing the Brain: A Guide to Neuroinformatics*, Academic Press, San Diego, CA.

Bedau, M.A. and Humphreys, P. (eds.) (2007) *Emergence – Contemporary Readings in*

*Philosophy and Science*, MIT Press, Cambridge, MA.

Bennett, M.R. and Hacker, P.M.S. (2003) *Philosophical Foundations of Neuroscience*, Blackwell, Malden, MA.

Bhave, S.H., Hornbaker, C., Phang, T.L., Saba, L., Lapadat, R., Kechris, K., Gaydos, J., McGoldrick, D., Dolbey, A., Leach, S., Sariano, B., Ellington, E., Jones, K., Mangion, J., Belknap, J.K., Williams, R.W., Hunter, L.E., Hoffmann, P.L., and Tabakoff, B. (2007) The PhenoGen Informatics website: tools for analyses of complex traits. *BMC Genet.*, **8**, 59.

Bleuler, E. (1911) *Dementia Präcox oder die Gruppe der Schizophrenien*, Deuticke, Leipzig.

Block, N. (1980) What is functionalism, in *Readings in Philosophy of Psychology VI* (ed. N. Block), Harvard University Press, Cambridge, MA, pp. 174–181.

Boccara, N. (2004) *Modeling Complex Systems*, Springer, Berlin.

Boogerd, F.C., Bruggeman, F.J., Hofmeyr, J.-H.S., and Westerhoff, H.V. (2007) *Systems Biology – Philosophical Foundations*, Elsevier, Amsterdam.

Bunge, M. (1998) *Philosophy of Science*, vol. 2, Transaction Publishers, London.

Carnap, R. (1928) *Der Logische Aufbau der Welt*, Meiner, Hamburg.

Carnap, R. (1932) Psychologie in physikalischer Sprache. *Erkenntnis*, **3**, 107–142.

Chalmers, D. (1995) Facing up to the problem of consciousness. *Journal of Consciousness Studies*, **2**, 200–219.

Churchland, P.M. (1981) Eliminative materialism and the propositional attitudes. *Journal of Philosophy*, **78**, 67–90.

Churchland, P.M. (1984) *Matter and Consciousness*, MIT Press, Cambridge, MA.

Churchland, P.S. (1986) *Neurophilosophy: Toward a Unified Science of the Mind–Brain*, MIT Press, Cambridge, MA.

Churchland, P.M. (2007) *Neurophilosophy at Work*, Cambridge University Press, New York.

Cooper, R. (2007) *Psychiatry and Philosophy of Science*, McGill-Queen's University Press, Montreal.

Crick, F.H.C. (1994) *The Astonishing Hypothesis*, Simon & Schuster, London.

Davidson, D. (1970) Mental events, in *Experience and Theory* (eds. L. Foster and J.W. Swanson), University of Massachusetts Press, Amherst, MA, pp. 79–101.

Dayan, P. and Abbott, L. (2005) *Theoretical Neuroscience. Computational and Mathematical Modeling of Neural Systems*, MIT Press, Cambridge, MA.

Dennett, D.C. (2006) *Sweet Dreams. Philosophical Obstacles to a Science of Consciousness*, MIT Press, Cambridge, MA.

Edelman, G.M. and Tononi, G. (2000) *A Universe of Consciousness How Matter Becomes Imagination*, Basic Books, New York.

Erdi, P. (2008) *Complexity Explained*, Springer, Berlin.

Finney, A., Hucka, M., Borstewin, J., Keating, S.M., Shapiro, B.E., Matthews, J., Kovitz, B.L., Schilstra, M.J., Funahashi, A., Doyle, J., and Kitano, H. (2006) Software infrastructure for effective communication and reuse of computational models, in *System Modeling in Cellular Biology: From Concepts to Nuts and Bolts* (eds. Z. Szallasi, V. Periwal, and J. Stelling), MIT Press, Cambridge MA, pp. 355–378.

Fishwick, P. A. (ed.) (2007) *Handbook of Dynamic System Modeling*, Chapman & Hall/CRC Press, Boca Raton, FL.

von Foerster, H. (2002) *Understanding Understanding: Essays on Cybernetics and Cognition*, Springer, Berlin.

Freeman, W.J. (2000) *Neurodynamics: An Exploration in Mesoscopic Brain Dynamics*, Springer, Berlin.

Gebicke-Haerter, P. (2008) Systems biology in molecular psychiatry. *Pharmacopsychiatry*, **41** (Suppl. 1), S19–S27.

Griesinger, W. (1882) *Mental Pathology and Therapeutics*, William Wood & Co., New York.

Griesinger, W. (1845) *Die Pathologie und Therapie der psychischen Krankheiten*, Krabbe, Stuttgart.

Heil, J. (2004) *Philosophy of Mind*, Oxford University Press, Oxford.

Hempel, C.G. and Oppenheim, P. (1948) Studies in the logic of explanation. *Philosophy of Science*, **15**, 135–175.

Hempel, C.G. (1965) *Aspects of Scientific Explanation*, Free Press, New York.

Hodgkin, A.L. and Huxley, A.F. (1952) Currents carried by sodium and potassium ions through the membrane of the giant axon of Loligo. *Journal of Physiology*, **117**, 500–544.

Hyman, St.E. (2007) Can neuroscience be integrated into the DSM-V? *Nature Reviews Neuroscience*, **8**, 725–732.

Jackson, F. (1982) Epiphenomenal qualia. *Philosophical Quarterly*, **32**, 127–136.

Kell, D.B. and Knowles, J.D. (2006) The role of modeling in systems biology, in *System Modeling in Cellular Biology: From Concepts to Nuts and Bolts* (eds. Z. Szallasi, J. Stelling, and V. Periwal), MIT Press, Cambridge, MA, pp. 3–18.

Kim, J. (1998) *Philosophy of Mind*, Westview Press, Boulder, CO.

Kim, J. (ed.) (2002) *Supervenience*, Cambridge University Press, Cambridge.

Kitano, H. (2002a) Systems biology: a brief overview. *Science*, **295**, 1662–1664.

Kitano, H. (2002b) Computational systems biology. *Nature*, **420**, 206–210.

Klipp, E., Herwig, R., Kowald, A., Wielring, C., and Lehrach, H. (2005) *Systems Biology in Practice*, Wiley-VCH Verlag GmbH, Weinheim.

Klipp, E., Liebermeister, W., Herwig, R., Wierling, C., Kowald, A., Lehrach, H., and Herwig, R. (2009) *Systems Biology: A Textbook*, Wiley-VCH Verlag GmbH, Weinheim.

Kraepelin, E. (1902) *Clinical Psychiatry*, Macmillan, New York.

Levine, J. (1983) Materialism and qualia: the explanatory gap. *Pacific Philosophy Quarterly*, **64**, 354–361.

Lewin, K. (1952) *Field Theory in Social Science: Selected Theoretical Papers*, Tavistock, London.

Lewis, D. (1966) An argument for the identity theory. *Journal of Philosophy*, **63**, 17–25.

Libet, B. (2005) *Mind Time*, Suhrkamp, Frankfurt.

Libet, B., Gleason, C.A., Wright, E., and Pearl, D. K. (1983) Time of conscious intention to act in relation to onset of cerebral activity (readiness-potential). The unconscious initiation of a freely voluntary act. *Brain*, **106**, 623–642.

Marin-Sangiuno, A. and Mendoza, E.R. (2008) Hybrid modeling in computational neuropsychiatry. *Pharmacopsychiatry*, **41** (Suppl. 1), S85–S88.

Metzinger, T. (2009) *The Ego Tunnel: The Science of the Mind and the Myth of the Self*, Basic Books, New York.

McGinn, C. (1989) Can we solve the mind–body problem? *Mind*, **98**, 349–366.

Nagel, T. (1974) What is it like to be a bat? *Philosophical Review*, **83**, 435–450.

Newton, R.G. (2007) *From Clockwork to Crapshoot: A History of Physics*, Pelknap Press of Harvard University Press, Cambridge, MA.

Noble, D. (2006) Multilevel modeling in Systems biology: from cells to whole organs, in *System Modeling in Cellular Biology: From Concepts to Nuts and Bolts* (eds. Z. Szallasi, V. Periwal, and J. Stelling), MIT Press, Cambridge, MA, pp. 297–312.

Noble, D. (2008) *Music of Life. Biology Beyond Genes*, Oxford University Press, New York.

Northoff, G. (2000) *Das Gehirn – Eine neurophilophische Bestandsaufname*, Mentis, Paderborn.

Northoff, G. (2004) *Philosophy of the Brain. The Brain Problem*, Benjamins, New York.

Palsson, B.O. (2006) *Systems Biology*, Cambridge University Press, Cambridge.

Periwal, V., Szallasi, Z., and Stelling, J. (2006) System modeling – why and how?, in *System Modeling in Cellular Biology: From Concepts to Nuts and Bolts* (eds. Z. Szallasi, V. Periwal, and J. Stelling), MIT Press, Cambridge, MA, pp. vii–viv.

Place, U.D. (1956) Is consciousness a brain process? *British Journal of Psychology*, **47**, 44–50.

Popper, K.R. and Eccles, J.C. (1977) *The Self and the Brain*, Springer, London.

Putnam, H. (1965) Brain and behavior, in *Analytical Philosophy* (ed. R.J. Butler), Blackwell, Oxford, pp. 1–19.

Quarteroni, A., Formaggia, L., and Veneziani, A. (eds.) (2006) *Complex Systems in Biomedicine*, Springer, Berlin.

Sadock, B.J., and Sadock, V.A. (2007) *Kaplan's and Sadock's Synopsis of Psychiatry*, 10th edn, Lippincott Williams & Wilkins, New York.

Savageau, M. (1976) *Biochemical Systems Analysis: A Study of Function and Design in Molecular Biology*, Addison-Wesley, Reading, MA.

Searle, J.R. (1997) *The Mystery of Consciousness*, Granta, London.

Searle, J.R. (2005) *Mind: A Brief Introduction*. Oxford University Press, New York.

Shiflet, A.B. and Shiflet, G.W. (2006) Introduction to computational science.

Modeling and Simulation for the Sciences. Princeton Univ. Press, Princeton, New Yersey.

Shoemaker, S. (1996) *The First-Person Perspective, and Other Essays*, Cambridge University Press, New York.

Singer, W. (1999) Neuronal synchrony: a versatile code for the definition of relations? *Neuron*, **24**, 49–65.

Sober, E. (2000) *Philosophy of Biology*, Westview Press, Boulder, CO.

Stadler, F. (2001) *The Vienna Circle*, Springer, Berlin.

Stelling, U.J., Doyle, F., and Doyle, J. (2006). Complexity and robustness of cellular systems. In: Szallasi, Z., Stelling, U.J., and Periwal, V. (eds.) System modeling in cellular biology. MIT Press, Cambridge, Mass. 19–39.

Stich, P. (1978) Autonomous psychology and the believe/desire thesis. *The Monist*, **61**, 573–591.

Strogatz, S.H. (2001): Nonlinear Dynamics and Chaos. With Applications to Physics, Biology, Chemistry and Engineering. The Perseus Books Group, New York.

Taylor, M.A. and Vaidya, N.A. (2009) *Descriptive Psychopathology*, Cambridge University Press, New York.

Tretter, F. (1989) Systemwissenschaft in der Medizin. *Deutsches Ärzteblatt*, **43**, 3198–3209.

Tretter, F. (2005) *Systemtheorie im klinischen Kontext*, Pabst, Lengerich.

Tretter, F. (2007) Die gehirn–geist-debatte. Wissenschaftstheoretische probleme in hinblick auf die psychiatrie [The brain–mind debate problems in the philosophy of science with regard to psychiatry]. *Der Nervenarzt*, **78**, 498, 501–504.

Tretter, F. and Albus, M. (2007) Computational neuropsychiatry of working memory disorders in schizophrenia: the network connectivity in prefrontal cortex – data and models. *Pharmacopsychiatry*, **40** (Suppl. 1), S2–S16.

Tretter, F. and Albus, M. (2008) Systems biology and psychiatry – modeling molecular and cellular networks of mental disorders. *Pharmacopsychiatry*, **41** (Suppl. 1), S2–S18.

Tretter, F. and Gruenhut, Ch. (2010) *Das Gehirn ist der Geist? Einführung in die Neurophilosphie*, Hogrefe, Göttingen.

Tretter, F., Müller, W. and Carlsson, A. (eds.) (2006) Systems science, computational science and neurobiology of schizophrenia. *Pharmacopsychiatry*, **39** (Suppl. 1).

Tretter, F., Gallinat, J. and Müller, W. (eds.) (2008) Systems biology and psychiatry. *Pharmacopsychiatry*, **41** (Suppl. 1).

von Bertalanffy, L. (1968) *General System Theory*, Braziller, New York.

Walter, S. and Heckmann, H.D. (2003) *Physicalism and Mental Causation*, Imprint Academic, Charlottesville, VA.

Watts, D.J. and Strogatz, S.H. (1998) Collective dynamics of 'Small World' networks. *Nature*, **393**, 440–442.

Wiener, N. (1948) *Cybernetics*, MIT Press, Cambridge, MA.

WHO (1992) *ICD-10: International Statistical Classification of Diseases and Related Health Problems*, 10th revision. WHO, Geneva.

Wundt, W. (1896) *Grundriss der Psychologie*, Engelmann, Leipzig.

# 2
# Neuropsychiatry – Subject, Concepts, Methods, and Computational Models

*Felix Tretter and Peter J. Gebicke-Haerter*

## 2.1
### Introduction

In this chapter, in Section 2.2 we give a brief overview on some basic issues of psychiatry: general issues, psychopathology, psychiatric diagnoses, and the special field of theoretical reasoning in psychiatry. We refer to aspects of schizophrenia in order to give an example of clinical psychiatry. In Section 2.3, we review some basic aspects of neurobiology that are relevant for psychiatry. Additionally, we refer to schizophrenia, demonstrating the neurobiology of this illness and describing some systemic modeling approaches. Where appropriate, references to depression, addiction, and dementia are made. For a deeper understanding of psychiatry, the reader is directed to textbooks of psychiatry (e.g., Sadock and Sadock, 2007).

## 2.2
### Psychiatric Fundamentals of Neuropsychiatry

#### 2.2.1
##### General Psychiatry

Psychiatry is a field of medicine devoted to the study, diagnosis, treatment, and prevention of mental disorders in humans (Sadock and Sadock, 2005). It encompasses investigations on the origin (etiology) and prognosis of those disorders as well as their psychopathology as a basis for diagnostic evaluations (Taylor and Vaidya, 2009), and pharmacology as a field of drug therapies (Stahl, 2008a, 2008b). Examples of mental disorders are schizophrenia, depression, anxiety, post-traumatic disorders, hyperactivity disorder, dementia, addiction, and so on (e.g., Sadock and Sadock, 2005). Schizophrenia, which is a complex illness with a world-wide prevalence of about 1% (Sadock and Sadock, 2007), will be a focus of this and some other chapters of this book.

*Systems Biology in Psychiatric Research.*
Edited by F. Tretter, P.J. Gebicke-Haerter, E.R. Mendoza, and G. Winterer
Copyright © 2010 WILEY-VCH Verlag GmbH & Co. KGaA, Weinheim
ISBN: 978-3-527-32503-0

## 2.2.2
### Psychopathology

Psychopathology is a special field in psychiatry that describes and classifies abnormal mental states, processes, and behavior. Mental disorders were already recognized and documented thousands of years ago. Psychotic and depressive states, but also dementive stages, were reported. Later on, more precise classifications (nosology) were developed. For instance, Emil Kraepelin distinguished "dementia praecox" from "depressive disorder" (Kraepelin, 1902; see also Chapter 5 by Kawohl and Hoff). Today, in order to standardize the classification of mental diseases, the *International Classification of Diseases*, 10th revision (ICD-10) has been established by the World Health Organization (WHO, 1992). These illnesses are diagnosed on the basis of psychopathological symptoms that are reported to or observed by the psychiatrist.

In clinical practice, the clinical interview conducted by a psychiatrist explores various mental functions such as consciousness, perception, thinking, memory, affective behavior, and so on (Table 2.1). For instance, acoustic hallucinations (e.g., hearing voices) and delusions (e.g., feeling to be Jesus) are symptoms related to, but not specific for, schizophrenia. Also, a deficiency in working memory may be indicative of schizophrenia. However, deficiencies in working memory may also be noticed in other disorders.

After gathering this data on the patient's mental state, a diagnosis has to be made (Table 2.2). The diagnostic classification is based on a brief checklist of symptoms that characterizes the disorder. The sum of several symptoms is called a syndrome, which is used to be subsumed under a diagnostic category. For instance, schizophrenia is characterized by so-called positive symptoms (e.g., hallucinations), negative symptoms (e.g., withdrawal), and cognitive symptoms (e.g., impaired working memory function). Depending on the individual symptom profile, different types of schizophrenia can be distinguished (see Section 2.2.3.1).

#### 2.2.2.1 Quantitative Psychopathology
In order to obtain a maximum of unbiased and quantitative evaluations of the psychopathological status, several rating scales were developed. They are used both in clinical practice and in research. The Hamilton Depression Scale is used for rating depressive states, the Hamilton Anxiety Scale for scaling of anxiety states, and the

**Table 2.1** Checklist of symptoms for psychiatric clinical examination (Sadock and Sadock, 2005, pp. 21–34).

| | |
|---|---|
| Consciousness | wakefulness, orientation, concentration, attention |
| Perception | illusions, hallucinations as pathological function |
| Thinking | thought disorder, delusions as pathological function |
| Speech | aphasia as pathological function |
| Memory | working memory, short-term memory, long-term memory (amnesia as pathological function) |
| Emotions | acute or sustained anxiety, depression, mania, dysphoria, and so on |
| Motivation | incentive, sexual interests, interests, hobbies |
| Motor behavior | coordination, goal-directed movements |

**Table 2.2** Some diagnoses of mental disorders (Sadock and Sadock, 2005, pp. 36–44).

Disorders usually first diagnosed in infancy, childhood, or adolescence (e.g., mental retardation, learning disorders, motor skills disorders)
Delirium, dementia, and amnestic and other cognitive disorders (e.g., delirium, dementia, amnestic disorders)
Substance-related disorders (e.g., abuse, dependence, intoxications, withdrawal as consequence of drug use)
Schizophrenia and other psychotic disorders (e.g., schizophrenia, schizoaffective disorders)
Mood disorders: bipolar disorders (manic and depressive episodes) and depressive disorders (e.g., major depression)
Anxiety disorders
Eating disorders
Sleep disorders
Personality disorder

Positive and Negative Syndrome Scale for quantification of symptoms of schizophrenia (Burt, Sederer and Isgack, 2002). Changes of the scores over time are considered as indicators of efficacy of the respective treatments.

Additionally, various psychological tests can be used. For instance, the computer-based Wisconsin Card Sorting Test can help to identify cognitive dysfunctions observed in schizophrenia. Four cards are presented to the subject and a fifth card is to be related to one of these cards regarding the features of color, shape or number. Speed and number of errors are measured, indicating semantic disorders. The Stroop Test evaluates the ability to counter-balance interferences. The subject is asked to read a series of words that describe a color. Wrong associations, like a word "green" in red color, have to be identified. The number of errors and the reaction time are quantitative measures indicating the degree of cognitive dysfunction.

### 2.2.2.2 Theoretical Psychopathology

Theory in psychiatry aims to "understand" and "explain" more detailed mechanisms, – cause–effect relationships – underlying the described clinical syndromes (see Chapter 1 by Tretter). In order to understand the underlying mechanisms of psychopathological phenomena, a functionalistic view could be developed. This is useful for computational modeling in psychiatry.

**Example – Working Memory Deficiency in Functionalist Language**  In many cognitive tasks it is necessary to retain briefly presented information for some time whilst other information is being processed. After recall of this stored information, the storage will be erased. This is based on the working memory. Working memory function can be tested by the response to a brief stimulus that must be remembered some seconds later (Figure 2.1). Typically, in view of the stimulus– response paradigm, memory shows as a prolonged rectangular response function to the respective stimulus. Pathological working memory function can be characterized by slow onset and/or premature decline, by weak and vague stimulus representation, or by prolonged persistence of the function.

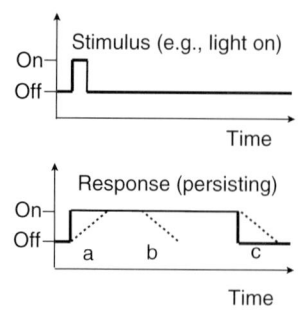

**Figure 2.1** Working memory. Functional definition in the framework of the stimulus–response paradigm. Prolonged response to a brief rectangular stimulus (e.g., a light flash) is the basis of a memory function. Pathological working memory function mainly shows as (a) a slow onset and/or (b) an earlier decline or (c) prolonged maintenance (slow offset) of the function. The respective response measure could be the discharge rate of "memory neurons" (see text).

This conceptualization of working memory can be used for computer-based modeling. A slow onset of storage could be due to a weak reciprocal activation of neuronal units. This will be discussed later in Section 2.3.1.7.

### 2.2.3
### Psychiatric Diagnoses

Mental dysfunctions are assessed in several global diagnostic categories as they are set-up in diagnostic taxonomies such as ICD-10 and the *Diagnostic and Statistical Manual of Mental Disorders*, text revision, 4th edn (DSM-IV; American Psychiatric Association, 2000) (Table 2.2).

#### 2.2.3.1 Diagnostic Criteria
Several criteria from ICD-10 or DSM-IV manuals have to be fulfilled to ascertain a respective disease. The focus are symptoms that have to occur several times a year and last for a certain time. For instance, for "schizophrenia," symptoms like delusions, hallucinations, thought disorders, ambivalence, and so on have to be observed to identify this illness (Table 2.3).

It is important to note that schizophrenia is not one well-defined disorder, but may manifest in several subtypes. For instance, if negative symptoms dominate, the disorganized type should be diagnosed; if mainly positive symptoms occur, the paranoid type can be taken as granted.

### 2.2.4
### Theoretical Psychiatry

Several approaches to explain mental disorders have been pursued after World War II – psychoanalytic, social psychiatric, and later biological concepts

**Table 2.3** Schizophrenia DSM-IV (American Psychiatric Association, 2000): Criteria to be fulfilled to diagnose schizophrenia.

A. *Characteristic symptoms:* two (or more) symptoms occurring during a 1-month period:
  (1) delusions
  (2) hallucinations
  (3) disorganized speech (e.g., frequent derailment or incoherence)
  (4) grossly disorganized or catatonic behavior
  (5) negative symptoms (i.e., affective flattening, alogia, or avolition)
B. *Social/occupational dysfunction:* disturbances in one or more major areas of functioning, such as work, interpersonal relations, or self
C. *Duration:* disturbances persist for several (e.g., 6) months
D. *Exclusion of other disorders:* brain disorders, intoxications, side effects of medications and so on

(see Chapter 5 by Kawohl and Hoff). Today, most clinical psychiatrists try to understand mental illness in the conceptual framework of a so-called "bio-psycho-social" model that integrates the action of biological, psychological, and social variables, and helps to design personalized therapeutic programs. Recently, biological psychiatry appears to be the most interesting analytical approach. There is hope that biological psychiatry can help to understand the complexity and dynamics of mental disorders, especially if computational methods are integrated into biological psychiatry.

### 2.2.4.1 "Computational Neuropsychiatry"

Some tools of modeling "natural" neural networks have been developed recently in neuroscience where "computational neuroscience" is established as a special approach of mathematical and computational modeling (Dayan and Abbott, 2005). Some authors also call this field "theoretical neuroscience." Computational neuroscience aims to develop formal models (e.g., artificial neural networks) in order to describe and explain various phenomena that are observed in experimental neurobiology. In an analogous way, "computational neuropsychiatry" can be understood as an approach in neuropsychiatry (or biological psychiatry) concerned with the development of data-based models that help to understand the (neuro)biological basis of mental disorders (Tretter, Müller, and Carlsson, 2006; Tretter and Müller, 2007). To achieve this, at present formal models are developed for only one symptom or syndrome, but not for the whole disease (Tretter and Scherer, 2006; Tretter and Albus, 2007). For instance, thought disorders in schizophrenia can be explained by a low reciprocal inhibition between nodes of a semantic network: activation of one node induces pathological coactivation of nearby nodes. Artificial neural network models allow us to explore such networks by computer-based experiments ("in silico" experiments; Cohen and Servan-Schreiber, 1993; Hoffman and McGlashan, 2006).

Currently, many circuit diagrams of "neuronal circuitries" in mental disorders are published in neuropsychiatric publications in order to demonstrate "the neuronal circuitry" of a certain mental disorder. These diagrams summarize empirical findings in a conceptual, visualized framework. Problems arise upon the introduction of the temporal component, because even with only three nodes it is hard to predict the behavior of a system. Computer-based models may help to animate these and more complex diagrams. Since it is hard to imagine the temporal development of a system even with only three components, these diagrams have to be animated by computer-based modeling and simulation.

The first step of computational modeling is to relate psychopathological terms to a functional language that represents the syndrome, as demonstrated above for the memory function that can be described as a sustained reaction to a stimulus. Next, the modalities of the respective circuit have to be defined (excitatory or inhibitory effects, strength of coupling, kinetics of processes, etc.). Until now, computational neuroscience has focused on electrical signaling in the brain, such as electroencephalography (EEG) (Freeman, 2000; see Chapter 6 by Freeman) and spiking of single neurons (see Chapter 7 by O'Donnell). Open questions are: what are the "meanings" of those signals, how do single impulses or field potentials of a group of neurons influence changes in networks and circuits? How can complex circuits (reciprocally connected pathways) be understood functionally? This is discussed later and in other papers in more detail (see Chapter 1 by Tretter and Chapter 11 by Mendoza). It should be kept in mind that modeling is always a selection from the complex reality of neurobiology. In addition, classifications of empirically observed effects of activities of some components (e.g., excitation or inhibition) are generalizations that may add skews and twists to biology, but they may help for orientation and exploratory purposes of modeling.

#### 2.2.4.2 "Systems Neuropsychiatry"

The term "systems biology" has been coined when aiming for a bottom-up synopsis of molecular biological data on the functional level of the cell, tissue, or whole organism (Kitano, 2002a, 2002b; Klipp *et al.*, 2005, 2009; Helms, 2008; Noble, 2006a, 2006b; see also Chapter 3 by Schulz and Klipp and Chapter 4 by Noble). Therefore, this approach seems to be the best basis for molecular neuropsychiatry that could be called "systems neuropsychiatry" (Tretter, Gallinat, and Müller, 2008; Tretter, Gebicke-Haerter, and Heinz, 2009; Gallinat *et al.*, 2007; Gebicke-Haerter, 2008). Several definitions of systems biology are presented in the literature, such as the framework of the National Institutes of Health (Table 2.4).

Systems biology aims to reconstruct the cell and higher-level systems on the computer (Palsson, 2006; Szallasi, Stelling, and Periwal, 2006). In this sense, psychiatry is an area of science encompassing these features in an ideal manner. Although the origins of psychiatric disorders can be associated with the brain, they can exert a marked influence on the whole organism and vice versa – physical disorders can heavily influence the brain functions.

In the following section we describe some basic neurobiological aspects of mental illnesses and present a multilevel view in modeling approaches in schizophrenia to advance the systemic view in psychiatry. Our focus is on working memory, neurochemistry, and molecular biological aspects.

**Table 2.4** Key features of a definition of systems biology (http://www.nigms.nih.gov/Initiatives/SysBio/).

- Is at the intersection of biology, mathematics, engineering, and the physical sciences
- Integrates experimental and computational approaches to study cells, tissues, and organisms
- Studies are of a quantitative nature in data collection and mathematical modeling
- Focuses on interactions among individual elements such as genes, proteins, and metabolites
- Integrates data from multiple levels of the biological information hierarchy
- Pays particular attention to dynamic processes
- Exerts successive iterations of experiment and theory development
- Generates models that are intended to predict physiological behavior to natural and artificial
- perturbations in order to understand and treat human diseases

## 2.3 Neurobiological Fundamentals of Neuropsychiatry

### 2.3.1 Basic Findings of (Neuro)biological Psychiatry

The German psychiatrist Wilhelm Griesinger had already proposed that "Mental disorders are brain disorders" (Griesinger, 1845, 1882). This view has gained more attention in recent decades with improved methods to study the brain. In line with this, psychiatry today is centered on the biological perspective, and focuses more and more on molecular biological data (Andreasen, 2004; Gebicke-Haerter, 2006). Genetic analyses (McGuffin, Owen, and Gottesman, 2004), and analyses of the transcriptome and of the proteome (Rohlff and Hollis, 2003; Reckow *et al.*, 2008), have become major topics of clinical research (Thome, 2005) and experimental animal research (Koch, 2006). These empirical approaches have gained additional power through the development of new molecular biological methods such as microarrays – techniques that generate huge amounts of quantitative data (Gebicke-Haerter, 2008). In this methodological orientation psychiatry is called "biological Psychiatry" when referring to the whole body or "neuropsychiatry" when focusing on the brain. Here, we want to briefly sketch some issues of neurobiology. For further information we suggest textbooks like Nicholls *et al.* (2001), Bear, Connors, and Paradiso (2001), and Webster (2002).

#### 2.3.1.1 Neuropsychopathology

Neuropsychopathology aims to relate the psychological level, such as mental symptoms, syndromes, and illnesses, to the neurobiological level (Taylor and Vaidya, 2009). Research questions are, for instance, how are auditory hallucinations in schizophrenia, which are categorized psychopathologically, related to a hyperactivity of cortical areas that are involved in hearing and speaking (auditory cortex)? On a biochemical level it is known through clinical psychopathology of drug consumption that hallucinations can be evoked by a high level of dopamine or serotonin or by a low level of acetylcholine. In order to obtain a more differentiated view on the neurobiology of auditory hallucinations, a distinction between different types of voices that are subjectively heard by the patient could be helpful. This view can be refined by determinations of endophenotypic markers. Endophenotypes are biological markers between the genotype and external phenotype that may indicate susceptibility to or manifest as early signs of a wide range of mental disorders. Such questions are subject to metatheoretical considerations in psychopathology that are not discussed very frequently (Meyer-Lindenberg and Weinberger, 2006; Puls and Gallinat, 2008; see Chapter 5 by Kawohl and Hoff).

#### 2.3.1.2 Neurobiological Methods

**Global Electrical Signals of the Brain** Communication between the various brain structures mentioned above is ensured by electrical and chemical signaling. These activities can be identified by the EEG recorded from the scalp via electrodes. The EEG is produced by the depolarization of neurons within the brain. Since electric potentials generated by single neurons are far too small to be picked by EEG, EEG activity always reflects the summation of the synchronous activity of thousands or millions of neurons that have a similar spatial orientation in the tissue.

Scalp EEG activity shows oscillations at a variety of frequencies. Several of these oscillations have characteristic frequency ranges and spatial distributions, and are associated with different states of brain functioning (e.g., waking and various sleep stages). These oscillations are supposed to represent synchronized activity over a network of neurons (see Chapter 6 by Freeman).

This hypothesis is based on the observation that during recognition, $\gamma$ waves of about 30 Hz can be identified. During mental activity, $\beta$ waves (15 Hz) are observed, and in the case of wakefulness, $\alpha$ waves (10 Hz) are present. In states of tiredness, $\theta$ waves of 7 Hz occur, whereas in sleep slow waves ($\delta$ waves of 2 Hz) can be recorded (Figure 2.2).

In schizophrenia the $\gamma$ waves are only weak and seldom. This is a sign that indicates the impaired cognitive functioning. It is also a hint that inhibitory mechanisms in schizophrenia are insufficient, because inhibitions are important for oscillations.

**Imaging Methods** The major neuroimaging techniques used in psychiatric research are positron emission tomography (PET), single photon emission computed tomography (SPECT), and magnetic resonance imaging (MRI). MRI uses magnetic fields and radio waves to produce high-quality two- or three-dimensional images of brain structures without injecting radioactive tracers (Schlösser, Koch, and Wagner, 2007). In contrast to MRI, PET detects emissions from radioactively labeled chemicals that have been injected into the bloodstream. The data are used to produce two- or three-dimensional images of the distribution of the chemicals throughout the brain and body. Similar to PET, radioactive tracers are also used in SPECT. The data are converted into two- or three-dimensional images of active brain regions by use of a computer. Together with EEG, these methods are noninvasive and measure biological activity through the skull *in vivo*. Each technique has its own advantages and each provides different information about the relationship between brain structure and function.

Gamma waves: recognition, attention, consciousness

Beta waves: mental activity

Theta waves: drowsiness

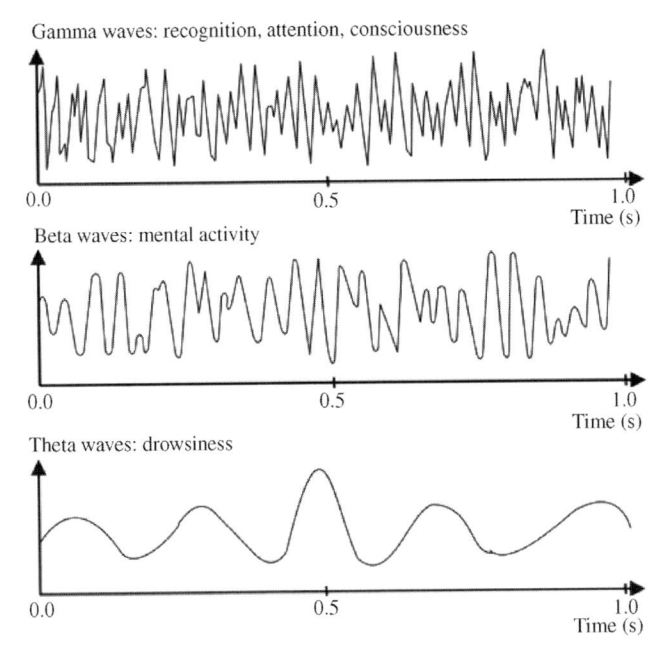

**Figure 2.2** EEG patterns during cognition (γ waves, 30 Hz), general mental activities (β waves, 15 Hz), and drowsiness (θ waves, 7 Hz) (modified from Tretter, 2005).

For this reason, studies that integrate two or more techniques are becoming more popular. For example, merging a PET scan image that shows activity at brain molecular sites, or receptors, with a highly detailed MRI image of brain structure can produce a composite image that makes it possible to identify more precisely where in the brain the activity is localized (see Chapter 9 by Vernalaken *et al.*). Suddenly increased, local neuronal activities and energy consumption evokes increased blood flow occurring after approximately 1–5 s. At these sites, hemoglobin, which is diamagnetic when carrying $O_2$, changes its configuration upon delivery of oxygen to those active neurons and becomes paramagnetic. Therefore, the magnetic resonance signal of blood is slightly different depending on the level of heme oxygenation and, hence, can be used to identify brain regions of increased neuronal activity. Some more recent reports have suggested that the increase in cerebral blood flow following neural activity is not causally related to the metabolic demands of the brain region, but rather is driven by the presence of neurotransmitters, especially glutamate.

### 2.3.1.3 Experimental Paradigms

Several experimental set-ups help to explore patients with mental disorders. First, psychophysiology and imaging methods are an important approach to study specific reactions of psychiatric patients according to standardized experimental situations. For example, the so-called prepulse inhibition is a test of a physiological response (e.g., eye blink) to two successive auditory stimuli. Schizophrenic patients do not show a reduced response to the second stimulus as observed with healthy subjects.

This is an example of an "endophenotype" (see Section 2.3.1.1) that might represent one biological basis of the clinical syndrome (see Chapter 5 by Kawohl and Hoff). Furthermore, functional MRI can be used to investigate changes in regional brain activities of alcohol-dependent patients upon exposure to cues of their addiction (e.g., a glass of beer). The degree of activation of cortical components of the limbic system (e.g., anterior cingulate cortex) is correlated with the degree of addiction (Heinz et al., 2004). It is shown also that compared to healthy subjects, depressive patients exert a stronger activation of the amygdala if they are exposed to pictures with sad or fearful faces (Canli et al., 2005).

**Animal Experiments**  Experiments in animal model systems that represent one or several features of the respective disease can be extremely helpful for medical research. In brain research, in particular, such kinds of experiments are often indispensable. Recently, a comprehensive book was published that describes several animal models of psychiatric disorders (Koch, 2006).

Here, we briefly refer to a fruitful neurobiological approach to understand impaired working memory function in schizophrenia.

**Example – Neurobiological Basis of Working Memory**  Working memory experiments were conducted in monkeys in the laboratory of Goldman-Rakic (Goldman-Rakic, 1995, 1999; Goldman-Rakic, Muly, and Williams, 2000). The animals had to fixate their eyes on a visual stimulus and were trained to retain the information of a visual cue that was briefly presented parafoveally. When the cue stimulus was switched off, the monkeys had to fixate further for some seconds and than had to look to the (absent) cue. Upon doing so, they obtained fruit juice as a reward. Simultaneous electrophysiological recordings from single neuronal units by electrodes that were inserted in the prefrontal cortex showed that those neurons exhibited a pattern of cellular discharge activity (spikes) that was related to the cognitive function of the animals (Figure 2.3). Therefore, these neurons can be interpreted as the neuronal substrate of working memory.

Investigations and experiments in human subjects showed that the working memory function depends on the level of dopamine transmission in the prefrontal cortex. Too much dopamine, as occurs after cocaine or amphetamine application, induces attention deficits, whereas low dopamine, as, observed in Parkinson's disease, diminishes the retention or intermediate storage of information (Figure 2.4). In situations of low dopamine, dopamine $D_1$ receptor agonists can improve the working memory function. In conclusion, it is not absolutely true that any component of neuronal networks such as dopamine synapses evokes linear, dose-dependent effects but rather that the effects are state-dependent. Mathematically, these observations can be summarized by an inverted U-function (Figure 2.4).

These findings are in concordance with the one very valid explanatory hypothesis for schizophrenia that is known as the "dopamine dysfunction hypothesis." This hypothesis encompasses a cortical hypofunction of dopamine-based signal transmission combined with the subcortical hyperfunction of dopamine transmission (Carlsson, 1988, 2006; Carlsson et al., 1999).

(a)

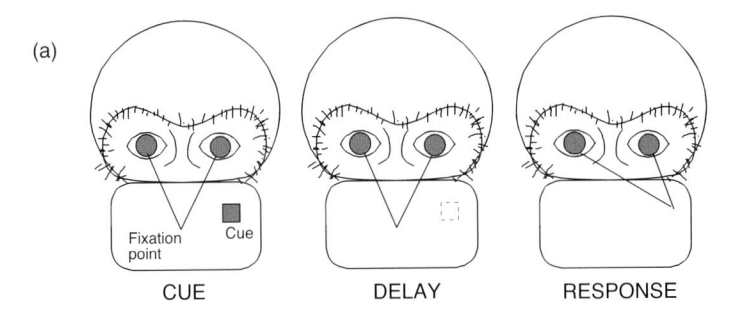

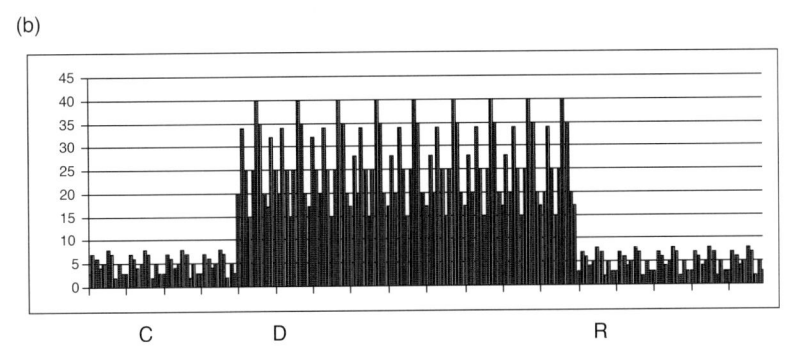

(b)

**Figure 2.3** Animal experiment on working memory function of neurons. At first, during fixation the neurons exert a low activity. When a cue is exposed (C) the discharge activity jumps up and persists although the cue is turned off (D). If fixation is switched to prior cue location the neuronal discharge activity drops back to baseline C: cue; D: delay; R: response. (After Goldman-Rakic, 1999; Wang, 2006.)

Working memory performance

$D_1$ agonist improves performance

$D_1$ agonist impairs performance

Optimum

e.g., Parkinson's disease

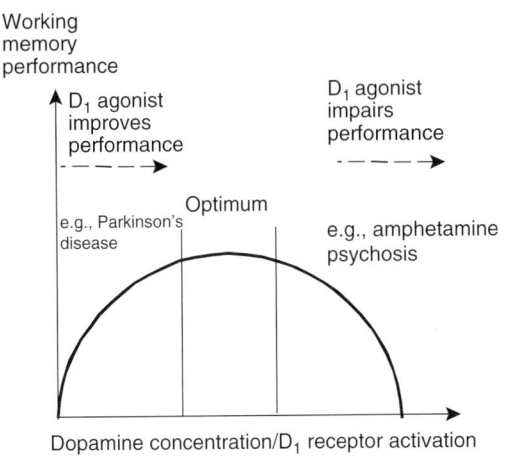

e.g., amphetamine psychosis

Dopamine concentration/$D_1$ receptor activation

**Figure 2.4** Inverted U-function that describes the relationship between level of dopamine and/or dopamine $D_1$ receptor type-based transmission and performance of working memory. This finding is relevant for understanding reduced working memory functions in schizophrenia.

**Table 2.5** Anatomical brain structures (Figure 2.5) and their mental functions.

| Structure | Function |
|---|---|
| Prefrontal cortex (PFC), at anterior pole of cerebrum | attention, planning, working memory |
| Orbitofrontal cortex (OFC), frontal area of cerebral cortex, localized above the orbita | behavioral inhibition |
| Visual cortex (VC) | cortical center of vision |
| Motor cortex (MC) | programs of motion |
| Anterior cingulated cortex (AC), structure of the limbic cortex | processes of emotional evaluation |
| Striatum (S, STR, STRI) | automatized behavior (e.g., procedural memory) |
| Thalamus (T, THA), relay nucleus in "dienecephalon" | Filter for perceptual processes |
| Nucleus accumbens (NA), area in the limbic system | states of pleasure, reward |
| Hippocampus (HIP) | episodic memory |
| Amygdaloid body, amygdala (A), area in the limbic system | anxiety |
| Substantia nigra (SN), complex of dopamine containing cells in midbrain (upper brain stem) | center of organization of movements |
| ventral tegmental area (VTA), complex of dopamine containing cells in midbrain | center of reward-related processes |
| Reticular formation (RF) | wakefullness |
| Basal forebrain (BF) | alertness |
| Brain stem (BS) | vegetative functions |

#### 2.3.1.4 Structure and Function of the Brain

Neuropsychological studies in healthy and mentally ill subjects have provided a huge amount of knowledge about the functional structure of the brain (Gazzaniga, 2004; Table 2.5 and Figure 2.5). Here we give a brief overview following a multilevel top-down procedure to describe the neurobiological findings relevant for mental disorders. The most investigated structures of the brain related to psychic and behavioral functions are the cortex cerebri with the occipital cortex that is involved in vision, the temporal cortex that generates auditory functions and is related to acoustic hallucinations, the prefrontal cortex that exhibits attention processes and working memory functions, and the orbitofrontal cortex that exerts behavior inhibition, control of emotions, and drives. A very prominent structure is the striatum that is composed of the putamen, pallidum externum, pallidum internum, and caudate nucleus. This structure is involved in the organization of automatic motor behavior. The thalamus is characterized as a sensory and motor relay unit. The hippocampus exhibits memory functions, the hypothalamus is involved in drives and vegetative functions. The nucleus accumbens is seen as a center for reward processes. The nucleus amygdala estimates relevance of stimuli, and induces anxiety and avoidance behavior. In the brain stem, the locus coeruleus is involved in alertness and general activation, the

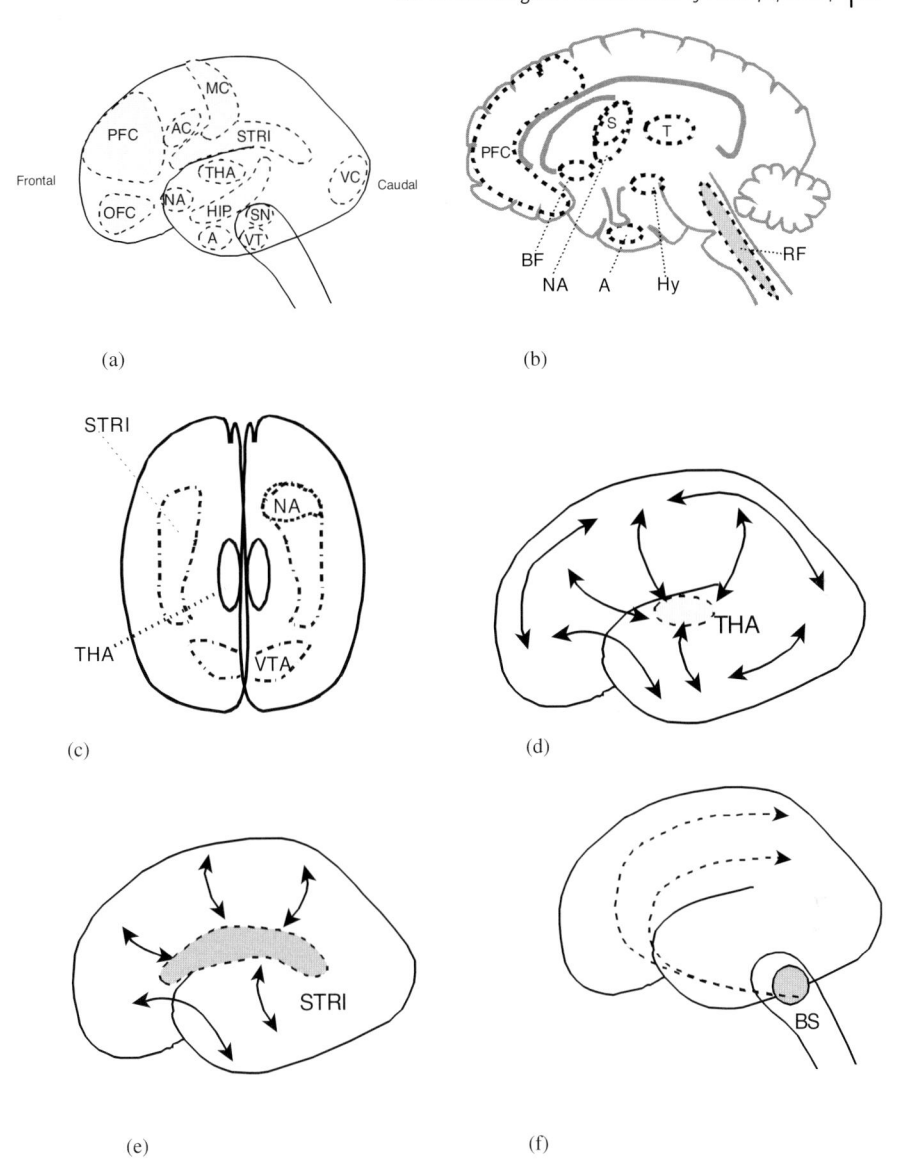

**Figure 2.5** Brain structures important for various mental functions (see text). (a) Cerebrum from the left side. (b) Right hemisphere from the medial side. (c) View of both hemispheres. (d) Circuits relevant for consciousness between the thalamus and cortex and with intracortical connections. (e) Circuits for subconscious information processing between the striatum and cortex. (f) Modulatory pathways from the brainstem to cortex (e.g., parts of dopamine systems). See Table 2.5 for abbreviations

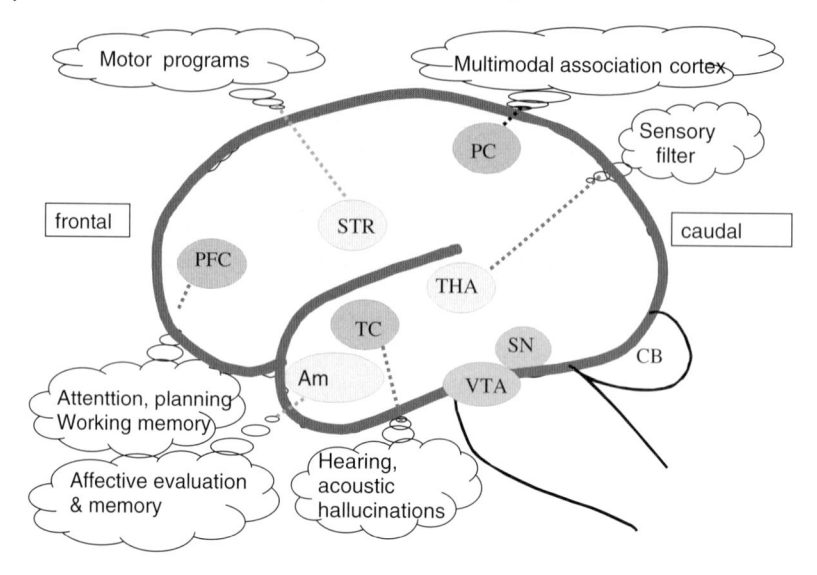

**Figure 2.6** Brain areas involved in the production of symptoms of schizophrenia. PC: parietal cortex; TC: temporal cortex; see also Table 2.5 for further abbreviations.

nucleus raphe is related to impulsiveness, the ventral tegmental area seems to be the source of the reward system, and the substantia nigra controls motor behavior by connections to the striatum.

**Brain Areas Relevant for Schizophrenia** From an anatomical point of view, mental disorders are defined as abnormalities of the morphological properties of the brain. In schizophrenia, imaging studies and clinical diagnostics often revealed enlarged ventricles. Imaging studies that analyzed working memory functions showed a hypofunction of the prefrontal cortex of these patients (Schlösser, Koch, and Wagner, 2007; Meisenzahl et al., 2007). This finding is related to the concept that a deficiency of prefrontal cortical dopamine transmission is the basis of this impairment. Several other areas in the brain are also involved in disturbed information processing in schizophrenia (Figure 2.6).

### 2.3.1.5 Global Circuits and their Connectivities

Neurons – as the atomic units of the brain – are connected to each other by short or longer fiber tracts. Short fiber connections form local neuronal networks, whereas long fiber bundles provide communication between brain regions. Cell–cell contacts between neurons are made by synapses, where electrical signals are transformed into chemical signals (transmitter) that affect activities of subsequent neurons (Table 2.6). As incoming signals are not just relayed to the next cell, but are modified in various ways, synapses are not simple transducers, but can be viewed as complex information processors. Details are discussed in Section 2.3.3.

Modification of incoming signals at synapses depends greatly on the neurotransmitter to be released into the synaptic cleft. Basically, there are activating (excitatory)

**Table 2.6** Some transmitter substances and usual abbreviations.

| | |
|---|---|
| Excitatory transmitters | glutamate (Glu) |
| | acetylcholine (ACh) |
| Inhibitory transmitters | γ-amino-butyric acid (GABA) |
| Modulators | dopamine (DA), |
| | norepinephrine (NE)/noradrenaline (NA) |
| | serotonin (5-hydroxytryptamine, 5) |

transmitters, such as glutamate, that depolarizes the membrane, and inhibiting (inhibitory) transmitters, such as γ-aminobutyric acid (GABA), that hyperpolarizes the membrane. Some transmitters can act in both ways because receptors with activating and inhibiting effects can be distinguished, even if there are some exceptions of this rule (Table 2.6). Some examples are mentioned here and will be discussed in detail in Section 2.3.3 (see also Chapter 7 by O'Donnell):

- *Excitatory subtypes of receptors* are dopamine $D_1$ receptors, serotonin (5-hydroxytryptamine) receptors of the 5-$HT_{2A}$ type, and norepinephrine $\alpha_1$ receptors.
- *Inhibitory subtypes of receptors* are dopamine $D_2$ receptors, serotonin 5-$HT_{1A}$ receptors, and norepinephrine $\alpha_2$ receptors.

On this basis of some aspects of molecular signaling mechanisms, a simple overview of the complex global circuitry in the brain that is important for each mental disorder can be given as in Figure 2.7, where activating (+) and inhibiting (−) connections between 14 components of the brain are depicted. This diagram summarizes some neurobiological evidence, but it is not able to provide an "explanation" of diseases. This can only be achieved by computational models of these circuits that simulate the respective processes.

**Global Circuits Relevant for Schizophrenia** Specifically, two main circuits require increased attention in schizophrenia research (Figure 2.8; Carlsson, 2006):

- One circuit that starts from the mid brain (ventral tegmental area) is composed of dopamine neurons that project to the medial prefrontal cortex (mPFC). This projection is affected by a *hypofunction of dopaminergic transmission* and may be an essential factor for the impairments of network functions in mPFC (see Section 2.3.1.7). Moreover, a weak glutamate pathway projecting back to the ventral tegmental area has to be taken into consideration. Owing to this "circular causality," it is hard to assess which component – the subcortical or the cortical component – is the main factor of pathology.
- Additionally, a second loop has to be considered – a *hyperactive dopaminergic projection* from the substantia nigra to inhibitory neurons in the striatal complex. This results in strong activation of the thalamus with resultant strong input into the cortex (e.g., auditory cortex with ensuing hallucinatory events). After intracortical processing, a weak activated glutamate projection signals to the substantia nigra by targeting local inhibitory neurons. Weak inhibitory coupling in substantia

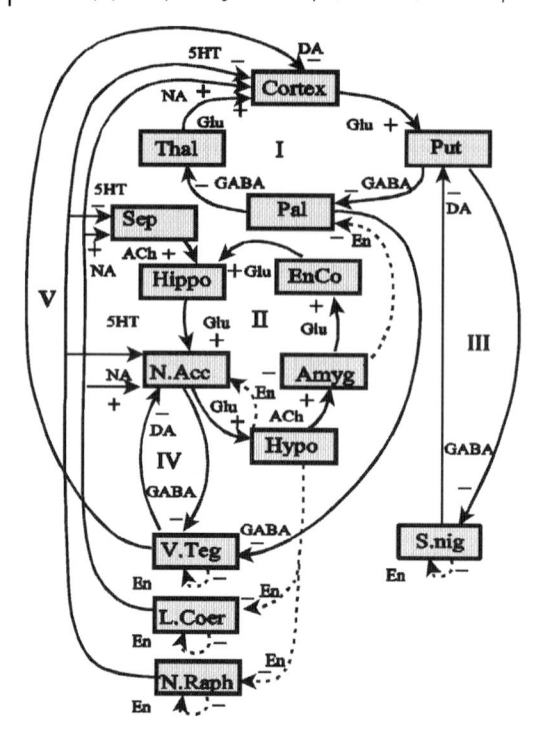

**Figure 2.7** Scheme of connectivity of global neuronal circuits (see text). Legend (see also Table 2.6): −: inhibiting action; +: activating, excitatory action; En: enkephalin; Put: putamen; Thal: thalamus; Pal: pallidum internum and externum; Sep: septal area; Hippo: hippocampus; EnCo: entorhinal cortex; Amyg: amygdala; N.Acc: nucleus accumbens; Hypo: hypothalamus; V.Teg: ventral tegmental area (dopamine center); S.nig: subtantia nigra (dopamine center); L.Coer: locus coeruleus (norepinephrine center); N.Raph: nucleus raphe (serotonin center).

nigra then results in a high inhibitory coupling by dopamine synapses in the striatal complex and so forth.

**Subcortical Circuits and Schizophrenia** One subsystem, circuit I of the global circuits of cerebral circuits of Figure 2.7, has already been very well studied. It is the system of the so called cortico-striato-cortical circuits that are relevant for symptoms of Parkinson's disease, schizophrenia, obsessive disorders, and addiction (Figure 2.9). Starting from the cortex, the feedback loop via the thalamus consists of two subloops: a direct loop via the putamen–pallidum internum–thalamus–cortex with two serially coupled inhibitions ("disinhibiton") that activates the thalamus ("accelerator") and a indirect loop via the putamen–pallidum externum–subthalamic nucleus and pallidum internum that exerts inhibition on the thalamus ("brake"). In schizophrenia, a dopamine hyperfunction via inhibitory $D_2$ receptors in functional terms can inhibit a serially coupled inhibitor with the consequence that the thalamus is disinhibited. Therefore, the thalamus might overstimulate the cortex, which could result in

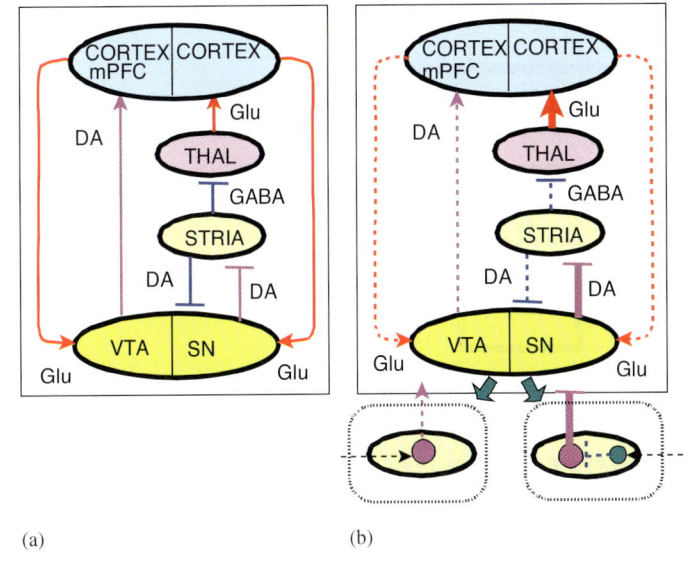

**Figure 2.8** Global circuitry relevant for schizophrenia. (a) Circuitry in a healthy subject. (b) Circuitry in a schizophrenia patient. An attenuated dopamine projection to the mPFC results in a weak glutamate feedback to the ventral tegmental area (VTA) with consequent reduced dopamine stimulation. Additionally, reduced inhibition of dopamine neurons is evoked in substantia nigra (SN) that increases the inhibitory tonus on the striatum (STRIA), with attenuated inhibition on the thalamus (THAL) and enhancement of the glutamate input to the cortex. Dashed lines: weak activity; thin lines: normal activity; thick lines: strong activity; DA: dopamine; Glu: glutamate.

hallucinations. Regarding neurological motor disorders, this circuit was animated by a computer simulation by Berns and Sejnowski (1998).

### 2.3.1.6 Local Networks of Neurons

In addition to global neuronal circuitries connecting specialized brain regions, and more easily amenable to identification and characterization, there are large numbers of local neuronal networks and local circuitries. In particular, the cortex is composed of incomprehensible networks of cells and their wiring that can be associated with different local areas (Figure 2.10a). Trials to reveal certain principles in these networks have largely failed. Attempts have been made to work with "modules" that represent one or two pyramidal cells and one inhibitory neuron. In these "canonical circuits," as introduced by the neurobiologist Gordon Shepherd (1994, 2004), the two excitatory neurons (e.g., pyramidal cells) and a inhibitory neuron are reciprocally interconnected (Figure 2.10b). Additionally, the neurons make recurrent loops onto themselves. Inhibitory feedback loops are of great importance for the behavior of local networks (Brunel and Wang, 2001; Lewis, Hashimoto, and Volk, 2005). This module also receives subcortical inputs. In this model, fast and slow processes are superimposed in a way that is poorly understood. Appropriate data

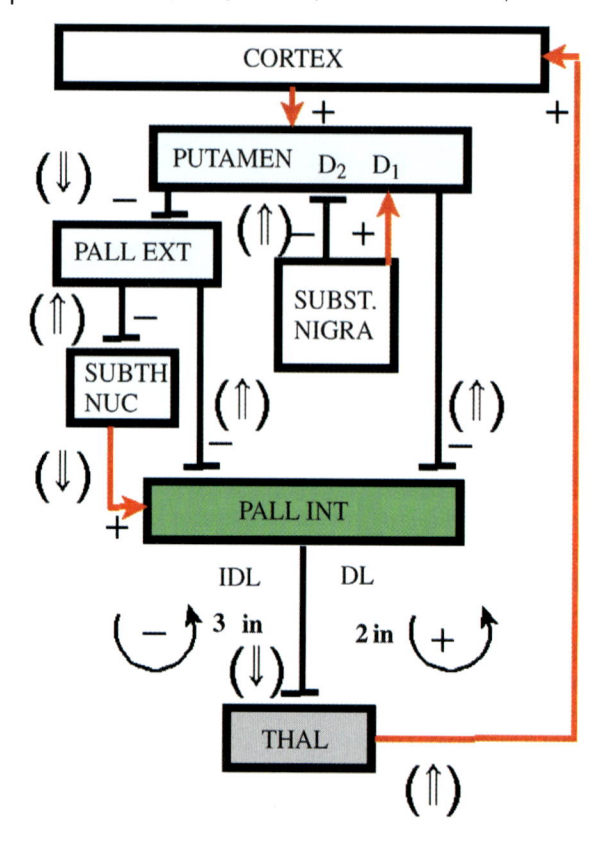

**Figure 2.9** Scheme of the functional structure of a selected subsystem of the global brain circuitry that is important for schizophrenia, obsessions, addiction, and Parkinson's disease. The cortex exhibits "antagonistic convergence", it activates (+) the thalamus via the direct loop (DL) with two inhibitions ("2 in," "accelerator") and inhibits (−) the thalamus via the indirect loop (IDL) with three inhibitions ("3 in," "brake"). The substantia nigra (SUBST NIGRA) as a center of the dopamine system can only interfere by activating $D_1$ receptors and via inhibiting $D_2$ receptors in a "synergistic convergence". In the case of schizophrenia, a hyperactivity of the dopamine pathway from substantia nigra via $D_2$ receptors (↑) inhibits the putamen so that the inhibition of the pallidum externum (PALL EXT) is reduced (↓). There is also a stronger inhibition on the subthalamic nucleus (SUBTH NUC) with subsequent attenuated effects on the pallidum internum (PALL INT). The putamen is strongly activated via $D_1$ receptors. As a consequence, the pallidum internum is strongly inhibited (↑) so that the thalamus (THAL) is disinhibited (↑). As a result, the sensory areas of the cortex could be overstimulated by the thalamus.

that describe the kinetic parameters of this network are still lacking. Therefore, we are presently far from a valid functional model of the (prefrontal) cortex that could explain mental functions and dysfunctions (Yang, Seamans, and Gorelova, 1999; Seamans and Yang, 2004; Yang and Chen, 2005).

**Development of Local Networks in Brain** During individual development there is a steady change of connections between neurons. Synapses are formed and eliminated. Connections between the neurons can be made via axon–dendrite, axon–axons,

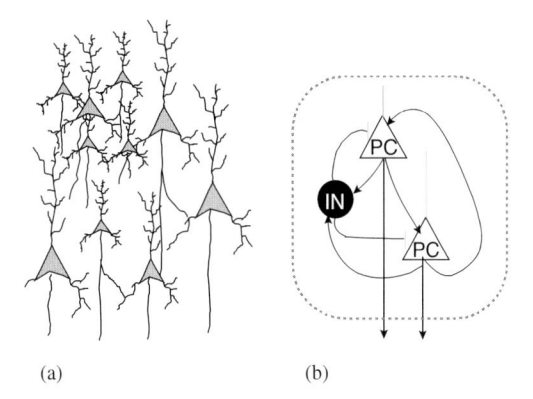

**Figure 2.10** Schemes of cortex and cortical "modules." (a) Scheme of a microscopic picture of the cortex with pyramidal cells. (b) Scheme of a cortical module, proposed by Shepherd (2004), with GABA-ergic inhibition by inhibitory interneurons (IN; bars) and glutamatergic excitation of pyramidal cells (PC; arrows). The pyramidal cells exhibit self-excitation and also the inhibitory neurons have self-inhibition (both not shown here).

axon–cell body, and dendrite–dendrite connections. An axon can branch and make contacts with more than one neuron. See Figure 2.11.

At some time in early childhood, the number of synapses reaches its maximum and then declines until some kind of optimal wiring has been established. This developmental process is called priming (Hoffman and Dobscha, 1989). It is controlled by cellular adhesion and repellant factors that are developmentally regulated. The optimization process can be disturbed by "unfavorable" conditions around birth, in early childhood, or even during adolescence. These disturbances may include stressful events like hypoxia, social deprivation, or drug abuse by mothers. For example, depressive patients show a reduced mass in the hippocampus (Bremner *et al.*, 2000) and cortical neuronal circuitries appear to be abnormal in schizophrenia, which gave rise to the "disconnectivity hypothesis" (Hoffman and McGlashan, 2006).

**Neuronal Excitation and Inhibition**  In the view of systems theory, the counteraction between inhibitory and excitatory impacts is a driving factor for the control of

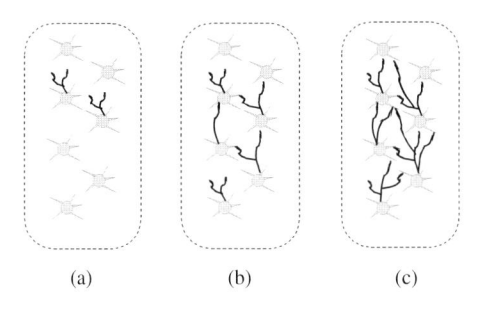

**Figure 2.11** Development of neuronal networks. From single neurons to networks (a–c; see text).

development and maintenance of neuronal circuitries. In terms of the electrophysiology (firing) of neurons, lateral inhibition and recurrent inhibition are well-known filtering mechanisms to exert directed movements or to focus one's attention (thinking, reading, writing, etc.). Therefore it is not surprising that inhibitory neurons are very frequent in cerebral cortex and that they express various receptor subtypes for inhibitory neurotransmitters to fine-tune their responses to excitation.

The present classification of inhibitory neurons in the cortex is based on morphological properties, connection properties, chemical properties and electrophysiological properties. However, there is still a low correspondence between the different methods, mainly based on animal experiments, and also in different species (Table 2.7 and Figure 2.12). Some of these findings are used for computational modeling (see below).

#### 2.3.1.7 Prefrontal Network in Schizophrenia

Recently, disturbances of information processing in schizophrenia have been related to a deficiency of inhibitory interneurons. The parvalbumin-containing interneurons that use GABA as a transmitter and that exhibit fast spiking in electrophysiological experiments seem to be crucial for the proper functioning of local cortical networks, especially in the prefrontal cortex (Figure 2.13) (Lewis, Hashimoto, and Volk, 2005). A reduction of GABA-ergic inhibition can be caused by weak GABA function and/or by weak glutamate transmission, and/or by reduced dopamine input into the cortex. These deficiencies, occurring in the dopaminergic, GABA-ergic, and glutamatergic connectivity of the prefrontal cortex, are believed to be crucial in schizophrenia

**Table 2.7** Simplified classification of inhibitory interneurons (Kawaguchi and Kubota, 1997; Lewis, Hashimoto, and Volk, 2005; Povysheva *et al.*, 2007).

| Cell morphology | Connectivity | Histochemistry | Electrophysiology |
| --- | --- | --- | --- |
| Large basket-cell type and chandelier cells | (peri)soma targeting (PTC/STC) | parvalbumin-containing c. (PV) | fast-spiking cells (FS) |
| Cells with narrow dendritic and axonal arbors; double bouquet cells, also Martinotti cells | dendrite targeting cells (DTC) | calbindin-containing cells (CB) | spike-frequency adaptation |
| | interneuron targeting cells (ITC) | calretinin-containing cells (CR) | irregular spiking patterns (often: bursting pattern, BS) |
| Nonpyramidal cells (NP) | cell body targeting | not specific | burst-spiking (BS) activity |
| Basket cells | cell body targeting | vasoactive intestinal peptide (VIP) | regular spiking cells (RS) |
| Neurogliaform cells | dendrite targeting | not specific | late spiking cells (LS) |

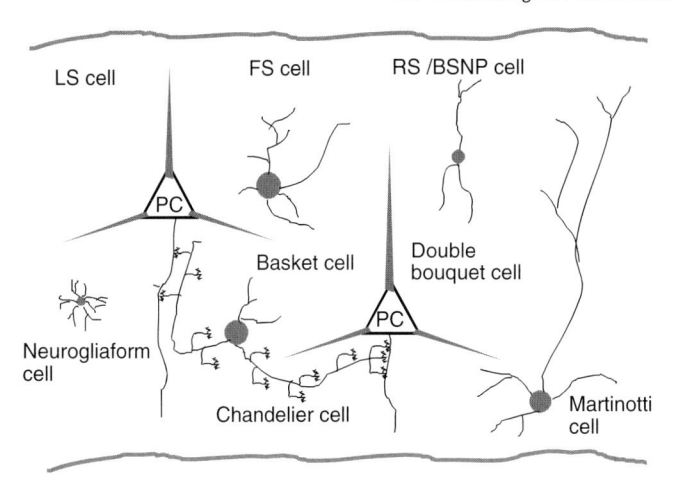

**Figure 2.12** Morphological appearance of some inhibitory neurons in the cerebral cortex (modified from Lewis, Hashimoto, and Volk, 2005). PC: pyramidal cell; see also Table 2.7 for further abbreviations.

(Figure 2.13) (Wassef, Baker, and Kochan, 2003; Winterer, 2006). Such theoretical questions can best be investigated by computer simulations.

**Computational Modeling of Working Memory Deficiency** Most computational models of the brain are based on the concept of artificial neural networks. They are constructed in analogy to biological neuronal networks (see Chapter 12 by Smith and Hütt). Each node of a network represents several, possibly some hundreds or thousands of neurons. When modeling the cortical network, decisions have to be made as to how the basic structure of the network should be designed. In most cases the structure of the networks entails inhibitory and excitatory actions. "Modules" are components of the network that exhibit a certain configuration of a circuit with excitatory and inhibitory neurons (as was discussed in Section 2.3.1.6). These modules can exhibit different functions that are equivalent to the *in vivo* processes (e.g., $\gamma$ oscillations). It should be kept in mind that each neuron also has to be modeled in detail as a multicompartment unit composed of the dendrite, cell body (soma), and axon. These compartments have to include the different ion channels, as well. Therefore, the elements of networks with increasing complexity have to be modeled as detailed as possible. The art of modeling is to end up with the most simple model that is able to display the maximum biophysical functions.

One of the most interesting computational models of schizophrenic symptoms is related to the impaired memory function, as published by Durstewitz, Seamans, and Sejnowski (2000). These authors used a network model based on hundreds of relatively simple modules with one type of inhibitory neuron and one type of excitatory neuron (Figure 2.14a). Models from other authors (e.g., Wang et al., 2004a; Wang, 2006) use more complex modules, such as with three types of inhibitory neurons (Figure 2.14b). Such models allow us to simulate more functions.

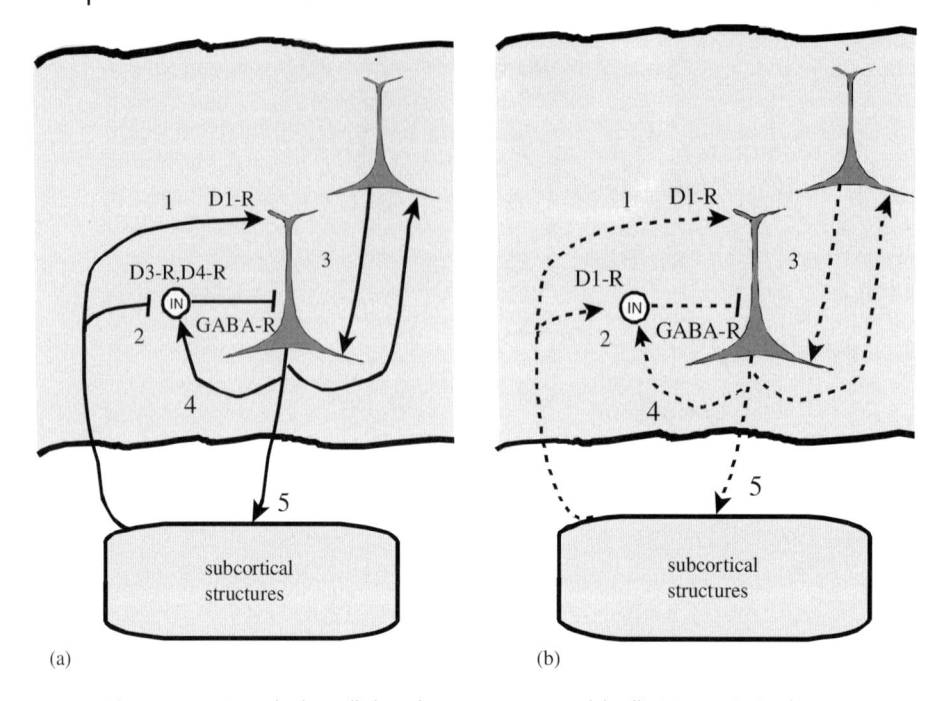

**Figure 2.13** Neurobiologically based hypothetical diagram of the "canonical" prefrontal cortical microcircuit and connections with the dopamine system (modified from Shepherd, 2004; Yang, Seamans, and Gorelova, 1999, 2004; Seamans and Yang, 2004; Yang and Chen, 2005; Lewis, Hashimoto, and Volk, 2005). (a) The balanced ratio of $D_1$ receptors (D1-R) dominating on pyramidal cells (1) versus $D_2$ ($D_3/D_4$) receptors (D3-R/D4-R) dominating on inhibitory neurons (2) may generate proper "cognitive" functions (e.g., working memory) in the subcircuit between pyramidal cells that is based on reciprocal excitatory interactions (3). Self-inhibition of pyramidal cells (4) via GABA-releasing inhibitory neurons (IN) and connections to dopamine cell centers in the mid brain (e.g., ventral tegmental area, nucleus accumbens) must also be taken into consideration for a functional understanding. (b) In the case of schizophrenia, a weak dopamine transmission is assumed with a compensatory $D_1$ receptor dominance on inhibitory neurons that can be related to a weak glutamate activity. A weak GABA transmission could also be essential. All results in a weak activation of pyramidal cells. Owing to "circular causality," appropriate modeling can only be provided by computational circuit models.

In any case, each node within the network represents a locus in the visual space. If a visual stimulus is supposed to be stored for some seconds, the respective node has to be activated for that period of time, whereas other nodes have to remain inhibited (Figure 2.15a). At low functionality of this network, any new stimulus could induce an activation of the respective node of the network (Figure 2.14b). This pathological condition occurs at low dopamine input into the network. This condition can be simulated by the model and allows us to perform computer experiments (*in silico* experiments) that stimulate new experiments in the neurobiological laboratory.

The dopamine transmission was conceptualized by Durstewitz, Seamans, and Sejnowski (2000) by modeling the $D_1$ receptors and $D_2$ receptors as the $D_1$ receptor/

(a) (b)

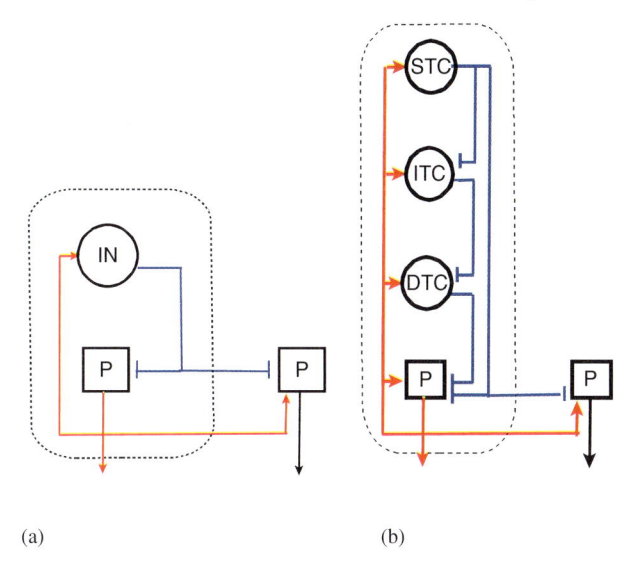

**Figure 2.14** Structure of different modules of artificial cortical network models. (a) Modules with one type of inhibitory neurons used by Durstewitz, Seamans, and Sejnowski (2000). (b) Module with three types of inhibitory neurons used by Wang et al. (2004a). P: pyramidal cell; IN: inhibitory neuron; DTC: dendrite-targeting cell; ITC: interneuron-targeting cell; STC: soma-targeting cell.

$D_2$ receptor ratio represents the filter function of the network. Under these conditions, the computer experiments showed that at a low $D_1$ receptor/$D_2$ receptor ratio the working memory function of the network was low. It results in a multifocal $D_2$ receptor-dominated state of the network. A unifocal $D_1$ receptor-dominated state of

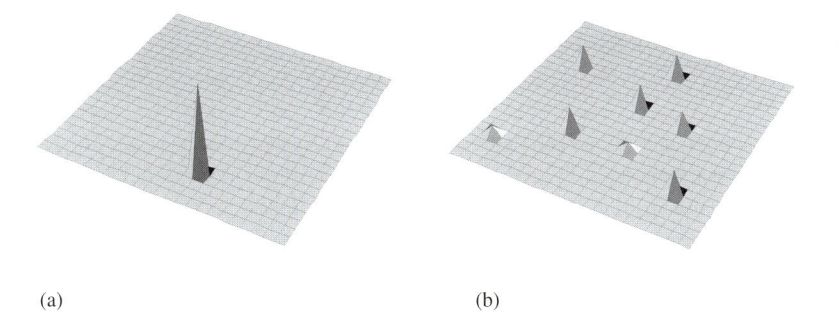

(a) (b)

**Figure 2.15** Display of states of artificial networks with two optional conditions in the prefrontal cortex in terms of processing of memory-related information. (a) Focused activation of nodes of the network is correlated with strong working memory function. This means that $D_1$ receptors prevail in the network, showing a single, strong and sustained centre of activity. (b) Multifocal activation of the network is caused by functional dominance of $D_2$ receptors. However, they show multiple, weaker, and short-lived centers of activation. This multifocal activation of the network is incompatible with strong working memory performance. (After Durstewitz, Seamans, and Sejnowski, 2000; Seamans et al., 2001; Winterer and Weinberger, 2004; see also Winterer, 2006; generated with Mathematica®.)

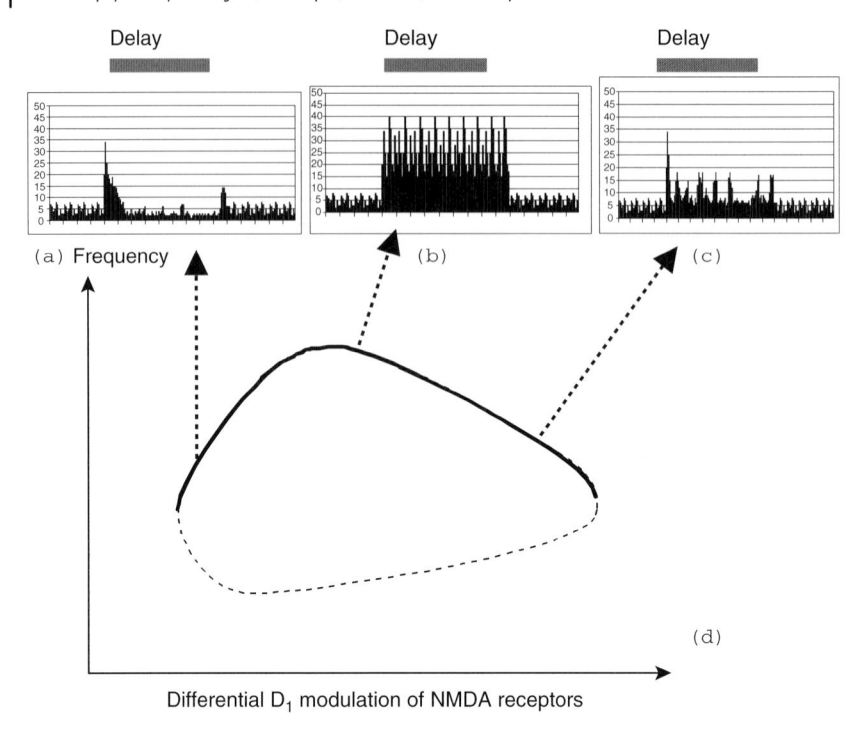

**Figure 2.16** Simulation of working memory task under various conditions. Scheme of validated simulation of neuronal computation of instable, stable, and weak working memory function depending on the $D_1$ modulation of NMDA receptors (modified from Brunel and Wang, 2001; Wang et al., 2004a). (a) Low persistent neuronal activity at low $D_1$ receptor modulation. (b) High persistent activity at medium $D_1$ receptor modulation. (c) Low persistent activity at high $D_1$ receptor modulation. (d) Resulting inverted U-function as a component of the phase portrait of the attractor.

the network corresponds to a good working memory performance. This can be caused by sufficient lateral inhibition between the nodes of the network. Dopamine could enhance this inhibition.

In the case of models that represent the biological diversity of inhibitory neurons (e.g., three types of inhibitory neurons), the experimental findings can be reconstructed in detail (Figure 2.16 and cf. Figure 2.3) and also γ oscillations can be generated in case of optimal cognitive function (Wang et al., 2004a; Wang, 2006).

### 2.3.2
### Neuron

In neuropsychiatry it is crucial to understand information processing by the functional brain pathology and, finally, by properties of neurons. For a deeper understanding of local neuronal networks, the structure and processes of the elementary units of the network, the neurons, have to be considered. This is in line with the concept of cellular pathology that was developed by Rudolf Virchow. He had already assumed that the cell is the atomic unit of pathology (Virchow, 1858, 1971).

Cells as autonomous but environment-related units and as parts of organs and organisms can be characterized by metabolism, growth, differentiation, reactivity, motility, reproduction, and regeneration. Specifically, the main task of neurons is information processing.

In terms of signal transmission, the neuron can be viewed as a complex processor that receives numerous inputs through its dendrites and cell body, and generates an output by its axon. The connection between neurons is called the synapse (Figure 2.17).

The cellular tasks can only be fulfilled by division of labor in a multicompartment fashion. In this context, the internal world of the neuron is composed of organelles, such as the mitochondria, the actin–tubulin cytoskeleton, lysosomes, and the Golgi apparatus, as well as the endoplasmic reticulum with polysomes and the cell nucleus containing the DNA (Figure 2.18). Within these compartments, molecules operate in functionally defined subsystems or modules that are interconnected with each other. Each of these molecular operations are controlled by cascades and networks of genetic and environmental factors. Some of these molecules and their functions are listed in Section 2.3.4.1.

At this point, we have to talk about the important function of glial cells, which are 10 times more numerous in brain than neurons and are largely neglected in modeling studies. Without oligodendrocytes, elaborating the myelin sheath around many neurons, the fast electrophysiological conductance would not be possible. Moreover, using microarray studies, it has been found in schizophrenia brains that genes involved in the lipid metabolism of myelin are specifically affected (Hakak *et al.*, 2001; Tkachev *et al.*, 2003). Furthermore, it has been known for a long time that glutamate neurotransmission is tightly controlled by astrocytes expressing glutamate transporters that efficiently remove the transmitter from the synaptic cleft. This exemplifies that very likely many more molecular mechanisms apart from the ones

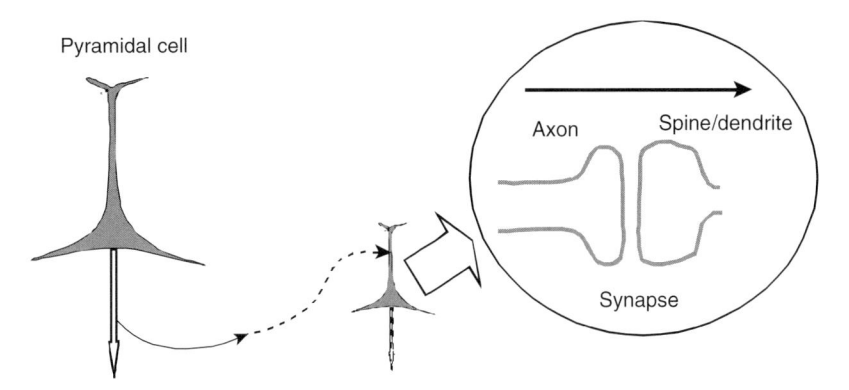

**Figure 2.17** Neuron and synapse.

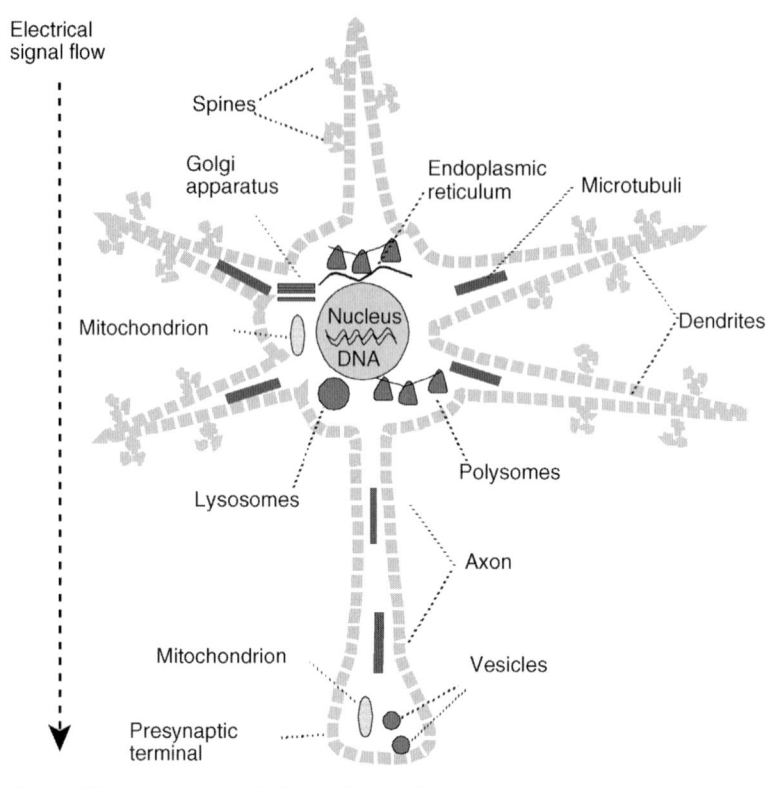

**Figure 2.18** Neuron – morphology and intracellular structures.

located in the neuron and related to dopamine are involved in the development and progression of schizophrenia.

#### 2.3.2.1 Electrical Signaling of the Neuron

Electrical signals in the brain are closely related to conscious experience and motor behavior. As has been described in Section 2.3.1.2, the frequency and synchrony of EEG signals are important indicators of the level of consciousness. These signals are generated by assemblies of neurons. Here, we focus on the electrical signals in the neuron (Figure 2.19).

The input for a neuron usually causes a slow and small variation of the membrane potential, either enhancing the potential (inhibitory postsynaptic potential) or reducing it (excitatory postsynaptic potential). The latter facilitates sudden discharges or action potentials (spikes). During the action potential, that starts at about −55 mV and reaches about + 30 to + 50 mV, sodium ions flow into the axon through sodium channels. Repolarization of the membrane potential is initiated by an outflow of potassium ions followed by extrusion of sodium ions. The resting potential is maintained at about −65 mV, hyperpolarization occurs at about −90 mV, and the discharge threshold is at about −55 mV. Specific ion pumps re-establish the stable

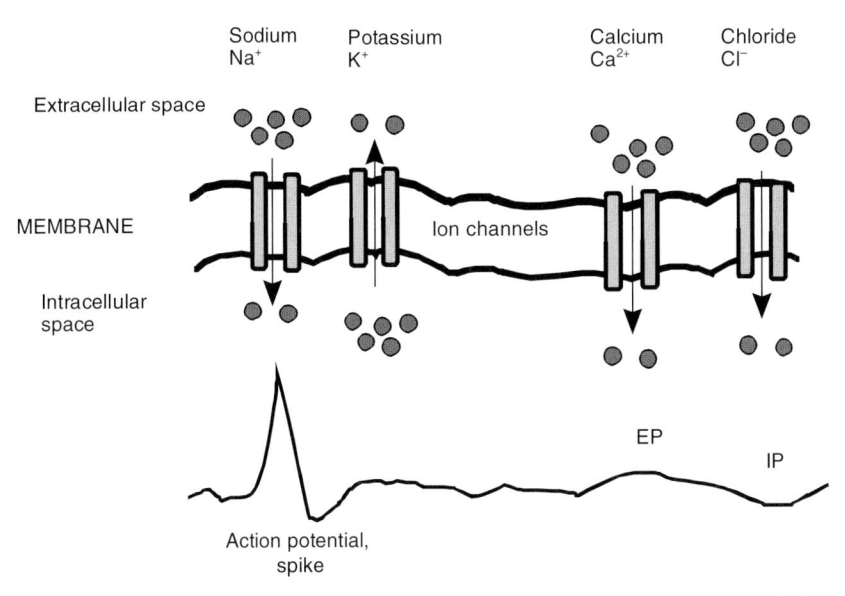

## 2.3 Neurobiological Fundamentals of Neuropsychiatry | 53

**Figure 2.19** Electrical signals of neurons. Plot of fluctuations of the membrane potential of the cell body and the dendrites of the cell with excitatory potentials (EP) and inhibitory potential (IP), and with action potentials (spikes) that usually start at the axon hillock at the beginning of the axon.

electrical "nonequilibrium" of the cell at the resting potential. From the point of discharge until restoration of the resting potential, the cell is in a refractory state that impedes or attenuates the following action potentials. This refractory state restricts the maximum spiking frequency of a neuron. The basis of electrical signaling of the neuron are gated ion channels for sodium ($Na^+$) and calcium ($Ca^{2+}$) for activating changes of the membrane potential (depolarization) and potassium ($K^+$) and chloride ($Cl^-$) for inhibiting effects on the membrane potential (hyperpolarization). Calcium ion influx into the cell is the major, excitatory driving force for eliciting spikes. The gating of the ion channels is made by receptors (ligand-gated) or caused by changes of the membrane potential (voltage-gated). Many subtypes of channels are identified; for instance, some sodium channels exert transient currents, others show persistent currents. Some calcium channels have low, others high threshold; some potassium channels are activated by strong depolarization, others by an increase of calcium.

It should be mentioned here that neurons in the cortex and striatum prefer two subthreshold membrane potentials: the up-state at about $-45$ mV and the down-state at about $-65$ mV. These states are related to slow oscillations (Wilson and Kawaguchi, 1996).

Fast neuron–neuron communication is ensured by myelinated axons. The myelin sheath is interrupted at regular intervals by unmyelinated (naked) sites, the nodes of Ranvier, where fast-conducting $Na^+$ channels are clustered to facilitate sodium influx and depolarization. This is the basis underlying the "saltatory" nerve conductance. The $Na^+$-triggered action potential leads to neurotransmitter release at the presynaptic nerve ending (axon terminal) either on the cell body of a neuron, on dendritic spines, or on a synapse of another neuron. In this way, a signal quickly traveling along an axon is slowed down at the presynaptic site by chemical transmission.

The neuronal discharge activity is supposed to be one level of neuronal coding – the higher the frequency, the higher the intensity of motor activity in the case of a motor neuron that activates muscles. Sensory signaling is also proportionally related to the frequency of discharge. It is important for modeling to distinguish tonic (sustained), phasic (transient) and burst activity.

### 2.3.3
**Synapse**

Contact sites between axons and dendrites (dendritic spines) or cell bodies of the following neurons are special structures, named synapses. Apart from axo-dendritic and axo-somatic synapses, dendrito-dendritic synapses also exist that are classified as transmitter-free gap junctions with direct electrical coupling that might evoke synchronization of spiking between neurons.

Signal transmission at the synapse takes place by the transformation of electrical signals (spikes) into chemical signals (transmitter substances). Action potentials arrive at the axon terminal and activate voltage-dependent calcium channels. Inflow of calcium stimulates vesicles where the transmitter substance is stored to fuse with the presynaptic membrane. The transmitter is released and can bind to postsynaptic receptors or to presynaptic inhibitory autoreceptors. Some transmitters are degraded by enzymes, others are eliminated by reuptake mechanisms (transporters). See Figure 2.20.

As already mentioned, the most thoroughly studied transmitter substances are norepinephrine, noradrenalin, serotonin, glutamate, GABA, dopamine, and acetylcholine. Each of these transmitters bind to specific pre- and postsynaptic receptors that can be grouped into several subtypes. For instance, dopamine-based transmission relies on $D_1$ type and $D_2$ type receptors. Dopamine receptors of the $D_1$ subtype are coupled with $G_s$-proteins that stimulate adenylyl cyclase, an enzyme that synthesizes cyclic AMP, whereas receptors of the $D_2$ subtype inhibit this pathway by inhibitory $G_i$-proteins. These receptors can also induce release of intracellular calcium. On the presynaptic site, $D_2$ receptors inhibit dopamine release. Glutamate receptors can also be divided into different subtypes: ionotropic glutamate receptors (subtypes: N-methyl-D-aspartate (NMDA)/α-amino-3-hydroxy-5-methyl-4-isoxazol-propionacid (AMPA)/kainate receptors) are permeable for calcium but also for sodium ions, whereas metabotropic receptors are coupled to G-proteins. GABA-A receptors are ionotropic receptors selective for chloride ions. Postsynaptic molecular mechanisms of receptor activation are discussed later. For a summary of receptor subtypes, see Table 2.8.

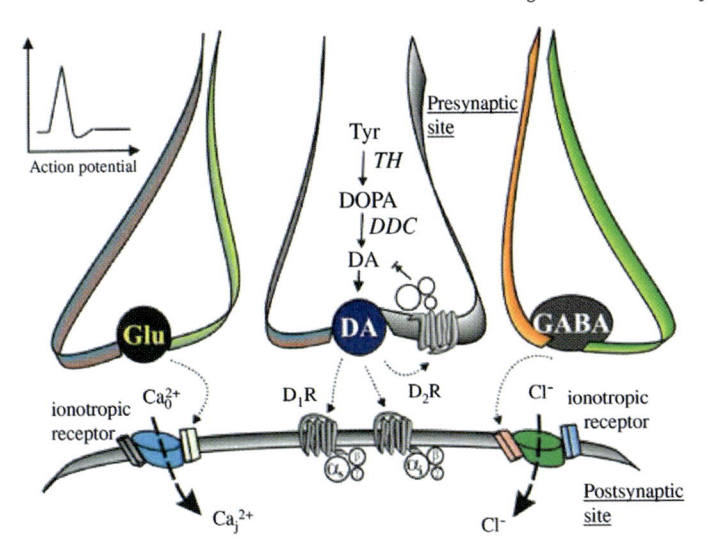

**Figure 2.20** Three axon endings releasing different neurotransmitters on one (postsynaptic) neuron. The neurotransmitters dopamine, glutamate, and GABA can activate a variety of postsynaptic receptors with α, β, and γ subunits of G-proteins (see text). Additionally, GABA and metabotropic glutamate receptors can be expressed on the postsynaptic site (not shown). DA: dopamine; Glu: glutamate; Tyr: tyrosine; DOPA: precursor of dopamine; DDC: DOPA decarboxylase; R: receptor.

To understand the synapse as a functional unit within neuronal networks, several processes have to be considered simultaneously: transmitter synthesis, transmitter transport, insertion into vesicles, storage of transmitter in vesicles, priming of vesicles, docking of vesicles, their fusion with the membrane, release of transmitter, diffusion, receptor binding, autoreceptor binding, decay, and so on, determine the processing dynamics of the synapse (Figure 2.21). The temporal dimension of these synaptic processes – "slow"/"fast" and "phasic"/"transient" processes – has to be considered in detail and should be represented in computational models (Abbott and Regehr, 2004).

Therefore, to understand the synapse as a system of interactive, molecular networks, functional consequences of presynaptic release of neurotransmitters and their postsynaptic effects have to be taken into account in studies on psychiatric disorders. In that respect, molecular systems biology could help to investigate molecular processes by computation and modeling techniques. Furthermore, actions of psychopharmaceutical drugs could be understood in a better way and this could markedly influence the development of new (and better) drugs. Some of these synaptic processes and structures are discussed here more in detail.

It should be mentioned that already some attempts have been made to establish a "systems biology of the synapse" (Grant, 2003; Pocklington et al., 2006; www.neurosciencecampus-amsterdam.nl).

### 2.3.3.1 Receptors

Neurotransmitter receptors are the most relevant structures in neuropsychiatric research and pharmacotherapy. Therefore, the biochemistry of these structures is of central importance.

The receptors can be roughly subdivided into ionotropic and metabotropic receptors. Ionotropic receptors are ion channels permeable for either positively

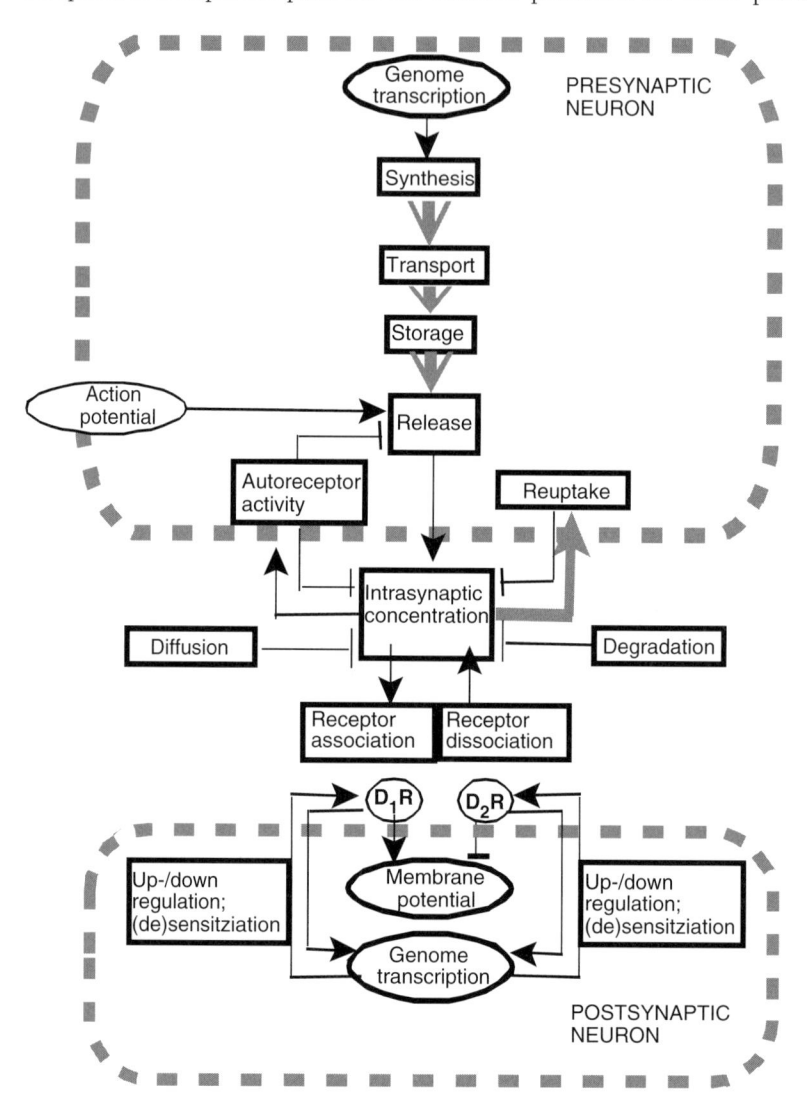

**Figure 2.21** Wiring diagram of processes related to dopamine-based neurotransmission: some molecular functional components that determine the dynamics of transmitter concentration. Legend: bold arrows: flow of molecules; thin arrows: elevation/activation of the targeted component; bars: reduction/inhibition; $D_1R$: receptor of dopamine $D_1$ family; $D_2R$: receptor of dopamine $D_2$ family.

**Table 2.8** Transmitters, receptors, mechanisms, and effects.

| Neurotransmitter | Receptor | Molecular mechanism | Effects |
|---|---|---|---|
| Acetylcholine | muscarinic $M_1$ $M_2$, $M_{3-5}$ | $G_s$, $G_{i/o}$ | de-/hyperpolarization |
| | nicotinic A, $B_{1-9}$ | $Na^+$, $K^+$, $Ca^{2+}$ channel | fast depolarization |
| Dopamine | $D_1$ type | $G_s$ | depolarization |
| | $D_2$ type | $G_{i/o}$ | hyperpolarization |
| GABA | GABA-$A_{\alpha 1-6, \beta 1-3, \gamma 1-3}$ | $Cl^-$ channel | fast hyperpolarization |
| | GABA-$B_{1,2}$ | $G_{i/o}$ | slow hyperpolarization, inhibits presynaptic GABA release |
| Glutamate | $AMPA_{1-4}$ | $Na^+$, $K^+$ channel | fast depolarization |
| | $kainate_{1-5}$ | | |
| | NMDA, $NR_{1,2A-D}$ | $Na^+$, $K^+$, $Ca^{2+}$ channel | slow depolarization, activation of CaM-protein kinase |
| | $NR_{3A,B}$ | | |
| | metabotropic $GluR_{1-8}$ | $G_s$, $G_{i/o}$ | activation of PKC |
| Serotonin | 5-$HT_{1A,B,D,E,F}$ | $G_{i/o}$ | hyperpolarization |
| | 5-$HT_{2A-C}$, 5-$HT_{4-7}$ | $G_s$ | de-/hyperpolarization |
| Norepinephrine, noradrenaline | $\alpha_{1A,B,D}$ | $G_s$ | depolarization |
| | $\alpha_{2-4, 6-10}$ | $G_{i/o}$ | hyperpolarization |
| | $\beta_{1-3}$ | $G_s$ | reduction of after hyperpolarization |
| Opioids | μ, δ, κ, σ | $G_{i/o}$ | reduced frequency of action potentials |
| Glycine | GlyR | $Cl^-$ channel | hyperpolarization? |
| Histamine | $H_1$ | $G_s$ | hyperpolarization |
| | $H_2$ | $G_s$ | reduction of after hyperpolarization |

($Na^+$) or negatively ($Cl^-$) charged ions. Metabotropic receptors are coupled with G-proteins that activate ($G_s$) or inhibit ($G_i$) consecutive molecular signaling cascades (Figure 2.20). Some effects of receptor activation are listed in Table 2.8. In line with this, a brief overview on some receptor types of the most important transmission systems may help to understand the complexity on this level of pivotal brain functions (Figure 2.22). For detailed information, see Cooper, Bloom, and Roth (2004), Webster (2002), and Stahl (2008a).

i) Receptor subtypes of the *norepinephrinergic/noradrenergic transmission system* are the excitatory $\alpha_1$, $\beta_1$, $\beta_2$, and $\beta_3$ receptors, and the inhibitory $\alpha_2$ receptors. These receptors are subdivided into:
   (a) $\alpha_{1A-1D}$ receptors that stimulate phospholipase C (PLC), which converts phosphatidylinositol bisphosphate ($PIP_2$) into inositol trisphosphate

**AMPA-R**
**NMDA-R**
**D₁-R**
**a1NA-R**
**5HT₂ₐ-R**
**nACh-R**
**GABA-B-R**

**GABA-R**
**D₂-R**
**a2NA-R**
**5HT₁ₐ-R**
**mACh-R**
**mGlu-R**

Action potential

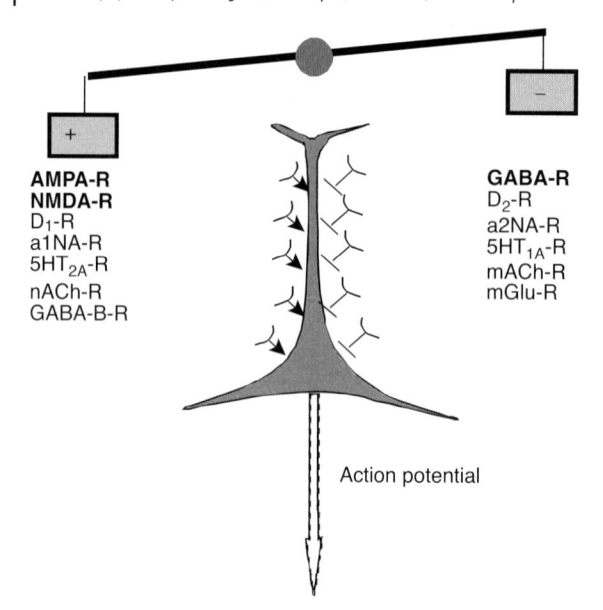

**Figure 2.22** Hypothetical dynamic equilibrium of excitation (+) and inhibition (−) by a balanced ratio of activating and inhibiting receptors, demonstrated by the scheme of the pyramidal cell in the cerebral cortex. Legend: Activating receptors (left): AMPA-R: AMPA receptor (glutamate receptor subtype); NMDA-R: NMDA receptor (glutamate receptor subtype); $D_1$-R: dopamine $D_1$ receptor; 5-$HT_{2A}$-R: serotonin receptor type 2A; a1NA-R: $\alpha_1$-norepinephrine receptor; nACh-R: nicotinic acetylcholine receptor; GABA-B-R: GABA-B receptors (inhibits presynaptically GABA release). Inhibiting receptors (right): GABA-A-R: GABA-A receptors; $D_2$-R: dopamine $D_2$ receptor; 5-$HT_{1A}$-R: serotonin receptor type 1A; a2NA-R: $\alpha_2$-norepinephrine receptor; mACh-R: muscarinic acetylcholine receptor; mGlu-R: metabotropic glutamate receptor (e.g., $mGlu_{2,3,4}$).

($IP_3$). $IP_3$ binds its receptor expressed on the endoplasmic reticulum and releases intracellular calcium.

(b) $\alpha_{2A}$, $\alpha_{2B}$ and $\alpha_{2C}$ receptors inhibit adenylyl cyclase, which reduces cellular activity. Presynaptic $\alpha_2$ receptors inhibit synthesis, vesicular storage, and release of noradrenaline. This autoreceptor mechanism is also used by antidepressant drugs such as mirtazapin.

ii) For the *dopamine transmission system*, a $D_1$ receptor family and a $D_2$ receptor family have been identified: $D_1$ family receptors ($D_1$, $D_5$) activate adenylyl cyclase, increase cAMP, and stimulate the reactivity of the respective cell. The $D_2$ family receptors ($D_2$, $D_3$, and $D_4$) inhibit adenylyl cyclase (inhibition of cAMP production). They also stimulate PLC, which converts $PIP_2$ into $IP_3$ (Figure 2.23). Furthermore, they activate potassium channels (outward rectifiers) with resultant hyperpolarization of the corresponding cell membrane. At the presynaptic site, they inhibit dopamine release.

iii) The *serotonergic transmission system* (5-HT) is made up of approximately 15 subtypes, such as 5-$HT_{1A,B}$, 5-$HT_{1D-F}$, 5-$HT_{2A-C}$, and 5-$HT_{3-7}$. They are summarized as the 5-$HT_1$ family and the 5-$HT_2$ family, including the 5-$HT_4$

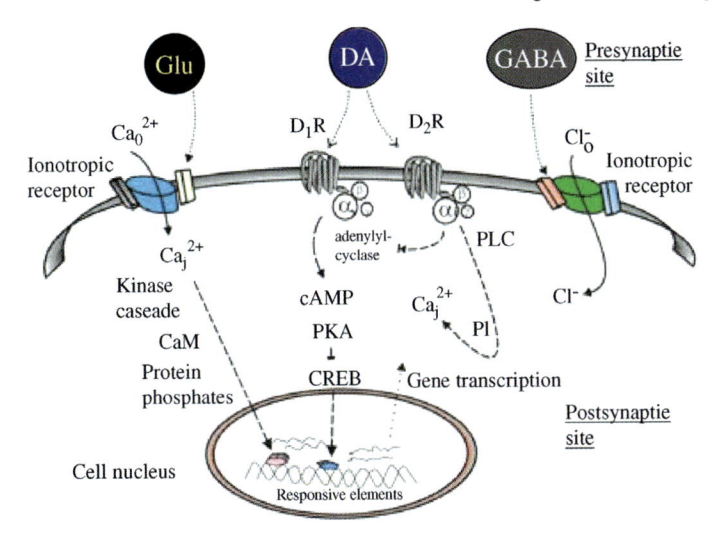

**Figure 2.23** Postsynaptic signaling cascade evoked by receptor activation.

receptors. $5\text{-HT}_1$ and $5\text{-HT}_5$ receptors inhibit adenylyl cyclase and cellular activity by activation of the above-mentioned potassium channel. They are mainly localized presynaptically and inhibit serotonin release. $5\text{-HT}_2$ receptors stimulate PLC and release of intracellular calcium. $5\text{-HT}_3$ receptors work as ligand-gated (excitatory) $Na^+/K^+$ channels. $5\text{-HT}_4$, $5\text{-HT}_6$, and $5\text{-HT}_7$ receptors are excitatory, as well as stimulating adenylyl cyclase activity.

iv) The *cholinergic transmission system* is composed of muscarinic and nicotinic receptors that depolarize the membrane potential ($M_1$ and $M_3$ receptors), whereas $M_2$ receptors hyperpolarize the membrane potential. $M_1$ receptors activate $G_q$-proteins that activate the PLC–$IP_3$ pathway. $M_2$ receptors open potassium channels and inhibit adenylyl cyclase by $G_i$-proteins. $M_3$ receptors are coupled to the $IP_3$ pathway via $G_q$-proteins. The neuronal nicotinergic receptors can be subdivided into $\alpha$ and $\beta$ families with eight and three members, respectively, and are ligand-gated $Na^+$ channels. (Some new data that are relevant for psychiatry are reported in Chapter 8 by Mobascher and Winterer.)

v) The *glutamatergic* system consists of two ionotropic receptor types: the NMDA and the non-NMDA (AMPA/kainate) receptors. NMDA receptors are activated via $AMPA_A$ receptors. They facilitate calcium influx with some minor selectivity for sodium, as well. This calcium influx may contribute to depolarization of postsynaptic cells. Additionally, glutamate-responsive metabotropic receptors $mGlu_{1-8}R$ have also been identified. Some of these have excitatory effects ($mGlu_{1-8}$ and $mGlu_5$) and some have inhibitory effects (e.g., $mGlu_2$, $mGlu_3$, $mGlu_4$, $mGlu_6$, $mGlu_7$, and $mGlu_8$).

vi) In the *GABA-ergic* system, two receptor types have been described: GABA-A and -B receptors. GABA-A receptors are expressed pre- and postsynaptically. They are ion channels composed of five subunits and facilitate influx of $Cl^-$, which induces hyperpolarization and inhibition of neuronal activity. GABA-B receptors are G-protein-coupled metabotropic receptors linked to $K^+$ channels (outward rectifiers) that induce hyperpolarization of the neuron. The $\alpha$ and $\gamma$ subunits of GABA-A receptors are binding sites for benzodiazepines.

Other receptors of importance are endorphin/enkephalin receptors. Here, four receptor subtypes have been identified: $\mu$, $\varkappa$, $\delta$, and $\sigma$ receptors. The $\mu$ receptors display inhibitory effects through hyperpolarization of the membrane potential. Hyperpolarization is induced by increase of the $K^+$ permeability. The $\mu$ receptors are subdivided into $\mu_1$ and $\mu_2$ receptors. $\varkappa$ receptors inhibit adenylyl cyclase and voltage-dependent calcium currents. This results in reduction of presynaptic activation of vesicles and of transmitter release. The $\delta$ receptors show similar effects. The functional relevance of the $\sigma$ receptors for psychopathology remains elusive.

Histamine receptors and glycine receptors are briefly described in Table 2.2.

Additionally, it has to be mentioned here that there is some evidence that the simple distinction of activating and inhibiting receptors is not fully appropriate: for dopamine receptors it has been shown that in a hyperpolarized state of the cell, $D_1$ receptor activation might activate potassium channels, thus leading to a prolonged state of hypoactivity of the cell, whereas at a low membrane potential, $D_1$ receptor activation would enhance calcium influx, thus facilitating discharge (cf. Durstewitz, 2006). These properties indicate that $D_1$ receptors act as amplifiers of the respective state of the cell.

Finally, noting that every neuron has thousands of receptors that can functionally modulate the activity and reactivity of the cell, it is feasible that these processes can be neither measured simultaneously nor can be imagined (Figure 2.22). Therefore, computer simulations could probably help us to understand these molecular processes of modulation of the activity of the cell.

**Pharmacotherapeutical Options for the Treatment of Schizophrenia** Receptor-centered explanatory hypotheses of mental disorders have become very popular (Cooper, Bloom, and Roth, 2004). They are based on findings that there are either more or less receptors expressed in certain brain regions compared to the respective receptor density in healthy subjects. The former is called upregulation of receptors; the later is called downregulation of receptors – persistent elevation of intrasynaptic transmitter concentration could result in adaptive downregulation of receptors.

For instance, in schizophrenia, dopamine $D_2$ receptor expression in subcortical structures seems to be more elevated than in healthy subjects, whereas dopamine $D_1$ receptors dominate in the cortex. Therefore, it is assumed that the predominance of the respective receptor type in these brain regions reflects an adaptive mechanism to an intrasynaptic oversupply of dopamine in the case of inhibiting $D_2$ receptors and a relative lack in the case of activating $D_1$ receptors. Although this interpretation might not be completely true, it can be useful for further explorations of synaptic functions.

Psychopharmaceutical drugs used in the treatment of psychiatric disorders act on different synaptic sites:

- Blockade of postsynaptic receptors: $D_2$ receptor blockade by antipsychotics.
- Blockade of presynaptic receptors: $D_2$ receptor blockade by antipsychotics.

- Activation of presynaptic receptors: $\alpha_2$ receptor activation by clonidine in the treatment of noradrenergic symptoms in alcohol withdrawal.
- Activation of postsynaptic receptors: serotonin 5-$HT_{2A}$ receptor activation by LSD.
- Blockade of transporters: serotonin uptake inhibitors, norepinephrine transporter inhibitors (antidepressants).
- Blockade of transmitter synthesis: $D_2$ autoreceptor blocks dopamine synthesis.
- Blockade of enzymes of transmitter decay: acetylcholine esterase blocker.

Since we are presently unable to provide a synopsis of the dynamic cooperation of these mechanisms, their kinetics are usually described separately. In particular, the processes of synaptic/cellular adaptation after repeated application of pharmaceuticals/drugs are far from being understood.

Antipsychotics or neuroleptics used to treat psychotic states in schizophrenia primarily inhibit pre- and postsynaptic $D_2$ receptors (Table 2.9). However, the first-generation antipsychotics (e.g., levopromazin, promethazin) mainly had blocking effects on norepinephrinergic and histaminergic receptors, whereas the more efficient second-generation (haloperidol) had a high affinity for $D_2$ receptors, blocking their action. New antipsychotics also inhibit serotonin 5-$HT_{2A}$ receptors, which results in less side-effects ("atypical antipsychotics").

A differentiated discussion of effects and side-effects of antipsychotics or antidepressants can help us to understand the molecular dynamics of the respective mental disease and its treatment.

In this receptor-centered view, depression is also understood as being caused by an upregulation of 5-$HT_2$ receptors and norepinephrinergic $\alpha_1$ receptors (Stahl, 2008b). Similar to what has been suggested for dopamine receptors in schizophrenia, this upregulation may also be a result of a compensation of low synaptic supply with the respective transmitter. When selective serotonin reuptake blockers are applied therapeutically, the increased supply of serotonin might evoke a compensatory downregulation of the receptors with the consequence that the normal function of the signal transmission in the respective neuronal networks is restored.

It is important to note that full and stable therapeutic effects induced by antipsychotics and antidepressants are observed only after about 10 days of delay, although the functions of the molecular machinery in synapses are already changed within minutes after drug application. Hence, the puzzling question of how to interpret the delayed response of the brain to pharmacotherapy remains elusive.

In this context, modeling of dynamic aspects of synaptic events and respective molecular networks could be very important for understanding the symptoms and endophenotypes of various mental diseases. Computational simulations of respective molecular networks would also be invaluable for the development of targeted pharmacotherapies. A concept that understands the cell and the cellular/subcellular networks as systems of molecular machines that are loosely coupled subsystems, and that are able to self-adapt and self-optimize their operations, may help us to get a hunch of what may be going on here.

**Table 2.9** Receptor binding profile of various antipsychotics (cf. Benkert and Hippius, 2008; Stahl, 2008a).

| Substance | $D_1$ | $D_2$ | $D_3$ | $D_4$ | $H_1$ | ACH | $\alpha_1$ | $\alpha_2$ | $5\text{-}HT_1$ | $5\text{-}HT_2$ |
|---|---|---|---|---|---|---|---|---|---|---|
| Clozapin[a] | ++ | + | + | +++ | ++ | + | ++ | ++ | + | +++ |
| Haloperidol | ++ | +++ | ++ | +++ | (+) | (+) | (+) | (+) | (+) | ++ |
| Olanzapin[a] | ++ | +++ | ++ | ++ | +++ | ++ | + | + | ++ | +++ |
| Promethazin | 0 | 0 | 0 | ? | +++ | 0 | ++ | ? | ? | 0 |
| Quetiapin[a] | + | ++ | + | − | ++ | − | ++ | ++ | ++ | + |
| Risperidon[a] | + | ++ | ++ | +++ | ++ | (+/−) | +++ | ++ | ++ | +++ |
| Sulpirid[a] | 0 | ++ | +++ | ? | 0 | 0 | 0 | ? | ? | 0 |

a) Atypical antipsychotic.

## 2.3.4
### The Cell as a System of Interacting Molecules

A view that is focused on molecular sites such as receptors is obviously not sufficient to understand the brain as the basis of mental disorders. In contrast, the brain should be seen as a cellular network of neurons. Therefore, the focus of molecular biology/biochemistry must be directed to the overall function of the cell.

A basic requirement for the biology of cells is the flow of energy and matter based on nonequilibrium states as conceptualized very early on by Ludwig von Bertalanffy (1968) by the term "flow equilibrium." Nonequilibrium situations can be found everywhere in living systems, such as in sustained and transient differences of ion concentration between extracellular and intracellular space, which are important for the electrical reactivity of the cell. Evidently, these differences of electric potential are well balanced out and, if running out of control, the reactivity of the cell is at stake, which may result in disease or even death.

For this reason, the regulation of the various intracellular molecular processes is of crucial importance. Divergence of local signal events as well as convergence of signaling cascades and their feedback loops must be considered in order to ensure proper function of the cells.

#### 2.3.4.1 Intracellular Signal Cascades – From Receptor to Genome

In the postsynaptic cell, multiple incoming signals are integrated into complex signal-processing events and "firing" of this cell only occurs if a critical threshold of excitation is overcome. Simultaneously, many metabolic processes are influenced in the postsynaptic cell, such as synthesis of retrograde second messengers (e.g., nitric oxide), oxygen and energy consumption, gene transcription and protein synthesis, apoptosis and cell proliferation, and transport of lipids and proteins throughout the cell, including all the fibers.

Typically, activation of postsynaptic receptors induces a variety of intracellular signaling pathways consisting of a bunch of messenger molecules (Table 2.10). For example, receptor activation can stimulate gene transcription via cAMP and PKA, and phosphorylation of transcription factors such as CREB and others. Then, phosphorylated CREB binds to CREB-responsive elements of some specific genes, which modifies the transcription process. Eventually, if expression of certain genes results in synthesis of proteins that are building blocks of membranes or ion channels, the reactivity of the cell may be altered (compare Figure 2.23).

The complexity of the networks that are involved in intracellular signal processing implies that only subsystems can be studied (Figure 2.24). Regarding this aspect, the time-courses of activation and inhibition, the dynamics of the topology of molecular density, and robustness of pathways in terms of interfering perturbances are extremely important mechanisms to be implemented in systems biological computations (Szallasi, Stelling, and Periwal, 2006). An exemplary graph-theoretical approach to study complexity has been proposed by Uri Alon, who is interested in basic design principles of biological circuits (Alon, 2007; see Chapter 12 by Smith and Hütt).

**Table 2.10** Signal transduction: some intracellular signaling molecules related to dopamine transmission (modified from Tretter and Albus, 2008).

| Notation | Name | Function |
| --- | --- | --- |
| AC | adenylyl cyclase | synthesis of cAMP second messenger isoenzymes exert different reactivity |
| Arc | signal protein | energy metabolism |
| cAMP | cyclic adenosine monophosphate | PKA activation |
| PKA | protein kinase A | protein phosphorylation (CREB) |
| $Ca^{2+}$ | calcium ion | control of enzyme activities |
| CaM | calmodulin (calcium-modulated protein) | regulator protein of calcium-dependent processes |
| CaMII | calmodulin II (calcium-modulated protein) | regulator protein of multiple processes |
| CaMK | calmodulin kinase | phosphorylation of proteins |
| Cdk5 | cyclin-dependent kinase | cell cycle regulation |
| CREB | cAMP-response element binding protein | transcription factor |
| DARPP-32 | dopamine and cAMP-regulated phosphoprotein-32 | integrating signal transduction pathways |
| ERK | extracellular signal-regulated kinase | phosphorylation |
| Fos (genes and proteins) | gene family (and protein) | transcription factor |
| Fras | Fos-related antigens | transcription factor |
| Homer | | NMDA receptor-associated protein |
| GTP | guanidine triphosphate | G-protein activation |
| MAPK | mitogen-activated protein kinase | phosphorylation in mitotic processes |
| MEK | MAPK/ERK kinase | phosphorylation of ERK |
| PP1 | protein phosphatase 1 | dephosphorylation of proteins |
| PLC | phospholipase C | controls phosphoinositol metabolism |
| PDE1 | phosphodiesterase 1 | degradation of cAMP and cGMP |
| PP2A | protein phosphatase 2A | dephosphorylation of proteins |
| PP2B | protein phosphatase 2B | dephosphorylation of proteins |
| Raf | protein | regulates MEK pathway |
| Ras | proto-oncogene | regulates MEK pathway |

#### 2.3.4.2 Modeling Signal Transduction Networks Relevant in Schizophrenia

The relevance of the dopamine system to understanding the pathology of schizophrenia is also emphasized by the specific analysis of dopamine-induced activation of $D_1$ and $D_2$ receptors, and the resultant intracellular signal transduction. At the first level of transduction (adenylyl cyclase), both receptor types have antagonistic effects; however, after about four steps at the PKA stage they exert synergistic effects. Moreover, effects on DARPP-32 as an integrator of different signaling cascades are evident (Figure 2.25). However, the graphic representation of the different diverging and converging pathways and feedback loops alone cannot describe the *in vivo* behavior of this molecular network, because the different kinetics and feedback loops influence the behavior in a way that may only be explored by computer-based

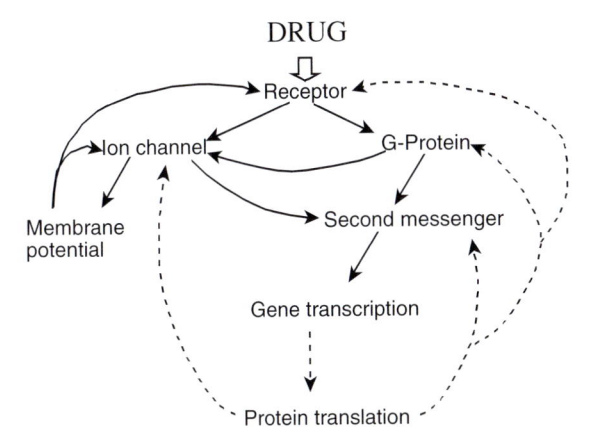

**Figure 2.24** Intracellular loops of signal transduction pathways that connect molecular signaling and electrical properties of the cell.

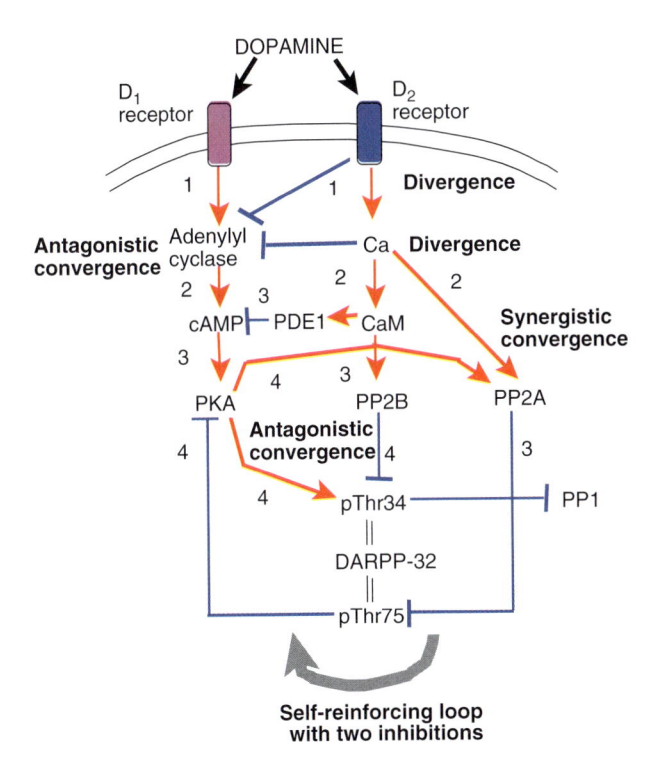

**Figure 2.25** Simplified scheme of the molecular network of signal transduction cascades of $D_1$ and $D_2$ receptor signaling, with various steps of processing converging on DARPP-32. Comment: PP2A acts as an (indirect) activator of PKA by double serial inhibition. See Table 2.10 for abbreviations.

Signaling pathway →

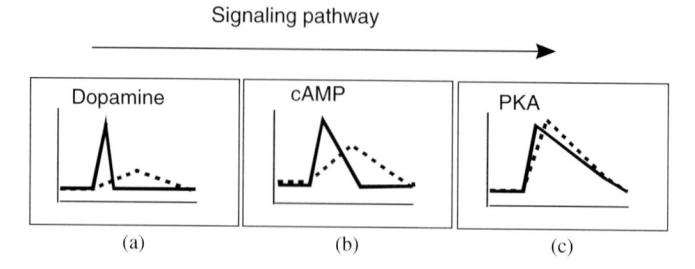

|Dopamine| |cAMP| |PKA|
(a) (b) (c)

**Figure 2.26** "*In silico*" test of the model of signal transduction. Different inputs can result in similar outputs (adapted from Lindskog et al., 2006). (I) Strong transformation of a fast (pulsatile) dopamine signal (solid line) over a triangular signal of cAMP to a saw-tooth-like signal of PKA. (II) Low transformation of a slow (triangular) dopamine signal (stippled line) with similar curve at stage of PKA.

modeling and simulation (Fernandez et al., 2006). Some years ago, Lindskog et al. (2006) constructed a computational model that also included effects of NMDA receptor activation. In their *in silico* experiments they could reproduce the above-described antagonistic and synergistic effects of dopamine receptor activation on the various levels of signal transduction. For instance, their computer experiments showed that a strong and fast pulsatile dopamine signal is transformed to a triangular signal step by step at the level of cAMP and PKA. By contrast, a slow and weak signal can be enhanced stepwise (Figure 2.26; *cf.* Lindskog, 2008).

This example shows how systems biology can advance our understanding of molecular events in the neuron by performing computer experiments on *in silico* cells.

Finally, additional immediate effects brought about by activation of the signaling cascade, like phosphorylation of various ion channels, ought to be included in computational systems biology of the neuron. This aspect would lead to an understanding of dynamic changes in membrane potential that, in turn, influence actions of the various receptors. In addition, thinking in multidimensional molecular networks should also include long-term effects of PKA-induced transcriptional and translational changes. This example of a systems biology approach to the neuron is believed to greatly improve our understanding of the dynamic action of agonists and antagonists as well as of the origin of mental disorders.

### 2.3.4.3 Genomics and Proteomics
Molecular psychiatry focuses on the world of genes and proteins – "genomics" and "proteomics." We will only briefly touch on some basic aspects.

**Genomics** The molecular genetics of mental disorders has made rapid progress by various high-throughput methodologies. However, the complexity of data has not allowed us to identify the genetic basic of mental disorders in detail. There is wide general agreement that there is no "master gene" generating the respective disorder. In schizophrenia, as in other diseases, some susceptibility genes have been identified. Their gene products are involved in events that entail neurotransmitter

**Table 2.11** Some susceptibility genes for schizophrenia (Thome, 2005; Harrison and Weinberger, 2005; Stahl, 2008a).

| | |
|---|---|
| *DTNBP1* (dysbindin or dystobrevin-binding protein-1) | *ERBB4* |
| *NRG1* (neuregulin) | *FEZ1* |
| *DISC1* (disrupted in schizophrenia-1) | *MUTED* |
| *DAOA* (D-amino acid oxidase activator; G72/G30) | *MRDS1* (*OFCC1*) |
| *DAO* (D-amino acid oxidase) | *BDNF* (brain-derived neurotrophic factor) |
| *RGS4* (regulator of G-protein signaling-4) | *Nur77* |
| *CHRNA7* ($\alpha_7$ nicotine cholinergic receptor) | *MAO-A* |
| *GAD1* (glutamic acid decarboxylase-1) | spinophylin |
| *GRM3* (mGlu$_3$R) | calcyon |
| *PPP3cc* | tyrosine hydroxylase |
| *PRODH2* | dopamine $D_2$ receptor |
| *AKT1* | dopamine $D_1$ receptor |

synthesis and degradation as well as neurotransmission (Table 2.11); however, this has not resulted in any good hypothesis of their disease-relevant functions. Nevertheless, it has been established for a long time, supported by twin studies, that most mental illnesses have a genetic causation of at least 50%. Obviously, more detailed concepts of the ongoing processes in the genome have to be developed. Problems in data-gathering, methodological strategies, and so on are discussed by in Chapter 10 by Tabakoff's group.

**Proteomics** Proteins are the "work horses" in all living systems, acting as building materials, energy stores, enzymes, and so on. However, they are also the most difficult molecules to be investigated, because their structural and functional variabilities are several orders of magnitude greater than those of DNA or RNA. Notwithstanding, recent research efforts are focused increasingly more on this level of molecular systems. High-throughput technologies have been established and are steadily being improved. Additionally, there are upcoming approaches trying to model smaller parts of this system. It is determined which protein interacts with which other proteins. This is a task for graph theory as described in Chapter 12 by Smith and Hütt.

### 2.3.4.4 Gene Regulation – Circular Signaling Pathways

The intracellular molecular signaling network downstream from the receptors and the world of genes and proteins were described in the previous sections. Now we turn to the circularity of molecular signaling cascades. In the 1950s, the dogma of the unidirectionality of genetic control of the organism was raised. In the 1960s, it had to be revisited when the Nobel laureates Francoise Jacob and Jacques Monod detected the circular regulatory mechanisms of gene expression (Jacob and Monod, 1961) – proteins synthesized after gene transcription can repress their own transcription or the transcription of other genes (negative feedback; Figure 2.27). Regulation and

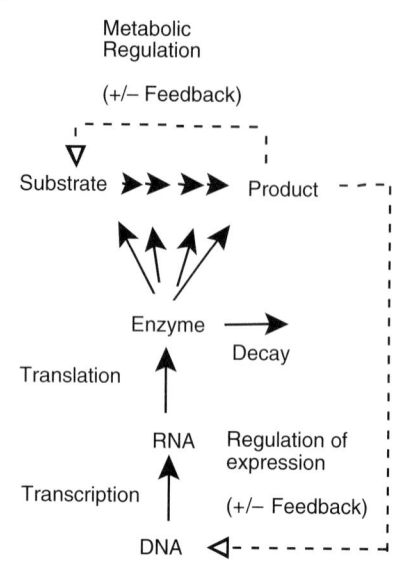

**Figure 2.27** Control loops of gene expression and of the production of proteins (after Palsson, 2006).

control of transcription, translation, metabolic pathways, and signal transduction, as well as the strict control of cell division and apoptosis, all integrate to ensure the successful coping of the organism with endogenous and exogenous disturbances. Various types of genes were identified, such as structural genes, promoter genes, operator genes and regulator genes. They constitute the operon as a functional subunit in the genome, as identified in bacteria by Jacob and Monod, and therefore it is also called the operon model (Jacob and Monod, 1961). Thus, in the systemic view, the simple concept that genes "determine" the structure and the function of the proteins and their expression is not precise enough. There are many exogenous influences that modify these parameters (e.g., epigenetic modifications).

**Regulation of Clock Genes** An excellent example of a biological feedback system, which is also subject to environmental modifications, is the molecular system of the biological clock that has been studied extensively (Ueda, 2007). Transcription of the *per* (period) and other genes of this system is high when the concentrations of the respective proteins are low (Albrecht, 2006). This results in translation of increasingly more protein that enters the nucleus and represses its own gene transcription or transcription of other "clock" genes (Hirayama and Sassone-Corsi, 2005). Examples of these negative feedback regulators are PER, CRY (cryptochrome), and REV-ERBα proteins, whereas positive (inducing) proteins are CLOCK and BMAL (brain and muscle Amt-like) (Figure 2.28). Typically, the cycle of protein accumulation and subsequent inhibition of transcription runs for approximately 25 h. Under those so-called "free-running" conditions in total darkness, there is a small but steady shift of the normal day/night rhythm. Therefore, daylight is important to reset the "clock" to 24 h (light entrainment). This example shows that there are intrinsic mechanisms controlling rhythmic expression of the "clock" and

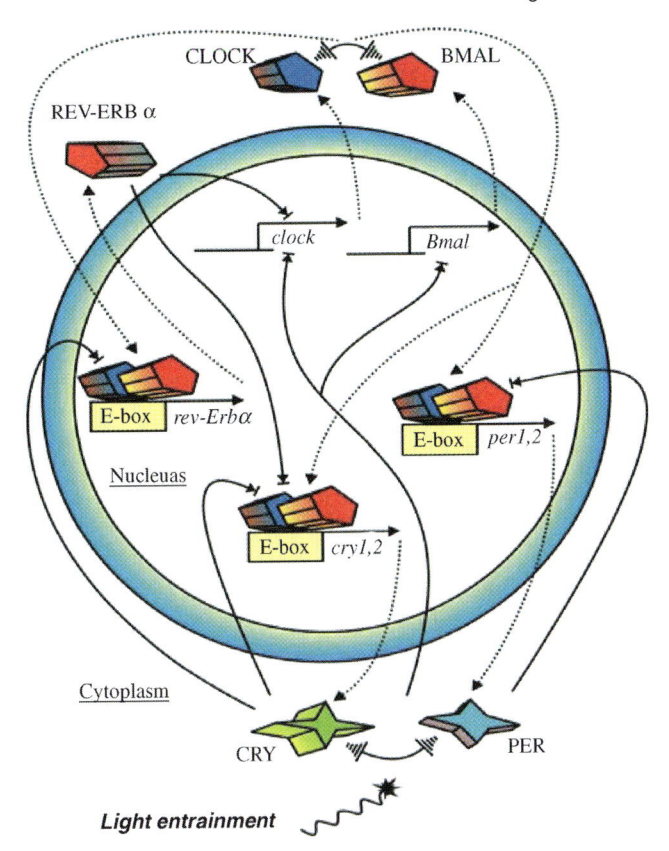

**Figure 2.28** Molecular feedback mechanisms of the "biological clock" (simplified). The cryptochrome (CRY) and period (PER) proteins may form heterodimers and repress gene transcription. The CLOCK and BMAL proteins can also act in concert, but typically induce gene transcription. Often, rhythmically expressed genes contain one or more E-Box elements in their promoter regions as binding sites for "clock" proteins.

many other genes. It also shows, however, that these molecular mechanisms rely on maintained inputs from the environment (Van der Zee, Boersma, and Hut, 2009). Hence, light and many other environmental factors can exert significant influences on molecular biological systems.

These findings could be relevant for the understanding of mechanisms of depression.

Considering these issues, the cell can be understood as a complex network of operating molecules that organize their proper way of functioning in a self-organized manner related to a complex spatio-temporal pattern of activation and inhibition.

### 2.3.4.5 Systems Biology of the Neuron

Summing up, the various components of the neuron and their molecular components as described above can be understood as a functionally closed system that

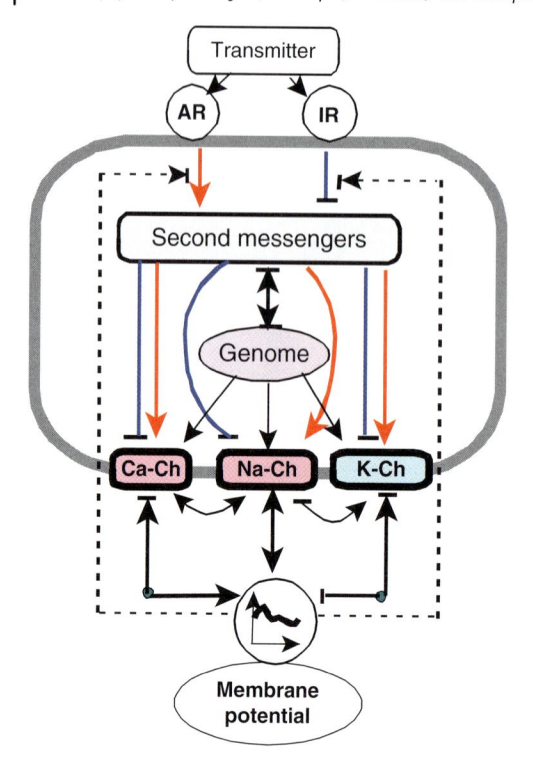

**Figure 2.29** Scheme of some components of the intracellular molecular signaling network controlled by antagonistic action of dopamine $D_1$ and $D_2$ receptors on cascades of second messenger molecules. AR: activating receptor; IR: inhibiting receptor; Na-Ch: sodium channel; K-Ch: potassium channel; Ca-Ch: calcium channel.

operates with electrochemical signaling processes. Variations of the membrane potential result in short- and long-term changes of the molecular networks of the cell, and these changes are followed by changes of the electrical reactivity of the cell (Figure 2.29).

## 2.3.5
### The Brain as a Neurochemical Oscillator

Considering the fact that the brain shows a basic sleep/wake cycle, it is feasible to conceptualize the brain as a dynamic network of coupled oscillators based on chemical transmission systems. The oscillations likely are not just sinusoidal waves, but impose as overlays of waves with varying amplitudes and wavelengths. Within this general conceptual framework, mental disorders can be characterized by abnormal synchrony of these oscillators (Singer, 1999).

For instance, in schizophrenia patients, low γ synchrony above the frontal cortex and high γ synchrony in parietal cortices is observed in EEGs (Symond et al., 2005).

During hallucinations, the EEG shows low $\gamma$ synchrony above frontal cortex and high $\gamma$ synchrony in parietal cortices, a reduction of $\gamma$ power after auditory stimulation, but a higher $\gamma$ and $\alpha$ synchrony. In the disorganized type of schizophrenia, reduced $\gamma$ synchrony can be observed.

Disturbances of the coupling of these neurochemical oscillators have also been identified in affective disorders, especially in bipolar disorders, (Stahl, 2008b); one first obvious reason is that depressive patients display irregular sleep, a reduced drive in the morning, but improved performance in the evening. These symptoms are related to a dysfunction of the interplay between norepinephrine, acetylcholine, serotonin, and dopamine. Interestingly, therapeutic sleep deprivation has been used with some success to improve the well-being of patients; however, it does not confer a long-lasting effect (several days only). Furthermore, depressive episodes can be recurrent. There are also patients who undergo transient manic phases alternating with depressive phases. These "bipolar" patients may experience low-frequency oscillations (months or even years) or pass through high-frequency ranges ("rapid cycling"). Finally, several dysfunctional molecular feedback loops have been identified in depression (e.g., the circadian regulation of cortisol exhibits an impaired circadian profile).

#### 2.3.5.1 Neurochemical Interaction Matrix

A serious problem when trying to develop an integrative model of brain dynamics that can be used to study the neurobiology of mental disorders is the enormous amount of data available in the literature – the more details are inserted in the model, the higher the model's complexity. This relation reduces the explanatory power, although the validity is high – the model might exceed the cognitive potential of the scientist. On the other hand, simple models may have great explanatory power, but they might be too simple. There is presently no resolution to this dilemma between the simplicity and validity of a model. Nevertheless, we want to present a simple heuristic concept based on findings that actions of various neurotransmitter systems are reciprocally connected. This view relates to the concept of a "neurochemical interaction matrix" (Figure 2.30) and can be transformed into a qualitative "toy model," called a "neurochemical mobile" ("mobile" being a metaphor for a hanging mobile ornament above a child's bed).

Experience from clinical practice and from laboratory research tells us that the major systems of neurotransmission that substantially contribute to and affect mental disorders are the dopamine, serotonin, acetylcholine, norepinephrine, noradrenaline, GABA, and glutamate systems. Schizophrenia, addiction, depression, and others can be characterized as disorders where the balanced interactions between these neurotransmitters are disturbed (Benes, 2009). Supposedly, the momentum of going astray arises by enhancement or blockade of transmitter release or by up/downregulation of respective receptors induced by environmental cues or drugs. If compensation of these events is insufficient, a whole cascade of inappropriate reactions may follow that spreads to and also jeopardizes other transmitter systems.

This can be demonstrated by the "neurochemical mobile" that is a simplification of this matrix.

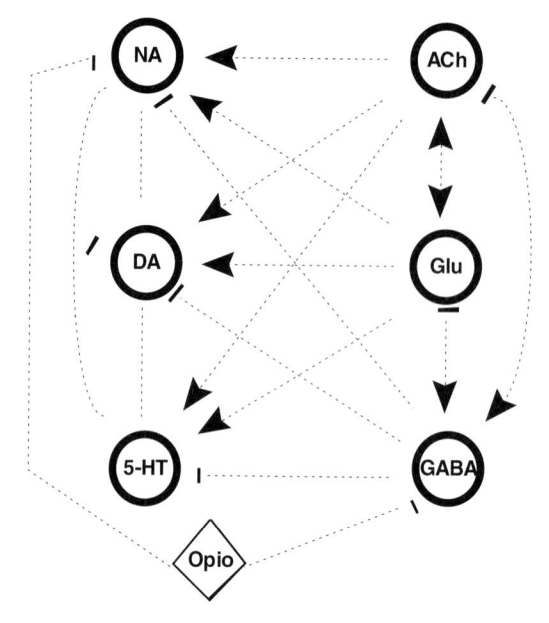

**Figure 2.30** Scheme of some interactions between neurochemical transmission systems. Opio: opiods.

#### 2.3.5.2 "Neurochemical Mobile"

The metaphorical framework of a "neurochemical mobile" represents the concept of a dynamic balance of neurotransmitter actions. In this framework, mental disorders can be described by a sustained imbalance of activity of the transmitter systems with focal dysfunctions. The basic constellation is that the norepinephrinergic system and the cholinergic system oppose each other in their effects on a common target (e.g., heart). On the side of the norepinephrinergic system, serotonin and dopamine are opponents. They also oppose each other with respect to their effects on local neuronal networks. On the side of acetylcholine, glutamate and GABA act as opponents that also have opposing effects (Figure 2.31). A complex picture of the global transmission arises when taking into account that the receptor subtypes determine the final function of the respective transmission system (Figure 2.31).

The heuristic value of this model is that most of the singular hypotheses of the neurochemical basis of mental disorders can be integrated. For instance, focusing on the neurochemistry of schizophrenia, several hypotheses can be seen:

- *Dopamine hyperfunction* (Carlsson, 1988). Some evidence for this hypothesis is based on the fact that cocaine consumption can evoke a psychotic state and that cocaine blocks the dopamine transporter resulting in a high dopamine transmission. Additionally, psychotic states can be treated by application of dopamine $D_2$ receptor blockers such as haloperidol.
- *Serotonin hyperfunction* (Aghajanian and Marek, 2000). LSD and mescaline that activate the serotonergic transmission can induce paranoid hallucinatory states.

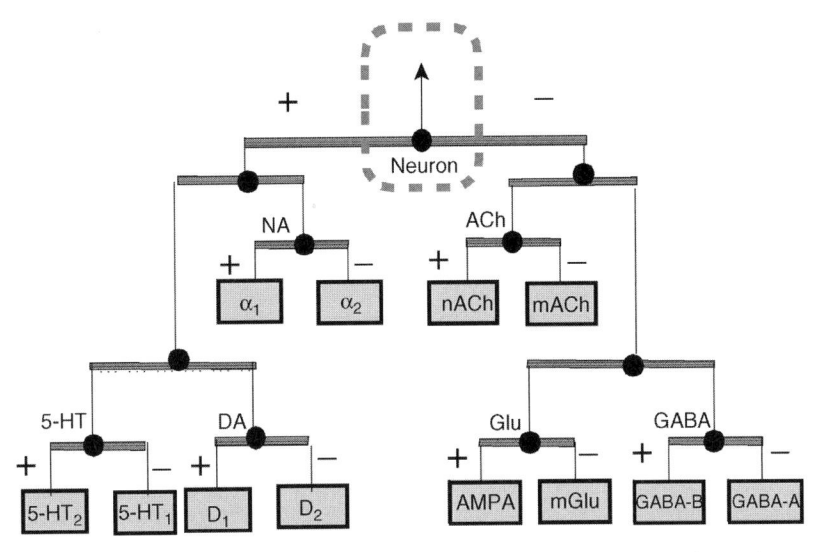

**Figure 2.31** Heuristic scheme of a "neurochemical mobile." Opposing neurochemical transmission systems and components with activating (+) and inhibitory (−) effects on cellular activities (see Table 2.8).

- *Glutamate hypofunction* (Moghaddam, 2003; Moghaddam and Homayoun, 2008). The drug phencyclidine (also know as "PCP") that acts as a glutamate antagonist can induce paranoid hallucinatory states.
- *GABA hypofunction* (Lewis, Hashimoto, and Volk, 2005). Withdrawal of benzodiazepines after chronic use induces psychotic states and epileptic seizures. These drugs induce a downregulation of GABA transmission.

The mobile allows us to represent these findings and treatment options in one model (Figure 2.32): antipsychotic action can be expected by antagonists of dopamine and/or serotonin transmission and by agonists of glutamate and/or GABA. Effective antipsychotic substances block $D_2$ receptors and $5-HT_2$ receptors. Benzodiazepines can also be useful in treatment in an acute schizophrenic episode by enhancing the GABA-ergic system (Figure 2.33).

Additionally, the framework of the mobile allows to understand depression as a hypofunction of the norepinephrine and serotonin system or as a hyperfunction of the acetylcholine system.

## 2.4
## Conclusions and Perspectives

The "traditional" single-component analyses in neurobiology have provided a huge amount of data, but failed to increase our understanding of the working of the whole system. Putting together all the pieces of knowledge on the molecular and cellular

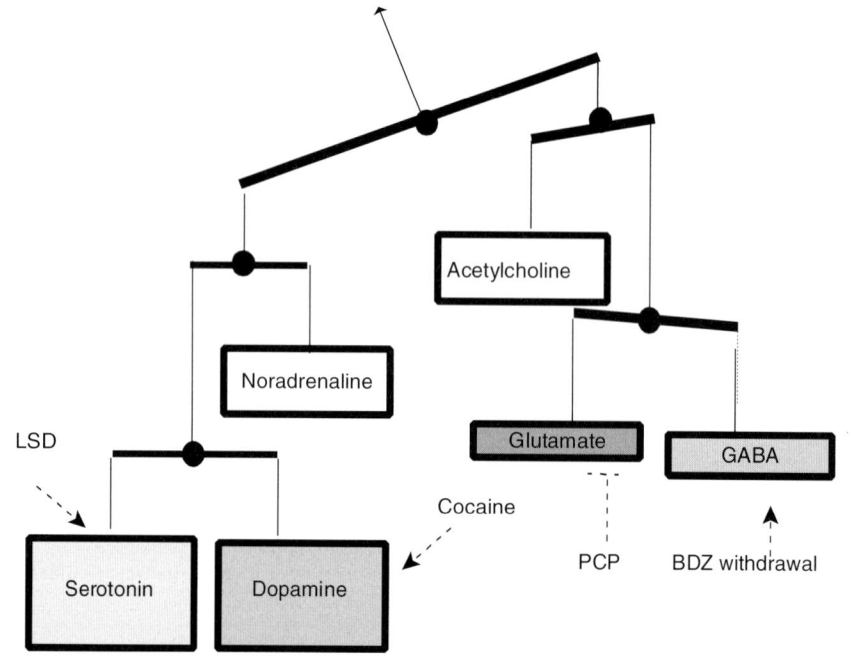

**Figure 2.32** Basic configuration of the neurochemical mobile: cases of schizophrenia, acute drug consumption, and of benzodiazepine (BDZ) withdrawal (see text).

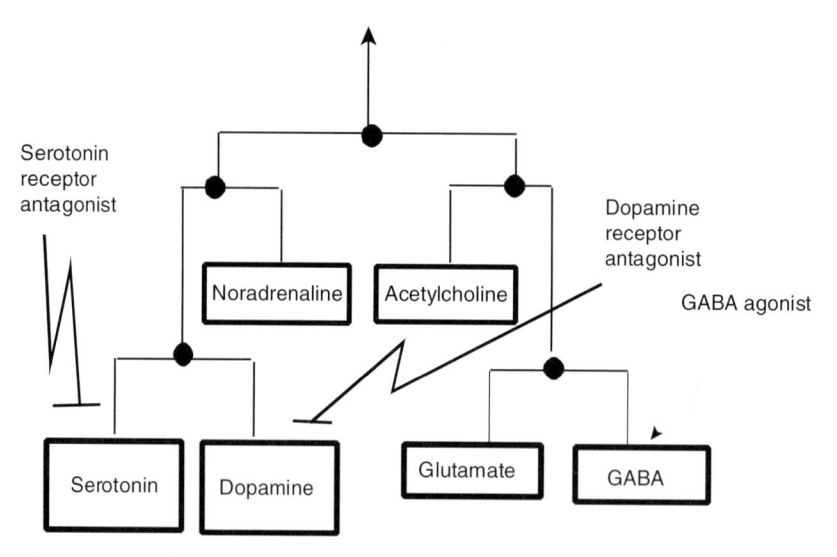

**Figure 2.33** Psychopharmaceutical treatment of productive psychosis induces a "healthy configuration" of the mobile (see text).

systems is far beyond the imagination of the human brain. This holds true also for the large amounts of data generated recently by high-throughput technologies at the transcription and protein levels. Although the development of appropriate computer algorithms to tackle these issues are well under way, they are still in their infancy. Molecular systems in single-cell organisms with several thousand molecules have been shown to be manageable for computer simulations (Conrad and Tyson, 2006; Paulsson and Elf, 2006). Extensions of those programs for multicellular organisms are being developed. Other programs have also been developed for higher-order levels, like region-specific communications in both the molecular and cell–cell interactive context. The more we try to refine our computational models, the more we recognize that there is extensive interaction between all levels. To understand the development of a disease requires a thorough understanding of all these interactions. Typically, the diagnosis of a brain disorder is zillions of steps of communication away from its onset. Systems biology can help, by aid of good computational tools, to zoom in into disrupted molecular/cellular events leading to disease (Gallinat, Obermayer, and Heinz, 2007).

Some next steps seem to be appropriate:

- *Global brain circuits* could be designed as computer models. Knowledge of systemic behavior gathered in first attempts of computational neuropsychiatry like addiction research (Ahmed, Graupner, and Gutkin, 2009) or schizophrenia research (Hoffman and McGlashan, 2006; Wang, 2006) can be integrated in such computer models (see Chapter 6 by Freeman, Chapter 14 by Lindskog *et al.*, and Chapter 15 by Loh *et al.*).
- *Small neuronal networks* and circuits can be modeled as it has been shown exemplarily by the prefrontal cortical working memory circuits that are dysfunctional also in schizophrenia (see Chapter 15 by Loh *et al.*).
- *Signaling pathways* and their interference with pathways that regulate cell growth can be studied systematically and quantitatively with regard to the kinetics of serial inhibitions (and disinhibitions). This is shown in principle in Chapter 14 by Lindskog *et al.*
- *Genetic analysis* can generate huge amount of data that has to be analyzed with regard to candidate genes, as proposed for addiction research in Chapter 10 by Tabakoff's group (see also Bhave *et al.*, 2007). In this field, strategies of data mining may be rewarding (see Chapter 13 by Gebicke-Haerter *et al.*).

The main challenges of understanding mental disorders from a system's perspective lie in the application of new mathematical tools for data mining (see Chapter 12 by Smith and Hütt) as well as for construction of simulation models (see Chapter 3 by Schulz and Klipp, and Chapter 11 by Mendoza). Competencies in this computational methodology are believed to stimulate the development of theoretical models explaining also the origins of psychiatric diseases. Investments of resources in such a "systems biology of the brain" appear to be promising.

In summary, by using systems biology approaches as described here, it will be possible to greatly improve our understanding of spatio-temporal interactions in the nervous (and other) systems on various levels, and to develop new types of psychiatric medications that are specifically tailored to the network dynamics and that therefore influence the different modules in nerve cells in a more efficient manner. For this reason, theoretical psychiatry could benefit from exchange with systems biology. However, major achievements can only be made if researchers from different disciplines like psychiatrists, neurobiologists, chemists, and computational scientists join hands to collaborate.

## References

Abbott, L. and Regehr, W.G. (2004) Synaptic computation. *Nature*, **431**, 796–803.

Aghajanian, G.K. and Marek, G.J. (2000) Serotonin model of schizophrenia: emerging role of glutamate mechanisms. *Brain Research Reviews*, **31**, 302–312.

Ahmed, S.H., Graupner, M., and Gutkin, B. (2009) Computational approaches to the neurobiology of drug addiction. *Pharmacopsychiatry*, **42** (Suppl. 1), S144–S152.

Albrecht, U. (2006) Orchestration of gene expression and physiology by the circadian clock. *Journal of Physiology*, **100**, 243–251.

Alon, U. (2007) *Systems Biology – Design Principles of Biological Circuits*, Chapman & Hall, New York.

Andreasen, N. (2004) *Brave New Brain – Conquering Mental Illness in the Era of the Genome*, Oxford University Press, New York.

American Psychiatric Association (2000) *Diagnostic and Statistical Manual of Mental Disorders*, text revision, 4th edn, American Psychiatric Association, Washington, DC.

Bear, M.F., Connors, B.W., and Paradiso, M.A. (2001) *Neuroscience: Exploring the Brain*, 2nd edn, Lippincott Williams & Wilkins, Baltimore, MD.

Benes, F.M. (2009) Neural circuitry models of schizophrenia: is it dopamine, GABA, glutamate or something else? *Biological Psychiatry*, **65**, 1003–1005.

Benkert, O. and Hippius, H. (2008) *Kompendium der Psychiatrischen Pharmakotherapie*, 7th edn, Springer, Berlin.

Berns, G.S. and Sejnowski, T.J. (1998) A computational model of how the basal ganglia produce sequences. *Journal of Cognitive Neuroscience*, **10**, 108–121.

Bhave, S.V., Hornbaker, C., Phang, T.L., Saba, L., Lapadat, R., Kechris, K., Gaydos, J., McGoldrick, D., Dolbey, A., Leach, S., Soriano, B., Ellington, A., Ellington, E.,

Jones, K., Mangion, J., Belknap, J.K., Williams, R.W., Hunter, L.E., Hoffman, P.L., and Tabakoff, B. (2007) The PhenoGen informatics website: tools for analyses of complex traits. *BMC Genetics*, **8**, 59.

Bremner, D.J., Narayan, M., Anderson, E.R., Staib, L.H., Miller, H.L., and Charney, D.S. (2000) Hippocampal volume reduction in major depression. *American Journal of Psychiatry*, **157**, 115–118.

Brunel, N. and Wang, X.-J. (2001) Effects of neuromodulation in a cortical network model of object working memory dominated by recurrent inhibition. *Journal of Computational Neuroscience*, **11**, 63–85.

Burt, T., Sederer, L., and Isgack, W.W. (eds.) (2002) *Outcome Management in Psychiatry. The Critical Review*, American Psychiatry Press, Washington, DC.

Canli, T., Cooney, R.E., Goldin, P., Shah, M., Sivers, H., Thomason, M.E., Whit, S., Whitfield-Gabrieli, S., Gabrieli, J.D.E., and Gotlib, I.H. (2005) Amygdala reactivity to emotional faces predicts improvement in major depression. *NeuroReport*, **16**, 1267–1270.

Carlsson, A. (1988) The current status of the dopamine hypothesis of schizophrenia. *Neuropsychopharmacology*, **1**, 179–186.

Carlsson, A. (2006) The neurochemical circuitry of schizophrenia. *Pharmacopsychiatry*, **39** (Suppl. 1), S10–S14.

Carlsson, A., Waters, N., and Carlsson, M.L. (1999) Neurotransmitter interactions in schizophrenia- therapeutic implications. *European Archives of Psychiatry and Clinical Neuroscience*, **4** (Suppl.), IV-37–IV-43.

Cohen, J.D. and Servan-Schreiber, D. (1993) A theory of dopamine function and its role in cognitive deficits in schizophrenia. *Schizophrenia Bulletin*, **19**, 85–104.

Conrad, E.D. and Tyson, J.J. (2006) Modeling molecular interaction networks with nonlinear ordinary differential equations, in *System Modeling in Cellular Biology: From Concepts to Nuts and Bolts* (eds. Z. Szallasi, J. Stelling, and V. Periwal), MIT Press, Cambridge, MA, pp. 97–124.

Cooper, J.R., Bloom, F.E., and Roth, R.H. (2004) *The Biochemical Basis of Neuropharmacology*, 8th edn, Oxford University Press, Oxford.

Dayan, P. and Abbott, L. (2005) *Theoretical Neuroscience. Computational and Mathematical Modeling of Neural Systems*, MIT Press, Cambridge, MA.

Durstewitz, D. (2006) A few important points about dopamine's role in neural network dynamics. *Pharmacopsychiatry*, **39** (Suppl. 1), S1–S88.

Durstewitz, D., Seamans, J.K., and Sejnowski, T.J. (2000) Dopamine-mediated stabilization of delay-period activity in a network model of prefrontal cortex. *Journal of Neurophysiology*, **83**, 1733–1750.

Fernandez, E., Schlappe, R., Girault, J.-A., and Le Novere, N. (2006) DARPP-32 is a robust integrator of dopamine and glutamate signals. *PLoS Computational Biology*, **2**, e176.

Freeman, W.J. (2000) *Neurodynamics: An Exploration in Mesoscopic Brain Dynamics*, Springer, Berlin.

Gallinat, J., Obermayer, K., and Heinz, A. (2007) Systems neurobiology of the dysfunctional brain. *Pharmacopsychiatry*, **40** (Suppl. 1), S40–S44.

Gazzaniga, M. (2004) *The Cognitive Neurosciences III*, 3rd edn, MIT Press, Cambridge, MA.

Gebicke-Haerter, P.J. (2006) Expression profiling in brain disorders and beyond. *Nihon Shinkei Seishin Yakurigaku Zasshi*, **26**, 1–10.

Gebicke-Haerter, P. (2008) Systems biology in molecular psychiatry. *Pharmacopsychiatry*, **41** (Suppl. 1), S19–S27.

Goldman-Rakic, P.S. (1995) Cellular basis of working memory. *Neuron*, **14**, 477–485.

Goldman-Rakic, P.S. (1999) The physiological approach: functional architecture of working memory and disordered cognition in schizophrenia. *Biological Psychiatry*, **456**, 650–661.

Goldman-Rakic, S.P., Muly, E.C. III, and Williams, G.V. (2000) $D_1$ receptors in prefrontal cells and circuits. *Brain Research Reviews*, **31**, 295–301.

Grant, S.G.N. (2003) Systems biology in neuroscience: bridging genes to cognition. *Current Opinion in Neurobiology*, **13**, 577–582.

Griesinger, W. (1845) *Die Pathologie und Therapie der psychischen Krankheiten*. Krabbe, Stuttgart.

Griesinger, W. (1882) *Mental Pathology and Therapeutics*, William Wood & Co., New York.

Hakak, Y., Walker, J.R., Li, C., Wong, W.H., Davis, K.L., Buxbaum, J.D., Haroutunian, V., and Fienberg, A.A. (2001) Genome-wide expression analysis reveals dysregulation of myelination-related genes in chronic schizophrenia. *Proceedings of the National Academy of Sciences of the United States of America*, **98**, 4746–4751.

Harrison, P.J. and Weinberger, D.R. (2005) Schizophrenia genes, gene expression, and neuropathology: on the matter of their convergence. *Molecular Psychiatry*, **10**, 40–68.

Heinz, A., Siessmeier, T., Wrase, J., Hermann, D., Klein, S., Grüsser, S.M., Flor, H., Braus, D.F., Buchholz, H.G., Gründer, G., Schreckenberger, M., Smolka, M.N., Rösch, F., Mann, K., and Bartenstein, P. (2004) Correlation between dopamine $D_2$ receptors in the ventral striatum and central processing of alcohol cues and craving. *American Journal of Psychiatry*, **161**, 1783–1789.

Helms, V. (2008) *Principles of Computational Cell Biology*, Wiley-VCH Verlag GmbH, Weinheim.

Hirayama, J. and Sassone-Corsi, P. (2005) Structural and functional features of transcription factors controlling the circadian clock. *Current Opinion in Genetics & Development*, **15**, 548–556.

Hoffman, R.E. and Dobscha, S.K. (1989) Cortical pruning and the development of schizophrenia: a computer model. *Schizophrenia Bulletin*, **15**, 477.

Hoffman, R.E. and McGlashan, T.H. (2006) Using a speech perception neural network computer simulation to contrast neuroanatomic versus neuromodulatory models of auditory hallucinations. *Pharmacopsychiatry*, **39** (Suppl. 1), S54–S64.

Jacob, F. and Monod, J. (1961) Genetic regulatory mechanisms in the synthesis of proteins. *Journal of Molecular Biology*, **3**, 318–356.

Kawaguchi, Y. and Kubota, Y. (1997) GABAergic cell subtypes and their synaptic connections in rat frontal cortex. *Cerebral Cortex*, **7**, 476–486.

Kitano, H. (2002a) Systems biology: a brief overview. *Science*, **295**, 1662–1664.

Kitano, H. (2002b) Computational systems biology. *Nature*, **420**, 206–210.

Klipp, E., Herwig, R., Kowald, A., Wielring, C., and Lehrach, H. (2005) *Systems Biology in Practice*, Wiley-VCH Verlag GmbH, Weinheim.

Klipp, E., Liebermeister, W., Wierling, Ch., Kowald, A., Lehrach, H., and Herwig, R. (2009) *Systems Biology: A Textbook*, Wiley-VCH Verlag GmbH, Weinheim.

Koch, M. (2006) *Animal Models of Neuropsychiatric Diseases*, Imperial College Press, London.

Kraepelin, E. (1902) *Clinical Psychiatry*, Macmillan, New York.

Lewis, D.L., Hashimoto, T., and Volk, D.W. (2005) Cortical inhibitory neurons and schizophrenia. *Nature Reviews Neuroscience*, **6**, 312–324.

Lindskog, M. (2008) Modelling of DARPP32 regulation to understand intracellular signaling in psychiatric disorder. *Pharmacopsychiatry*, **41** (Suppl. 1), S99–S104.

Lindskog, M., Kim, M.S., Wikstrom, M.A., Blackwell, K.T., and Kotaleski, J.H. (2006) Transient calcium and dopamine increase PKA activity and DARPP-32 phosphorylation. *PLoS Computational Biology*, **2**, e119.

McGuffin, P., Owen, M.J., and Gottesman, I.I. (eds.) (2004) *Psychiatric Genetics and Genomics*, Oxford University Press, New York.

Meisenzahl, E.M., Scheuerecker, J., Schmitt, G.J.S., and Möller, H.-J. (2007) Dopamine, prefrontal cortex and working memory functioning in schizophrenia. *Pharmacopsychiatry*, **40** (Suppl. 1), S62–S72.

Meyer-Lindenberg, A. and Weinberger, D.R. (2006) Intermediate phenotypes. and genetic mechanisms of psychiatric disorders. *Nature Reviews Neuroscience*, **10**, 818–827.

Moghaddam, B. (2003) Bringing order to the glutamate chaos in schizophrenia. *Neuron*, **40**, 881–884.

Moghaddam, B. and Homayoun, H. (2008) Divergent plasticity of prefrontal cortex networks. *Neuropsychopharmacology Reviews*, **33**, 42–55.

Nicholls, J.G., Martin, A.R., Wallace, B.G., and Fuchs, P.A. (2001) *From Neuron to Brain: A Cellular and Molecular Approach to the Function of the Nervous System*, 4th edn, Sinauer Associates, New York.

Noble, D. (2006a) *The Music of Life: Biology Beyond the Genome*, Oxford University Press, New York.

Noble, D. (2006b) Multilevel modelling in systems biology: from cells to whole organs, in *System Modelling in Cellular Biology: From Concepts to Nuts and Bolts* (eds. Z. Szallasi, V. Periwal, and J. Stelling), MIT Press, Cambridge, MA, pp. 297–312.

Palsson, B.O. (2006) *Systems Biology*, Cambridge University Press, Cambridge.

Paulsson, J. and Elf, J. (2006) Modeling molecular of intracellular kinetics, in *System Modelling in Cellular Biology: From Concepts to Nuts and Bolts* (eds. Z. Szallasi, V. Periwal, and J. Stelling), MIT Press, Cambridge, MA, pp. 149–176.

Pocklington, A.J., Cumiskey, M., Armstrong, J.D., and Grant, S.G. (2006) The proteomes of neurotransmitter receptor complexes form modular networks with distributed functionality underlying plasticity and behaviour. *Molecular Systems Biology*, **2**, 2006.0023.

Povysheva, N.V., Zaitsev, A.V., Kröner, S., Krimer, O.A., Rotaru, D.C., Gonzalez-Burgos, G., Lewis, D.A., and Krimer, L.S. (2007) Electrophysiological differences between neurogliaform cells from monkey and rat prefrontal cortex. *Journal of Neurophysiology*, **97**, 1030–1039.

Puls, I. and Gallinat, J. (2008) The concept of endophenotypes in psychiatric diseases – meeting the expectations? *Pharmacopsychiatry*, **41** (Suppl.), S37–S43.

Reckow, S., Gormanns, P., Holsboer, F., and Turck, C.W. (2008) Psychiatric disorders biomarker identification: from proteins to systems biology. *Pharmacopsychiatry*, **41** (Suppl.), S70–S77.

Rohlff, C. and Hollis, K. (2003) Modern proteomic strategies in the study of complex neuropsychiatric disorders. *Biological Psychiatry*, **53**, 847–853.

Sadock, B.J., and Sadock, V.A. (2005) *Kaplan's and Sadock's Synopsis of Psychiatry*, 9th edn, Lippincott Williams & Wilkins, Philadelphia, PA.

Sadock, B.J., and Sadock, V.A. (2007) *Kaplan's and Sadock's Synopsis of Psychiatry*, 10th edn, Lippincott Williams & Wilkins, Philadelphia, PA.

Schlösser, R.G.M., Koch, C., and Wagner, G. (2007) Assessing the state space of the brain with fMRI: an integrative view of current methods. *Pharmacopsychiatry*, **40** (Suppl. 1), S85–S92.

Seamans, J.K., and Yang, C.R. (2004) The principal features and mechanisms of dopamine modulation in the prefrontal cortex. *Progress in Neurobiology*, **74**, 1–57.

Seamans, J.K., Durstewitz, D., Christie, B., Stevens, C.F., and Sejnowski, T.J. (2001) Dopamine $D_1/D_5$ receptor modulation of excitatory synaptic inputs to layer V prefrontal cortex neurons. *Proceedings of the National Academy of Sciences of the United States of America*, **98**, 301–306.

Shepherd, G. (1994) *Neurobiology*, Oxford University Press, New York.

Shepherd, G. (2004) *The Synaptic Organization of the Brain*, Oxford University Press, New York.

Singer, W. (1999) Neuronal synchrony: a versatile code for the definition of relations? *Neuron*, **24**, 49–65.

Stahl, S. (2008a) *Stahl's Essential Psychopharmaocology*, Cambridge University Press, Cambridge.

Stahl, S. (2008b) *Depression and Bipolar Disorder: Stahl's Essential Psychopharmacology*, 3rd edn, Cambridge University Press, New York.

Symond, M.P., Harris, A.W., Gordon, E., and Willams, L.M. (2005) "Gamma synchrony" in first-episode schizophrenia : a disorder of temporal connectivity? *Am. J.Psychiatry*, **162**, 459–465.

Szallasi, Z., Stelling, J., and Periwal, V. (eds.) (2006) *System Modelling in Cellular Biology: From Concepts to Nuts and Bolts*, MIT Press, Cambridge, MA.

Taylor, M.A. and Vaidya, N.A. (2009) *Descriptive Psychopathology*, Cambridge University Press, Cambridge.

Thome, J. (2004) *Molekulare Psychiatrie*, Huber, Bern.

Tkachev, D., Mimmack, M.L., Ryan, M.M., Wayland, M., Freeman, T., Jones, P.B., Starkey, M., Webster, M.J., Yolken, R.H., and Bahn, S. (2003) Oligodendrocyte dysfunction in schizophrenia and bipolar disorder. *Lancet*, **362**, 798.

Tretter, F. (2005) *Systemtheorie im klinischen Kontext*, Pabst, Lengerich.

Tretter, F. and Albus, M. (2007) "Computational neuropsychiatry" of working memory disorders in schizophrenia: the network connectivity in prefrontal cortex – data and models. *Pharmacopsychiatry*, **40** (Suppl. 1), S2–S16.

Tretter, F. and Albus, M. (2008) Systems biology and psychiatry – modeling molecular and cellular networks of mental disorders. *Pharmacopsychiatry*, **41** (Suppl. 1), S2–S18, Review.

Tretter, F. and Müller, W. (eds.) (2007) Computational neuropsychiatry volume 2: the functional architecture of working memory networks in schizophrenia – data and models. *Pharmacopsychiatry*, **40** (Suppl. 1).

Tretter, F. and Scherer, J. (2006) Schizophrenia, neurobiology and the methodology of systemic modeling. *Pharmacopsychiatry*, **39** (Suppl. 1), S26–S32.

Tretter, F., Müller, W., and Carlsson, A. (eds.) (2006) Systems science, computational science and neurobiology of schizophrenia. *Pharmacopsychiatry*, **39** (Suppl. 1).

Tretter, F., Gallinat, J., and Müller, W.E. (eds.) (2008) Systems biology and psychiatry: the functional architecture of molecular networks in mental disorders – data and models. *Pharmacopsychiatry*, **41** (Suppl. 1).

Tretter, F., Gebicke-Haerter, P., and Heinz, A. (eds.) (2009) Systems biology of addiction. *Pharmacopsychiatry*, **42** (Suppl. 1).

Ueda, H.R. (2007) Systems biology of mammalian circadian clocks. *Cold Spring Harbor Symposia on Quantitative Biology*, **72**, 365–380.

Van der Zee, E.A., Boersma, G.J., and Hut, R.A. (2009) The neurobiology of circadian rhythms. *Current Opinion in Pulmonary Medicine*, **15**, 534–539.

Venter, J.C., Adams, M.D., Myers, E.W., Li, P.W., Mural, R.J. et al. (2001) The sequence of the human genome. *Science*, **291**, 1304–1351.

Virchow, R. (1971) *Die Cellularpathologie in ihrer Begründung auf physiologische und pathologische Gewebelehre* (transl. F. Chance), Dover, New York.

von Bertalanffy, L. (1968) *General System Theory*, Braziller, New York.

Yang, C.R. and Chen, L. (2005) Targeting prefrontal cortical dopamine $D_1$ and N-methyl-D-aspartate receptor interactions in schizophrenia treatment. *Neuroscientist*, **11**, 452–470.

Yang, C.R., Seamans, J.K., and Gorelova, N. (1999) Developing a neuronal model for the pathophysiology of schizophrenia based on the nature of electrophysiological actions of dopamine in the prefrontal cortex. *Neuropsychopharmacology*, **21**, 161–194.

Wang, X.J., Tegnér, J., Constandinidis, C., and Goldman-Rakic, P.S. (2004) Division of labor among distinct subtypes of inhibitory neurons in a cortical microcircuit of working memory. *Proceedings of the National Academy of Sciences of the United States of America*, **101**, 1368–1373.

Wang, X.J. (2006) Toward a microcircuit model for cognitive deficits in schizophrenia. *Pharmacopsychiatry*, **39** (Suppl. 1), S80–S87.

Wassef, A., Baker, J., and Kochan, L.D. (2003) GABA and schizophrenia: a review of basic science and clinical studies. *Journal of Clinical Psychopharmacology*, **23**, 601–640.

Webster, R. (ed.) (2002) *Neurotransmitters, Drugs and Brain Function*, John Wiley & Sons, Inc., New York.

WHO (1992) *ICD-10: International Statistical Classification of Diseases and Related Health Problems*, 10th revision. WHO, Geneva.

Wilson, C.J. and Kawaguchi, Y. (1996) The origins of two-state spontaneous membrane potential fluctuations of neostriatal spiny neurons. *Journal of Neuroscience*, **16**, 2397–2410.

Winterer, G. (2006) Cortical microcircuits in schizophrenia – the dopamine hypothesis revisited. *Pharmacopsychiatry*, **39** (Suppl. 1), S68–S71.

Winterer, G. and Weinberger, D.R. (2004) Genes, dopamine and cortical signal-to-noise-ratio in schizophrenia. *Trends in Neurosciences*, **27**, 683–690.

Zaccalo, M. and Pozzon, T. (2003) cAMP and $Ca^{2+}$ interplay: a matter of oscillation patterns. *Trends in Neuroscience*, **26**, 53–55.

# 3
# Introduction to Systems Biology
*Marvin Schulz and Edda Klipp*

## 3.1
## Introduction

### 3.1.1
### What is Systems Biology?

Long ago, biology started as a purely descriptive discipline. Scientists categorized organisms based on their appearance or formulated theories about how organisms evolve, which were at their time unprovable. Later, researchers focused on the exploration of the molecular building blocks of the cell. They started describing the structures of DNA, RNA, and proteins, and gave qualitative explanations of how fundamental processes like translation and transcription work. While this approach yielded detailed insights, it often neglected the interplay between different components and thus failed to give a holistic picture of the cell.

The link between the global behavior of cells or even complete organisms to cellular processes is provided by the emerging field of "systems biology" (Kitano, 2001). This new area integrates methods from biology, chemistry, physics, mathematics, computer science, and even philosophy in order to give an integrative view on certain processes. The major challenge in systems biology, however, is to make use of the various types of information coming from different sources. General properties of the underlying system have to be extracted from this information, which can afterwards be used for the formulation of qualitative or quantitative models. These properties are in most cases not assessable from knowledge concerning single parts of a system.

### 3.1.2
### Purpose of Modeling

In general, a model is an abstract description of a certain biological process. It is neither restricted to any mathematical formalism nor limited in the described processes. The ultimate goal of modeling is to describe a large system by a network

*Systems Biology in Psychiatric Research.*
Edited by F. Tretter, P.J. Gebicke-Haerter, E.R. Mendoza, and G. Winterer
Copyright © 2010 WILEY-VCH Verlag GmbH & Co. KGaA, Weinheim
ISBN: 978-3-527-32503-0

of simple representations of its fundamental processes, such as a pathway, which is a network of reactions fulfilling a certain task, by its single reactions. Such a description should be accurate enough that not only experimental results can be reproduced, but also that predictions about the system's behavior under different conditions can be made. An illustrative application is the identification of potent drugs for a certain disease. Given a disease affecting a certain cellular pathway, a model of this pathway can be used to predict the effects various drugs would have on this system. Following this approach, effective drugs could be identified more efficiently. Predictions made by these models can be verified or falsified in new, more directed experiments, which will lead to further refinements of the models. This cycle of model construction, hypotheses generation, and experimental validation is the fundamental core of systems biology. Eventually, it is supposed to result in a description of a certain system that is able to explain all observations in a series of experiments. However, as the cycle implies, the systems biology approach will never lead to one final model being able to explain every property of a system. Every model has boundaries and a certain level of detailedness limiting its generality. Therefore, modeling will always be a trade-off between simplicity and comprehensiveness.

However, the huge amounts of various kinds of data and the size of the developed models render the use of computational methods necessary. Depending on the mathematical framework in which the model is built, different tools can be used to analyze available data, construct or refine models, or generate testable hypotheses from them.

### 3.1.3
### Levels of Modeling

In systems biology, the modeled networks can represent different types of chemical entities – genes being successively expressed in a certain condition, small molecules being converted into each other, or proteins transmitting an external signal into the cell in a cascade. Data of different quality is available for the different kinds of experimentally analyzed entities. Also, the available knowledge from the literature and the questions a model is supposed to answer will not be of the same quality. The most popular frameworks are introduced in the following.

*Boolean networks* are a mathematical representation often used to describe genetic regulations by transcription factors. The variables in such models can take two different states, on and off, and refer to whether the expression of a gene is turned on or not. At each discrete time step the new states of the variables are calculated from the states in the previous step.

*Discrete models* extend the Boolean case by allowing the variables to take more than two different discrete values, which makes it possible to describe more than two states of a gene.

In *ordinary differential equation (ODE) models* the variables as well as the time become continuous. The change in the variable's values over time is given in the form of ODEs. This means that the state of a variable at a certain timepoint cannot be calculated easily, but requires an integration of the system over time.

*Stochastic models* can be used when the number of simulated molecules is low and random effects could have a large influence on the system's behavior. In this framework, the next reaction event and the time until it is happening are chosen at random from certain distributions defined by the model. These reactions events represent a single reaction converting a small set of molecules according to its stoichiometry.

## 3.2
## Data Analysis

The quality of the models developed in systems biology greatly depends on the quality of the biological data available. In the following we will give a short overview of different types of data and possible data processing steps.

### 3.2.1
### Types of Data

#### 3.2.1.1 Purification
The identification of individual molecules like DNAs, RNAs, or proteins from the extracts of a cell is possible with different techniques like gel electrophoresis and chromatography. During gel electrophoreses the samples inserted into pockets in a gel made of a cross-linked polymer (e.g., polyacrylamide) are separated by an electric field. Depending on the charge and size of the molecule, and the size of the pores in the polymer gel, the molecules travel in the electric field. Since in most cases the samples consist of certain molecule groups in which the charge of a molecule is approximately proportional to its mass, the distance the molecules travel is mainly dependent on their mass and three-dimensional structure. In the separation of proteins the relation of mass to charge can be ensured by the addition of sodium dodecylsulfate. This surfactant disrupts the structure of a protein and binds to it at every second amino acid, giving it a negative charge according to its length.

A second class of techniques for the separation of molecules constitutes the various types of chromatographies. In column chromatography, for example, a mixture of compounds is washed with a buffer (the mobile phase/eluent) through a column filled with a solid carrier material (the stationary phase/adsorbent). The type of the adsorbent and the composition of the eluent determine the retention of the different molecules in the column, and therefore which molecules leave the column at a certain time.

#### 3.2.1.2 Detection
Different blotting techniques can be used for the verification of a molecule's presence in a cell extract. After a gel electrophoresis step the molecules are transferred (blotted) to a membrane (made of nitrocellulose, for example) and hybridized with a radioactively or fluorescently labeled molecule. The presence and relative amount of

the labeled molecule can later be detected easily. With this technique it is also possible to attain quantitative information on the investigated molecule. Depending on the type of substance examined, the type of the labeled molecule and the name of the blot are different. In western blotting, proteins are hybridized with antibodies binding to them; in Southern blotting, DNA with complementary DNA; in northern blotting, RNA with DNA or RNA. It is also possible to detect the location of a certain DNA or RNA in intact tissues or organisms with *in situ* hybridization. The expression of a gene in a certain location can be shown with this technique by the addition of a labeled complementary nucleic acid strand. The investigated gene is supposed to be expressed in places where this labeled molecule accumulates after a washing step.

### 3.2.1.3 Large-Scale Analyses

Large-scale analyses of the expression of genes under different conditions can be performed with microarrays. A microarray is a matrix built of up to a million spots in which a small amount of a certain short nucleotide strand is fixed. Complementary strands can bind to these short nucleotide strands. In an analysis one first extracts mRNA, which is reverse transcribed to cDNA and labeled with a fluorescent dye. Then the labeled cDNA is hybridized with the DNA on the microarray. Finally, the amount of fluorescence, which is measured for the individual spots, corresponds to the relative amount of mRNA that binds to the spot. Similar techniques also exist for the identification of protein expression, protein–protein or protein–compound interactions, or the research on the composition of cell surfaces.

### 3.2.1.4 Identification of Components

The determination of the chemical composition of a cell extract can be achieved by mass spectrometry (MS). In this analysis the individual masses of the components are obtained, which in a second step can be used to identify them. During MS analysis the molecules of interest are ionized, accelerated in an electric field, and then, for example, the time of flight (TOF) in a vacuum is measured. This TOF corresponds to the mass over charge ($m/z$) ratio for the individual molecules. In the identification of proteins these can additionally be cut into small fragments by enzymes before the MS analysis is performed, which leads to peptide mass fingerprinting. With multiple subsequent analyses, as in the tandem MS, it is also possible to perform *de novo* peptide sequencing.

### 3.2.2
**Working with Data**

In many applications the huge amounts of data produced can no longer be overseen by humans. Therefore, computational techniques have to be applied in order to extract important features. Two examples of such techniques are explained in the following, but the reader is reminded that many more are available.

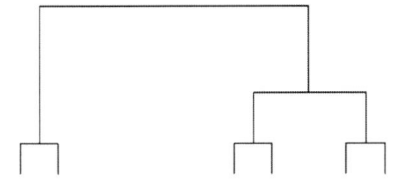

**Figure 3.1** Example of a hierarchical clustering.

### 3.2.2.1 Different Clustering Approaches

Clustering is an approach to group different data objects according to their similarity, such as genes with certain expression levels under different conditions. The similarity between data objects is judged by some kind of distance measure giving more correlated objects a smaller distance. Clustering methods can be divided into two categories. Depending on whether a training data set is available from which a clustering tool would be able to learn to which cluster a certain object is associated, the clustering is called supervised or unsupervised. Supervised clustering can be performed with various techniques from computer science (e.g., neuronal nets or support vector machines), while different approaches have to be distinguished in unsupervised clustering. In partitional clustering, one starts with a fixed set of clusters. During successive steps objects are assigned to the cluster with the center closest to them and then the centers are repositioned according to the middle of the objects currently associated with the cluster. This approach is guarantied to terminate, but is sensitive to the initial position of the cluster centers. In hierarchical clustering (Figure 3.1), one starts with one cluster (divisive) or clusters containing single objects (agglomerative) and successively divides or combines them according to the distance of the clusters. The result will vary depending on whether the distance of clusters is judged by their closest (single linkage) or most distant objects (complete linkage), or by their center (average linkage),

### 3.2.2.2 Principal Component Analysis

A different approach for the reduction of the dimensionality of a dataset is principal component analysis. In this approach orthogonal directions in which the variability is maximal in a dataset and which are uncorrelated throughout the experimental results are identified. These directions are linear combinations of, for example, gene expression levels in different individuals called principal components.

As an example one can consider a set of two genes expressed at the same level and a set of experiments with a varying common expression level. While the difference in the gene's expression levels does not change, the sum of the levels would vary greatly between experiments. Therefore, the sum would be the direction with the maximum variation between experiments, which would be the output of the analysis.

From a mathematical point of view the principal components are the eigenvectors of the empirical covariance matrix ordered by their eigenvalues. The eigenvalues can give a hint as to how many dimensions are necessary to explain the data to a certain degree of accuracy.

## 3.3
## ODE Modeling

The use of ODEs has become quite popular among modelers trying to explain certain phenomena. In this section we explain some basics of working with ODEs. The information we need to construct an ODE model for a certain system is knowledge about the initial concentrations of the participating substances, the reactions in which these substances participate, and differential equations for the rates of the reactions. From this information we can determine the state of the system at later points in time.

### 3.3.1
### Differential Equations

Vector $\vec{x}$ is an ordered set of concentrations of substances in a certain unit. Their change in time is determined by the reactions they participate in and their velocity. For example, if $x_1$ (the concentration of the first substance) is degraded in reaction $v_1$ that has a rate proportional to $x_1$, then the differential equation for its concentration change can be written as:

$$\frac{dx_1}{dt} = -v_1(x_1) = -a \cdot x_1.$$

The problem of computing the concentrations in the vector $\vec{x}$ for any time $t$ given ODEs and initial concentrations $\vec{x}(0)$ is called an initial value problem. Only for very simple examples can this problem be solved analytically – the analytical solution for the above-mentioned problem is $\vec{x}(t) = \vec{x}(0) \cdot e^{-a \cdot t}$. For larger, more complex problems the solution can only be computed with the help of numerical approximations.

### 3.3.2
### Stoichiometric Matrix

When dealing with large systems of reaction equations it becomes handy to divide the differential equations into a vector of reaction velocities and the so-called stoichiometric matrix. The entries in this matrix determine how many molecules of a certain substance are consumed or produced in a certain reaction. For example, in a system (as shown in Figure 3.2) where $v_1$ consumes one $x_1$ and produces two $x_2$ at a rate proportional to $x_1$ and $x_2$ is constantly degraded in reaction $v_2$, the differential equations can be written as:

$$\frac{d\vec{x}}{dt} = \frac{d}{dt}\begin{pmatrix} x_1 \\ x_2 \end{pmatrix} = \begin{pmatrix} -v_1(\vec{x}) \\ 2 \cdot v_1(\vec{x}) - v_2(\vec{x}) \end{pmatrix} = \begin{pmatrix} -1 & 0 \\ 2 & -1 \end{pmatrix} \cdot \begin{pmatrix} v_1(\vec{x}) \\ v_2(\vec{x}) \end{pmatrix} = N \cdot v(\vec{x}),$$

where:

$$v(\vec{x}) = \begin{pmatrix} a \cdot x_1 \\ b \end{pmatrix},$$

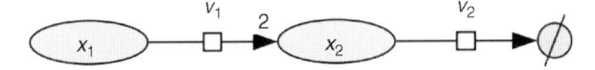

**Figure 3.2** Model described in Section 3.3.2.

is the vector of reaction rates. This notation makes it simple to develop even very large models because the structure of the reaction network, which is determined by the matrix N, and the reaction rates (or kinetics) are separated from each other.

### 3.3.3
### Reaction Kinetics

The rate of a reaction depends on the molecular mechanism of the reaction. Apart from constant rates, mass action kinetics are the most simple rate laws. The mass action kinetics describing an uncatalyzed reaction consist of a constant factor times a product of the concentrations of the participating reactants to the power of their stoichiometry. In the case of a reversible reaction the rate of the backward direction is subtracted from the forward rate. For example, a reaction:

$$S + R \leftrightarrow 2P,$$

would have a mass action kinetics of:

$$a \cdot s \cdot r - b \cdot p^2,$$

where $a$ and $b$ denote constants, and $s$, $r$, and $p$ denote concentrations.

Enzymatic reactions lead to more complex kinetics. It has, for example, to be considered that the amount of enzyme available is limited. Since this also limits the maximal rate, the kinetics have to be saturable with respect to the substrate concentration.

The most simple formula for an irreversible enzymatic reaction with one substrate S is the Michaelis–Menten kinetics:

$$v = \frac{v_{max} \cdot s}{k_m + s},$$

where $k_m$ is the substrate concentration at which the rate takes half of its maximum value.

For enzymes binding more than one substrate at a time the kinetics get more complex. This results from the fact that binding of the first substrate can change the structure of the enzyme making the binding of the next substrate more easy. Such an effect is called cooperativity and can, for example, be expressed by Hill kinetics:

$$v = \frac{v_{max} s^n}{k_d^n + s^n},$$

where $n$ and $k_d$ are the Hill constant and the dissociation constant for this reaction. The relationship between reaction rates and substrate concentrations for the different kinetics are depicted in Figure 3.3.

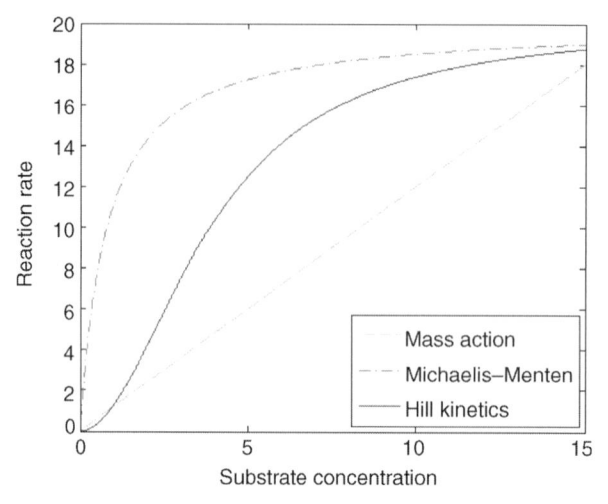

**Figure 3.3** Examples of reaction rates for different kinetics.

### 3.3.4
### Steady States

Once a model of ODEs has been constructed it can be analyzed for certain properties. The first interesting property is whether the system has stable or unstable steady states. A stable steady state is an interesting extreme condition because the system will tend to converge to it for large times without external influences.

Formally, a vector $\vec{x}^{ss}$ is a steady state if it fulfils the condition:

$$N \cdot v(\vec{x}^{ss}) = \frac{d\vec{x}^{ss}}{dt} = \vec{0},$$

where $\vec{0}$ is a vector containing only zeros. This means that the system will not leave this state without external influences. It is called stable if a system starting in a nearby state is drawn towards it and it is called stable unstable if it is pushed away.

In most models $\vec{x} = \vec{0}$ will be a steady state, because without a constant production of a substance all reaction kinetics will become zero. This state is called the trivial steady state. Other steady states in which $v(\vec{x}) \neq \vec{0}$ will only exist if the matrix $N$ has a certain structure (if it is column-rank deficient, certain nonzero vectors yield $\vec{0}$ if multiplied with this matrix from the right). Since $N$ depends only on the structure of the network, whether the system can have such nontrivial steady states or not can be seen from the network itself.

### 3.3.5
### Metabolic Control Analysis

In many applications one faces the problem of how a system can be perturbed in order to produce a certain outcome. One biotechnological example is the production of a particular substance by a certain organism. The question here is how the organism can be influenced to maximize the yield. Metabolic control analysis (Kacser and Burns, 1973; Heinrich and Rapoport, 1974) (MCA) has been developed for this purpose.

MCA is a powerful framework inferring changes in the steady state behavior of a system from changes in the systems parameters or initial concentrations. With this theory one can quantitatively infer which parameters or reaction rates have a large influence on the concentrations and fluxes through the reactions in steady state. It provides linear approximations of the changes in a concentration or the flux through a reaction in steady state resulting from changes in either a parameter or an initial concentration. These global approximations are easily calculable from simple local properties of the system – the elasticities.

When a system is in steady state this does not necessarily mean that all reaction rates are also equal to zero. Such a dynamic equilibrium can for example be taken by a metabolic pathway, if the substrate of the pathway is constantly available. In a real organism this pathway will of course in most cases not be in equilibrium because of for example a changing environment. Nevertheless, the system will often operate close to this state which makes it interesting for further considerations.

The $\varepsilon$ elasticity is the derivative of the reaction velocities with respect to substance concentrations.

$$\varepsilon_i^k = \frac{\partial v_k}{\partial x_i}$$

Storing these derivatives in a matrix $\varepsilon$, such that for each row the reaction from which the derivative is taken and for each column the concentration with respect to which the derivative is taken stays the same, makes it easy to write down formulas for the global approximations in steady state, like the concentration control coefficient:

$$C_v^x = \frac{\partial \vec{x}^{ss}}{\partial v} = -(N \cdot \varepsilon)^{-1} \cdot N$$

### 3.3.6
**Simulating Models**

Since in most cases the models are too complex to be solved explicitly, numerical methods for the computation of substance concentrations at different timepoints have to be applied. The most basic way of solving an ODE model is the explicit Euler method. In this scheme the concentrations at a timepoint $t + h$ are computed from the concentrations at time $t$ with the formula:

$$\vec{x}(t+h) = \vec{x}(t) + h \cdot f(\vec{x}(t)).$$

The idea behind this method is that $f$ will not change significantly on the interval between $t$ and $t + h$ if $h$ is sufficiently small. Therefore, one can take $f$ at time $t$ as an approximation for $f$ on the whole interval which makes the determination of $\vec{x}$ at time $t+h$ very simple.

This brief explanation should serve as a short introduction to numerical solutions to differential equations. Numerical methods currently implemented in most tools are much more complex.

### 3.3.7
### Parameter Estimation

In many modeling cases not all parameters of a systems are available from the literature. With the help of concentration time-courses of the substances in the system gained from biological experiments, we can estimate these parameters such that the simulated time-courses are close to the measured ones. For this purpose, we define a so-called objective function $f$ that measures the deviation of the model to the measured data. It is dependent on the measured ($\bar{x}_i$) and the simulated concentrations ($x_i$) at distinct time points $t_1, \ldots, t_P$:

$$f = \sum_{i=1}^{I} \sum_{p=1}^{P} (\bar{x}_i(t_p) - x_i(t_p))^2.$$

This function, called the sum of squared residuals, can be minimized with local and global optimizers like BFGS (Shanno, 1970) and simulated annealing (Kirkpatrick, 1984). In each function evaluation the system is simulated with a new set of parameters and the time-courses are afterwards compared to the experimentally measured ones. In case the objective function is sufficiently close to zero, the optimizers terminate. For the corresponding parameter set the simulated time-courses will approximate the measured ones (as shown in Figure 3.4).

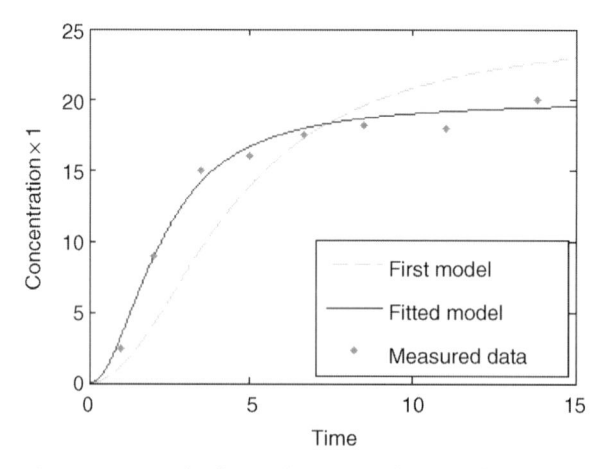

**Figure 3.4** Example of a simple system before and after fitting to experimental results.

## 3.4
## Results Gained from Systems Biology

Systems biology has provided research with results on many different topics (e.g., in the understanding of metabolic pathways, signal transduction, or genetic regulation). In the following, one example of a theoretical prediction will be given, which was later confirmed in living organisms.

### 3.4.1
### Just-in-Time Transcription

Klipp, Heinrich, and Holzhütter (2002) studied the optimal gene expression in a linear pathway with linear kinetics. The objective of the pathway was to convert an initial substrate, of which a certain amount is given into the environment, into the product of the chain as fast as possible. It was supposed that the enzymes in this pathway can instantly be produced and degraded, but that the total enzyme concentration is limited for economic reasons. Also, the number of times gene expression is allowed to change is restricted to the number of enzymes involved.

In the optimal gene expression profile the genes are first expressed exclusively starting with the first enzyme in the pathway. For the first gene the expression time is relatively long, but these times become shorter along the chain. In the last step, instead of the last gene getting activated exclusively, all genes are expressed with increasing strength from the front to the back of the chain.

These theoretical results have later been shown to hold in the amino acid biosynthesis in *Escherichia coli* (Zaslaver *et al.*, 2004). Under certain conditions it could be demonstrated that the genes involved in the unbranched part of the pathway are expressed in their functional order with a delay of about 10 min. It has also been found that, as predicted, the expression levels are higher for the genes involved in early positions in the pathway.

## 3.5
## Standard Formats, Databases, and Tools

In this section we discuss practical aspects of modeling. We give a short overview on standardized formats for the description of ODE models, introduce databases containing various information useful for modeling, and show tools with which models can be constructed and simulated.

### 3.5.1
### XML-Based Formats for ODE Models

In order to promote the exchange of ODE and other types of models between different research groups, standardized formats for their description have been developed. Along various approaches, like CellML (Lloyd, Halstead, and Nielsen, 2004), the

Systems Biology Markup Language (SBML) (Hucka et al., 2007) has become the quasi-standard with more than 160 tools and databases supporting it. SBML is an Extensible Markup Language (XML) format based on the biochemical structure of a model rather than on the mathematical type of description. In an SBML model, the concentrations or amounts of substances can be described as species in a certain compartment. These species can participate in reactions with arbitrary kinetics or their numerical values can be altered by general rules or events.

A machine-readable connection to the biological meaning behind a certain modeled entity can be given in SBML with annotations compliant to the MIRIAM standard (Le Novère et al., 2005). This standard defines a format describing how SBML elements like species or reactions can be linked to database entries providing detailed biological information. In particular, for the use of annotating biochemical models, a new ontology has been created – Systems Biology Ontology. With its help, the role of parameters in, for example, a reaction kinetic can be annotated.

### 3.5.2
**Databases**

Since the process of collecting information for ODE models can be quite laborious, the use of different kinds of databases providing various kinds of information becomes necessary. In the following we will introduce a small selection of databases and ontologies containing information important to modelers.

Gene Ontology (www.geneontology.org) (Harris et al., 2004) provides a vocabulary describing biological processes, molecular functions, and cellular components. This vocabulary is useful in modeling because it uses directed relations to connect the terms (e.g., one term being a generalization of another term).

KEGG (Kanehisa et al., 2008), the Kyoto Encyclopedia of Genes and Genomes (www.genome.jp/kegg), is a large database containing different information. One part of it contains nucleotide sequences, links between similar genes in different organisms, and protein similarities for amino acid sequences. A second part contains cross-linked information about more than 15 000 chemical compounds, over 7000 reactions, and more than 5000 enzymes. The third part connects all information by providing maps of metabolic or signaling pathways containing links to the other parts.

Reactome (Vastrik et al., 2007) (www.reactome.org) does not focus on single chemical entities, but rather gives information about complex interactions between proteins, nucleic acids, and other molecules. This information is divided into physical entities, reactions, and pathways. One interesting feature of Reactome is the so-called "sky," which is a graphical representation of the human reaction network. Each edge in the graph is linked to a page giving detailed information on the corresponding reaction. The pathways described in Reactome can be a good starting point for modeling, since they can be exported to SBML.

After the structure of a network has been modeled with the help of the previous databases, the reaction kinetics have to be determined. This can be done with the help of Brenda (Chang et al., 2008) (www.brenda-enzymes.org), which gives information about kinetic constants (e.g., $k_m$ values or turnover numbers) that are hard to gather from publications manually.

Also available on the web are resources for complete models in the SBML format like BioModels (Le Novere et al., 2006) (www.ebi.ac.uk/biomodels-main) and JWS online (Olivier and Snoep, 2004) (www.jjj.bio.vu.nl). These models can either be used as a starting point for new, more advanced models or they can be attached to existing models, with tools like semanticSBML (Schulz et al., 2006), in order to produce more comprehensive descriptions of biochemical processes.

### 3.5.3
### Tools for the Construction, Simulation, and Analysis of ODE Models

Various tools can be used for the construction and analysis of ODE models. Here, two SBML compatible tools will be described. CellDesigner (Funahashi et al., 2008) is a process diagram editor that can be used to easily construct models by drawing. Boxes of different shapes representing different types of molecule classes can be placed on a map and connected by reactions. Properties like the kinetic law of a reaction or the initial concentration of a molecule can be changed via dialog boxes. The user is also able to retrieve information from various databases and calculate time-courses for the constructed models.

Another popular tool among the systems biology community is Copasi (Hoops et al., 2006) (COmplex PAthway SImulator). This tool is not only able to construct ODE models and simulate their temporal behavior, but also gives the user the opportunity to analyze them with the help of MCA, to scan parameter ranges, or to estimate parameters in order to fit the simulation to given measured time-courses.

Descriptions of many similar tools can be found on the SBML website (sbml.org).

### 3.6
### Future Directions in Systems Biology

Two main paradigms exist in biological research. The research conducted under the first paradigm focuses on small, isolated topics, (e.g., the behavior of a certain enzyme under different conditions). For these small topics, hypotheses can easily be formulated and tested. Results gained from this approach are usually presented in text form. According to the second paradigm, researchers are focusing on large-scale systematic analyses (e.g., gene expression data), which are afterwards collected in big databases.

Systems biology tries to bridge the gap between those two paradigms by integrating data from both fields. Results from detailed small-scale research can be incorporated into the construction of models for a certain process, while large-scale data can be used for the refinement and support of the model. Thus, in this way, a more holistic view can be gained.

Systems biology has proven to be valuable in the planning of research experiments. Instead of giving just one hypothesis, which can be falsified or supported, models, once they have been constructed, can be used for the generation of multiple hypotheses. In case some of them can be falsified, this may yield hints how to refine the model, which would lead to more testable experiments. In this way, different alternatives for the mechanism underlying a certain phenomenon can also be tested.

Systems biology will most probably become more and more important in medical research. Possible applications are the identification of promising drug targets, the pharmacokinetic testing of drugs, or personalized medicine. When developing new drugs the pharma companies often focus on creating drugs against already known targets with less adverse effects. Targeting new proteins bears a huge financial risk as drugs against those fail trials more often. Good targets could be identified more efficiently with the help of models created by systems biology. Furthermore, side-effects of the drugs might be ruled out and good drug combinations could be found without the financial risk of clinical trials. Using pharmacokinetics for modeling the administration, distribution, metabolism, and excretion of a drug candidate could also become more valuable for pharma companies. Testing candidates *in silico* would reduce the costs of clinical trials or animal tests and might give hints for the development of new, more effective drugs. A third medical application of systems biology in medicine is individualized therapies. Here, data from a patient (e.g., genomic data) can be used to select appropriate drugs or drug combinations. In this way, cancer patients could get the most appropriate therapy for their type of cancer or severe side-effects could be ruled out.

Another field for the application of systems biology is biotechnology. In many processes, microorganisms are used to produce a certain substance. An analysis of their metabolic network can give ideas that can improve the product yield per resources spent (e.g., a change of the environmental conditions or a gene knockout).

A total enumeration of all possible applications of systems biology is impossible within the scope of this chapter, but it should have become clear that the list of mentioned applications for this expanding field is far from being comprehensive.

## References

Chang, A., Scheer, M., Grote, A., Schomburg, I., and Schomburg, D. (2008) BRENDA, AMENDA and FRENDA the enzyme information system: new content and tools in 2009. *Nucleic Acids Research*, **37**, D588–D592.

Funahashi, A., Matsuoka, Y., Jouraku, A., Morohashi, M., Kikuchi, N., and Kitano, H. (2008) CellDesigner 3.5: a versatile modeling tool for biochemical networks. *Proceedings of the IEEE*, **96**, 1254–1265.

Harris, M.A., Clark, J., Ireland, A., Lomax, J., Ashburner, M., Foulger, R., Eilbeck, K., Lewis, S., Marshall, B., Mungall, C. *et al.* (2004) The Gene Ontology (GO) database and informatics resource. *Nucleic Acids Research*, **32**, D258–D261.

Heinrich, R. and Rapoport, T.A. (1974) A linear steady-state treatment of enzymatic chains. General properties, control and effector strength. *European Journal of Biochemistry*, **42**, 89–95.

Hoops, S., Sahle, S., Gauges, R., Lee, C., Pahle, J., Simus, N., Singhal, M., Xu, L., Mendes, P., and Kummer, U. (2006) COPASI – a complex pathway simulator. *Bioinformatics*, **22**, 3067.

Hucka, M., Finney, A., Hoops, S., Keating, S., and Le Novere, N. (2007) Systems biology markup language (SBML) Level 2: structures and facilities for model definitions. *Nature Precedings*, **58** (2); http://hdl.handle.net/10101/npre.2007.58.2.

Kacser, H. and Burns, J.A. (1973) The control of flux. *Symposia of the Society for Experimental Biology*, **27**, 65–104.

Kanehisa, M., Araki, M., Goto, S., Hattori, M., Hirakawa, M., Itoh, M., Katayama, T., Kawashima, S., Okuda, S., Tokimatsu, T. *et al.* (2008) KEGG for linking genomes to life and the environment. *Nucleic Acids Research*, **36**, D480–D484.

Kirkpatrick, S. (1984) Optimization by simulated annealing: quantitative studies. *Journal of Statistical Physics*, **34**, 975–986.

Kitano, H. (2001) Systems biology: toward system-level understanding of biological systems, in *Foundations of Systems Biology*, MIT Press, Cambridge, MA, pp. 1–29.

Klipp, E., Heinrich, R., and Holzhütter, H.G. (2002) Prediction of temporal gene expression. *FEBS Journal*, **269**, 5406–5413.

Le Novère, N., Finney, A., Hucka, M., Bhalla, U.S., Campagne, F., Collado-Vides, J., Crampin, E.J., Halstead, M., Klipp, E., Mendes, P. *et al.* (2005) Minimum information requested in the annotation of biochemical models (MIRIAM). *Nature Biotechnology*, **23**, 1509–1515.

Le Novere, N., Bornstein, B., Broicher, A., Courtot, M., Donizelli, M., Dharuri, H., Li, L., Sauro, H., Schilstra, M., Shapiro, B. *et al.* (2006) BioModels Database: a free, centralized database of curated, published, quantitative kinetic models of biochemical and cellular systems. *Nucleic Acids Research*, **34**, D689–D691.

Lloyd, C.M., Halstead, M.D.B., and Nielsen, P.F. (2004) CellML: its future, present and past. *Progress in Biophysics and Molecular Biology*, **85**, 433–450.

Olivier, B.G. and Snoep, J.L. (2004) Web-based kinetic modelling using JWS online. *Bioinformatics*, **20**, 2143–2144.

Schulz, M., Uhlendorf, J., Klipp, E., and Liebermeister, W. (2006) SBMLmerge, a system for combining biochemical network models. *Genome Informatics Series*, **17**, 62–71.

Shanno, D.F. (1970) Conditioning of quasi-Newton methods for function minimization. *Mathematics of Computation*, **24**, 647–656.

Vastrik, I., D'Eustachio, P., Schmidt, E., Joshi-Tope, G., Gopinath, G., Croft, D., de Bono, B., Gillespie, M., Jassal, B., Lewis, S., Matthews, L., Wu, G., Birney, E., and Stein, L. (2007) Reactome: a knowledge base of biologic pathways and processes [Correction in *Genome Biology* 2009, **10**, 402]. *Genome Biology*, **8**, R39.

Zaslaver, A., Mayo, A.E., Rosenberg, R., Bashkin, P., Sberro, H., Tsalyuk, M., Surette, M.G., and Alon, U. (2004) Just-in-time transcription program in metabolic pathways. *Nature Genetics*, **36**, 486–491.

# 4
# Mind Over Molecule: Systems Biology for Neuroscience and Psychiatry
*Denis Noble*

## 4.1
### Introduction: Mind and Molecule Meet

The following is based on a philosophical anecdote first published in Noble (2008d).

Imagine a world in some *m*th dimension in which Mind and Molecule meet. Molecule, the little David, surveys Mind, the big Goliath, and teases him: "You think yourself so big. In fact you have no dimensions at all. And when you do, you just turn out to be a bunch of molecules like me!" Mind recalls this taunt from their days in the world of four dimensions. "You jumped-up tiny midget! Where do you get this idea that your level of causality is privileged?" Molecule likes this challenge. "It's obvious isn't it? I am what the whole of life reduces to. In my most extended form, as DNA, I am the great Genie. As one of our Earthlings put it, *I created you*, body and mind." Mind: "But he also called you selfish!" Molecule: "That's why I always win!"

Mind is tired of this tautologous nonsense, so he climbs Meditation Mountain and thinks. He thinks so deeply that he becomes empty of form. Even in *m* dimensions he finds that when he meditates profoundly he seems to disappear. As he returns from Nirvana he has his Eureka moment. "I've got it," he exclaims, "we are *all* empty, even that little Molecule. He's just . . . he's just a bunch of tangled strings. He is a molecule precisely because he interacts with other molecules; he is nothing on his own."

Meditation Mountain has two large tablets of stone. Mind carves into them his 10 commandments: the principles of Systems Biology. I have outlined these principles from different perspectives in three recent articles (Noble, 2008a, 2008c, 2008d). They are listed in Table 4.1. Here, I will explain how they have been applied to modeling the heart, how they might be applied to modeling neurons and the brain, and what their implications may be for psychiatry.

However, first of all, I should explain an important clarification of the 10 principles since they were published. The sixth principle, "There is no genetic program," has been the subject of considerable confusion and debate. Much depends on how one defines a gene (Keller, 2000; Noble, 2008c). Defined as a particular stretch of DNA, the statement is correct. Those sequences alone do not form a program that could be analyzed syntactically. However, it is valid to refer to what many people call

Table 4.1 Ten principles of systems biology (Noble, 2008a).

| | |
|---|---|
| 1 | Biological functionality is multilevel |
| 2 | Transmission of information is not one way |
| 3 | DNA is not the sole transmitter of inheritance |
| 4 | The theory of biological relativity – there is no privileged level of causality |
| 5 | Gene ontology will fail without higher-level insight |
| 6 | There is no genetic program |
| 7 | There are no programs at any other level |
| 8 | There are no programs in the brain |
| 9 | The self is not an object |
| 10 | There are many more to be discovered – a genuine "theory of biology" does not yet exist |

"developmental programs," which necessarily include much more than the DNA. For a clear discussion of the distinction between "genetic program" and "developmental program," the reader is referred to (Keller, 2000, pp. 80–82).

## 4.2
### First Steps: Modeling Excitable Cells

The first model of a cardiac cell was developed in 1960 (Noble, 1960) following the discovery of two types of potassium channel in the heart (Hutter and Noble, 1960). The equations for one of these channels – the delayed rectifier – was based on using a slowed-down version of the Hodgkin–Huxley squid nerve potassium channel (Hodgkin and Huxley, 1952), while the sodium channel equations were also modified from their work. The nervous system therefore provided the paradigm for the early cardiac work. Later developments incorporated many more ionic currents and related mechanisms: several calcium channels, more sodium and potassium channels, ion exchangers like the Na–K pump, the Na–Ca exchanger, intracellular calcium signaling, internal and external ion concentrations, excitation–contraction coupling, to name a few. The cardiac cell models are now therefore highly developed and very complex (Noble, 2002a, 2007a; Noble and Rudy, 2001), and they are used extensively in the study of drugs and arrhythmia (Fink et al., 2008; Noble, 2007b, 2008b).

At the same time as the early development of such computational models, considerable progress was made in the application of mathematics both to cardiac and to neuronal structures. This work was summarized in the book by Jack, Noble, and Tsien (1975). This book became the standard text on cable theory and conduction in excitable cells, and it includes extensive treatment of neurons and dendrite function, including the mathematics of rhythmic firing in neurons. Meanwhile, modeling of nerve cells using ion channel models became progressively more sophisticated (Koch and Segev, 1998).

An important general conclusion from this extensive work on heart and nerve is that causality in biological systems works in two directions between two levels of organization. The molecular (protein) mechanisms generate the electric current and other processes (such as changes in ion concentrations) essential for the system to work at all, while the integrated result of their activity, including the electric potential of the cell as a whole, in turn acts as the driver of the gating processes at the protein level. There is therefore feedback from the higher level, the cell, down onto the lower level, the protein transporters, as well as upward causation from the transporters to the cell level, thus forming a two-level cycle, originally called the Hodgkin cycle (Figure 4.1).

Consider, as an example, the cardiac pacemaker mechanism. This depends on ionic current generated by a number of protein channels carrying sodium, calcium,

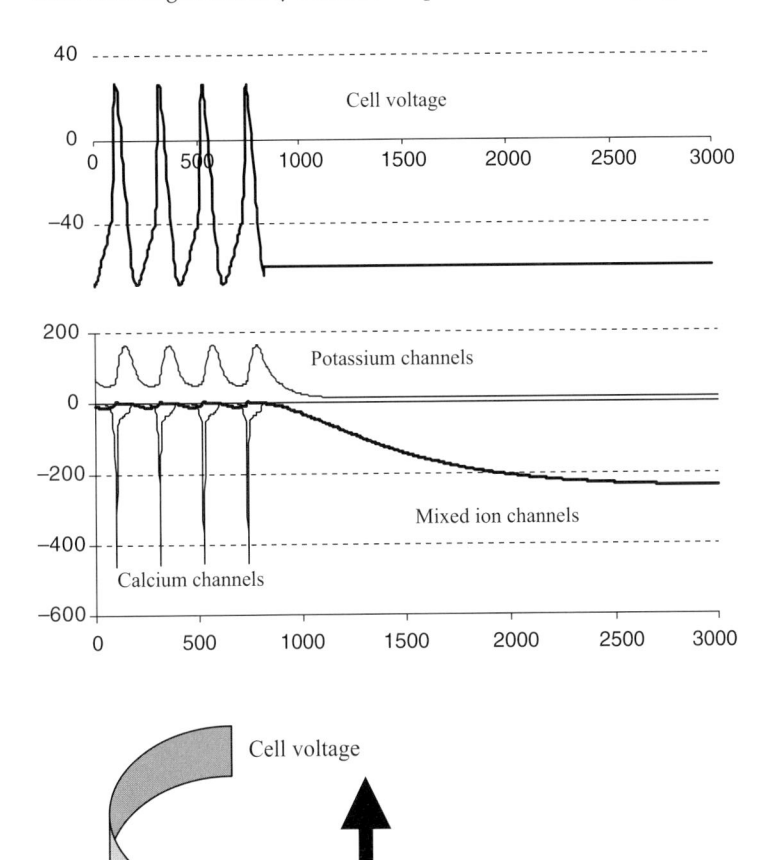

**Figure 4.1** Computer model of pacemaker rhythm in the heart (Noble and Noble, 1985). For the first four beats the model is allowed to run normally and generates rhythm closely similar to a real heart. Then the feedback from cell voltage to protein channels is interrupted. All the protein channel oscillations then cease. They slowly change to steady constant values. The diagram shows the causal loop involved. Protein channels carry current that changes the cell voltage (upward arrow), while the cell voltage changes the protein channels (downward arrow). In the simulation, this downward arrow was broken at 800 ms. (From Noble, 2006.)

potassium, and other ions. The activation, deactivation, and inactivation of these channels proceed in a rhythmic fashion in synchrony with the pacemaker frequency. We might therefore be tempted to say that their oscillations generate that of the overall cell electrical potential (i.e., the higher-level functionality). However, this is not the case. The kinetics of these channels varies with the electrical potential. There is therefore feedback between the higher-level property, the cell potential, and the lower level property, the channel kinetics (see chapter 5 in Noble, 2006). If we remove the feedback (e.g., by holding the potential constant, as in a voltage clamp experiment), the channels no longer oscillate (Figure 4.1). The oscillation is therefore a property of the system as a whole, not of the individual channels or even of a set of channels unless they are arranged in a particular way in the right kind of cell.

Nor can we establish any priority in causality by asking which comes first, the channel kinetics or the cell potential. This fact is also evident in the differential equations we use to model such a process. The physical laws represented in the equations themselves, and the initial and boundary conditions, operate *at the same time* (i.e., during every integration step however infinitesimal), not sequentially.

It is simply a prejudice that inclines us to give some causal priority to lower-level, molecular events. The concept of level in biology is itself metaphorical. There is no literal sense in which genes and proteins lie *underneath* cells, tissue and organs. It is a convenient form of biological classification to refer to different levels and we would find it very hard to do without the concept. However, we should not be fooled by the metaphor into thinking that "high" and "low" here have their normal meanings. From the metaphor itself, we can derive no justification for referring to one level of causality as privileged over others. That would be a misuse of the metaphor of level.

We can generalize this principle of multilevel causality and the existence of downward causation to all levels of biological organization (Figure 4.2).

## Downward Causation

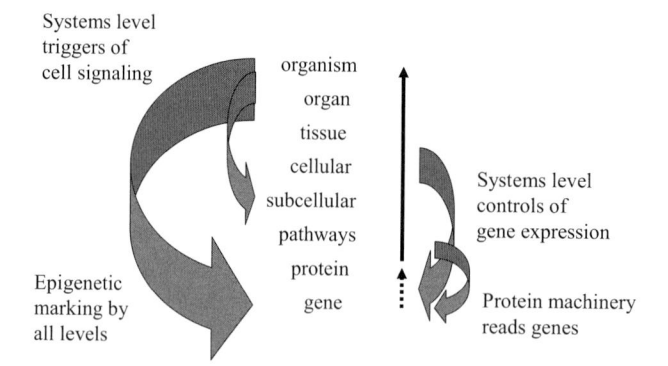

**Figure 4.2** Some of the various forms of downward causation, such as higher levels triggering cell signaling and gene expression. Loops of interacting downward and upward causation can be built between all levels of biological organization. As emphasized in the text, this diagram is metaphorical since it uses the concepts of "level," "upward," and "downward." (Modified from Noble, 2006.)

One of the consequences of the existence of downward causation is that transmission of information in biological systems is not one way. The central dogma of molecular biology has been misapplied to suggest that it is and that there is a genetic program controlling the unfolding of the organism. I prefer the metaphor of genes as the pipes of a vast organ (there are pipe organs with roughly the same number of pipes as genes in the human genome!) which are played by the various types of cells and tissues of the body to produce the very different patterns of gene expression that are involved.

The central dogma has a correct, but limited, application, which is the original one proposed by Crick (1958, 1970) – that while DNA codes for amino acid sequences in proteins, the amino acid sequence does not code for DNA. However, there are two respects in which the central dogma is incomplete. The first is that it defines the relevant information uniquely in terms of the DNA code – the sequence of C, G, A, T bases. Although the most that this information can tell us is which protein (or proteins in the case of genes with multiple exons) will be made. The DNA sequence alone does not tell us how much of each protein will be made. Yet, this is one of the most important characteristics of any protein-producing machinery. I first encountered this kind of problem when working on the mathematics of excitable cell conduction (Jack, Noble, and Tsien, 1975). The speed of conduction of a nerve or muscle impulse depends on the density of rapidly activated sodium channels – the larger the density, the greater the ionic current and the faster the conduction. However, this rule applies only up to a certain optimum density. As the channel gating also contributes to the cell capacitance, which itself slows conduction, there is a point at which adding more channel proteins is counter-productive (Jack, Noble, and Tsien, 1975, p. 432). This optimum density was also shown theoretically in the same year by Alan Hodgkin who noted that the actual density in an unmyelinated nerve such as the squid giant axon is close to this optimum (Hodgkin, 1975). This means that a feedback mechanism must operate between the electrical properties of the nerve and the expression levels of the sodium channel protein. We now refer to such feedback mechanisms in the nervous system, which take many forms, as electro-transcription coupling (Deisseroth et al., 2003).

Similar processes must occur in the heart. One of the lessons I have learnt from many attempts to model cardiac electrophysiology (Noble, 2002b) is that, during the slow phases of repolarization and pacemaker activity, the ionic currents are so finely balanced that it is inconceivable that nature arrives at the correct expression levels of the relevant proteins without some kind of feedback control. We do not yet know what that control might be, but we can say that it must exist. Nature cannot be as fragile as our computer models are! Robustness is an essential feature of successful biological systems.

## 4.3
## Higher-Level Simulation

Neither the heart nor the nervous system can be analyzed in terms of cells alone. Modeling therefore needs also to be done at the tissue and organ levels, and to be extended to the systems and whole organism levels, as in the Virtual Physiological Human project. The approach though has to be very different. The heart is a syncytial structure. All the electrically excitable cells are interconnected via gap junctions. In principle it is therefore possible to build a virtual heart from interconnected virtual cells (Crampin et al., 2004; Hunter, Robbins, and Noble, 2002; Noble, 2002a). In practice, compromises have to be made. Computations involving tens of millions of grid points representing cellular level events can be performed. However, computations at the level of hundreds of millions of cells – the actual number involved is around 1 billion – are beyond the capacity of even the largest supercomputers planned. This problem is circumvented by lumping numbers of cell elements together at each grid point. Since neighboring cells have much the same properties, this can be done without too much loss of detail.

However, this approach cannot be used so readily, if at all, for the brain. First, the numbers of cells involved is much larger. Second, the connections between them are not usually electrical gap junctions, but rather chemical synapses with chemical specificity and usually complex geometry. Each cell may connect in this way with thousands of others from many different parts of the brain, rather than the 6–10 neighboring cells in a cardiac syncytium. Third, the form and activity of the different types of nerve cell differ widely. Finally, we really do not know to what extent the functioning of the brain is dependent on fine detail at the cellular and molecular levels. Where exactly is memory stored? If it really were at the molecular level, as some have postulated, the difficulties in modeling would be daunting indeed. My own guess is that the molecular level is probably too low. Stochasticity and nonspecificity would make precise coding difficult and probably impossible. After all, even gene expression is strongly stochastic (Kupiec, 2009).

Despite all the technical difficulties, in both cases, tissue and organ level modeling are necessary. Cardiac arrhythmia, for example, is a property of large numbers of cells, although the genesis of such arrhythmia may often have a cellular basis. In the case of the brain, it is obvious that many physiological functions, and certainly those of the mind, are high-level ones.

There is therefore a considerable literature over many decades on modeling the brain at levels higher than single cells (e.g., David and Friston, 2003; Jansen and Rit, 1995; Wilson and Cowan, 1972, 1973).

A fundamental property of systems involving multiple levels between which there are feedback control mechanisms is that there is no privileged level of causality. All levels can be involved in the causal loops involved. As we have seen there is no reason, other than metaphorical ones, to give causal priority to lower-level, molecular events.

One of the aims of my book, *The Music of Life* (Noble, 2006), was to explore the limitations of biological metaphors. This is a form of linguistic analysis that is rarely applied in science, though a notable exception is Steven J. Gould's monumental work on the theory of evolution in which he analyses the arguments for the multiplicity of levels at which natural selection operates (see chapter 8 in Gould, 2002).

These points can be generalized to any biological function. The only sense in which a particular level might be said to be privileged is that, in the case of each function, there is a level at which the function is integrated and it is one of our jobs as scientists to determine what that level may be. These will generally be higher, not lower, levels. The idea that there is no privileged level of causality is the basis of what, in the 10 principles (Figure 4.1), I call one of them a theory of biological relativity (Noble, 2008a).

A related principle is that gene ontology will fail without higher-level insight. Genes, as defined by molecular genetics to be the coding regions of DNA, code for proteins. Biological function then arises as a consequence of multiple interactions between different proteins in the context of the rest of the cell machinery. Each function therefore depends on many genes, while many genes play roles in multiple functions. What then does it mean to give genes names in terms of functions? The only unambiguous labeling of genes is in terms of the proteins they code for. Thus, the gene for the sodium-calcium exchange protein is usually referred to as *ncx*. Ion channel genes are also often labeled in this way, as in the case of sodium channel genes being labeled *scn*.

This approach, however, naturally appears unsatisfactory from the viewpoint of a geneticist, since the original question in genetics was not which proteins are coded for by which stretches of DNA (in fact, early ideas on where the genetic information might be found (Schrödinger, 1944) favored the proteins), but rather what is responsible for higher-level phenotype characteristics. There is no one-to-one correspondence between genes or proteins and higher-level biological functions. Thus, there is no "pacemaker" gene. Cardiac rhythm depends on many proteins interacting within the context of feedback from the cell electrical potential.

Another good example of this approach is the discovery of what are called clock genes, involved in circadian rhythm. Mutations in a single gene (now called the Period (*per*) gene) are sufficient to abolish the circadian period of fruit flies (Konopka and Benzer, 1971). This discovery of the first "clock gene" was a landmark since it was the first time that a single gene had been identified as playing such a key role in a high-level biological rhythm. The expression levels of this gene are clearly part of the rhythm generator. They vary (in a daily cycle) in advance of the variations in the protein that they code for. The reason is that the protein is involved in a negative feedback loop with the gene that codes for it (Hardin, Hall, and Rosbash, 1990). The idea is very simple. The protein levels build up in the cell as the *per* gene is read to produce more protein. The protein then diffuses into the nucleus where it inhibits further production of itself by binding to the promoter part of the gene sequence. With a time delay, the protein production falls off and the inhibition is removed so that the whole cycle can start again. So, we not only have a single gene capable of regulating the biological clockwork that generates circadian rhythm, it is itself a key component in the feedback loop that forms the rhythm generator.

However, such rhythmic mechanisms do not work in isolation. There has to be some connection with light-sensitive receptors (including the eyes). Only then will the mechanism lock on to a proper 24-h cycle rather than free-running at say 23 or 25 h. In the mouse, for example, many other factors play a role. Moreover, the CLOCK gene itself is involved in other functions. That is why Foster and Kreitzman (Foster and Kreitzman, 2004) have written:

... what we call a clock gene may have an important function within the system, but it could be involved in other systems as well. Without a complete picture of all the components and their interactions, it is impossible to tell what is part of an oscillator generating rhythmicity, what is part of an input, and what is part of an output. In a phrase, it ain't that simple!

As Foster and Kreitzman emphasize, there are many layers of interactions overlaid onto the basic mechanism. So much so that it is possible, surprisingly, to knock out the CLOCK gene in mice and retain circadian rhythm (Debruyne et al., 2006).

The point is obvious. We should not be misled by gene ontology. The first function a gene is found to be involved in is rarely if ever the only one and may not even be the most important one. Gene ontology will require higher-level insight to be successful in its mission.

## 4.4
### Genetic Programs?

It is time to return to the Mind and Molecule story in my introduction. For, I can already sense that, out there in my imagined $m$th dimension, Molecule is getting impatient with Mind. As we have developed the ideas of a systems approach to biology, Mind has been running all over him like a steamroller flattening everything in its path. Surely, Molecule thinks, I must win on the question of genetic programs. They are all over the place. They are the crown jewels of the molecular genetic revolution, invented by none other than the famous French Nobel Prize winners, Monod and Jacob (Jacob, 1970; Monod and Jacob, 1961). Their enticing idea was born during the early days of electronic computing, when computers were fed with paper tape or punched cards coded with sequences of instructions. I knew those computers well since this was the kind of machine on which I carried out my first modeling computations in 1960 (see chapter 5 in Noble, 2006). The instructions on the tape were clearly separate from the machine itself that performed the operations. They dictated those operations. Moreover, the coding on the tape, as in the stored programs in today's computers, was digital. The analogy with the digital code of DNA is obvious. So, are the DNA sequences comparable to the instructions of a computer program?

An important feature of such computer programs is that the program is separate from the activities of the machine that it controls. Originally, the separation was physically complete, with the program on the tape or cards only loaded temporarily into the machine. Nowadays, the programs are stored within the memory of the machine, and the strict distinction between the program, the data, and the processes controlled may be breaking down. Perhaps computers are becoming more like living systems, but in any case the concept of a genetic program was born in the days when programs were separate identifiable sets of instructions.

So, what do we find when we look for genetic programs in an organism? I suggest that we find no genetic programs! There are no sequences of instructions in the genome that could possibly play a role similar to that of a computer program. The reason is very simple. A database is not a program. DNA is a database. It is used like an organ of the cell, and by the tissues and organs of the body. To find anything comparable to a program we *always* have to extend our search well beyond the genome itself. Thus, as we have seen above, the sequence of events that generates circadian rhythm includes the *per* gene, but it necessarily also includes the protein it codes for, the cell in which its concentration changes, the nuclear membrane across which it is transported with the correct speed to effect its inhibition of transcription. This is a gene–protein–lipid–cell network, not simply a gene network. In fact, it involves all the other levels as well: the interactions underlying circadian rhythm involve the systems as a whole. The nomenclature matters. Calling it a gene network fuels the misconception of genetic determinism. In the generation of a 24-h rhythm, none of these events in the feedback loop is privileged over any other. Remove any of them, not just the gene, and you no longer have circadian rhythm.

Moreover, it would be strange to call this network of interactions a program. The network of interactions is itself the circadian rhythm process. As Enrico Coen, the distinguished plant geneticist, put it "Organisms are not simply manufactured according to a set of instructions. There is no easy way to separate instructions from the process of carrying them out, to distinguish plan from execution" (Coen, 1999). In short, the concept of a program here is completely redundant. Organisms are interaction machines, not Turing machines; biology achieves efficiency through highly parallel processing, and by combining analog and digital responses to endogenous and exogenous impacts.


## 4.5
## Programs in the Brain?

This leads me to what is possibly the most controversial of my conclusions: there are no programs in the brain! However, before I explain what I mean by this statement, let us note that, by this stage, Molecule is getting more than impatient. We are moving onto what ought to be Mind's territory – the brain. Suspecting that Mind might have an easy ride at this level, in desperation Molecule calls in help from his brilliant discoverer, the Nobel Prize winner Francis Crick. In his book *The Astonishing Hypothesis*, Crick proclaimed "You, your joys and your sorrows, your memories and your ambitions, your sense of personal identity and free will, are in fact no more than the behavior of a vast assembly of nerve cells and their associated molecules" (Crick, 1994). As Molecule savors the implications of this sweeping statement he does a *pas de deux* with his alter ego – the other strand of the double helix.

Meanwhile, Mind is at last in some kind of trouble of his own. Many biologists tell him that the solution to the old mind–brain problem is simple. In some sense or other, the mind is just a function of the brain. The pancreas secretes insulin, endocrine glands secrete hormones ... and the brain "secretes" consciousness! All that's left is to find out how and where in the brain that happens. In one of his last statements, Crick has even hinted at where that may be: "I think the secret of consciousness lies in the claustrum" (Francis Crick, 2004, quoted by V.S. Ramachandran; http://www.edge.org/documents/archive/edge147.html). This structure is a thin layer of nerve cells in the brain. It is very small and it has many connections to other parts of the brain, but the details are of no importance to the argument. The choice of brain location for the "secret of consciousness" varies greatly according to the author. Descartes even thought that it was in the pineal gland. The mistake is always the same, which is to think that in some way or other the brain is a kind of performance space in which the world of perceptions is reconstructed inside our heads and presented to us as a kind of Cartesian theatre. However, that way of looking at the brain leaves open the question: where is the "I," the conscious self that sees these reconstructions? Must that be another part of the brain that views these representations of the outside world?

We are faced here with a mistake similar to that of imagining that there must be programs in the genomes, cells, tissues, and organs of the body. There are no such programs, even in the brain. The activity of the brain and of the rest of the body simply *is* the activity of the person, the self. Once again the concept of a program is superfluous. Thus, when I play my guitar and my fingers rapidly caress the strings at an automatic speed that comes from frequent practice, there is no separate program that is making me carry out this activity. The patterns and processes in my nervous system and the associated activities of the rest of my body simply *are* me playing the guitar. The practice is also itself a formative influence, resulting in gradual modifications of the "self" over time.

Similarly, when we deliberate intentionally there is no nervous network "forcing" us to a particular deliberation. The nervous networks, the chemistry of our bodies, together with all their interactions within the social context in which any intentional deliberation makes sense, *are* us acting intentionally. We, ourselves, our "selves," simply are the processes of the body that might be called programs. And so I arrive at the highest of the principles, and the one that has the greatest relevance to psychiatry: the self is not an object. It is a process. Moreover, it is a process that, like all the parts of a biological system, is subject to development. From DNA (Barbara McClintock in her Nobel Prize lecture called the genome a "highly sensitive organ of the cell" (Keller, 1983)) through all the levels, to the nervous system and beyond, development is continuous.

In terms of our story, Mind wins yet again, not because he is a separate object competing for activity and influence with the molecules of the body. Thinking in that way was originally the mistake of the dualists, like Sherrington and Eccles, led by their interpretation of the philosophy of Descartes. Modern biologists have abandoned the separate substance idea, but many still cling to a materialist version of the same mistake (Bennett and Hacker, 2003), based on the idea that somewhere in the brain the self is to be found as some neuronal process. The reason why that level of integration is too low is that the brain, and the rest of our bodies which are essential for attributes like consciousness to make sense (chapter 9 in Noble, 2006), are tools (back to the database idea again) in an integrative process that occurs at a higher level involving social interactions. We cannot attribute the concept of self-ness to ourselves without also doing so to others (Strawson, 1959). Contrary to Crick's view, our selves are indeed much "more than the behavior of a vast assembly of nerve cells and their associated molecules" precisely because the social interactions are essential even to understanding what something like an intention might be. I analyze an example of this point in much more detail in chapter 9 of *The Music of Life* (Noble, 2006). This philosophical point is easier to understand when we take a systems view of biology since it is in many ways an extension of that view to the highest level of integration in the organism.

I was taught medicine under one of the great anatomists and zoologists of the nervous system, J.Z. Young, who did much of his seminal work on the brain of the octopus. His book, *A Model of the Brain* (Young, 1964), was a seminal work for me. He wrote (p. 27) "it does not seem likely that the brain operates with a detailed programme of logical instructions." After 44 years of further reflection on this question I find myself agreeing strongly with his conclusion and I admire his insight all those years ago.

## 4.6
## Conclusions

It is time in this chapter to leave Mind and Molecule arguing it out in their *m*th dimension. Perhaps I have convinced you that in our world, at least, the argument is sterile. It will be an interesting subject for future historians of science to determine why we were misled for so long by the idea of molecular determinism. But we can move on. We need to understand all those molecules, but we need even more to understand their interactions. The parameter space in which those interactions occur is mind-bogglingly vast. "There wouldn't be enough material in the whole universe for nature to have tried out all the possible interactions even over the long period of billions of years of the evolutionary process" (chapter 2 in Noble, 2006).

The systems approach works well together with another important insight, which is that it is somewhat artificial to view objects and their relationships as separate aspects of reality. An object that had no relationship to any other would be undetectable. That is why, when in my story Mind cuts himself off from the world as he meditates, he literally becomes empty in the sense in which I used the term in the Introduction. In the limit everything, even Molecule, becomes empty. Nothing and everything meet out there in the *m*th dimension.

## Acknowledgments

Work in the author's laboratory is supported by EU FP6 BioSim network, EU FP7 PreDiCT project, Biotechnology and Biological Sciences Research Council, and the Engineering and Physical Sciences Research Council. Some of the material used in this chapter has been published in different form in my book, *The Music of Life* (Noble, 2006), and the story of Mind and Molecule is taken from Noble (2008d).

## References

Bennett, M.R. and Hacker, P.M.S. (2003) *Philosophical Foundations of Neuroscience*, Blackwell, Oxford.

Coen, E. (1999) *The Art of Genes*, Oxford University Press, Oxford.

Crampin, E.J., Halstead, M., Hunter, P.J., Nielsen, P., Noble, D., Smith, N., and Tawhai, M. (2004) Computational physiology and the physiome project. *Experimental Physiology*, **89**, 1–26.

Crick, F.H.C. (1958) On protein synthesis. *Symposia of the Society for Experimental Biology*, **12**, 138–163.

Crick, F.H.C. (1970) Central dogma of molecular biology. *Nature*, **227**, 561–563.

Crick, F.H.C. (1994) *The Astonishing Hypothesis: The Scientific Search for the Soul*, Simon & Schuster, London.

David, O. and Friston, K.J. (2003) A neural mass model for MEG/EEG: coupling and neural dynamics. *Neuroimage*, **20**, 1743–1755.

Debruyne, J.P., Noton, E., Lambert, C.M., Maywood, E.S., Weaver, D.R., and Reppert, S.M. (2006) A clock shock: mouse CLOCK is not required for circadian oscillator function. *Neuron*, **50**, 465–477.

Deisseroth, K., Mermelstein, P.G., Xia, H., and Tsien, R.W. (2003) Signaling from synapse to nucleus: the logic behind the mechanisms. *Current Opinion in Neurobiology*, **13**, 354–365.

Fink, M., Noble, D., Virag, L., Varro, A., and Giles, W. (2008) Contributions of HERG $K^+$ current to repolarization of the human ventricular action potential. *Progress in Biophysics and Molecular Biology*, **96**, 357–376.

Foster, R. and Kreitzman, L. (2004) *Rhythms of Life*, Profile Books, London.

Gould, S.J. (2002) *The Structure of Evolutionary Theory*, Harvard University Press, Cambridge, MA.

Hardin, P.E., Hall, J.C., and Rosbash, M. (1990) Feedback of the *Drosophila* period gene product on circadian cycling of its messenger RNA levels. *Nature*, **343**, 536–540.

Hodgkin, A.L. (1975) The optimum density of sodium channels in an unmyelinated nerve. *Proceedings of the Royal Society B*, **270**, 297–300.

Hodgkin, A.L. and Huxley, A.F. (1952) A quantitative description of membrane current and its application to conduction and excitation in nerve. *Journal of Physiology*, **117**, 500–544.

Hunter, P.J., Robbins, P., and Noble, D. (2002) The IUPS human physiome project. *Pflügers Archiv – European Journal of Physiology*, **445**, 1–9.

Hutter, O.F. and Noble, D. (1960) Rectifying properties of heart muscle. *Nature*, **188**, 495.

Jack, J.J.B., Noble, D., and Tsien, R.W. (1975) *Electric Current Flow in Excitable Cells*, Oxford University Press, Oxford.

Jacob, F. (1970) *La Logique du vivant, une histoire de l'hérédité*, Gallimard, Paris.

Jansen, B.H. and Rit, V.G. (1995) Electroencephalogram and visual evoked potential generation in a mathematical model of coupled cortical neurons. *Biological Cybernetics*, **73**, 357–366.

Keller, E.F. (1983) *A Feeling for the Organism: The Life and Work of Barbara McClintock*, Freeman, New York.

Keller, E.F. (2000) *The Century of the Gene*, Harvard University Press, Cambridge, MA.

Koch, C. and Segev, I. (1998) *Methods in Neuronal Modeling*, 2nd edn, MIT Press, Cambridge, MA.

Konopka, R.J. and Benzer, S. (1971) Clock mutants of *Drosophila melanogaster*. *Proceedings of the National Academy of Sciences*, **68**, 2112–2116.

Kupiec, J.-J. (2009) *The Origin of Individuals: A Darwinian Approach to Developmental Biology*, World Scientific, London.

Monod, J. and Jacob, F. (1961) Teleonomic mechanisms in cellular metabolism, growth and differentiation. *Cold Spring Harbor Symposia on Quantitative Biology*, **26**, 389–401.

Noble, D. (1960) Cardiac action and pacemaker potentials based on the Hodgkin–Huxley equations. *Nature*, **188**, 495–497.

Noble, D. (2002a) Modelling the heart: from genes to cells to the whole organ. *Science*, **295**, 1678–1682.

Noble, D. (2002b) Modelling the heart: insights, failures and progress. *BioEssays*, **24**, 1155–1163.

Noble, D. (2006) *The Music of Life*, Oxford University Press, Oxford.

Noble, D. (2007a) From the Hodgkin–Huxley axon to the virtual heart. *Journal of Physiology*, **580**, 15–22.

Noble, D. (2007b) Heart simulation, arrhythmia and the actions of drugs, in *Biosimulation in Drug Development* (eds. M. Bertau, E. Mosekilde, and H. Westerhoff), Wiley-VCH Verlag GmbH, Weinheim, pp. 259–272.

Noble, D. (2008a) Claude Bernard, the first systems biologist, and the future of physiology. *Experimental Physiology*, **93**, 16–26.

Noble, D. (2008b) Computational models of the heart and their use in assessing the actions of drugs. *Journal of Pharmacological Sciences*, **107**, 107–117.

Noble, D. (2008c) Genes and causation. *Philosophical Transactions of the Royal Society*, **A366**, 3001–3015.

Noble, D. (2008d) Mind over molecule: activating biological demons. *Annals of the New York Academy of Sciences*, **1123**, xi–xix.

Noble, D. and Noble, S.J. (1985) A model of sino-atrial node electrical activity based on a modification of the DiFrancesco–Noble (1984) equations. *Proceedings of the Royal Society*, **B222**, 295–304.

Noble, D. and Rudy, Y. (2001) Models of cardiac ventricular action potentials: iterative interaction between experiment and simulation. *Philosophical Transactions of the Royal Society*, **A359**, 1127–1142.

Schrödinger, E. (1944) *What is Life?*, Cambridge University Press, Cambridge.

Strawson, P.F. (1959) *Individuals*, Routledge, London.

Wilson, H.R. and Cowan, J.D. (1972) Excitatory and inhibitory interactions in localized populations of model neurons. *Biophysical Journal*, **12**, 1–24.

Wilson, H.R. and Cowan, J.D. (1973) A mathematical theory of the functional dynamics of cortical and thalamic nervous tissues. *Kybernetik*, **13**, 55–80.

Young, J.Z. (1964) *A Model of the Brain*, Oxford University Press, Oxford.

# Part Two
# Basics

In Chapter 5, special aspects of psychopathology are presented by Wolfram Kawohl and Paul Hoff. Major changes coming up in categorizing and diagnosing mental disorders under the influence of neurobiological research in psychiatry are described and discussed. Categories such as endophenotypes appear to be suitable to relate neurobiological findings to mind functions.

Considering mental disorders as being related to brain processes, electrical signaling is an essential issue. Therefore, Chapter 6 focuses on electrophysiological parameters that may provide insights into psychopathological symptoms and syndromes. Electroencephalography has already been used successfully in both human and animal experimentation. In this chapter, Walter Freeman interprets his electrophysiological findings as oscillating circuits that are able to display self-enhancing activities from which to learn how neurons elaborate and maintain the unity of intentional action. These electrical activities are described in the olfactory system on three levels: (i) microscopic local field potentials, (ii) the mesoscopic, intracranial electrocorticogram, and (iii) the macroscopic electroencephalogram. Many reverberating circuits represent circular causality that generates self-enhancing activity. This is reflected by the frequency and degree of coherence that can be correlated with the degree of cognitive functioning.

The following Chapter 7 by Patricio O'Donnell gives an insight into the world of single unit recordings regarding the dopaminergic system and the structures of cortical areas. Here, we see the difficulties for experimental neurobiology to establish appropriate links between neurochemical issues and electrophysiological findings in an experimental set-up.

# 5
# Neuropsychiatry, Psychopathology, and Nosology – Symptoms, Syndromes, and Endophenotypes
*Wolfram Kawohl and Paul Hoff*

## 5.1
### Introduction

This chapter will focus on a central and controversial issue in psychiatry: the structure of the diagnostic process and the question what kind of information is guiding this process – psychopathological symptoms and syndromes or neurobiological findings such as endophenotypes. This cannot be done without a conceptual and historical introduction into the specific qualities and problems of *psychiatric* diagnosis and nosology, as compared to other medical fields. Then, we illustrate the debate on the diagnostic relevance of neurobiological data by a more detailed analysis of the concept of endophenotypes. The last part addresses the question of what the future role of psychopathology might be in psychiatry, both from a clinical and from a research point of view.

## 5.2
### Conceptual and Historical Introduction

The diagnostic process in psychiatry has always been a controversial issue. This has to do with certain special features of our field. Of all medical specialties, psychiatry and psychotherapy are probably most intensively connected with political, historical, and social developments (Hoff, 2009). Some landmarks of psychiatric thinking from the late eighteenth century to the present time will be mentioned here.

Psychiatry, as we know and recognize it today – as a medical discipline closely connected with neurobiological, psychological, sociological, and philosophical issues – may with good reasons be called a product of the era of enlightenment. From this time on, the view of mentally ill people as *persons* gained influence, as persons with indispensable individual rights and with a personal autonomy that may be diminished, but not eradicated by whatever illness they might have. Psychiatry began to emerge as a medical discipline, rooted in scientific research and dedicated to the treatment of individual persons. Prominent figures in this context are Philippe Pinel,

William Tuke, or John Conolly, who later became known as the leader of the "no-restraint movement."

The first decades of the nineteenth century saw an influential group of psychiatric authors, mainly in German-speaking countries, who were part of romanticism with its emphasis on affectivity, even irrationality and vagueness, as opposed to enlightenment's strong focus on rationality and preciseness. In psychiatry, this led to a major interest in the subjective perspective of the single person and his/her *idiographic* development before the onset of a mental illness, again as opposed to the *nomothetic* approach of enlightenment (Benzenhoefer, 1993; Marx, 1990, 1991). Prominent authors of this time were J.C.A. Heinroth and K. Ideler (Schmidt-Degenhard, 1985).

One of the most remarkable figures in the history of modern psychiatry, Wilhelm Griesinger, marked the turning point from romanticism in psychiatry to the rise of modern empirical and especially neurobiological research into mental illness. He postulated (e.g., Griesinger, 1845) that psychiatry should deal with the mind–body relationship empirically (i.e., by clinical and psychophysiological research) and not metaphysically. *Philosophically*, he advocated a methodological, not a metaphysical materialism. *Nosologically,* he postulated the existence of only one psychotic illness that clinically may appear in different stages ("unitary psychosis" or "Einheitspsychose") from affective syndromes to paranoid-hallucinatory and catatonic syndromes and, finally, to chronic states with severe cognitive deficits, nowadays called dementia.

In the second half of the nineteenth century, neuroanatomical and biological research brought new insights into the structure and the function of the brain. Many authors regarded mental illness as biological disorder of the brain. For example, the Viennese psychiatrist Theodor Meynert chose "illnesses of the forebrain" ("Erkrankungen des Vorderhirns") as the subtitle of his psychiatric textbook from 1884.

Emil Kraepelin postulated that psychotic disorders may be classified in a "natural" (i.e., primarily biological) system, no matter which scientific method is applied. Anatomy, etiology and symptomatology, if developed sufficiently, will necessarily converge in the same "natural disease entities." The most prominent result of this basic idea was Kraepelin's nosological dichotomy, which – as will be shown below – is debated with much skepticism nowadays. It divided the field of major psychotic illnesses in the two areas of "dementia praecox" (markedly bad prognosis) versus "manic-depressive insanity" (markedly better prognosis) (Kraepelin, 1902).

The Swiss psychiatrist Eugen Bleuler published his highly influential work *Dementia praecox oder die Gruppe der Schizophrenien* in 1911 (Bleuler, 1911). He agreed with Kraepelin in some important respects, such as the clinical dichotomy between schizophrenia (dementia praecox) and manic-depressive illness or the generally naturalistic attitude towards mental illness. In marked contrast to Kraepelin, Bleuler integrated the psychological (also in the sense of hermeneutical and psychoanalytical) perspective into clinical psychiatry. He acknowledged the highly heterogeneous course of schizophrenic illness, thus departing definitely from Kraepelin's very pessimistic point of view. His main argument for switching from dementia praecox to schizophrenia was that the disease does not always become a *dementia* and it does not always appear *praecociter.*

Karl Jaspers' *General Psychopathology*, originally published in 1913 (Jaspers, 1963), still must be called a cornerstone of psychiatric conceptualization. He regarded psychopathology as a central practical and research tool for the psychiatrist, that he tried to establish as both an empirical and theoretical scientific field. For Jaspers, it will not be possible to completely describe or even explain human mental life only by objective and quantitative procedures. An essential point here is that our access to mental events of other persons is never a direct one. It is indirect and necessarily involves intersubjectivity insofar as we depend on this person's expressions by language, nonverbal communication, behavior patterns, even by his or her literary or other pieces of art.

As Karl Jaspers, Kurt Schneider was part of the Heidelberg psychopathological tradition, and explicitly acknowledged that neurobiological factors play a major role in the etiology and pathogenesis of mental disorders. He additionally claimed that this does not rule out other factors, such as psychological, and social ones. As for psychiatric diagnoses, he insisted that these are not objective, "naturalistic" statements, but conceptual constructs based on empirical data (Schneider, 1992). In his attempt to differentiate and sharpen the diagnostic process, such as by his subtle description of "first- and second-rank symptoms of schizophrenia," Kurt Schneider may well be regarded a precursor of our present-day operationalized diagnostic manuals, *International Statistical Classification of Diseases and Related Health Problems*, 10th revision (ICD-10) (WHO, 1992) and *Diagnostic and Statistical Manual of Mental Disorders*, text revision, 4th edn (DSM-IV TR) (American Psychiatric Association, 2000)

Another important approach to the concept and diagnosis of mental illness is anthropological psychiatry, although in recent decades this has lost a lot of influence. However, until the early 1960s this perspective was prominent, especially in academic psychiatry. It was decisively oriented at existential philosophy, and focused on the idiographic and biographical aspect in the pathogenesis and etiology of mental disorders. In particular, the school of "Daseinsanalyse," founded by Ludwig Binswanger, declined any elementaristic approach (e.g., as in association psychology), and tried to get access to the complete mental act and its inner structure ("Ganzheit"). In this perspective, psychosis, for example, is not only the appearance of isolated symptoms such as delusions and hallucinations, but a specifically human disorder of shaping one's life. This disorder, on the one hand, may severely diminish degrees of freedom and personal autonomy. On the other hand, to view psychotic (and other psychiatric) states not only as mere deficits, but also – albeit being pathological and creating significant suffering – as carrying meanings with regard to the person's life and self-understanding, may open up psychotherapeutic options.

In recent decades enormous progress in neuroscientific research has been made and rapidly entered the field of psychiatric practice, especially through the emergence of effective pharmacological treatments for people with severe mental disorders like schizophrenia or bipolar disorder. Our diagnostic tools began to reach far beyond the psychopathological level by adopting neurophysiological (e.g., electroencephalography (EEG), evoked potentials), neuropsychological (e.g., executive functions), biochemical (e.g., genetic polymorphisms), and imaging techniques (e.g., functional magnetic resonance imaging (fMRI)).

However, two major problems arose exactly here that have been challenges and sometimes obstacles for psychiatric research:

i) The practical problem that there was no generally accepted terminology, not to say language, in psychiatry to guide research activities and, most important, to enable the international research community to communicate in a reliable way.
ii) The theoretical question whether the psychopathological level will be scientifically necessary at all in psychiatric research in the era of neuroscience.

As for the practical terminological issue, high expectations were raised by the concept of operationalized psychiatric diagnosis, as nowadays represented by ICD-10 (WHO, 1992) and DSM-IV TR (American Psychiatric Association, 2000). Situated in the epistemological tradition of logical empiricism and analytical philosophy, these diagnostic approaches put the emphasis on descriptive psychopathological elements that are delineated by explicit criteria and (wherever possible) stay clear of etiological presuppositions. This critical, even puristic attitude towards psychiatric (and especially diagnostic) terms has its merits, given the many incompatible systems our field has seen during its history. However, it also has its limitations. If quantification and reliability on the level of operationally defined single symptoms become the only points of reference for the diagnostic process, complex, albeit therapeutically relevant psychopathological and intersubjective phenomena might be overlooked, underestimated or even regarded as unscientific (e.g., patient–doctor relationship, complex delusional experiences, specific affective qualities in severe depression). This may lead to unjustified restrictions and simplifications of psychopathology.

In the next part of this chapter we take a closer look at the arguments within neuroscience that focus on the correlation between biological and mental phenomena, trying to minimize the often quoted "explanatory gap" between the two areas as far as possible. In recent years, some – but not all, probably not even the majority of – neuroscientific researchers explicitly denied any difference between mental and biological phenomena, thus changing correlation to identity and adopting the long-standing philosophical position of (eliminative) materialism. This position is by far not "only" of academic interest, but could possibly develop – via the notions of mental illness and of personal responsibility – a marked influence on our clinical and forensic practice (Gruen, Friedman, and Roth, 2008; Kroeber, 2007).

In our context we focus on a less fundamental and therefore more pragmatic issue: the question whether "endophenotypes" might serve as missing link between the psychopathological level, encompassing experiences and behavior of the individual person, and the neurobiological level, encompassing, for example, gene expression and basic physiological activities of the neuron.

## 5.3
### Finding the "Atomic Unit" in Psychopathology: Endophenotypes

The term endophenotype was introduced to the literature by the American schizophrenia researcher Irving Gottesman (Gottesman and Gould, 2003). This concept is presently very popular in psychiatric research (Zobel and Maier, 2004). Endophenotypes are neurobiological correlates of disorders. They are genetically influenced and stable over time. Environmental influences on endophenotypes are lower compared to the phenotype of a disorder. The genetic determination of an endophenotype is, due to its current definition, presumably less complex than the one of the phenotype.

Endophenotypes are used in the search for genes that cause a higher susceptibility for the development of a disorder – so-called susceptibility genes.

Various neurophysiological methods such as EEG and evoked potentials are promising techniques in terms of their ability to determine endophenotypes. EEG characteristics are especially promising with regard to the investigation of endophenotypic properties because of their extensive interindividual variability, their high intraindividual stability, and the proven heritability of more than 80%. E.g., various psychiatric disorders are associated with a reduced amplitude of the P300 in the oddball paradigm (e.g., alcoholism and schizophrenia).

### 5.3.1
### Susceptibility Genes

The search for susceptibility genes – genes that are associated with a susceptibility or sensitivity for the development of a disorder (e.g., schizophrenia) – has proven to be very difficult.

It is only recently that large genetic studies have revealed the first successful identification of possible susceptibility genes, such as *DTNBP1*, *NRG1*, and *DAOA* (G30/G72) (Maier, 2008; also see Table 2.11 in Chapter 2 by Tretter and Gebicke-Haerter).

A major problem in the search for susceptibility genes is the large variety in the phenomenology of schizophrenia. It can be assumed that "schizophrenia" is not one single disorder with a single underlying genetic basis, but a cluster of different disorders with different underlying influencing factors. Therefore, it is not surprising that the search for susceptibility genes in groups of schizophrenic patients (i.e., patients with the diagnosis of schizophrenia, but probably different, albeit similar, disorders) has been of such great difficulty up to now. The search for endophenotypes is a consequence of this problem. Endophenotypes take an intermediate position between the genetical and the phenomenological level. In contrast to a diagnosis that is located on the phenomenological level, endophenotypes represent changes on a more basal level (e.g., measurable changes in brain functioning). With their help the number of patients controlled for possible susceptibility genes can be limited and homogenized. Thus, patients can be examined who suffer from a disorder with similar underlying mechanisms. When different persons of a population show similar endophenotypical characteristics, the search for the underlying gene variant is more promising than in a group of patients that has been built according to a similar diagnosis because the endophenotype is influenced by less genetic variants than the disorder or even the group of symptoms leading to the denomination of a syndrome and, subsequently, a diagnosis.

Examples of endophenotypes can be certain changes in fMRI parameters, blood parameters, evoked potentials, EEG and electrocardiography abnormalities, and deviant neuropsychological parameters.

### 5.3.2
### Requirements for Endophenotypes

Some requirements for endophenotypes are listed in the literature. These requirements are considered to be necessary for a successful search for susceptibility genes:

i) The endophenotype is genetically influenced.
ii) The endophenotype is a possible neurobiological component that contributes to the disorder.
iii) The genes of which variants influence the endophenotype are the susceptibility genes.
iv) The endophenotype is determined by fewer genes than the disorder.
v) For the endophenotype, the penetrance is higher and environmental influences are lower; therefore, the endophenotype is influenced more directly than the disorder (phenotype) itself.

Those criteria are quasi-ideal criteria that cannot be tested in advance. More practical criteria for endophenotypes could be the more frequent occurrence of an endophenotype in patients with a certain disorder in contrast to healthy controls and stability over time including traceability before the beginning of the disorder. Another important factor is a more frequent occurrence of the endophenotype in relatives of the patients than in a normal population as a hint for genetic influences.

### 5.3.3
### Identified and Possible Endophenotypes

An example of a successfully identified endophenotype is the higher cholesterol level in patients with Alzheimer's disease. The cholesterol level is influenced by the enzyme apolipoprotein E (ApoE). The genetic variant *ApoE4* is associated with a higher risk for this disorder.

Possible endophenotypes in schizophrenia research are changes in evoked potentials and slow eye movements, impairments of working memory, and attention deficits. For the auditory evoked potential P50, an association with a variant of the gene coding for a subunit of the $\alpha_7$ nicotinic acetylcholine receptor has been reported (Adler *et al.*, 1998). Another example of a probable endophenotype is the elevated β activity in the EEG of alcoholics. In contrast to the normal population, this higher β activity can be found in the siblings and children of alcoholics, too. There is a clear coupling signal to a region on chromosome 4 enclosing several genes for subunits of the γ-aminobutyric acid-A receptor (Porjesz *et al.*, 2002).

A reduction of the P50 amplitude in the prepulse inhibition paradigm also shows intrafamilial similarities (Freedman, Adler, and Leonard, 1999). There are hints for couplings with a gene locus on chromosome 15q. These results have not yet been replicated yet.

In the group of event-related potentials, the loudness dependence of acoustical evoked potentials (LDAEPs) may be a promising candidate for another endophenotype. The LDAEP is related to central serotonergic activity (Juckel, Mendlin, and Jacobs, 1999) and exhibits an excellent test–retest reliability (Hegerl and Juckel, 2000). A central serotonergic function imbalance is hypothesized to be connected to a vast number of psychiatric disorders such as depression, anxiety disorders, and obsessive-compulsive disorders. It is considered a reliable method to gain information about the central serotonin level in human *in vivo* experiments (Kawohl et al., 2008).

### 5.3.4
### Endophenotypes and the Role of Psychopathology

The question has been raised whether psychopathology may still play a role in the era of endophenotypes. The promise of inexpensive genetic testing for all kinds of disorders seems to make psychopathological approaches dispensable. Furthermore, operational diagnostics appear comparatively time-consuming and, compared to an exact genetic assessment, imprecise. However, different arguments underline the role of psychopathology in the future of psychiatric research and therapy.

The screening and classification of participants for genetic association studies has to be done by psychopathological means. In addition, the presently known susceptibility genes have been identified without the help of endophenotyping. The identification of more susceptibility genes with the help of endophenotypes can be expected; nevertheless, the correlation of identified genes will still refer to the phenotype identified by means of operationalized diagnostics due to the nature of a psychiatric disorder as a cluster of different psychopathological symptoms and not a single symptom. Even the existence of a gene variant determining for a susceptibility to a psychiatric disorder does not mean that this disorder is existent, at least at present, in the carrier of the variant. The diagnosis has to be stated on the basis of a clinical psychopathological assessment. Thus, an exact psychopathological assessment of the subjects in such association studies is and will be inevitable. For example, in the search of genetic variants associated with aggression, the existence of aggressive behavior is tested by a psychopathological assessment (Rujescu et al., 2008). This also applies for the endophenotypes and their "in between" position.

### 5.4
### Basic Methodological Problem: Time/Spatial Resolution

The spatial and temporal resolution of the methods applied in neurobiological and neuropsychological experiments are an important theoretical and practical issue. Neuroimaging methods such as fMRI are largely considered to be of good spatial resolution. A poor spatial and good temporal resolution is attributed to electrophysiological methods such as evoked potentials. These statements can be put into perspective by considering the facts that fMRI data reflects blood flow and not neuronal activity (albeit blood flow in close relation to current neuronal activity), and that by source localization of somatosensory evoked potentials, even small structures such as the cortical somatosensory hand area can be determined (Kawohl et al., 2007). A combination of methods with good spatial and temporal resolution provides further insight in to the where and when of neuronal activity (Mulert et al., 2008). However, fMRI and electrophysiological experiments only provide a small insight into complex neural processes such as decision making due to their direct stimulus dependence. Thus, an even more important problem is the deviation between the time frame of mental processes and the temporal resolution of the methods that are applied for the investigation of these processes.

### 5.4.1
### An Example: Libet's Experiment

A good example for this is the famous trial of Benjamin Libet. In 1983, Benjamin Libet set up an experiment to control the time-course of movements and the preceding decision (Libet et al., 1983).

In this experimental set-up, subjects were connected to an EEG system with six recording channels placed on the scalp plus two linked mastoid electrodes as reference electrodes, and an electromyography (EMG) system with electrodes attached to the forearm. The subjects were asked to move their wrist whenever they freely decided to do so. The movement was recorded by the EMG system that sent a trigger signal to the EEG-recorder. The EEG sweeps corresponding to each trigger were averaged. During the experiment, the subjects watched the screen of a cathode ray oscilloscope. On the screen, a spot of light revolved in a clockwise circle with one revolution being completed in 2.56 s. The subjects were asked to report the position of the spot at the moment of their awareness of wanting to move. This point was called "W" by Libet, the moment of the movement was called "M." Forty trials were averaged.

The potential that is measured in this experimental set-up is the so-called readiness potential (RP). In 1965, Kornhuber and Deecke had reported on this phenomenon – a potential occurring several hundred milliseconds before a voluntary movement (Kornhuber and Deecke, 1965). This potential had been interpreted as an epiphenomenon (in the neurophysiological sense of the word) of the decisional process itself. Interestingly, in most subjects in Libet's trial, the RP occurred *before* the point in time of the decision to move (approximately 400 ms on average in advance of W, the time of a "conscious awareness of 'wanting' to perform a given self-initiated movement"). W occurred approximately 200 ms before M.

There are two basic problems in the interpretation of this data regarding the decisional process: (i) the correctness of the estimation of W and (ii) the explanatory power of the experiment as a whole. The discussion whether Libet's results rule out the existence of a free will is the subject of extensive literature (Kawohl and Habermeyer, 2007) and is not addressed in this chapter.

## 5.4.2
### The First Problem: The Estimation of W

The subject reports the time of wanting to move by watching a rotating spot on an oscillograph's screen. According to neurophysiological findings, the timepoint when the person consciously sees the spot on the screen at position X is not identical with the timepoint of the spot being at position X. Van de Grind thoroughly reviews the literature on time perception, and points out the problem that Libet would tie perceptual and EEG phenomena together by "time juggling" (van de Grind, 2002). Thus, physical, neural, and mental time differ from each other. This must be taken into account concerning the accordance of the time perception of the revolving spot, the moment of W, and the real time. Experiments on the so-called flash-lag phenomenon shed further light on this problem (Nijhawan, 1994–2001). In the flash-lag experiments, perception of a moving object and a nonmoving flash that is presented at the same time the object is at the position where the flash occurs is tested. Interestingly, the moving object is seen at a point in space that coincides with a point of time in the future. Nijhawan interprets this as a prediction of the motion by the nervous system. This hypothesis has also been questioned, because a change of direction of the object at the moment of the flash would still allow to perceive the movement in the correct direction (Whitney and Murakami, 1988).

It must also be taken into account that Zeki and Moutoussis (Moutoussis and Zeki, 1997) found that changes in color are seen earlier than changes in orientation and that the perception of movements takes the longest time. They conclude that there is nothing like a time-compensating mechanism. Interestingly, this is in strong contrast to the central latencies of motion and color perception, which appear in reverse order (Zeki, 2001). The temporal binding problem, which is the subject of a vast field in neuroscience, is involved here.

## 5.4.3
### The Second Problem: The Explanatory Power in the Light of a Questionable Time Resolution

Different authors (van de Grind, 2002; Kroeber, 2004) state that in Libet's experiment highly automated functions are studied under artificial conditions. Zhu (2003) emphasizes that the role of volition can still be maintained – the decision to move had already been formed at the beginning of the experiment when the subjects agreed to the experimental instructions. An experiment by which the whole timing of decision making and not only the last seconds before a movement can be clarified has not yet been conducted. With today's techniques, it is difficult if not impossible to design. Such an experiment would require the monitoring of a time frame of approximately 1 h, including the decision to participate in the experiment. Such an electrophysiological measurement is, for the time being, not conceivable without the event-related potential technique. This technique is based on the averaging of numerous EEG sweeps around similar events and has no possibility for monitoring complex processes like listening to instructions given at one time in advance of a trial, not to mention the following decisional process. It must also be emphasized that the decision to act or not to act is, as every other decision, based on complex interactions of contexts of the past such as biographical factors and experiences, cognitive selections, affective, social, and moral influences (Emrich, 2002; Geyer, 2004). An experimental set-up to cover these elements and, in particular, their interactions is even more complex and presently unimaginable. This extends to psychological methods such as interviews or computerized tests, and also to other neuroimaging methods such as fMRI, positron emission tomography, or near-infrared spectroscopy with their comparatively poor temporal resolution.

Thus, this experiment may serve as a good example of the vast difference between the resolution of experimental methods and the character of the mental processes and their influencing factors that are targeted by neuroscientific experiments in the time domain.

## 5.5
### Future New Diagnostic Schedules and Research

Given the long-standing problems of theoretical heterogeneity and low reliability of psychiatric diagnosis, on the one hand, and the growing scientific impact of neuroscience in psychiatry, on the other hand, it is understandable that many authors suggest neurobiological measures as new diagnostic criteria. Indeed, as has already been discussed above, there is a growing literature on this issue with special emphasis on endophenotypes and their "in between" position. This discussion typically centers around the following questions:

i) Should we abandon Kraepelin's (categorical) dichotomy of schizophrenia and affective disorders, and replace it – no matter what kind of diagnostic criteria will be applied – by a dimensional model?
ii) Are neuroscientific findings in major psychiatric disorders already robust enough to practically serve as diagnostic criteria in addition to, or even instead of, psychopathological criteria?
iii) Are nonempirical aspects of psychiatric classification (e.g., the notion of value) relevant or even indispensable elements in both the theoretical debate on nosology and the practical diagnostic process or should they be labeled as unscientific and therefore be avoided wherever possible?

To sum up the present status of this discussion, there is, on the one hand, increasing skepticism against understanding different mental disorders as categorically distinct "natural kinds," as Kraepelin had suggested. On the other hand, the existing neurobiological data are often regarded as not yet sufficient to establish a new and valid nosological approach, even by authors working in the field of neuroscience themselves. As a consequence of such a rather cautious attitude, some consider it premature to totally leave the Kraepelinian dichotomy because there still is no generally accepted alternative available (Gaebel and Zielasek, 2008; Moeller, 2008). These arguments focus on evidence-based, mainly quantitative and algorithm-oriented diagnostic procedures. In contrast to this, nonempirical, qualitative and value-oriented aspects are addressed in a markedly smaller, but rising number of publications (Fulford, Thornton, and Graham, 2006; Zachar and Kendler, 2007). We will come back to this topic in Section 5.6.

Practically speaking, there will have to be decisions on these issues during the ongoing development of the 11th edition of the WHO's *International Classification of Diseases* (ICD-11) and of the fifth edition of American Psychiatric Association's *Diagnostic and Statistical Manual of Mental Disorders* (DSM-V). These decisions will, of course, deeply influence psychiatric diagnosis both in the clinical and the research area (Krueger and Bezdjian, 2009).

## 5.6
## On the Future Role of Psychopathology

Finally, psychopathology's possible future role in psychiatry is discussed. Regarded as a central point of reference in psychiatry in earlier times (Janzarik, 1979), psychopathology finds itself in a much more defensive position nowadays not only from a scientific, but also from a clinical point of view (Hoff, 2007). This is not only due to the growing impact of neuroscience. There are also critical voices arguing that since the work of Kurt Schneider, the Heidelberg psychopathologist already mentioned above, there has not been any substantial progress in descriptive psychopathology apart from the formal aspect of stricter operationalization by ICD-10 and DSM-IV. In addition, psychopathological data are thought to be not objective and reliable enough. Even more radical, some authors in the era of neuroscientific psychiatry predict that in the long run there will not be any room left for an enterprise such as psychopathology that necessarily deals with subjective and interpersonal phenomena. In their view, psychopathology (and mentalism in general) will then be completely replaced by quantitative neurobiological criteria (Churchland, 1986; Roth, 2003). Again, others have argued that clinical psychopathology should be developed in the direction of functional psychopathology (van Praag, 1988). The core idea of this approach is to focus on the correlation between psychopathological phenomena, on the one hand, and neurobiological dysfunctions, on the other hand. The search for endophenotypes, as exemplified above, is also in this tradition.

Remarkably, however, in the last few years there are indicators of a renewed interest in psychopathology. Given the complexity of *any* scientific perspective on mental illness, unidimensional approaches lose credit due to their simplifying and over-reductionistic attitudes. Of course, there still are serious philosophical theories strongly supporting neurobiological materialism, such as John Bickle's (2003) book *Philosophy and Neuroscience*, subtitled *A Ruthlessly Reductive Account*. However, within the psychiatric and neuroscientific research community itself, many authors nowadays agree that we need a concept beyond single methodological perspectives. In this line of thought the notion of person encompassing the areas of subjectivity, personal autonomy, and responsibility comes into focus again. Two examples may illustrate this:

i) The World Psychiatric Association's Institutional Program for Psychiatry of the Person, operative since 2005 (Mezzich, 2006), to a great extent tackles the question how the theoretical notion of "person centeredness" can be implemented practically into the future diagnostic process in psychiatry. In its structure, this debate resembles the neurobiological arguments mentioned above on how endophenotypes could play a significant role in diagnosis.

ii) There is a philosophically substantial *and* widely recognized debate on the relationship between evidence- and value-based decision making in psychiatry. These topics in themselves are, of course, not new for the field, but, as mentioned above, they have attracted remarkable interest and generated fruitful discussions, especially since Fulford's (1989) book on moral theory and medical practice.

Generally speaking, the interface between psychiatry, neurosciences, and philosophy, often simply (and partly misleadingly) called "neurophilosophy," is at present an active, albeit highly heterogeneous field. This dialogue between scientific disciplines that were neatly separated before has gained momentum in recent years. One reason for this is the fact that, in contrast to their earlier stages of development, newer research tools such as fMRI or neuropsychological tests of executive functions nowadays not only deal with "simple" cognitive tasks, but with highly complex areas such as the interaction between cognitive and affective phenomena, or even the notions of "self," "identity," or "consciousness" (Frith, 1992; Kircher and David, 2003; Vogeley, 2007).

To jump to the conclusion directly, one of psychopathology's major tasks in the future should be to prevent dogmatic simplification in psychiatry – and there is such a risk in *any* theoretical framework that came up since psychiatry slowly emerged as a medical discipline in the era of enlightenment. The realistic neurobiological approach to mental illness, nowadays most radically represented by eliminative materialism, is at risk of *naturalistic reductionism* by straightforwardly identifying mental illness with disturbed neurobiological processes. ("Realistic" in an epistemological sense – the assumption that the real world exists fully independently from our conceptualization or our mental acts; Emil Kraepelin's concept of "natural disease entities in psychiatry" is *the* example for such a realistic approach to psychiatric nosology (Hoff, 1994).) The nominalistic definitions of ICD-10 and DSM-IV may become dogmatic in the sense of a *formalistic reductionism*. This happens, as has been mentioned above, if users of operationalized diagnostic algorithms presume that the entire phenomenon of psychosis is covered or even explained by these operationalized procedures. However, biographical or hermeneutical approaches may also run into dogmatism. To believe that the etiology, pathogenesis, and clinical symptomatology of a given mental illness may be fully understood by the methods of empathy and interpretation constitutes a *heuristical reductionism*.

It is exactly at this crucial point that psychopathology in a broader sense might again become a relevant point of reference for psychiatry. However, in order to succeed in this, it will have to fulfill a number of quite demanding conditions:

i) Combining operational with "open" descriptions of phenomena. This, for example, means to respect complex qualitative phenomena such as personal and interpersonal traits or biographical information that are not easily, if at all, represented by the usual, reliability-oriented diagnostic criteria.
ii) Closely cooperating with neurosciences and social sciences. For example, by further developing research designs for the interface between social cognition, empathy, or altruism, on the one side, and brain function, on the other side, without adopting any naive reductionism.
iii) Acknowledging subjectivity as a scientific topic, and implementing conceptual and historical knowledge about psychiatric theories into the current debate on psychiatry's identity, especially regarding basic concepts like mental illness itself, the mind–body relationship and interpersonality.

Essentially, this debate is focused on the relationship between three areas in psychiatric research: *empirical data*, no matter whether from neurobiology, social sciences, or psychopathology, the *notion of mental illness* in general (including nonempirical issues such as values) and the *anthropological framework* guiding our research. If psychopathology in a differentiated and nondogmatic way as outlined above will gain influence in the future, this without doubt will also be fruitful for neuroscientific research into mental illness.

## References

Adler, L.E., Olincy, A., Waldo, M., Harris, J.G., Griffith, J., Stevens, K., Flach, K., Nagamoto, H., Bickford, P., Leonard, S., and Freedman, R. (1998) Schizophrenia, sensory gating, and nicotinic receptors. *Schizophrenia Bulletin*, 24, 189–202.

American Psychiatric Association (2000) *Diagnostic and Statistical Manual of Mental Disorders, text revision*, 4th edn, American Psychiatric Association, Washington, D.C.

Benzenhoefer, U. (1993) *Psychiatrie und Anthropologie in der ersten Haelfte des 19. Jahrhunderts*, Pressler, Stuttgart.

Bickle, J. (2003) *Philosophy and Neuroscience. A Ruthlessly Reductive Account*, Kluwer, Dordrecht.

Bleuler, E. (1911) *Dementia Praecox oder die Gruppe der Schizophrenien*, Deuticke, Leipzig.

Churchland, P.S. (1986) *Neurophilosophy: Towards a Unified Theory of the Mind–Brain*, MIT Press, Cambridge, MA.

Emrich, H.M. (2002) Towards a neuropsychology of autonomy: re-presenting forgetting, subliminality and freedom. *Fortschritte Der Neurologie Psychiatrie*, 70, 511–519.

Freedman, R., Adler, L.E., and Leonard, S. (1999) Alternative phenotypes for the complex genetics of schizophrenia. *Biological Psychiatry*, 45, 551–558.

Frith, C.D. (1992) *The Cognitive Neuropsychology of Schizophrenia*, Psychology Press, New York.

Fulford, K.W.M. (1989) *Moral Theory and Medical Practice*, Cambridge University Press, Cambridge.

Fulford, K.W.M., Thornton, T., and Graham, G. (2006) Values, ethics, and mental health, in *The Oxford Textbook of Philosophy and Psychiatry* (eds K.W.M. Fulford, T. Thornton, and G. Graham), Oxford University Press, Oxford, part IV.

Gaebel, W. and Zielasek, J. (2008) The DSM-V initiative "deconstructing psychosis" in the context of Kraepelin's concept on nosology. *European Archives of Psychiatry and Clinical Neuroscience*, 258 (Suppl. 2), 41–47.

Geyer, C. (2004) *Hirnforschung und Willensfreiheit. Zur Deutung der neuesten Experimente*, Suhrkamp, Frankfurt am Main.

Gottesman, I.I. and Gould, T.D. (2003) The endophenotype concept in psychiatry: etymology and strategic intentions. *American Journal of Psychiatry*, **160**, 636–645.

Griesinger, W. (1845) *Die Pathologie und Therapie der psychischen Krankheiten*. Krabbe, Stuttgart.

Gruen, K.J., Friedman, M., and Roth, G. (eds.) (2008) *Entmoralisierung des Rechts. Massstaebe der Hirnforschung fuer das Strafrecht*, Vandenhoeck & Ruprecht, Goettingen.

Hegerl, U. and Juckel, G. (2000) Identifying psychiatric patients with serotonergic dysfunctions by event-related potentials. *World Journal of Biological Psychiatry*, **1**, 112–118.

Hoff, P. (1994) *Emil Kraepelin und die Psychiatrie als klinische Wissenschaft. Ein Beitrag zum Selbstverstaendnis psychiatrischer Forschung*, Springer, Berlin.

Hoff, P. (2007) Ueber die zukuenftige rolle der psychopathologie: Grundlagen- oder hilfswissenschaft?, in *Subjektivitaet und Gehirn* (eds. T. Fuchs, K. Vogeley, and M. Heinze), Pabst, Lengerich, pp. 195–209.

Hoff, P. (2009) Historical roots of the concept of mental illness, *Psychiatric Diagnosis: Challenges and Prospects* (eds. I.M. Salloum and J.E. Mezzich), John Wiley & Sons, Ltd, Chichester, pp. 3–14.

Jaspers, K. (1963) *General Psychopathology* (transl. J. Hoenig and M.W. Hamilton), University of Chicago Press, Chicago, IL.

Janzarik, W. (1979) *Psychopathologie als Grundlagenwissenschaft*, Enke, Stuttgart.

Juckel, G., Mendlin, A., and Jacobs, B.L. (1999) Electrical stimulation of rat medial prefrontal cortex enhances forebrain serotonin output: implications for electroconvulsive therapy and transcranial magnetic stimulation in depression. *Neuropsychopharmacology*, **21**, 391–398.

Kawohl, W. and Habermeyer, E. (2007) Free will: reconciling German civil law with Libet's neurophysiological studies on the readiness potential. *Behavioral Sciences and the Law*, **25**, 309–320.

Kawohl, W., Hegerl, U., Muller-Oerlinghausen, B., and Juckel, G. (2008) Insights in the central serotonergic function in patients with affective disorders. *Neuropsychiatry*, **22**, 23–27.

Kawohl, W., Waberski, T.D., Darvas, F., Norra, C., Gobbele, R., and Buchner, H. (2007) Comparative source localization of electrically and pressure-stimulated multichannel somatosensory evoked potentials. *Journal of Clinical Neuroscience*, **24**, 257–262.

Kircher, T. and David, A. (2003) *The Self in Neuroscience and Psychiatry*, Cambridge University Press, Cambridge.

Kornhuber, H. and Deecke, L. (1965) Hirnpotentialänderungen bei Willkürbewegungen und passiven Bewegungen des Menschen: Bereitschaftspotential und reafferente Potentiale. *Pflugers Archiv*, **284**, 1–17.

Kraepelin, E. (1902) *Clinical Psychiatry*, Macmillan, New York.

Kroeber, H.L. (2004) Structural dynamics and delinquent behavior. *Fortschritte Der Neurologie Psychiatrie*, **72** (Suppl. 1), S40–S44.

Kroeber, H.L. (2007) The historical debate on brain and legal responsibility – revisited. *Behavioral Sciences & the Law*, **25**, 251–261.

Krueger, R.F. and Bezdjian, S. (2009) Enhancing research and treatment of mental disorders with dimensional concepts: toward DSM-V and IVD-11. *World Psychiatry*, **8**, 3–6.

Libet, B., Gleason, C.A., Wright, E.W., and Pearl, D.K. (1983) Time of conscious intention to act in relation to onset of cerebral activity (readiness-potential). The unconscious initiation of a freely voluntary act. *Brain*, **106**, 623–642.

Maier, W. (2008) Common risk genes for affective and schizophrenic psychoses. *European Archives of Psychiatry and Clinical Neuroscience*, **258** (Suppl. 2), 37–40.

Marx, O.M. (1990) German romantic psychiatry. Part I. *History of Psychiatry*, **1**, 351–381.

Marx, O.M. (1991) German romantic psychiatry. Part II. *History of Psychiatry*, **2**, 1–25.

Meynert, Th. (1884) *Psychiatrie. Klinik der Erkrankungen des Vorderhirns*. Braumueller, Wien.

Mezzich, J.E. (2006) Institutional consolidation and global impact: towards a psychiatry for the person. *World Psychiatry*, **5**, 65–66.

Moeller, H.J. (2008) Systematic of psychiatric disorders between categorical and dimensional approaches. Kraepelin's dichotomy and beyond. *European Archives of Psychiatry and Clinical Neuroscience*, **258** (Suppl. 2), 48–73.

Moutoussis, K. and Zeki, S. (1997) Functional segregation and temporal hierarchy of the visual perceptive systems. *Proceedings of the Royal Society of London B*, **264**, 1407–1414.

Mulert, C., Seifert, C., Leicht, G., Kirsch, V., Ertl, M., Karch, S., Moosmann, M., Lutz, J., Moller, H.J., Hegerl, U., Pogarell, O., and Jager, L. (2008) Single-trial coupling of EEG and fMRI reveals the involvement of early anterior cingulate cortex activation in effortful decision making. *Neuroimage*, **42**, 158–168.

Nijhawan, R. (1994) Motion extrapolation in catching. *Nature*, **370**, 256–257.

Nijhawan, R. (1997) Visual decomposition of colour through motion extrapolation. *Nature*, **386**, 263–282.

Nijhawan, R. (2001) The flash-lag phenomenon: object motion and eye movements. *Perception*, **30**, 263–282.

Porjesz, B., Almasy, L., Edenberg, H.J., Wang, K., Chorlian, D.B., Foroud, T., Goate, A., Rice, J.P., O'Connor, S.J., Rohrbaugh, J., Kuperman, S., Bauer, L.O., Crowe, R.R., Schuckit, M.A., Hesselbrock, V., Conneally, P.M., Tischfield, J.A., Li, T.K., Reich, T., and Begleiter, H. (2002) Linkage disequilibrium between the beta frequency of the human EEG and a GABA$_A$ receptor gene locus. *Proceedings of the National Academy of Sciences of the United States of America*, **99**, 3729–3733.

Roth, G. (2003) *Fuehlen, Denken, Handeln*, Suhrkamp, Frankfurt am Main.

Rujescu, D., Giegling, I., Mandelli, L., Schneider, B., Hartmann, A.M., Schnabel, A., Maurer, K., Moller, H.J., and Serretti, A. (2008) NOS-I and -III gene variants are differentially associated with facets of suicidal behavior and aggression-related traits. *American Journal of Medical Genetics*, **147B**, 42–48.

Schmidt-Degenhard, M. (1985) Zum melancholiebegriff JCA heinroths, in *Psychiatrie auf dem Wege zur Wissenschaft* (eds. G. Nissen and G. Keil), Thieme, Stuttgart, pp. 12–18.

Schneider, K. (1992) *Klinische Psychopathologie*, 14th edn, Thieme, Stuttgart.

van de Grind, W. (2002) Physical, neural, and mental timing. *Consciousness and Cognition*, **11**, 241–264, discussion 308–213.

van Praag, H.M. (1988) Serotonin disturbances in psychiatric disorders. Functional versus nosological interpretation, in *Selective 5-HT-Reuptake Inhibitors: Novel or Commonplace Agents?* (eds. M. Gastpar and J. Wakelin), Advances in Biological Psychiatry, Karger, Basel, pp. 52–57.

Vogeley, K. (2007) Disturbances of time consciousness from a phenomenological and a neuroscientific perspective. *Schizophrenia Bulletin*, **33**, 157–165.

Whitney, D. and Murakami, I. (1988) Latency difference, not spatial extrapolation. *Nature Neuroscience*, **1**, 656–657.

WHO (1992) *ICD-10: International Statistical Classification of Diseases and Related Health Problems*, 10th revision. WHO, Geneva.

Zachar, P. and Kendler, K.S. (2007) Psychiatric disorders: a conceptual taxonomy. *American Journal of Psychiatry*, **164**, 557–565.

Zeki, S. (2001) Localization and globalization in conscious vision. *Annual Review of Neuroscience*, **24**, 57–86.

Zhu, J. (2003) Reclaiming volition: an alternative interpretation of Libet's experiment. *Journal of Consciousness Studies*, **10**, 61–77.

Zobel, A. and Maier, W. (2004) Endophenotype – a new concept for biological characterization of psychiatric disorders. *Nervenarzt*, **75**, 205–214.

# 6
# System Properties of Populations of Neurons in Cerebral Cortex

*Walter J. Freeman*

Brains are notable for their immense number of cells, approaching a trillion glia and neurons together, and for their unity in intentional action. The unity is achieved in the main by synaptic interaction among the neurons at short and long distances. The interaction is created by the parts, yet it transcends them by imposing order on the neurons' activities. That circular causality creates an entity that is called an order parameter, which is a system property that can only be defined, understood, and measured by summation over large populations of neurons. The electric fields of potential known as the intracranial electrocorticogram (ECoG) and scalp electroencephalogram (EEG) are the sums of potentials caused by the flows of ionic current from synaptic potentials. They are not the agency of enforcing order; they provide indices of order parameters at three levels of clinical observation: microscopic local field potentials, the mesoscopic ECoG, and the macroscopic EEG. These indices of system order, despite their present obscurity and complexity, offer a prolific source of novel information from which to learn how neurons create and maintain the unity of intentional action. Here my aim is to introduce the distinction between microscopic activity and mesoscopic order, mainly using the properties of three-layered allocortex in the olfactory system as a model that is simpler than the six-layered neocortices of the sensory and limbic systems.

## 6.1
### Introduction

Brain waves are doubtlessly the least understood and yet the most tantalizing among physiological signals from the body. Coming as they do from the most phylogenetically advanced of all organs, the neocortex, the signals offer a direct window onto the global expanse of the machinery of the mind, spread like a map below the scalp and skull. Their enigmatic fluctuations look like noise, and in fact they largely are noise

(Freeman, 2006), but they also contain indefinite amounts of information about all levels of function in brains, and that information is locked into the morass of noise like precious elements waiting to be discovered and shaped into the service of medicine and psychiatry. In my view, brain waves offer the most significant and difficult challenge in contemporary bioscience.

On the basis of methods of observation, I define three levels of cortical function. Microscopic synaptic potentials from dendrites and spikes (action potentials) from axoms are viewed with microelectrodes. Mesoscopic field potentials that are used as indices of order parameters are seen with arrays of electrodes on cortical surfaces (Freeman and Kozma, 2010). Macroscopic observations are noninvasive. The ionic currents generated by mass actions of cortical dendrites provide the basis for brain waves. These currents flow intracellularly in dendritic shafts and extracellularly across the relatively fixed resistance of cortical tissue, mainly within the low resistance paths of capillaries feeding cortex. The ionic currents are accompanied by electric and magnetic fields, seen in electroencephalograms (EEGs) and magnetoencephalograms. The flows are supported by metabolic energy that is provided by the cardiovascular system, which directs blood flow to regions roughly related to the intensities of current. The areas are visualized with various techniques, the most widely used being functional magnetic resonance imaging combined with blood oxygen level depletion. Thus, the three main types of energy used for imaging brain activity are intrinsically linked to each other through dendritic currents. However, they come from differing neural populations, and have differing scales of time and space; the data they offer are complementary and are not directly mutually supportive (Freeman, Ahlfors, and Menon, 2009).

Below the microscopic level are properties at the molecular and atomic level that determine the ionic currents at synapses and trigger zones; above the macroscopic level are the cognitive and social determinants of intentional action. These levels lie beyond the scope of this essay. Here, I describe and explain the electrical manifestations of brain waves as *system* properties of cortex, because neuron populations generate them by mass interactions through dendrites and axons, synapses and gap junctions. My focus is on the micro–meso interface. I simplify my description in three ways. (i) I concentrate on brains of small mammals (rats, rabbits, and cats) as models of human cortex. Their relative simplicity facilitates studies of intracranial electric fields: local field potentials (LFPs) from depth electrodes and electrocorticograms (ECoG) from electrode arrays on the pial surface. Such studies are a prelude to the far more difficult task of making sense of brain waves on the scalp – the EEG. (ii) Rather than beginning with the six-layered neocortex found only in mammals, I begin with the three-layered allocortex that comprises the olfactory, hippocampal, and simpler limbic systems found in submammalian species as well as mammals. (iii) I concentrate on the system properties of the olfactory bulb and cortex, because their function in sensation and perception is more easily controlled and dominantly important in most animals. Furthermore, the somatic, visual, and auditory neocortices phylogenetically evolved from the allocortex, so that olfaction provides a model for the population dynamics of all four systems, which are alike in that they enable perception of sources of information at distances, allowing time for prediction and planning of action by the brain before threats like teeth and fire arrive at the body surface (Freeman, 2001). Finally, the output signals of all four cortices have the same spatio-temporal forms, so that brains readily combine them in the limbic system into multisensory percepts (gestalts) that occupy all cortices simultaneously and the motor cortices as well.

## 6.2
## Spatial Structure of Brain Waves

My first encounter with brain waves was in the hypothalamus while studying the thermoregulatory system. The prevailing dogma was that an anterior "heat center" and a posterior "cold center" controlled brain temperature. I disproved this by eliciting vasodilation and vasoconstriction, respectively, with heating and cooling only in the anterior hypothalamus. This finding also disproved the prevailing dogma of the insensitivity of the brain to stimuli; brains have receptors for temperature, osmolarity, and many other chemical parameters essential to their well being, seen now not as a collection of "centers," but as systems with sensors, effectors, networks of connections, and modifiable set points for homeostatic feedback control, which are regulated by cortical, subcortical and limbic systems in health and disease.

I next sought to detect the action potentials (spikes, pulses) of the thermoreceptor neurons. I did not succeed with the prevailing technology, so I turned to what was then called the "hypothalamogram," now called LFP, which was the record of potential differences between an extracellular electrode placed in the hypothalamus and a distant reference electrode. I used diathermy to apply heat while controlling local temperature with a thermistor. Not only was there no change in brain waves, the signals from the diathermy electrodes likewise showed no change, despite the increase in local temperature in excess of $60\,°C$. Clearly the "hypothalamogram" was coming from elsewhere. By triangulation, I tracked the sources of hypothalamic LFP oscillations to the nearby hippocampus, the septum, and by far the most powerful, the olfactory cortex. The olfactory brain waves became the focus of my research, and the self-regulatory system exemplified by thermoregulation became my model in the search for the origin, regulation, and function of brain waves as system properties (Freeman, 2006).

The reason that the olfactory cortex was such a powerful generator turned out to be its systemic cytoarchitecture. The olfactory cortex has a high packing density of pyramidal cells with an unusually high degree of lamination of the cell bodies with alignment of the dendrites above and axons below the cell body layer. The ionic currents flow in one direction inside the neurons and oppositely outside the neurons, where they sum in the volume conductor of the forebrain. During excitation cations ($+$) flow into the dendrites and out at the axons, and vice versa for anions ($-$). During inhibition the directions are reversed. The excitatory extracellular electric potential is negative at cation inflow/anion inflow (a "sink") and positive elsewhere (a "source"), giving the now classical dipole field (Figure 6.1), which reverses during inhibition. The dipole field of current extends throughout the brain. The inflows and outflows

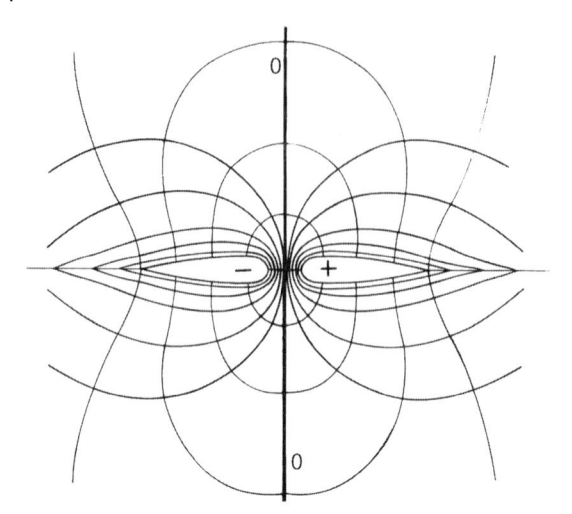

**Figure 6.1** Light lines show the field of ionic current flowing from the source (+) in the axonal tree to the dendritic sink (−) in the extracellular tissue of the brain during excitation at the dendrites promoting the discharge of spikes at the axon. Curves at right angles to the current lines show the field of potential measured with respect to a distant reference point. Note that the zero isopotential surface lies at or near the layer of cell bodies, from which the dendrites and axon extend on either side. (From Freeman, 1975.)

are equal, because current always flows in closed loops. Neurons neither create nor destroy electric charge.

The spatial separation of source and sink is necessary for the dipole field. Pyramidal cells have the necessary cytoarchitecture; stellate interneurons do not, because their branches are radially symmetric. Their fields of potential do not extend outside the dendritic arbors; they are true LFPs. If the neurons are axially symmetric but neither laminated nor polarized and instead are randomly oriented, the sources and sinks are still present, but the fields of extracellular potential tend to cancel and fail to appear at the pial surface (ECoG) or scalp (EEG). The several dipole fields of the neighboring cortices explain the hypothalamogram; the neurons in the basal ganglia and reticular nets of the forebrain lack the lamination and palisade of cortex, so their LFP are over-ridden by the cortical dipole fields. Brain waves appear in the basal ganglia the same way that they appear in the scalp EEG, but on opposite sides of the cortical dipoles. In summary, subcortical neurons also have dendritic ionic currents that determine their intracellular synaptic potentials and firing rates, but the lack of structural symmetry in their populations precludes formation of sufficiently high LFPs to over-ride the brain-wide components of the dipole fields that form the ECoG and EEG at recording surfaces, well outside dendritic arbors. The subcortical nuclei do not provide indices of order parameters.

It is important to understand that EEG and ECoG are mass action system properties of populations of cortical neurons. I call them *mesoscopic* as distinct from the spikes of cortical neurons that are *microscopic* and the synchronized signals from multiple areas of brains that are *macroscopic*. The basis for the distinction is that the alignment of neurons supports spatial *summation* of the electric (and magnetic) fields of dendritic potentials of all the active neurons with shared orientation in the neighborhood of an electrode on the scalp or pial surface. The summing makes the field potential a mesoscopic variable, typically from tens to hundreds of thousands of neurons under each electrode and tens of millions under an array of electrodes involving billions of synapses. Microelectrode recordings of axonal spikes give microscopic variables, typically from one or a few. Even if records are taken from hundreds of neurons simultaneously, they are insufficient to reach the population level. In any case the EEG and ECoG are decidedly not the "envelope" of summed axonal and dendritic spikes, because the spikes lack the spatial and spectral structural organization that is necessary for spatial summation.

This distinction between microscopic and mesoscopic is fundamental in coming to understand brain function, and it necessitates thinking about cortex as a multilevel system. The summation gives access to mesoscopic properties that do not exist at the microscopic level of spikes. In the same sense that temperature and pressure are properties of a mass of molecules, each having kinetic energy but neither temperature or pressure, the neural population has a pulse density and a wave density at each point in time and space, which are related to the spike frequency and membrane polarization of the individual neurons, but the order parameters are accessed only through summing or averaging the microscopic energies from very large numbers of neurons (Freeman and Vitiello, 2008).

The distinction becomes important when it comes to considering the forms taken by cortical output. If, on the one hand, the output axons are organized so as to project onto targets with preservation of point-to-point ordering ("topographic mapping"), then the output is microscopically organized. If, on the other hand, the cortical output axons from each local neighborhood diverge broadly to targets, and conversely each receiving neighborhood of neurons receives from a broad spatial distribution of neurons ("divergent–convergent projection"), then the output is a two-dimensional mesoscopic spatio-temporal pattern that cannot be observed by use of limited numbers of microelectrodes, but can be observed and measured as a spatial pattern, preferably by use of a surface electrode array to record the ECoG.

## 6.3
## Temporal Structure of the EEG/ECoG

Every area of cortex generates observable fields of electric potential in proportion to the degrees of palisade and lamination. These fields enable spatial mapping of EEG/ECoG potentials. What is required for summation beyond the cytoarchitectural organization is temporal synchrony, which is provided in the main by synaptic interaction. The key word is interaction. Neurons being driven feedforwardly by an external source show microscopic synchrony. In contrast, neurons in large numbers driving each other in feedback show mesoscopic synchrony. Feedback is the mesoscopic foundation for the ubiquitous background oscillations in brains, not the feedforward property of microscopic neurons equipped with pacemaker potentials.

Neuronal feedback is either positive or negative. Excitatory neurons typically greatly outnumber inhibitory neurons in cortex and most of their input is from each other, not extracortical (Braitenberg and Schüz, 1998). Typically, each neuron transmits to $10^4$ neighbors, these to another $10^4$, and they again to $10^4$, so that even in the first three synapses the number may involve virtually all cortical neurons. Multisynaptic distributed distances and delays of transmission randomize the excitatory feedback to the originating neuron, so that each neuron is immersed in spikes that resemble a thermal bath (Freeman, 1975). The sum of the microscopic extracellular spikes in auditory display gives a roar like a waterfall, manifesting so-called "white noise," because its spectrum is flat (a slope of zero) and it contains all frequencies at equal power. The mesoscopic information that is buried in the noise is not directly accessible.

The summation by dendrites smoothes the white noise, so that the high frequencies appear attenuated in the ECoG and the spectrum approaches the form of so-called "brown noise" (Figure 6.2) from its occurrence in Brownian motion. It is important to note that power at all frequencies is still present, but the log power decreases in linear proportion to the log frequency. For brown noise the slope is $-2$. This spectral form epitomizes the simple, basic form of the ECoG at rest. It is low-dimensional noise. Moreover, it is self-stabilized everywhere locally. When the anatomical density of excitatory connections reaches a threshold above which each

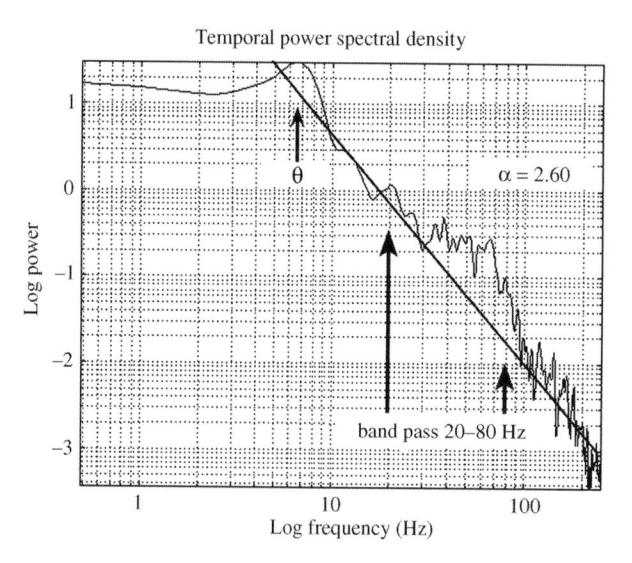

**Figure 6.2** The canonical form of the spectrum of the ECoG at rest in logarithmic coordinates is a straight line, which is the form of random noise. When the subject engages the environment, peaks of power appear above the line, which indicate the emergence of nonrandom structure, here in the θ and γ ranges. (From Freeman, 2001.)

neuron receives as many pulses on average as it transmits, there the population of neurons maintains random firing at a steady level (Freeman, 2006). Mesoscopic stability is maintained by refractory periods, not by inhibition. Every neuron has a near linear relation between dendritic current density and pulse frequency, but that range is bounded by two firm nonlinearities: the threshold at zero pulse frequency under inhibition, and a maximal firing rate, beyond which it cannot fire without stopping to rest. It cannot sustain firing without opportunity to recover. If it has just fired and gets new excitatory input, it simply fails to fire. Every cortical population is locally stabilized everywhere at a mean firing level that is controlled by neurohormones that adjust the refractory periods and thereby the set point of the level of background activity, as in temperature regulation. The effect of the refractory periods is to increase the slope of the spectrum of the ECoG by blocking the highest frequencies, making it steeper than $-2$. In deep slow wave sleep the slope goes to $-3$ or steeper, so-called "black noise" that exemplifies the modifiability of the set point of the noise level (Freeman et al., 2006).

The role of inhibition is to enhance oscillation in narrow bands of the spectrum, which can lead to bursts of oscillation at various frequencies. Two types of transmission delay determine the main frequency ranges of the oscillations in the $\beta$ (12.5–25 Hz) and $\gamma$ (25–100 Hz) ranges. Local oscillations are determined mainly by the passive membrane time constants of the excitatory and inhibitory neurons, which in the olfactory system are close to 5 ms, plus the dendritic synaptic and cable delays that sum to near 1.25 ms. Each cycle of oscillation by negative feedback requires four steps: excitation of pyramidal cells, excitation of inhibitory interneurons, inhibition of pyramidal cells, and inhibition of inhibitory interneurons (Figure 6.3). That last step allows re-excitation of pyramidal cells, but only when the background activity is present to provide excitation during disinhibition by release from inhibition. The four steps add to 25 ms, which is the wavelength of 40 Hz oscillations – the center frequency of $\gamma$ rhythms. Mutual inhibition and mutual excitation, both of which are forms of positive feedback, can and do modify the frequency transiently over the range from 25 to 100 Hz.

Large domains of synchronized oscillation depend on axonal coupling that introduces longer delays, which lower the frequencies of oscillation into the $\beta$ range. For both types of oscillation the emergence of observable EEG and ECoG requires synchronization both by convergence to a common frequency and to a common phase. A rule of thumb is that oscillations having the same frequency, but that diverge more than $\pm 45°$ of phase (a range of a quarter cycle), fail to sum effectively, because the shared power is less than half the total power ($\cos^2 = 0.5$). Typically the bursts of $\beta$ and $\gamma$ oscillation observed in the EEG last much less than three cycles of the current frequency, indicating that even when oscillation is established, it cannot last long, because the transmission delays that determine the frequencies are distributed, and therefore the intrinsic frequencies in cortical populations from inhibitory feedback are distributed. That is why sustained oscillations at single frequencies such as "40 Hz" are rarely seen. However, bursts of $\beta$ and $\gamma$ that are correlated with cognitive functions last three to five cycles of the carrier frequency, so that they manifest very strong departure from randomness.

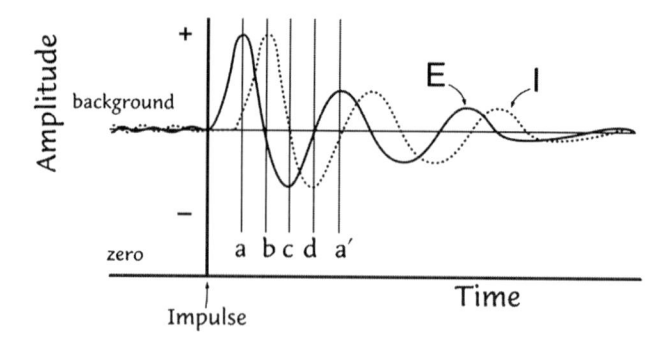

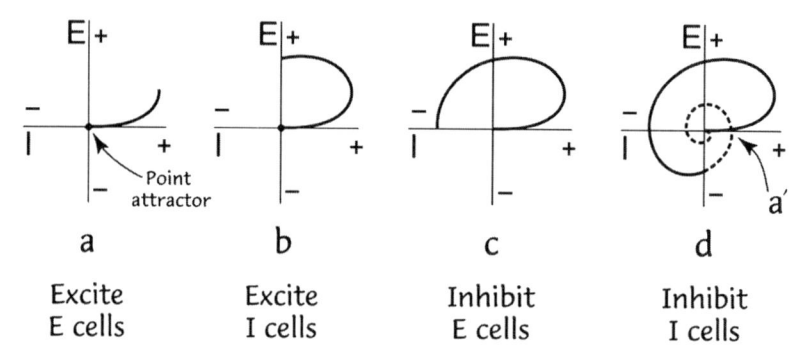

**Figure 6.3** Sequence of the four temporal steps in negative feedback between excitatory and inhibitory neuron populations in cortex. When each step takes 6.25 ms, four steps take 25 ms, which is the 40-Hz wavelength. (From Freeman, 2001.)

## 6.4
### Behavioral Correlates in Spatio-Temporal Patterns of the EEG

It should now be apparent that the difference between the microscopic pulses of neurons and the mesoscopic dendritic waves, despite their interdependence in the control of ionic currents at synapses by spikes and the control of spikes by ionic currents at trigger zones, has profound consequences for the relations of these two observable forms of cortical activity. The spike firing rates of individual neurons are often directly dependent on sensory stimulus parameters of location, intensity, rate of change, modality, and so on. Extracellular dendritic waves are emergent properties of interacting populations of neurons and their amplitudes depend not directly on inducing stimuli, but on the strengths of the synapses through which the interactions take place. Learning under reinforcement modifies many of these synapses; therefore the spatial patterns of the induced oscillations reflect the memory store of prior experience. In a word, the microscopic spikes are measures of sensation; the mesoscopic EEG waves are measures of perception. Examples of ECoG patterns from the olfactory system show that the spatial patterns of amplitude modulation (AM) of a shared frequency in the $\gamma$ range can be changed by

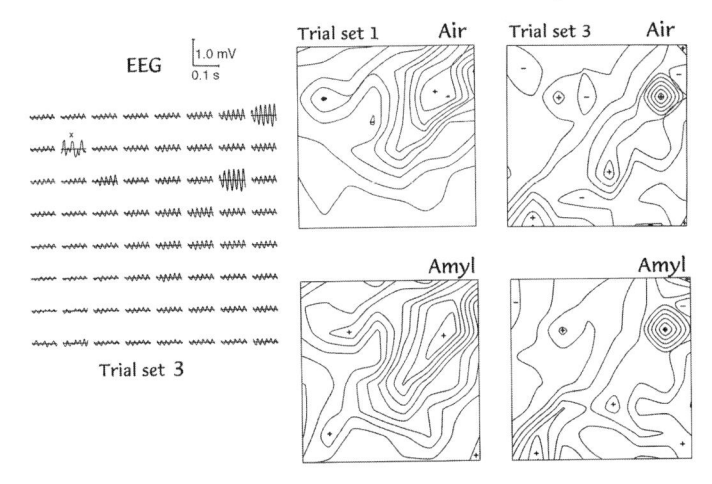

**Figure 6.4** The 64 ECoG waveforms on the left show the AM of the carrier oscillation. The contour plots on the right show vertically the short-term change with learning (upper and lower left frames) and horizontally the long-term change with the consolidation (left to right, both pairs). The lack of invariance with respect to fixed stimuli shows that perception is a creative process, not the processing, storage, and retrieval of fixed quantities of information. (From Freeman, 2001.)

reinforcement learning, but that the resulting patterns lack invariance with respect to fixed stimuli (Figure 6.4). Instead, the AM patterns change with the addition of new learned stimuli or with changes in reinforcement (Freeman, 2005a). They are not representations of the sensory input; that is the property of the spikes. Instead, AM patterns show the retrieved knowledge about the stimuli, which is immediately put to use by the subjects in the guidance of behavior. The bursts carrying AM patterns are not representations of stimuli; they are operators that retrieve and implement the knowledge that is triggered by stimuli.

Synaptic changes with learning occur in two ways. First, in the stage of acquisition by multiple samples of the stimulus from repeated sniffing, integration of Hebbian increments of synaptic strength between coexcited excitatory neurons creates a nerve cell assembly for that stimulus. The assembly generalizes over the collection of receptor inputs to that stimulus, because input to any small or large subset of the assembly activates all of the assembly. Analysis of the system properties of the olfactory bulb shows that this Hebbian change occurs among the projection neurons and not at the synapses delivering the receptor input, and that the effect is to increase up to 50-fold the oscillatory output of bulb to the cortex and other parts of the brain. This finding highlights the great value of system analysis at the mesoscopic level, because the microscopic analysis gives no hint of this degree of amplification of microscopic input by the mesoscopic population.

The second type of change with learning is in the stage of consolidation, which extends to the entire bulb the learning beyond the Hebbian nerve cell assembly that is estimated to contain on the order of 0.5% of the projection neurons in the bulb. Both synaptic changes are manifested in altered AM patterns, the first right after the initial accommodation and creation of a new AM pattern, the follow-up assimilation in days after as the new knowledge is incorporated into memory and the AM pattern continues to evolve (Figure 6.4). The synaptic changes are inferred to be widely and evenly distributed, because the information by which AM patterns are classified derives from all signals; as in a holographic image (Pribram, 1971), no channel has any more or less value than any other.

In addition to the γ AM patterns in olfactory perception, the γ AM patterns are found in the primary sensory areas relating to somatic, visual, and auditory perception. In rabbits the modal diameters of the neocortical regions of synchrony are much larger than those in the bulb (around 10 mm circumference), on average 15 mm, which suffice to include all subareas in each primary receiving area. The size indicates that the emergence of mesoscopic γ AM patterns is the mechanism of "binding," in which the features of complex inputs are integrated not only with each other but also with past experience. The AM patterns appear within 100 ms of learned stimulus onset and usually recur in the next 200 ms. As early as 300 ms after stimulus

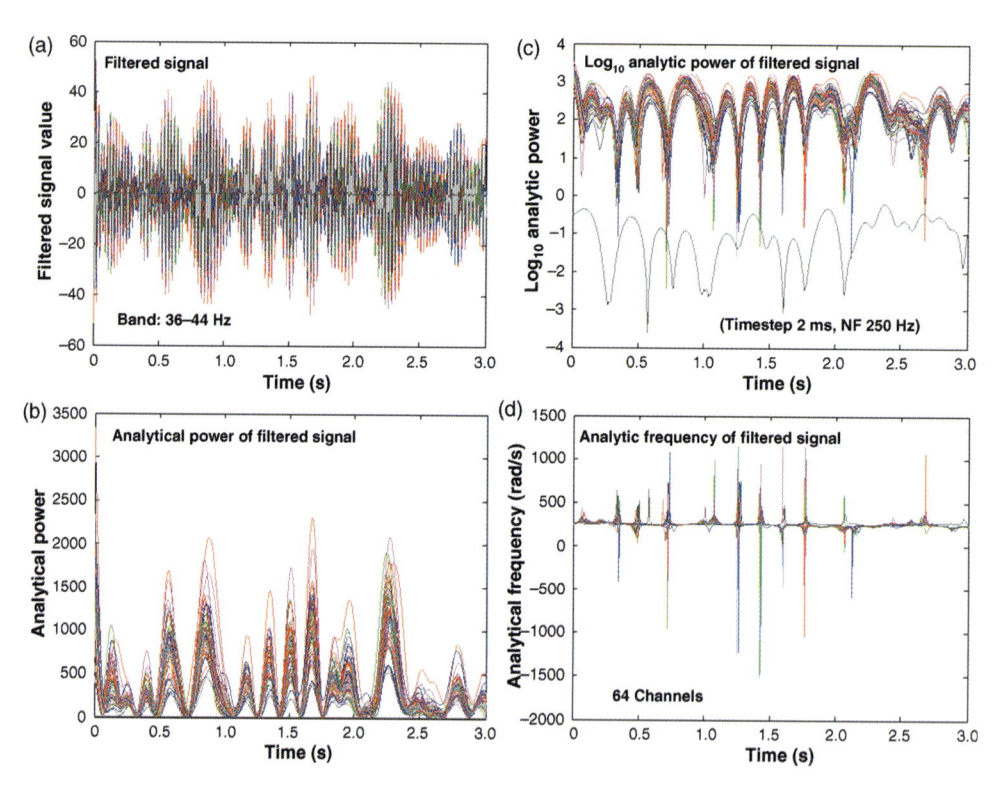

**Figure 6.5** (a) Band pass filtering the EEG and ECoG gives bursts of oscillation at the center frequency of the pass band. (b) The amplitude and power approach zero between bursts in beats. (c) The logarithm of the power shows loss of power by a factor of $10^{-4}$ or more in null spikes. (d) At these points in time the neural activity is disorganized, as shown by the transiently high spatial and temporal variances of the analytic frequency (phase in rad/digitizing step in s). (From Freeman, 2009.)

onsets and through the conditioned responses, bursts of oscillation emerge with frequencies in the β range. The diameters are larger than those of γ AM patterns and may often cover most of each hemisphere. They recur at intervals in the low θ range, prior to conditioned responses. The β AM patterns emerge that include all of the primary sensory areas and the entorhinal cortex (Freeman and Burke, 2003), and are based in synchronization of β oscillations in all areas (Freeman and Rogers, 2003). These findings suggest that the ECoG β bursts enact the stage beyond primary perception in each modality, at which multisensory fusion into gestalts takes place, together with synthesis in the hippocampal system of the time and place of gestalts in sequences. Whether such multicortical synchronized AM patterns in the ECoG might be detected in the scalp EEG, despite the distortions introduced by the intervening tissues and the gyrification, is to be determined.

The intermittency of the bursts of oscillation carrying AM patterns suggests that perception is episodic, even "cinematographic" (Sacks, 2004). If so, the frame rate should be expected to be in the high θ range (5–7 Hz) for γ frames and in the low θ range (3–5 Hz) for β frames, corresponding to the observed rates of repetition of classifiable AM patterns (Freeman, 2005a). It turns out that these rates can be explained as a property of the mechanism that maintains the background ECoG activity. Narrow band filtering of brown noise, and likewise of the ECoG from behaving rabbits, gives Rayleigh noise that is characterized by beats, in which the power transiently approaches zero in null spikes. The beats are due to interference on the summing of oscillatory signals from a population of neurons with a distribution of transmitting frequencies. The repetition rate of the null spikes is proportional to the band width independently of the center frequency by a factor of 0.641 (Rice, 1950; Freeman, 2009). I estimate the width of the distribution of carrier frequencies by calculating the minimal spatial standard deviation of carrier frequencies in the 64 signals that are recorded between null spikes. These estimates range from 5 to 10 Hz, which by Rice's conversion factor give repetition rates in the θ range. Thus, the interference phenomenon can explain the widely observed linkage of θ and γ oscillations in the ECoG and EEG.

To summarize, cortex of all kinds consists of interactive populations of excitatory and inhibitory neurons that sustain three kinds of feedback. Mutual excitation is a form of positive feedback that provides self-sustained, self-regulated background activity. The spectrum of that activity shows that it contains all frequencies in the clinical range (Freeman, 2009). Mutual inhibition is also a form of positive feedback that enhances spatial contrast in opposition to mutual excitation that smoothes spatial difference. Interaction between excitatory and inhibitory neurons is a form of negative feedback that gives the possibility of enhanced oscillations in the β and γ ranges. The negative feedback acts as a band pass filter, in which the characteristic center frequency depends on the axonal and synaptic delays in the feedback loop. The variation in component neurons and connections introduces distributions of delays and therefore of characteristic frequencies. The negative feedback acts as a narrow band filter applied to the background activity, which can be observed as Rayleigh noise (Freeman, 1975, p 190; Rice, 1950). The impact of sensory input that drives cortex away from its resting balance releases power in excess of the resting level. A sequence of such events is revealed by peaks in the ECoG spectrum in the $\gamma$, $\beta$, and $\theta$ ranges. It turns out that the coupling between $\theta$ and $\theta$–$\gamma$ is inherent in the homeostatic mechanism by which background activity and its cinematographic segmentation are autonomously provided and stabilized.

One further feature that Rice predicted is of special physiological value. Rice proposed that a narrow band signal added to white noise would cause emergence of beats in the Rayleigh noise at intervals nearly twice the duration of the intervals between beats in the absence of the signal. Indeed, an important property of the AM patterns that are classifiable with respect to conditioned stimuli and responses is that they have durations that substantially exceed expectations for random events. The classifiable AM patterns last three to five cycles of the carrier frequency; that prolongation of durations can reasonably be inferred to facilitate the effective transmission and reception of the information carried by the oscillatory bursts. In this regard it is important to recognize that strong evidence has been provided at the cellular level for tuning of local networks of neurons at adaptive frequencies in the $\gamma$ range (Whittington et al., 2000). The mesoscopic distributed connectivity provides the generation and transmission of $\gamma$; the membrane conductances that set the frequencies of target neurons are crucial for reception at the microscopic level.

## 6.5
### Synthesis of Two Levels of Function in the Cortical System

The significance of the exceptionally deep beats (null spikes) lies in the inter-relations of the microscopic and mesoscopic levels of function in the cortical system that supports perception. In essence the AM patterns in the $\beta$ and $\gamma$ ranges of the ECoG are observed in the sums of potential fields owing to the flow of ions driven by the dendrites of large populations of neurons. The action potentials drive dendritic currents for both excitation and inhibition, and after summation the dendritic currents determine the firing rates, so the spatial and temporal patterns of spike firing rates are highly correlated with dendritic potentials observed in the ECoG wherever both can be measured, although it must be again stated that the ECoG and EEG are not the "envelopes" of spikes. The neurons provide the power of the activity; the interactions provide the patterns. The neurons that provide the power that sustains the interactions are constrained into ordered states by those interactions. The ECoG is determined by both the mean degree of activation and the degree of synchrony of oscillation. The sum of extracellular currents is too weak to affect the population directly, but that same current passing across the high resistance of the membranes of axon trigger zones does control the firing, so the ECoG is an index of the degrees of power and synchrony that characterize the interaction producing the order. Therefore, I designate the ECoG as an index of a mesoscopic *order parameter*, in accord with usage by physicists (Haken, 1983).

This *circular causality* is the mechanism by which sensory information at the microscopic level selects perceptual patterns through the mediation of Hebbian nerve cell assemblies that constitutes the memory store. It is easy to see how a learned conditioned stimulus excites a Hebbian nerve cell assembly that represents that stimulus in the cortex, but how does the assembly capture the entire cortex in order to form an AM pattern? The proposed answer is that a null spike that occurs by the interference among a distribution of carrier frequencies suddenly and briefly reduces the order parameter. The microscopic neurons continue to fire and the microscopic dendritic power consumption continues unabated, but the degree of organization through synchrony drops and the neurons become disordered. In that brief interlude an active nerve cell assembly may push the mesoscopic trajectory of the cortical mechanism into the basin of its associated attractor, which expresses the appropriate memory store in its AM pattern when the power returns as the null spike ends. A sequence of null spikes may serve as a cinematographic shutter, enabling the staccato flow of images that constitutes perception and thought.

## 6.6
## Conclusions and Applications

This hypothesis raises many questions, which are too complex to be dealt with in this brief chapter. The essential point here and now is that cortex is a system of neuron populations and its operations can best be understood through studies of the novel properties that are imposed by mass action (Freeman, 2009; Freeman and Kozma, 2010). These studies require observation and measurement of the patterns of order that are imposed by mass interactions. That requires the definitions of the state variables that are to be measured, the control parameters such as arousal, and the order parameters that evaluate the degree of interaction. The EEG is the most valuable resource available to us as a source of indices of order parameters. However, the conceptual gap is very wide between brain system theory and clinical applications. Psychiatrists may well question whether mastering the theory will give sufficient advantage to justify the necessary investment of resources, particularly because the theory described in this chapter deals only with the relatively simple interactions by which sensory cortices perform the most basic operations of perception, which are barely above the threshold of human cognition. In my brief clinical experience, requests for EEG consultation to support psychiatric diagnosis yielded such jejune comments as "diffuse abnormalities of a non-specific kind." However, in my research experience, several insights have emerged that have bearing on psychiatric practice. I list them here with citations to full descriptions:

- I predicted by system theory that a population of γ-aminobutyric acid (GABA)-ergic interneurons in the input layer of the olfactory bulb, which was long considered to be inhibitory, was actually excitatory (Freeman, 1975). Physiological proof was by demonstration that these neurons have high intracellular concentrations of chloride ion (Siklós *et al.*, 1995), so that the effect of GABA on their GABA-A receptors is depolarizing. System analysis shows that the degree of excitation manifested in the set point of the background activity is enhanced by histamine (Freeman, 2005b). Histamine is the principal neurohormone medi-

ating arousal and the bulbectomized rat is the best available animal model for clinical depression. The implication from system theory is that the mode of action of anxiolytic drugs may depend primarily on their antihistaminic properties (Freeman, 1993).
- Most forms of epilepsy are commonly thought to result from runaway excitation. System analysis of the complex partial seizures with 3/s spikes and *absence* that are triggered by electrical stimulation of the olfactory tract shows that the seizures are manifestations of runaway inhibition. I proved this by demonstrating that the spikes are inhibitory postsynaptic potentials driven by firing bursts of inhibitory interneurons (Freeman, 1986). The implication from system theory is that prevention and mitigation of such seizures is best done using GABA blockers such as valproic acid, not by GABA mimetics.
- The lack of invariance of mesoscopic AM patterns in perception demonstrates at the mechanistic level of neural populations that perception is a creative process, not information processing. This experimental proof means that system analysis has profound implications for the philosophy (Dreyfus, 2007) and practice (Freeman, 2003) of psychiatry, including a new appreciation of neural mechanisms of unconscious dynamics and creative perceptual distortions such as hallucinations.
- Brain system analysis reinstates the concept of intentionality in its original form, as the process by which brains imagine desired future states, devise the strategic and tactical actions needed to achieve those states, predict by preafference the sensory consequences of the actions to be taken, and modify the brain and body to assimilate and adapt by learning (Freeman, 2001, 2007). System analysis further posits the necessity for a process I call unlearning, by which socialization takes place through the selective dissolution of past learning by oxytocin and its replacement with new knowledge gained in social environments. In the dyadic relation between analyst and patient, the unlearning has the form of transference (Pincus, Freeman, and Modell, 2007).
- Three years after completing his unpublished "Project for a Scientific Psychology" (Pribram and Gill, 1976), Sigmund Freud on 22 September 1898 wrote to his colleague Wilhelm Fliess: "I am not at all in disagreement with you, not at all inclined to leave psychology hanging in the air without an organic basis. But apart from this conviction, I do not know how to go on, neither theoretically nor therapeutically, and therefore must behave as if only the psychological were under consideration. Why I cannot fit it together [the organic and the psychological] I have not even begun to fathom" (Blumberg, personal communication).

The reason for the impasse, I believe, was the inadequacy of nineteenth century physics to support brain theory. System research using EEG and ECoG now reopens Freud's Project, so beautifully described and interpreted by Pribram and Gill, with tools of twenty-first century physics, including nonlinear dynamics (Freeman, 2001), nonequilibrium thermodynamics (Freeman, 2008), random graph theory (Freeman *et al.*, 2008), and renormalization group theory (Freeman and Cao, 2008).

# References

Braitenberg, V. and Schüz, A. (1998) *Cortex: Statistics and Geometry of Neuronal Connectivity*, 2nd edn, Springer, Berlin.

Dreyfus, H. (2007) Why Heideggerian AI failed and how fixing it would require making it more Heideggerian. *Philosophical Psychology*, **20**, 247–268.

Freeman, W.J. (1975) *Mass Action in the Nervous System*, Academic Press, New York.

Freeman, W.J. (1986) Petit mal seizure spikes in olfactory bulb and cortex caused by runaway inhibition after exhaustion of excitation. *Brain Research Reviews*, **11**, 259–284.

Freeman, W.J. (1993) Valium, histamine and neural networks [Editorial]. *Biological Psychiatry*, **34**, 1–2.

Freeman, W.J. (2001) *How Brains Make Up Their Minds*, Columbia University Press, New York.

Freeman, W.J. (2003) Neurodynamic models of brain in psychiatry. *Neuropsychopharmacology*, **28** (Suppl. 1), 54–63.

Freeman, W.J. (2005a) Origin, structure, and role of background EEG activity. Part 3. Neural frame classification. *Clinical Neurophysiology*, **116**, 1118–1129.

Freeman, W.J. (2005b) NDN, volume transmission, and self-organization in brain dynamics. *Journal of Integrative Neuroscience*, **4**, 407–421.

Freeman, W.J. (2006) Origin, structure, and role of background EEG activity. Part 4. Neural frame simulation. *Clinical Neurophysiology*, **117**, 572–589.

Freeman, W.J. (2007) Intentionality. *Scholarpedia*, **2**, 1337; http://www.scholarpedia.org/article/Intentionality.

Freeman, W.J. (2008) A pseudo-equilibrium thermodynamic model of information processing in nonlinear brain dynamics. *Neural Networks*, **21**, 257–265.

Freeman, W.J. (2009) Deep analysis of perception through dynamic structures that emerge in cortical activity from self-regulated noise. *Cognitive Neurodynamics*, **3**, 105–116.

Freeman, W.J. and Burke, B.C. (2003) A neurobiological theory of meaning in perception. Part 4. Multicortical patterns of amplitude modulation in gamma EEG. *International Journal of Bifurcation and Chaos*, **13**, 2857–2866.

Freeman, W.J. and Cao, Y. (2008) Proposed renormalization group analysis of nonlinear brain dynamics at criticality, in *Advances in Cognitive Neurodynamics: Proceedings of the 1st International Conference on Cognitive Neurodynamics* (eds. R. Wang, F. Gu, and E. Shen), Springer, Heidelberg, pp. 147–158.

Freeman, W.J. and Kozma, R. (2010) Freeman's mass action. *Scholarpedia*, **5** (1), 8040; http://www.scholarpedia.org/article/Freeman's_Mass_Action.

Freeman, W.J. and Rogers, L.J. (2003) A neurobiological theory of meaning in perception. Part 5. Multicortical patterns of phase modulation in gamma EEG. *International Journal of Bifurcation and Chaos*, **13**, 2867–2887.

Freeman, W.J. and Vitiello, G. (2008) Dissipation and spontaneous symmetry breaking in brain dynamics. *Journal of Physics A: Mathematical and Theoretical*, **41**, 304042.

Freeman, W.J., Holmes, M.D., West, G.A., and Vanhatalo, S. (2006) Fine spatiotemporal structure of phase in human intracranial EEG. *Clinical Neurophysiology*, **117**, 1228–1243.

Freeman, W.J., Kozma, R., Bollobás, B., and Riordan, O. (2008) Scale-free cortical planar networks, in *Handbook of Large-Scale Random Networks* (eds. B. Bollobás, R. Kozma, and D. Miklös), Bolyai Society Mathematical Studies, vol. 18, Springer, New York, pp. 277–324.

Freeman, W.J., Ahlfors, S.M., and Menon, V. (2009) Combining fMRI with EEG and MEG in order to relate patterns of brain activity to cognition. *International Journal of Psychophysiology*, **73**, 43–52.

Haken, H. (1983) *Synergetics: An Introduction*, Springer, Berlin.

Pincus, D., Freeman, W.J., and Modell, A. (2007) A neurobiological model of perception: considerations for transference. *Journal of Psychoanalytic Psychology*, **24**, 623–640.

Pribram, K.H. (1971) *Languages of the Brain: Experimental Paradoxes and Principles in Neuropsychology*, Prentice-Hall, Englewood Cliffs, NJ.

Pribram, K.H. and Gill, M.M. (1976) *Freud's 'Project' Re-Assessed. Preface to Contemporary Cognitive Theory and Neuropsychology*, Basic Books, New York.

Rice, S.O. (1950) *Mathematical Analysis of Random Noise. Technical Publications Monograph B-1589*, Bell Telephone, New York.

Sacks, O. (2004) In the river of consciousness. *New York Review of Books*, **51** (1), January 15.

Siklós, L., Rickmann, M., Joó, F., Freeman, W.J., and Wolff, J.R. (1995) Chloride is preferentially accumulated in a subpopulation of dendrites and periglomerular cells of the main olfactory bulb in adult rats. *Neuroscience*, **64**, 165–172.

Whittington, M.A., Faulkner, H.J., Doheny, H.C., and Traub, R.D. (2000) Neuronal fast oscillations as a target site for psychoactive drugs. *Pharmacology and Therapeutics*, **86**, 171–190.

ature-driven. This pleasant, open-source tool called *Inkscape* is a vector graphics editor that can be used for a variety of tasks, including the creation of logos, illustrations, and even simple web design. It's known for its versatility and ease of use, making it a popular choice among both beginners and experienced designers.

# 7
# Dopamine and the Electrophysiology of Prefrontal Cortical Networks
*Patricio O'Donnell*

## 7.1
## Introduction

To gain a comprehensive view on the role of dopamine (DA) modulation of prefrontal cortical activity, it is essential to consider not only receptor subtypes involved and the cell type DA acts upon, but also the sometimes dramatic postnatal changes that occur in these systems. DA actions on the prefrontal cortex (PFC) have been extensively studied with a variety of approaches, including recordings in cultured neurons and brain slices, as well as in anesthetized and freely moving animals. Such diversity in levels of analysis has contribute to a fair understanding of the complexity of DA actions in the PFC and other cortical circuits (Seamans and Yang, 2004). However, many divergent views persist and some of them arise from different techniques having studied animals at different postnatal developmental stages. For example, whole-cell recordings typically rely on slices from very young animals, in many cases obtained before weaning, whereas anatomical and behavioral studies more likely assess adult animals. In this chapter, DA modulation of cortical circuits is reviewed, with an emphasis on the roles on different cell types, including pyramidal neurons and interneurons, and the maturation of these actions during adolescence.

The anatomical and molecular organization of cortical microcircuits does change during adolescence. Connectivity changes during puberty and adolescence can be detected in human subjects with studies using diffusion tensor imaging, and these changes correlate with cognitive maturation (Casey, Giedd, and Thomas, 2000). In nonhuman primates, the cortical DA innervation, measured with the density of tyrosine hydroxylase-positive axon terminals, peaks during puberty and declines during adolescence (Rosenberg and Lewis, 1995). Furthermore, DA receptor mRNAs peak during adolescence in human PFC (Weickert *et al.*, 2007). In rodents, the density of DA receptors increases postnatally, with $D_1$ receptors reaching adult levels by postnatal day (PD) 60 (Tarazi and Baldessarini, 2000). Markers of $\gamma$-aminobutyric acid (GABA) transmission within the primate PFC also change during adolescence. Parvalbumin (PV)-containing fibers rise rapidly before being pruned to adult levels (Lewis, Hashimoto, and Volk, 2005). Dramatic processes of cell overproduction and elimination take place during adolescence in cortical regions (see Andersen, 2003, for review), and cognitive functions that depend on prefrontal DA, such as decision making and working memory (Funahashi, 2001), do evolve with the transition to adulthood (Bunge and Wright, 2007). Neurophysiological measures such as error-related negativity and event-related potentials also mature well into late adolescence (Segalowitz and Davies, 2004), suggesting that the neural substrate of cognitive functions is being refined at that time. Despite extensive information highlighting changes in PFC DA, GABA, and glutamate neurotransmission during adolescence, little is known about how the modulation of cortical physiology matures during this critical period.

## 7.2
### Electrophysiological Actions of DA in Prefrontal Cortical Circuits

The electrophysiological actions of DA in the PFC remain controversial, with both excitatory and inhibitory effects reported. Many factors can account for such diversity of responses, including which synaptic processes and cell types are modulated by DA (Seamans and Yang, 2004). *In vivo* intracellular recordings from adult rats reveal that intra-PFC administration of DA with iontophoresis depolarizes pyramidal neurons while suppressing firing in the vast majority of neurons (Bernardi *et al.*, 1982). Endogenous phasic DA release by electrical stimulation of the ventral tegmental area (VTA) with trains of pulses mimicking burst firing or by intra-VTA injection of *N*-methyl-D-aspartate (NMDA) – a procedure that induces DA cell burst firing – yields a similar change in PFC pyramidal neurons (Lewis and O'Donnell, 2000). The reduction of firing in most PFC pyramidal neurons has been interpreted as phasic DA reducing overall activity, allowing only strongly activated neurons to overcome such inhibition and fire action potentials; in short, DA has been proposed to increase signal-to-noise ratio in the system (DeFrance, Sikes, and Chronister, 1985; O'Donnell, 2003), thereby highlighting reward-related or salient stimuli.

DA acts primarily by modulating fast synaptic responses such as those provided by glutamate and GABA. In PFC slice recordings, $D_1$ receptors primarily enhance NMDA currents (Chen, Greengard, and Yan, 2004; Seamans *et al.*, 2001a) and potentiate NMDA effects on pyramidal cell excitability (Tseng and O'Donnell, 2004; Wang and O'Donnell, 2001) in slice recordings. This action of $D_1$ receptors can be blocked by protein kinase A (PKA) antagonists and by interfering with $Ca^{2+}$ signaling (Tseng and O'Donnell, 2004; Wang and O'Donnell, 2001), suggesting a dependence on $G_s$ activation and $Ca^{2+}$. $D_2$ receptors, on the other hand, reduce pyramidal cell excitability and attenuate α-amino-3-hydroxy-5-methyl-4-isoxazol-propionacid (AMPA) and NMDA responses in PFC pyramidal neurons (Gulledge and Jaffe, 1998; Tseng and O'Donnell, 2004). Several mechanisms could mediate the $D_2$ inhibition of AMPA/kainate synaptic transmission (Figure 7.1), including direct postsynaptic activation of phospholipase C (PLC) and inositol trisphosphate ($IP_3$), as well as inhibition of PKA signaling pathways (Tseng and O'Donnell, 2004). The $D_2$

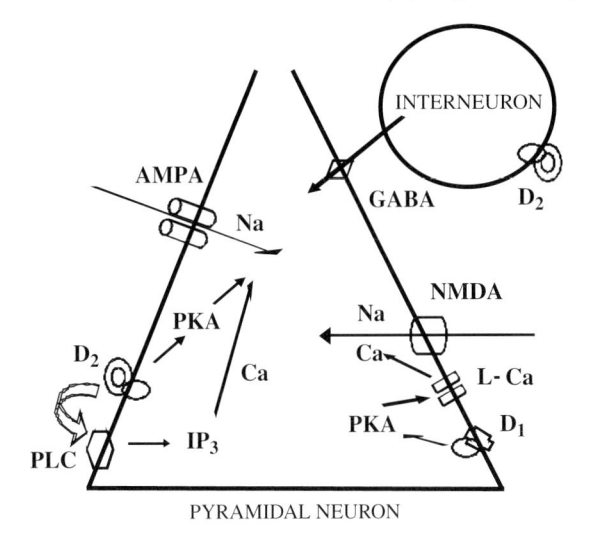

**Figure 7.1** Cartoon depicting the primary cellular mechanisms involved in DA modulation of AMPA and NMDA responses in PFC pyramidal neurons. $D_2$ receptors (left) can attenuate AMPA responses via both an inhibition of PKA activity and by engaging PLC and $IP_3$. The attenuation of NMDA responses by $D_2$ agonists (right) involves instead activation of GABA-A receptors, suggesting that $D_2$ receptors activate local inhibitory interneurons. NMDA responses are enhanced by $D_1$ agonists via potentiation of PKA and recruitment of L-type calcium channels (L-Ca).

inhibition of NMDA responses in pyramidal neurons, on the other hand, does not involve PKA blockade or calcium; it is blocked instead by GABA-A antagonists (Tseng and O'Donnell, 2004). Thus, $D_2$ receptor activation in slices from adult rats can recruit GABA synapses in the PFC. Indeed, the $D_2$ agonist quinpirole increases levels of GABA (Grobin and Deutch, 1998) and enhances GABA interneuron excitability in slices from adult rats (Tseng and O'Donnell, 2004, 2007b). So, we are faced again with the yin and yang of DA actions. On the positive side, a potentiation of NMDA responses by $D_1$ receptors would allow an excitatory action of DA only on already depolarized PFC neurons, thereby reinforcing ongoing behaviorally relevant cortical activity. *In vivo*, $D_1$ agonists enhance PFC long-term potentiation (LTP) (Gurden, Takita, and Jay, 2000), and suppression of VTA activity impairs hippocampal–PFC LTP (Gurden, Tassin, and Jay, 1999). Furthermore, $D_1$ agonists improve memory retrieval and working memory performance (Floresco and Phillips, 2001; Seamans, Floresco, and Phillips, 1998), and $D_1$–NMDA coactivation in the PFC is required for appetitive instrumental learning in adult rats (Baldwin, Sadeghian, and Kelley, 2002). These results suggest that a $D_1$ potentiation of NMDA responses is critical for several PFC-dependent functions. On the other hand, if phasic DA encounters pyramidal neurons at their resting membrane potential, a state in which NMDA receptors are not effectively activated, the dominant effect may be a $D_2$-mediated reduction of glutamate responses. Thus, DA actions in the PFC seem to be a combination of excitatory and inhibitory effects, with the net result being a $D_1$ reinforcement of strongly activated pyramidal neurons and a $D_2$-dependent attenuation of weakly driven pyramidal neurons.

As indicated above, DA also modulates local inhibitory interneurons in the PFC, many of which have a strong impact on whether pyramidal neurons fire action potentials. Juxtacellular recordings *in vivo* reveal that the reduction in pyramidal cell firing is accompanied by an increase in firing by fast-spiking interneurons with a similar time-course (Tseng et al., 2006). This finding suggests that the strong inhibitory effect of VTA stimulation on pyramidal neurons *in vivo* may involve activation of local interneurons. Whole-cell recordings in PFC slices show that DA modulates GABA inputs to pyramidal neurons (Seamans et al., 2001b), with a strong $D_1$ excitation of interneurons in slices from young animals (Gorelova, Seamans, and Yang, 2002). In slices from adult rats, $D_2$ receptors also increase interneuron excitability and attenuate NMDA-induced increased excitability in pyramidal neurons (Tseng and O'Donnell, 2004, 2007b). The $D_2$-mediated attenuation of NMDA responses can be blocked by GABA-A antagonists, suggesting that a $D_2$ recruitment of fast-spiking interneurons can shunt NMDA-dependent events. Thus, the combination of DA actions on pyramidal neurons and interneurons may contribute to the DA-dependent increase in signal-to-noise ratio in PFC information processing.

## 7.3
### Changes in DA Modulation of Pyramidal Neurons during Adolescence

DA effects on pyramidal neurons become refined during adolescence. We have shown age differences in the modulation of pyramidal cell excitability by AMPA, NMDA and $D_1$ agonists, as well as in $D_1$–NMDA interactions. In slices obtained from prepubertal rats (PD < 35), AMPA, NMDA, and the $D_1$ agonist SKF38393 enhance pyramidal cell excitability in response to intrasomatic current injection (Wang and O'Donnell, 2001). Similar recordings in slices from late adolescent or adult rats (i.e., older than 55 days) reveal similar effects, but with dose–response curves shifted to the left (Tseng and O'Donnell, 2004), an indication of higher potency of these agents in the adult brain. Furthermore, dendritic $Na^+$ and $Ca^{2+}$ regenerative potentials in pyramidal neurons, which are important players in synaptic plasticity, become effective in coupling distal apical dendrites with somata at PD 42 (Zhu, 2000), a time in which NMDA receptor subunit expression changes (Monyer et al., 1994). These observations indicate that DA and glutamate become more efficient in driving pyramidal cell firing as the animals mature through adolescence. The interactions between DA and glutamate also change during this critical period. In slices from juvenile rats, a $D_1$ agonist potentiates NMDA responses in a synergistic manner (Wang and O'Donnell, 2001). In slices from young adult rats, such synergism is capable of yielding persistent depolarizations similar to the *up states* that are observed *in vivo* (Lewis and O'Donnell, 2000). Coadministering SKF38393 and NMDA causes a series of plateau depolarizations lasting hundreds of milliseconds, but only in slices from adult rats (Tseng and O'Donnell, 2005). In slices from prepubertal rats, all $D_1$–NMDA induced depolarizations are in the range of tens of milliseconds, suggesting simple synaptic responses and not persistent activity (Tseng and O'Donnell, 2005). The $D_1$–NMDA plateaus observed in adult slices are likely driven by enhanced glutamatergic activity in the local network, as they disappear with tetrodotoxin or the AMPA antagonist 6-cyano-2,3-dihydroxy-7-nitro-quinoxaline (Tseng and O'Donnell, 2005). A $D_1$ facilitation of plateau depolarizations in the PFC can provide a temporal window during which context-relevant inputs can drive pyramidal neuron firing and NMDA-dependent synaptic plasticity would be enabled. As mesocortical DA activation is context-dependent and related to attention and salient stimuli (Cohen, Braver, and Brown, 2002; Horvitz, 2000; Tobler, Fiorillo, and Schultz, 2005), the relevant ongoing activity in the PFC (i.e., that mediated by AMPA and NMDA receptors) would therefore become enhanced. Thus, the maturation of $D_1$–NMDA interactions during adolescence renders them in a state in which persistent activity can be more readily driven and reinforced by DA.

## 7.4
## Changes in DA Modulation of GABA Interneurons during Adolescence

The DA modulation of local interneurons changes even more dramatically during adolescence. In slices from prepubertal rats, the $D_1$ agonist SKF38393 increases interneuron excitability (Gorelova, Seamans, and Yang, 2002) while the $D_2$ agonist quinpirole does not have a major effect (Gorelova, Seamans, and Yang, 2002; Tseng and O'Donnell, 2004). These actions are balanced by a DA-dependent attenuation of GABA synaptic responses in pyramidal neurons (Seamans *et al.*, 2001b; Trantham-Davidson *et al.*, 2004). On the other hand, whole-cell recordings in slices from adult rats reveal that quinpirole increases spontaneous firing of fast-spiking interneurons (Tseng and O'Donnell, 2004) and enhances interneuron excitability (Tseng and O'Donnell, 2007b). These effects are only observed in slices from rats older than PD 45 (Tseng and O'Donnell, 2007b), indicating that during adolescence DA becomes strongly excitatory on interneurons by virtue of both $D_1$ and $D_2$ receptors increasing their excitability (Figure 7.2). The cellular or synaptic changes responsible for this late maturation remain to be determined. It can be speculated that they could depend on changes in the receptor subtypes expressed by interneurons ($D_2$ versus $D_4$), the G-protein they are coupled to ($G_i$ versus $G_q$), or the dimerization with other receptors (Franco *et al.*, 2006). The outcome of such switch in the $D_2$ modulation of interneurons is a much stronger recruitment of local inhibition with phasic DA release in the adult PFC.

The changes in $D_2$ modulation of interneurons during adolescence affect the DA modulation of cortico-cortical information. The emergence of a $D_2$ upregulation of interneuron excitability during adolescence contributes to the $D_2$ attenuation of NMDA effects on pyramidal cell excitability that we observed in slices from adult rats, as blocking GABA-A receptors prevented the $D_2$ modulation of NMDA responses (Tseng and O'Donnell, 2004). Furthermore, a $D_2$ recruitment of interneurons has an impact on intracortical synaptic activity. Electrical stimulation of cortico-cortical fibers by placing an electrode in superficial layers (I or II) about 1 mm lateral to the

## (a) Prepubertal PFC

$D_1 \uparrow$ NMDA
$D_2 \downarrow$ AMPA

PYR

$D_2 -/\downarrow$
$D_1 \uparrow\uparrow$

FSI

excitability

$D_1 \uparrow$   $D_2 \downarrow$

## (b) Adult PFC

$D_1 \uparrow\uparrow\uparrow$ NMDA
$D_2 \downarrow$ AMPA

PYR

$D_2 \uparrow\uparrow$
$D_1 \uparrow\uparrow$

FSI

excitability

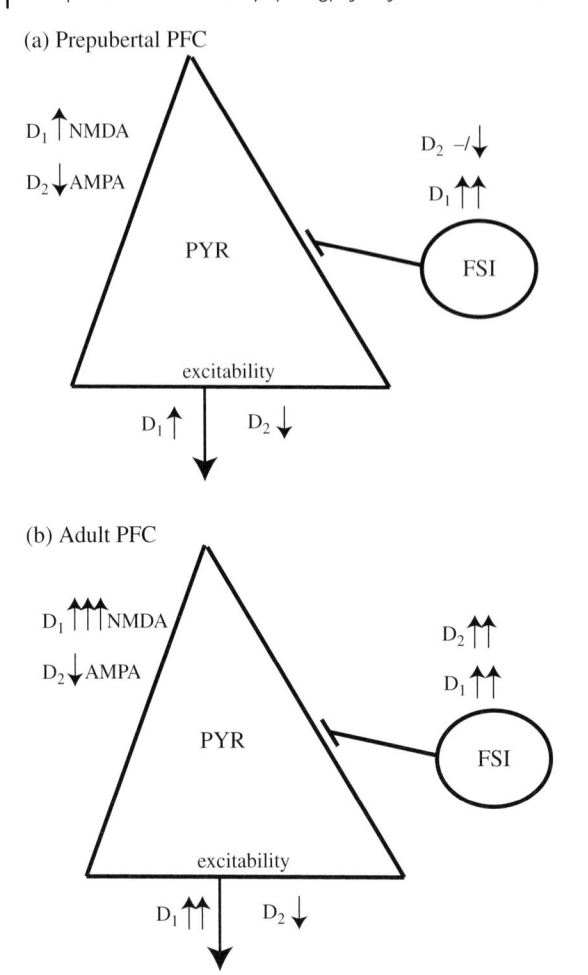

$D_1 \uparrow\uparrow$   $D_2 \downarrow$

**Figure 7.2** Model of periadolescent changes in DA modulation of pyramidal (PYR) neuron and interneuron excitability. (a) In prepubertal PFC circuits, $D_2$ receptors have a mild, if any, effect on fast-spiking interneuron (FSI) excitability. In the transition to an adult circuit (b), the $D_1$ upregulation of pyramidal cell excitability and its potentiation of NMDA responses in pyramidal neurons is enhanced. In fast-spiking interneurons, $D_2$ receptors become strongly excitatory.

apical dendrite of the deep layer pyramidal neuron being recorded evokes AMPA-dependent excitatory postsynaptic potentials (EPSPs). Adding quinpirole attenuates the EPSPs by a dual mechanism in slices from adult rats: (i) an early component that is not blocked by GABA-A antagonists and therefore may be due to a direct effect on $D_2$ receptors on the pyramidal neuron being recorded, and (ii) a slow component that lasts several minutes and is blocked by GABA-A antagonists (Tseng and O'Donnell, 2007a). In juvenile rats, only the early, direct component is observed

(Tseng and O'Donnell, 2007a), consistent with the notion that $D_2$ receptors activate interneurons in late adolescent or adult slices. The maturation of DA actions on interneurons is therefore important for appropriate information processing in the PFC, and may balance the increase in responsivity to $D_1$ and NMDA activation. Thus, the excitation–inhibition balance responsible for proper PFC processing of salient information becomes refined during adolescence and such refinement could contribute to establishing a more efficient PFC in the transition to adulthood.

## 7.5
## Abnormal Periadolescent Maturation of DA Actions in Developmental Animal Models of Schizophrenia

Adolescence is a critical period for several psychiatric disorders that are characterized by PFC and DA alterations. In schizophrenia, for example, although there are some early cognitive traits (Nuechterlein, Dawson, and Green, 1994), the full onset of hallucinations and delusions does not occur until late adolescence or early adult stages (Thompson, Pogue-Geile, and Grace, 2004). On the other hand, there is a clear genetic predisposition for this disorder (Harrison and Weinberger, 2005), suggesting that early developmental anomalies should be present. How can early developmental deficits cause such delayed symptom onset? Several animal models were developed to directly assess this issue. Perhaps one of the most extensively studied is the neonatal ventral hippocampal lesion (NVHL), established by Lipska and Weinberger (Lipska, Jaskiw, and Weinberger, 1993) to determine whether early hippocampal deficits would have impact on adult behaviors. Indeed, rats with a NVHL present behavioral, molecular, and electrophysiological anomalies, most of which emerge during adolescence. Specifically, adult NVHL rats are hyperactive (Lipska and Weinberger, 1995), show enhanced reactivity to stress, psychostimulants and NMDA antagonists (Al-Amin, Weinberger, and Lipska, 2000; Lipska, Jaskiw, and Weinberger, 1993; Swerdlow et al., 1995), and exhibit sensorimotor gating deficits in the form of reduced prepulse inhibition of the acoustic startle response (Lipska et al., 1995). Social interactions are also altered (Sams-Dodd, Lipska, and Weinberger, 1997) along with an increased liability for addictive behaviors (Brady et al., 2008; Chambers and Self, 2002). Furthermore, nucleus accumbens (NA) neurons respond to VTA stimulation with an abnormal increase in firing, instead of the normal decrease (Goto and O'Donnell, 2001), in adult but not prepubertal rats with a NVHL (Goto and O'Donnell, 2002). Thus, the NVHL is a useful tool to study periadolescent changes secondary to earlier developmental manipulations.

Several findings point to the PFC, and more specifically PFC interneurons, as being affected in this model. A PFC lesion in adult rats with a NVHL blocks hyperlocomotion (Lipska, al-Amin, and Weinberger, 1998) and the abnormal responses of NA neurons to VTA stimulation (Goto and O'Donnell, 2004). Furthermore, several PFC-dependent behavioral measures, such as working memory, are affected in NVHL rats (Chambers et al., 1996; Lipska et al., 2002) and primates (Bachevalier, Alvarado, and Malkova, 1999), and there is a reduction in GAD67 in the

PFC of NVHL rats (Lipska et al., 2003). Many anomalies in the NVHL model cannot be reproduced if the lesion is produced when the animals are already adults (O'Donnell et al., 2002), suggesting that the altered responses may reflect abnormal postnatal developmental changes within the mesocortical–PFC pathway. Thus, impaired PFC information processing is likely to underscore alterations in the NVHL model.

The DA modulation of glutamate and GABA responses is altered in the PFC of NVHL rats. In vivo intracellular recordings reveal that VTA stimulation with bursts of pulses caused the transition to an up state in pyramidal neurons from adult NVHL rats, but instead of the normal decrease in firing (Lewis and O'Donnell, 2000), pyramidal neurons increase their firing (O'Donnell et al., 2002), suggesting the possibility that interneuron activation by DA was impaired in this model. Whole-cell recordings reveal that adult PFC pyramidal neurons are hyperexcitable in response to NMDA and $D_1$ activation (Tseng et al., 2007) and fast-spiking interneurons are less active in slices from adult rats with a NVHL (Tseng et al., 2008). Furthermore, the periadolescent maturation of DA effects on PFC interneurons fails to occur in NVHL rats. Quinpirole increases interneuron excitability in slices from adult sham-treated rats, as it does in naïve rats, but does not yield an increase in excitability in slices from adult NVHL rats (Tseng et al., 2008). In many interneurons from adult NVHL rats, quinpirole actually reduces excitability. The NVHL procedure therefore causes an alteration in interneuron development such as the maturation of responses to DA during adolescence either does not occur or takes a wrong direction. Thus, even though the neonatal lesion may have caused abnormal development of PFC circuits, the functional impact of such anomaly is minimal in the immature brain; it is only when the normal periadolescent maturation fails to occur that symptoms become florid.

Other models also point to a deficit in cortical interneurons. Raising rats in social isolation also yields abnormal responses in adult PFC pyramidal neurons to VTA stimulation (Peters and O'Donnell, 2005). Treating pregnant rats with the antimitotic agent methylazoxymethanol decreases the number of PV-expressing neurons in the hippocampus (Penschuck et al., 2006) and PFC (Lodge, Behrens, and Grace, 2008), which is associated with loss of γ oscillations in the electroencephalogram. Some of the emerging genetic models also display interneuron deficits. For example, a dominant-negative form of DISC1 shows a reduction in PV interneurons (Hikida et al., 2007). An immune challenge in pregnant rats has been proposed to mimic the impact of maternal infection. In our hands, injecting the bacterial endotoxin lipopolysaccharide (LPS) in the ventral hippocampus at the same age as the lesions are typically conducted also causes an abnormal maturation of PFC interneurons during adolescence. In slices from LPS-treated rats, quinpirole fails to increase interneuron excitability (O'Donnell, Tseng, and Feleder, 2008). This indicates that the deleterious effects of the NVHL on interneuron development are not related to the lesion, but to abnormal activity or inactivation in the ventral hippocampus impairing development of PFC circuits. The convergence in interneuron deficits across several different models is quite remarkable and highlights the possibility that several different mechanisms may share a common interference with postnatal maturation of local inhibition in cortical circuits.

## 7.6
## Implications for Schizophrenia Pathophysiology and Novel Treatments

The periadolescent maturation of prefrontal cortical circuits is likely to play a major role in the pathophysiology of schizophrenia and related disorders. Although several candidate genes that confer predisposition for schizophrenia have been identified (Harrison and Weinberger, 2005), most symptoms do not emerge until late adolescence or early adulthood. Our work with NVHL rats suggests that abnormal PFC circuits as the result of early manipulations may become evident only when the late periadolescent maturation fails to occur. It is possible that a combination of predisposing genes and epigenetic factors contributes to establishing abnormally wired cortical circuits, perhaps characterized by altered interneuron function. It is with the critical maturation during adolescence that the impact of such abnormal circuitry on behavior becomes evident. Thus, the protracted maturation of inhibitory circuits in the cortex may serve as a bridge between the early developmental nature of predisposing factors and the late onset of symptoms.

The involvement of a protracted maturation of inhibitory circuits in symptom onset offers opportunities for new drug treatment approaches for schizophrenia. The traditional way to treat this disease has been DA antagonists, mostly targeting $D_2$ receptors. There is indeed evidence that the clinical efficacy correlates with the ability to block $D_2$ receptors (Seeman, 1987). Both classical and atypical antipsychotic drugs reverse some of the abnormal behaviors and electrophysiological deficits associated with the NVHL (Goto and O'Donnell, 2002; Le Pen and Moreau, 2002; Lipska and Weinberger, 1994), indicating that the lesion model may be reproducing pathophysiological changes that are targeted by antipsychotic drugs. As neuroleptics present disabling side-effects and compliance is poor, there has been an intense search for novel therapeutic approaches. For many years, this search has focused on different compounds that retained the $D_2$ antagonism. More recently, however, the conceptualization that excitation–inhibition balance in cortical regions may be a critical factor and DA dysregulation may occur downstream to altered cortical function led the field to consider approaches targeting glutamate and GABA receptors. A recent clinical trial revealed that restoring such balance with a metabotropic glutamate-2/3 agonist (which may reduce the levels of glutamate release) has similar efficacy as olanzapine (Patil et al., 2007). This was the first nondopaminergic compound with proven efficacy. The consideration that excitation–inhibition balance matures during adolescence should guide drug discovery, and in particular calls for consideration of external factors for which adolescents seem vulnerable and may contribute to triggering symptom onset, such as stress. In short, the maturation of DA effects in the PFC and other cortical areas is likely to determine whether a particular component in those circuits is vulnerable and may settle into an abnormal configuration that may lead to symptoms.

## References

Al-Amin, H.A., Weinberger, D.R., and Lipska, B.K. (2000) Exaggerated MK-801-induced motor hyperactivity in rats with the neonatal lesion of the ventral hippocampus. *Behavioural Pharmacology*, **11**, 269–278.

Andersen, S.L. (2003) Trajectories of brain development: point of vulnerability or window of opportunity? *Neuroscience and Biobehavioral Reviews*, **27**, 3–18.

Bachevalier, J., Alvarado, M.C., and Malkova, L. (1999) Memory and socioemotional behavior in monkeys after hippocampal damage incurred in infancy or in adulthood. *Biological Psychiatry*, **46**, 329–339.

Baldwin, A.E., Sadeghian, K., and Kelley, A.E. (2002) Appetitive instrumental learning requires coincident activation of NMDA and dopamine $D_1$ receptors within the medial prefrontal cortex. *Journal of Neuroscience*, **22**, 1063–1071.

Bernardi, G., Cherubini, E., Marciani, M.G., Mercuri, N., and Stanzione, P. (1982) Responses of intracellularly recorded cortical neurons to the iontophoretic application of dopamine. *Brain Research*, **245**, 268–274.

Brady, A.M., McCallum, S.E., Glick, S.D., and O'Donnell, P. (2008) Enhanced methamphetamine self-administration in a neurodevelopmental rat model of schizophrenia. *Psychopharmacology*, **200**, 205–215.

Bunge, S.A. and Wright, S.B. (2007) Neurodevelopmental changes in working memory and cognitive control. *Current Opinion in Neurobiology*, **17**, 243–250.

Casey, B.J., Giedd, J.N., and Thomas, K.M. (2000) Structural and functional brain development and its relation to cognitive development. *Biological Psychology*, **54**, 241–257.

Chambers, R.A., Moore, J., McEvoy, J.P., and Levin, E.D. (1996) Cognitive effects of neonatal hippocampal lesions in a rat model of schizophrenia. *Neuropsychopharmacology*, **15**, 587–594.

Chambers, R.A. and Self, D.W. (2002) Motivational responses to natural and drug rewards in rats with neonatal ventral hippocampal lesions: an animal model of dual diagnosis schizophrenia. *Neuropsychopharmacology*, **27**, 889–905.

Chen, G., Greengard, P., and Yan, Z. (2004) Potentiation of NMDA receptor currents by dopamine $D_1$ receptors in prefrontal cortex. *Proceedings of the National Academy of Sciences of the United States of America*, **101**, 2596–2600.

Cohen, J.D., Braver, T.S., and Brown, J.W. (2002) Computational perspectives on dopamine function in prefrontal cortex. *Current Opinion in Neurobiology*, **12**, 223–229.

DeFrance, J.F., Sikes, R.W., and Chronister, R.B. (1985) Dopamine action in the nucleus accumbens. *Journal of Neurophysiology*, **54**, 1568–1577.

Floresco, S.B. and Phillips, A.G. (2001) Delay-dependent modulation of memory retrieval by infusion of a dopamine $D_1$ agonist into the rat medial prefrontal cortex. *Behavioral Neuroscience*, **115**, 934–939.

Franco, R., Casado, V., Mallol, J., Ferrada, C., Ferre, S., Fuxe, K. *et al.* (2006) The two-state dimer receptor model: a general model for receptor dimers. *Molecular Pharmacology*, **69**, 1905–1912.

Funahashi, S. (2001) Neuronal mechanisms of executive control by the prefrontal cortex. *Neuroscience Research*, **39**, 147–165.

Gorelova, N., Seamans, J.K., and Yang, C.R. (2002) Mechanisms of dopamine activation of fast-spiking interneurons that exert inhibition in rat prefrontal cortex. *Journal of Neurophysiology*, **88**, 3150–3166.

Goto, Y. and O'Donnell, P. (2001) Network synchrony in the nucleus accumbens *in vivo*. *Journal of Neuroscience*, **21**, 4498–4504.

Goto, Y. and O'Donnell, P. (2002) Delayed mesolimbic system alteration in a developmental animal model of schizophrenia. *Journal of Neuroscience*, **22**, 9070–9077.

Goto, Y. and O'Donnell, P. (2004) Prefrontal lesion reverses abnormal mesoaccumbens response in an animal model of schizophrenia. *Biological Psychiatry*, **55**, 172–176.

Grobin, A.C. and Deutch, A.Y. (1998) Dopaminergic regulation of extracellular gamma-aminobutyric acid levels in the prefrontal cortex of the rat. *Journal of Pharmacology and Experimental Therapeutics*, **285**, 350–357.

Gulledge, A.T. and Jaffe, D.B. (1998) Dopamine decreases the excitability of layer V pyramidal cells in the rat prefrontal cortex. *Journal of Neuroscience*, **18**, 9139–9151.

Gurden, H., Takita, M., and Jay, T.M. (2000) Essential role of $D_1$ but not $D_2$ receptors in the NMDA receptor-dependent long-term potentiation at hippocampal–prefrontal cortex synapses *in vivo*. *Journal of Neuroscience*, **20**, RC106.

Gurden, H., Tassin, J.-P., and Jay, T. (1999) Integrity of the mesocortical dopaminergic system is necessary for complete expression of *in vivo* hippocampal–prefrontal cortex long-term potentiation. *Neuroscience*, **94**, 1019–1027.

Harrison, P.J. and Weinberger, D.R. (2005) Schizophrenia genes, gene expression, and neuropathology: on the matter of their convergence. *Molecular Psychiatry*, **10**, 40–68.

Hikida, T., Jaaro-Peled, H., Seshadri, S., Oishi, K., Hookway, C., Kong, S. *et al.* (2007) Dominant-negative DISC1 transgenic mice display schizophrenia-associated phenotypes detected by measures translatable to humans. *Proceedings of the National Academy of Sciences of the United States of America*, **104**, 14501–14506.

Horvitz, J.C. (2000) Mesolimbocortical and nigrostriatal dopamine responses to salient non-reward events. *Neuroscience*, **96**, 651–656.

Le Pen, G. and Moreau, J.L. (2002) Disruption of prepulse inhibition of startle reflex in a neurodevelopmental model of schizophrenia: reversal by clozapine, olanzapine and risperidone but not by haloperidol. *Neuropsychopharmacology*, **27**, 1–11.

Lewis, B.L. and O'Donnell, P. (2000) Ventral tegmental area afferents to the prefrontal cortex maintain membrane potential "up" states in pyramidal neurons via $D_1$ dopamine receptors. *Cerebral Cortex*, **10**, 1168–1175.

Lewis, D.A., Hashimoto, T., and Volk, D.W. (2005) Cortical inhibitory neurons and schizophrenia. *Nature Reviews Neuroscience*, **6**, 312–324.

Lipska, B.K. and Weinberger, D.R. (1994) Subchronic treatment with haloperidol and clozapine in rats with neonatal excitotoxic hippocampal damage. *Neuropsychopharmacology*, **10**, 199–205.

Lipska, B.K. and Weinberger, D.R. (1995) Genetic variation in vulnerability to the behavioral effects of neonatal hippocampal damage in rats. *Proceedings of the National Academy of Sciences of the United States of America*, **92**, 8906–8910.

Lipska, B.K., Jaskiw, G.E., and Weinberger, D.R. (1993) Postpuberal emergence of hyperresponsiveness to stress and to amphetamine after neonatal excitotoxic hippocampal damage: a potential animal model of schizophrenia. *Neuropsychopharmacology*, **90**, 67–75.

Lipska, B.K., Swerdlow, N.R., Geyer, M.A., Jaskiw, G.E., Braff, D.L., and Weinberger, D.R. (1995) Neonatal excitotoxic hippocampal damage in rats cause post-pubertal changes in prepulse inhibition of startle and its disruption by apomorphine. *Psychopharmacology*, **132**, 303–310.

Lipska, B., al-Amin, H., and Weinberger, D. (1998) Excitotoxic lesions of the rat medial prefrontal cortex. Effects on abnormal behaviors associated with neonatal hippocampal damage. *Neuropsychopharmacology*, **19**, 451–464.

Lipska, B.K., Aultman, J.M., Verma, A., Weinberger, D.R., and Moghaddam, B. (2002) Neonatal damage of the ventral hippocampus impairs working memory in the rat. *Neuropsychopharmacology*, **27**, 47–54.

Lipska, B.K., Lerman, D.N., Khaing, Z.Z., Weickert, C.S., and Weinberger, D.R. (2003) Gene expression in dopamine and GABA systems in an animal model of schizophrenia: effects of antipsychotic drugs. *European Journal of Neuroscience*, **18**, 391–402.

Lodge, D., Behrens, M., and Grace, A.A. (2008) A loss of parvalbumin containing interneurons is associated with diminished gamma oscillatory activity in an animal model of schizophrenia. *Schizophrenia Research*, **102**, 112.

Monyer, H., Burnashev, N., Laurie, D.J., Sakmann, B., and Seeburg, P.H. (1994) Developmental and regional expression in the rat brain and functional properties of four NMDA receptors. *Neuron*, **12**, 529–540.

Nuechterlein, K.H., Dawson, M.E., and Green, M.F. (1994) Information-processing abnormalities as neuropsychological vulnerability indicators for schizophrenia. *Acta Psychiatrica Scandinavica*, **90** (Suppl. 384), 71–79.

O'Donnell, P. (2003) Dopamine gating of forebrain neural ensembles. *European Journal of Neuroscience*, **17**, 429–435.

O'Donnell, P., Lewis, B.L., Weinberger, D.R., and Lipska, B.K. (2002) Neonatal hippocampal damage alters electrophysiological properties of prefrontal cortical neurons in adult rats. *Cerebral Cortex*, **12**, 975–982.

O'Donnell, P., Tseng, K.Y., and Feleder, C. (2008) Periadolescent emergence of impaired dopamine modulation of prefrontal GABA circuits in developmental animal models of schizophrenia. *International Journal of Neuropsychopharmacology*, **11**, 37.

Patil, S.T., Zhang, L., Martenyi, F., Lowe, S.L., Jackson, K.A., Andreev, B.V. et al. (2007) Activation of mGlu2/3 receptors as a new approach to treat schizophrenia: a randomized Phase 2 clinical trial. *Nature Medicine*, **13**, 1102–1107.

Penschuck, S., Flagstad, P., Didriksen, M., Leist, M., and Michael-Titus, A.T. (2006) Decrease in parvalbumin-expressing neurons in the hippocampus and increased phencyclidine-induced locomotor activity in the rat methylazoxymethanol (MAM) model of schizophrenia. *European Journal of Neuroscience*, **23**, 279–284.

Peters, Y.M. and O'Donnell, P. (2005) Social isolation rearing affects prefrontal cortical response to ventral tegmental area stimulation. *Biological Psychiatry*, **57**, 1205–1208.

Rosenberg, D.R. and Lewis, D.A. (1995) Postnatal maturation of the dopaminergic innervation of monkey prefrontal and motor cortices: a tyrosine hydroxylase immunohistochemical analysis. *Journal of Comparative Neurology*, **358**, 383–400.

Sams-Dodd, F., Lipska, B.K., and Weinberger, D.R. (1997) Neonatal lesions of the rat ventral hippocampus result in hyperlocomotion and deficits in social behavior in adulthood. *Psychopharmacology*, **132**, 303–310.

Seamans, J.K., and Yang, C.R. (2004) The principal features and mechanisms of dopamine modulation in the prefrontal cortex. *Progress in Neurobiology*, **74**, 1–58.

Seamans, J.K., Floresco, S.B., and Phillips, A.G. (1998) $D_1$ receptor modulation of hippocampal–prefrontal cortical circuits integrating spatial memory with executive functions. *Journal of Neuroscience*, **18**, 1613–1621.

Seamans, J.K., Durstewitz, D., Christie, B.R., Stevens, C.F., and Sejnowski, T.J. (2001a) Dopamine $D_1/D_5$ receptor modulation of excitatory synaptic inputs to layer V prefrontal cortex neurons. *Proceedings of the National Academy of Sciences of the United States of America*, **98**, 301–306.

Seamans, J.K., Gorelova, N., Durstewitz, D., and Yang, C.R. (2001b) Bidirectional dopamine modulation of GABAergic inhibition in prefrontal cortical pyramidal neurons. *Journal of Neuroscience*, **21**, 3628–3638.

Seeman, P. (1987) Dopamine receptors and the dopamine hypothesis of schizophrenia. *Synapse*, **1**, 133–152.

Segalowitz, S.J., and Davies, P.L. (2004) Charting the maturation of the frontal lobe: an electrophysiological strategy. *Brain and Cognition*, **55**, 116–133.

Swerdlow, N.R., Lipska, B.K., Weinberger, D.R., Braff, D.L., Jaskiw, G.E., and Geyer, M.A. (1995) Increased sensitivity to the sensorimotor gating-disruptive effects of apomorphine after lesions of medial prefrontal cortex or ventral hippocampus in adult rats. *Psychopharmacology*, **122**, 27–34.

Tarazi, F.I. and Baldessarini, R.J. (2000) Comparative postnatal development of dopamine $D_1$, $D_2$ and $D_4$ receptors in rat forebrain. *International Journal of Developmental Neuroscience*, **18**, 29–37.

Thompson, J.L., Pogue-Geile, M.F., and Grace, A.A. (2004) Developmental pathology, dopamine, and stress: a model for the age of onset of schizophrenia symptoms. *Schizophrenia Bulletin*, **30**, 875–900.

Tobler, P.N., Fiorillo, C.D., and Schultz, W. (2005) Adaptive coding of reward value by dopamine neurons. *Science*, **307**, 1642–1645.

Trantham-Davidson, H., Neely, L.C., Lavin, A., and Seamans, J.K. (2004) Mechanisms underlying differential $D_1$ versus $D_2$ dopamine receptor regulation of inhibition in prefrontal cortex. *Journal of Neuroscience*, **24**, 10652–10659.

Tseng, K.Y. and O'Donnell, P. (2004) Dopamine-glutamate interactions controlling prefrontal cortical pyramidal cell excitability involve multiple signaling mechanisms. *Journal of Neuroscience*, **24**, 5131–5139.

Tseng, K.Y. and O'Donnell, P. (2005) Post-pubertal emergence of prefrontal cortical up states induced by $D_1$–NMDA co-activation. *Cerebral Cortex*, **15**, 49–57.

Tseng, K.Y. and O'Donnell, P. (2007a) $D_2$ dopamine receptors recruit a GABA component for their attenuation of excitatory synaptic transmission in the adult rat prefrontal cortex. *Synapse*, **61**, 843–850.

Tseng, K.Y. and O'Donnell, P. (2007b) Dopamine modulation of prefrontal cortical interneurons changes during adolescence. *Cerebral Cortex*, **17**, 1235–1240.

Tseng, K.Y., Mallet, N., Toreson, K.L., Le Moine, C., Gonon, F., and O'Donnell, P. (2006) Excitatory response of prefrontal cortical fast-spiking interneurons to ventral tegmental area stimulation *in vivo*. *Synapse*, **59**, 412–417.

Tseng, K.Y., Lewis, B.L., Lipska, B.K., and O'Donnell, P. (2007) Post-pubertal disruption of medial prefrontal cortical dopamine-glutamate interactions in a developmental animal model of schizophrenia. *Biological Psychiatry*, **62**, 730–738.

Tseng, K.Y., Lewis, B.L., Hashimoto, T., Sesack, S.R., Kloc, M., Lewis, D.A. *et al.* (2008) A neonatal ventral hippocampal lesion causes functional deficits in adult prefrontal cortical interneurons. *Journal of Neuroscience*, **28**, 12691–12699.

Wang, J. and O'Donnell, P. (2001) $D_1$ dopamine receptors potentiate NMDA-mediated excitability increase in rat prefrontal cortical pyramidal neurons. *Cerebral Cortex*, **11**, 452–462.

Weickert, C.S., Webster, M.J., Gondipalli, P., Rothmond, D., Fatula, R.J., Herman, M.M. *et al.* (2007) Postnatal alterations in dopaminergic markers in the human prefrontal cortex. *Neuroscience*, **144**, 1109–1119.

Zhu, J.J. (2000) Maturation of layer 5 neocortical pyramidal neurons: amplifying salient layer 1 and layer 4 inputs by $Ca^{2+}$ action potentials in adult rat tuft dendrites. *Journal of Physiology*, **526**, 571–587.

# Part Three
# Research in Molecular Psychiatry

Chapter 8 by Arian Mobascher and Georg Winterer demonstrates the crucial role of the cholinergic system in mental disorders. This system is one of the six major neurochemical transmission systems and displays a special relationship to nicotine addiction. Special focus has been devoted to nicotinic cholinergic receptor signaling in the synapse.

Chapter 9 by Ingo Vernaleken *et al.* provides an excellent overview on the technology of molecular imaging. Examples of the specific anatomical localizations of serotonin, dopamine, and the respective receptors of serotonin, dopamine, and so on, are discussed. Step by step, the molecular network involved in mental disorders can be discovered by these molecular imaging methods. They may open up new perspectives of personalized drug treatments.

Chapter 10 by Boris Tabakoff *et al.* describes some current strategic problems of genetic psychiatry, with a focus here on the genetics of addiction.

# 8
# Nicotinic Cholinergic Signaling in the Human Brain – Systems Perspective

*Arian Mobascher and Georg Winterer*

## 8.1
## Introduction

Brain function in health and disease can be studied on multiple levels, like the molecular, cellular, neuronal network and behavioral or symptom levels. Over recent decades, neuroscience has provided a huge amount of information about brain function on these various levels. At the same time, neuroscientists have found it increasingly difficult to put all the pieces of information together that they have generated about neurotransmitters, their receptors, intracellular signaling, cellular electrophysiology, brain development and anatomy, and brain diseases. Therefore, the ultimate goal of neuroscience – to understand how the brain works – is far from being accomplished. It has been argued that a systemic view – looking at the different levels of brain function – is needed to integrate all the pieces of information together to accomplish that goal. The emerging fields of systems neuroscience and computational neuroscience are trying to address this issue of different levels of complexity that make it hard to put a certain piece of scientific information into the broader context of brain/neuronal network function. In systems neuroscience, the brain is conceptualized as a system that consists of various subsystems. One of the major goals of this research area is to generate computer models of these (sub)systems to study the effects of manipulating one subsystem on the higher-order systems.

From a more general, biological point of view: "Systems biology is the study of the behavior of complex biological organization and processes in terms of the molecular constituents" (Kirschner, 2005). It addresses the fundamental question of how the phenotype is generated from the genotype and acknowledges the fact that the classic paradigm of molecular biology (i.e., one gene–one enzyme–one function) is a simplistic view that is not sufficient to understand the biological function of many if not most genes and gene products. In other words, in systems biology it is recognized that "the diversity of genes cannot approximate the diversity of functions within an organism" (Kirschner, 2005). Instead, systems biologists emphasize the importance of the biological context when a certain genetic variation or pharmacological manipulation is studied. This is also a very important consideration for neuropsychiatric research. The genetic contribution to mental illness is a good example. Many major psychiatric diseases have a high heritability (e.g., schizophrenia and bipolar disorder). Schizophrenia shows several, quite distinct (clinical) phenotypes and in bipolar disorder, another phenotype is seen. At the same time, however, bipolar disorder and schizophrenia (regardless of the clinical appearance) seem to share susceptibility genes, suggesting that it is the context (neurodevelopmental, other biological, or psychosocial context) that may shape the ultimate phenotype – at least to some extent. This example points out that future research on the pathogenesis and therapy of psychiatric diseases will benefit from a systems approach as a clinical syndrome or disease (phenotype) or drug effect will in most cases not be accurately and comprehensively predicted by the genotype or the molecular/pharmacodynamic effects of a compound alone.

In the subsequent sections of this chapter we look at the nicotinic cholinergic neurotransmission in the brain. It exemplifies how (genetic or pharmacological) manipulation of a system can look quite differently, depending on which level of complexity is studied. We will first summarize the present knowledge about pharmacological manipulation of nicotinic acetylcholine receptors (nAChRs). We then describe the (intra)cellular effects of receptor stimulation by nicotine and look at three psychophysiological systems that are modulated by nicotinic cholinergic neurotransmission: the reward system, the stress response system, and cognition, namely learning and memory and attention. Finally, we summarize how genetic variation in the nicotinic system may affect genetic liability to human disorders. First, however, the clinical relevance of this transmitter system needs to be pointed out.

## 8.2
### Epidemiological Relevance of the Nicotinic Cholinergic System

The physiological neurotransmitter acetylcholine can bind to two classes of receptor: (i) ligand-gated ion channels that can be pharmacologically activated by nicotine, an alkaloid derived from the tobacco plant, and (ii) to G-protein-coupled receptors that can be stimulated by muscarine, a compound that is derived from a poisonous mushroom species. Only the nicotinic system will be further discussed in the following sections.

Nicotine is the major psychoactive compound of tobacco smoke. In most western societies, 25–30% of the population are regular tobacco smokers (Andreas and Loddenkemper, 2007; Andreas et al., 2007; Raupach et al., 2007). Smoking cigarettes on a regular basis reduces the average life expectancy by about 10 years (Raupach et al., 2007; Andreas et al., 2007). Every second smoker dies from smoking-related causes such as lung cancer, emphysema, and coronary artery disease (Raupach et al., 2007; Andreas et al., 2007). Therefore, smoking is the number one cause of early, avoidable deaths in developed countries (Welte, König, and Leidl, 2000). Nicotine addiction is the major reason for smokers to keep smoking despite their desire to quit. About one-third of smokers try to quit each year, but only a minor fraction of attempts is successful due to the addictive effects of nicotine (Batra, 2004).

There is even some evidence that nicotine is as addictive as some "hard" drugs like heroine. Twin studies suggest a genetic impact on smoking behavior with a heritability of generally over 60% (Carmelli *et al.*, 1992;Edwards, Austin, and Jarvic, 1995). Like in other behavioral traits and psychiatric disorders, the genetics of smoking behavior and nicotine addiction is complex-polygenic – a number of genes, each explaining a small fraction of the total variance, contribute together to the genetic liability of nicotine addiction (Mineur and Picciotto, 2008). At the same time, environmental factors are considered to be important. Some data point towards a leading role of environmental factors in the early stages of nicotine addiction, whereas genetic factors are more important later on (Heath *et al.*, 1993; Heath and Martin, 1993). However, there is also evidence that neuroplastic changes in the brain induced by nicotine may already start in the early stages of nicotine addiction (Mansvelder and McGee, 2000). In any case, nicotine is highly addictive. The aggressive antismoking campaigns and rigorous nonsmoker protection laws are starting to pay off in that the overall fraction of smokers is starting to fall in most western societies. However, a fraction of smokers – mostly heavy smokers – does not seem to benefit from these campaigns or the cessation programs and treatments that are available today. This is particularly true for a subgroup of smokers with clinical (or subclinical) psychiatric symptoms. Schizophrenia is the most striking example in this context. The proportion of smokers in schizophrenia is about 80% in most studies compared to the 25–30% in the general population (e.g., de Leon *et al.*, 1995). Acetylcholine – the physiological ligand of nAChRs – is known to play a critical role in cognitive processes such as memory and learning. Therefore, it has been proposed that schizophrenic smokers may self-medicate symptoms of their disease using nicotine. The notion of an important role of nicotinic cholinergic neurotransmission in schizophrenia is supported by several lines of evidence: (i) nicotine improves cognitive function in laboratory animals and humans that have not been exposed to nicotine before (Levin, McClernon, and Rezvani, 2006; Mansvelder *et al.*, 2006), (ii) the abundance of nAChRs in postmortem brain tissue of schizophrenics seems to be altered (Freedman *et al.*, 1995; Durany *et al.*, 2000; Freedman, Adams, and Leonard, 2000), and (iii) there is some genetic evidence linking the $\alpha_7$ subunit of nAChR subunits to schizophrenia (Harrison and Weinberger, 2005; Leonard *et al.*, 2002; Adams and Stevens, 2007). Therefore, a better understanding of the effects of nicotine on neurons and neuronal networks may not only lead to better cessation programs and pharmacological aids to quit smoking, but may also help to develop future treatments for schizophrenia.

## 8.3
## nAChrRs and the Cellular Effects of Nicotine

The effects of nicotine on neurons are mediated by nAChRs. They are pentameric protein complexes with a central channel for cations. Multiple genes (*CHRNA2–10* and *CHRNB2–4*) code for the respective receptor subunits $\alpha_2$–$\alpha_{10}$ and $\beta_2$–$\beta_4$ (Gotti, Zoli, and Clementi, 2006). The pentameric receptors are either homomeric (meaning

$\alpha_4\beta_2$nAChR
→ Depolarizing current

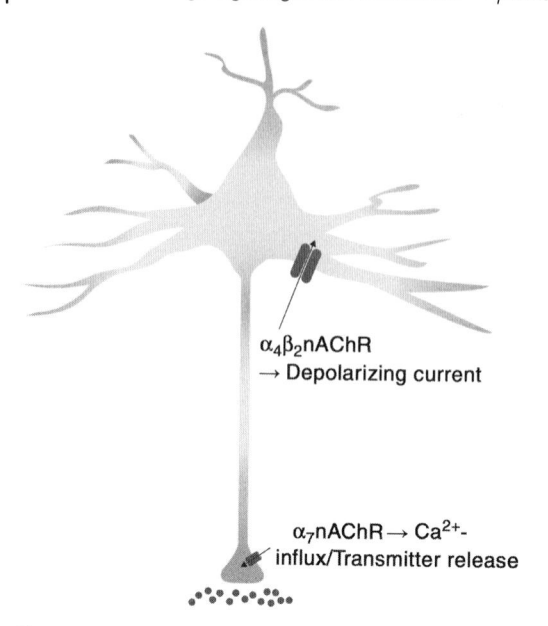

$\alpha_7$nAChR → $Ca^{2+}$-influx/Transmitter release

**Figure 8.1** Typical localization of common nAChR subtypes and net effect of their stimulation.

they are composed of only one type of α receptor subunits) or heteromeric (meaning that they contain α as well as β subunits). The most abundant receptors in the central nervous system are the homomeric $\alpha_7$ and the heteromeric $\alpha_4\beta_2^*$ receptors (Mineur and Picciotto, 2008). The latter consist of at least two $\alpha_4$ subunits and two $\beta_2$ subunits, and may contain a third subunit type completing the pentamer. nAChR are located at different sites of a neuron (i.e., presynaptic terminals, cell bodies, and dendrites) (Figure 8.1). Three functional states of nAChRs can be differentiated: (i) at rest, (ii) activated (ion flow through the channel), and (iii) desensitized (a state in which the receptor cannot be activated by its ligand or an agonist like nicotine). Upon stimulation of nAChRs the central channel opens for the cations sodium, potassium and calcium. Thus, nAChRs are nonselective. Upon stimulation of presynaptic nAChRs a $Ca^{2+}$ influx occurs that triggers an increased release of neurotransmitters (Figure 8.1). A stimulation of somatic or dendritic receptors induces a depolarizing current that is mostly driven by sodium influx and to a lesser extent by calcium influx (depending on the subunit composition of the receptor). The result is an increase in the neuron's excitability and firing rate (Wonnacott, Sidhpura, and Balfour, 2005) (Figure 8.2). By these mechanisms nAChRs modulate numerous neuronal systems, and cellular and cognitive processes such as the reward system, the attentional network, learning and memory, brain development, and neuroprotection (Mineur and Picciotto, 2008). Thus, the physiological effects of nicotinic cholinergic neurotransmission as well as the effects of exogenous nicotine depend on the receptor subtype, the localization of the stimulated receptor on the neuron (presynaptic, somatic, dendritic), the type of neuron that is stimulated (nAChRs are found on

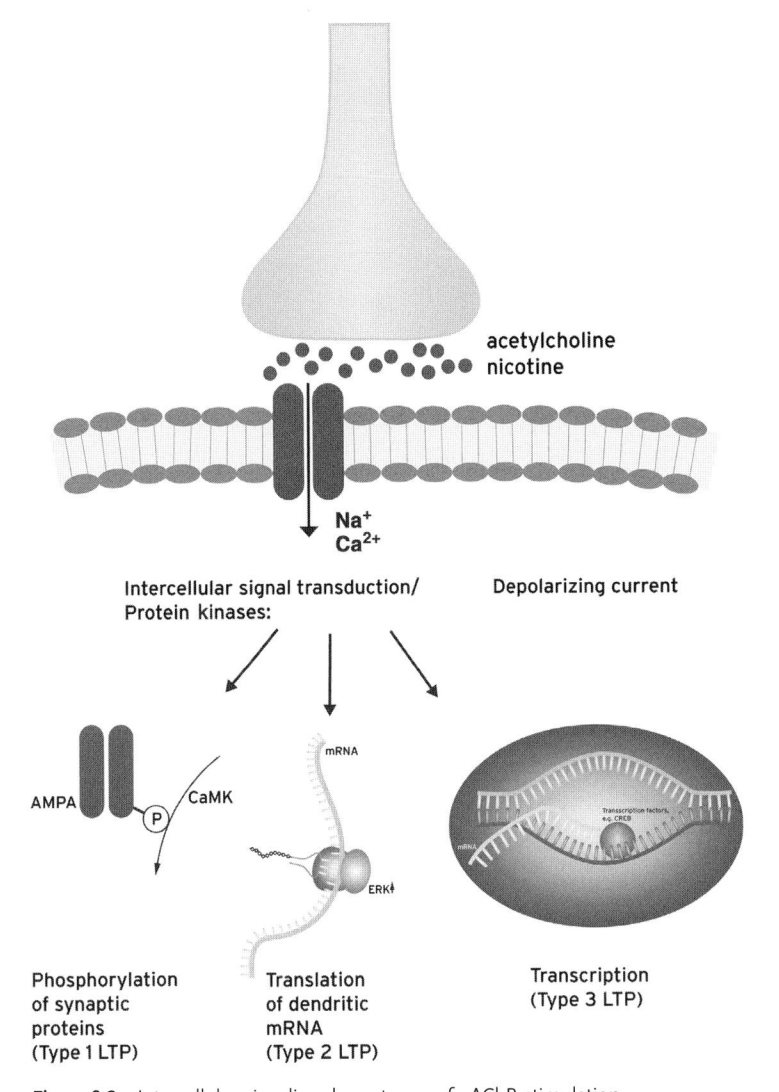

**Figure 8.2** Intracellular signaling downstream of nAChR stimulation.

various neurons such as γ-aminobutyric acid (GABA)-ergic, glutamatergic, dopaminergic, and cholinergic neurons), and the functional-anatomical system the stimulated neuron is integrated in (e.g., reward system, attentional network, learning and memory, and stress response system).

The stimulation of nAChRs by the physiological ligand acetylcholine is characterized by rapid exposure to the ligand, high peak concentrations, and termination of the stimulation within milliseconds because acetylcholine is rapidly cleaved and thereby inactivated by the enzyme cholinesterase. Under these physiological conditions nAChRs are either in their resting state or they are activated. Stimulation of nAChRs by nicotine shows substantially different kinetics. Smoking of one cigarette causes a longer exposure of the receptor to lower doses of the agonist compared to physiological stimulation with acetylcholine because (i) nicotine is provided for several minutes and (ii) nicotine is not inactivated by the enzyme cholinesterase. This nonphysiological, continuous, low-dose stimulation of nAChRs facilitates the inactivated/desensitized conformation of the receptor. This receptor desensitization is the most common model to explain tolerance to nicotine on the physiological and behavioral level (Robinson et al., 2007).

Long-term exposure of a receptor to elevated levels of its physiological ligand or a pharmacological agonist usually leads to receptor endocytosis and downregulation. The effect of long-term exposure to nicotine on nAChRs is a remarkable exception to this rule. In fact, heteropentameric $\beta_2$-containing (mostly $\alpha_4\beta_2^*$) nicotinic receptors are upregulated upon chronic nicotine exposure. This may go along with a potentiation of the effects of nicotine on cellular, network, and behavioral levels dependent on the receptor's conformation and activation/desensitization state.

The phenomenon of receptor upregulation has been explained in several ways. Most data point towards a post-translational mechanism (i.e., stabilization of nAChR protein complexes) as mRNA levels of nAChR subunits remain unchanged when a neuron is exposed to nicotine. Peng et al. (1994) suggested that desensitized receptors on the cell surface may be stabilized. An alternative mechanism has recently been postulated by Sallette et al. (2005). Unlike the physiological ligand acetylcholine, nicotine can penetrate the cell membrane and may therefore have intracellular effects, too. Recently, the authors showed that intracellular nicotine promotes proper folding and assembly of heteromeric, $\beta_2$-containing nAChR subunits. Thereby, the fraction of properly folded and assembled pentameric receptor complexes that are able to pass the "quality check" in the endoplasmic reticulum (ER) and leave the rough ER via the medial Golgi apparatus is increased. This means that more $\beta_2$-containing nAChRs eventually reach their destination – the cell membrane. It is estimated that under physiological conditions only about 15% of synthesized AChR subunits escape from ER-associated proteasomal degradation and end up in functional pentameric nAChRs on the cell surface. Sallette et al. (2005) suggest that the chaperone-like activity (chaperones are molecules that help proteins to fold properly) of nicotine increases the fraction of functional receptors on the cell surface. This phenomenon of nAChR upregulation upon chronic nicotine exposure may contribute to some aspects of nicotine addiction such as sensitization and withdrawal symptoms during abstinence from nicotine. In this scenario, an increased number of nAChRs may cause cholinergic hyperexcitability that may be associated with withdrawal symptoms (Dani and De Biasi, 2001; Staley et al., 2006). Yet another level of complexity to the problem of the effects of nicotine on neurons was recently added by Rezvani et al. (2007). The authors provided evidence that nicotine inhibits the proteasome (a multisubunit molecular machine that is the major protease for cellular proteins) and thereby stabilizes a number of synaptic proteins like $\alpha_7$ nAChRs as well as $\alpha$-amino-3-hydroxy-5-methyl-4-isoxazol-propionacid (AMPA), N-methyl-D-aspartate (NMDA), and metabotropic glutamate receptor subunits, and PSD95, which is a scaffolding protein of the postsynaptic density.

In addition to the nicotine effect on nAChR abundance, chronic nicotine causes further plastic changes that may alter the neurophysiological properties of neurons. Phenomena like synaptic plasticity (e.g., long-term potentiation (LTP) and long-term depression (LTD) upon nicotine exposure) have been observed in several neuronal networks (Mansvelder and McGee, 2000; Ji, Lape, and Dani, 2001; Couey et al., 2007). Network-specific aspects (reward system, learning and memory, and attention) are discussed in the subsequent sections of this book. However, basic cellular aspects of nicotine-mediated neuronal/synaptic plasticity need to be addressed first.

The majority of plastic changes in neurons like LTP are affected by intracellular $Ca^{2+}$-dependent signaling (Raymond, 2007). With respect to LTP, three different types can be distinguished that are regulated by intracellular $Ca^{2+}$ in different ways: type 1 = neither transcription- nor translation-dependent, type 2 = dependent on translation but not on transcription, and type 3 = dependent on transcription and translation. In type-1 LTP, neuroplastic changes are induced solely by protein–protein interactions, mostly phosphorylation events that are catalyzed by protein kinases. In type 2, new proteins are translated from pre-existing dendritic mRNA. In type 3 plastic changes, genomic DNA is transcribed into mRNA and new proteins are subsequently synthesized. In the latter, the signal transduction machinery downstream of nicotine involves activation of transcription factors that regulate gene expression. Nicotine can increase intracellular $Ca^{2+}$ in different ways and thereby induce the above-mentioned processes. (i) Activation of nAChRs, especially of $\alpha_7$ receptors that show high $Ca^{2+}$ influx through the open channel, may directly trigger $Ca^{2+}$-dependent intracellular processes. (ii) nAChR (e.g., $\alpha_4\beta_2^*$ subtype) receptor-mediated depolarization events may open voltage-gated ion channels. $Ca^{2+}$-influx through these channels may then trigger $Ca^{2+}$-dependent plastic events (Wonnacott, Sidhpura, and Balfour, 2005). (iii) Activation of presynaptic nAChRs at glutamatergic and dopaminergic synapses may induce release of these transmitters, and thereby induce plastic changes at this synapse and in the postsynaptic neuron (Mansvelder and McGee, 2000; Chen et al., 2007). For instance, dopamine-mediated (i.e., by means of nicotine-induced activation of dopaminergic neurons) plastic effects of nicotine are relevant in the nucleus accumbens (NAcc, a central structure in the reward system) (Raymond, 2007; Inoue et al., 2007; Di Chiara et al., 2004). $Ca^{2+}$-dependent signal transduction processes that induce the above-mentioned plastic changes are mediated by a number of protein kinases (Raymond, 2007) and downstream transcription factors (Raymond, 2007; Nakayama et al., 2001). Stimulation of nAChRs can activate at least three different signal transduction cascades: (i) the calcium/calmodulin-dependent kinase (CaMK) pathway, (ii) the extracellular signal-regulated kinase (ERK)/mitogen-activated protein kinaseMAPK) pathway, and (iii) the phosphoinositide 3-kinase(PI3K)/protein kinase B(Akt)/glycogen synthase kinase-3β (GSK-3β) pathway (Sugano et al., 2006; Steiner, Heath, and Picciotto, 2007). These three pathways are involved in synaptic plasticity (the neuronal correlate of cognitive processes like learning and memory) and they have been linked to schizophrenia directly or to other human diseases that are accompanied by cognitive deficits, such as certain forms of inherited mental retardation (Weeber and Sweatt, 2002). Nicotine activates CaMKs – a process that is $Ca^{2+}$-dependent. The CaMK pathway is involved in early- and late-phase LTP (Raymond, 2007; Steiner, Heath, and Picciotto, 2007). For instance, CaMKII phosphorylates AMPA-type glutamate receptors that results in increased conductance/ion flow through the receptor, a form of type 1 LTP. Furthermore, the CaMK pathway may induce type 2 LTP via ERK-mediated translational activation (Kelleher, Govindarajan, and Tonegawa, 2004). Finally, CaMKs are involved in late LTP (type 3) via activation of transcription factors that are involved in synaptic plasticity. cAMP-response element binding protein (CREB) is a prominent example (Schmitt et al., 2005). Nicotine induces $Ca^{2+}$-dependent ERK phosphorylation/activation (Sugano et al., 2006; Steiner, Heath, and Picciotto, 2007). Dependent on the cell type, protein kinase A, protein kinase C, and CaMKs have been shown to be upstream activators of ERK. It seems that ERK functions in at least two distinct neuronal compartments, Activated ERK has been shown to be an activator of the general translational machinery, thereby promoting dendritic mRNA translation – the cell biological correlate of type 2 LTP. A second pool of phosphorylated ERK may activate downstream transcription factors such as CREB and thereby promote type 3 LTP (Weeber and Sweatt, 2002). Of note, dysregulation of the ERK/CREB pathway has been implicated in mental retardation associated with neurofibromatosis type 1 and Rubinstein–Taybi Syndrome (Weeber and Sweatt, 2002). At this point it is quite unclear which nAChR subtypes are most important for activation of the ERK pathway. However, it has recently been shown by Bitner et al. (2007) that $\alpha_7$ nAChRs (stimulated by an $\alpha_7$-selective agonist) are sufficient to activate the ERK/CREB pathway. Interestingly, in this study the agonist also improved a variety of cognitive domains in laboratory animals such as short-term memory, long-term memory consolidation, and sensory gating. Furthermore, nicotine activates the PI3K/Akt/GSK-3$\beta$ pathway (Sugano et al., 2006; Kihara et al., 2001). It has been suggested that this pathway may be activated via ERK-mediated upregulation of the insulin receptor substrate-I and -II proteins that function as activators of the PI3K and ERK pathways (Sugano et al., 2006). Alternatively, PI3K may be activated more directly by nicotine as Kihara et al. (2001) have shown that $\alpha_7$ nAChRs and PI3K can be found in the same protein complex. One of the major downstream substrates of PI3K is Akt. This protein kinase has a variety of cellular substrates (Manning and Cantley, 2007). GSK-3$\beta$ is the first protein that was identified as a target of Akt, and is the best studied Akt substrate in the context of synaptic plasticity, cognition, and psychiatric disease (Jope and Roh, 2006). GSK-3$\beta$ has high basal activity. It phosphorylates more than 40 substrates and is inhibited by Akt-mediated phosphorylation (Jope and Roh, 2006). The PI3K/Akt/GSK-$\beta$ pathway has been shown to be involved in a variety of fundamental cellular processes in neurons, such as cell survival, neuronal polarity, and synaptic plasticity (Jope and Roh, 2006; Wang et al., 2003; Peineau et al., 2007). It is important to note that this cascade is involved in intracellular signaling downstream of several neurotransmitters, such as dopamine (Beaulieu et al., 2007), glutamate (Yoshii and Constantine-Paton, 2007; van der Heide, Ramakers, and Smidt, 2006), and acetylcholine (Sugano et al., 2006; Kihara et al., 2001), that have all been implicated in schizophrenia.

Furthermore, it may act downstream of neuregulin/Erb signaling (Kanakry et al., 2007). Genes coding for components of the neuregulin/Erb pathway (i.e., *NRB1* and *ERBB4*) are among the most promising susceptibility genes for schizophrenia (Harrison and Weinberger, 2005). Consistent with a role of the PI3K/Akt/GSK-β pathway in schizophrenia, Akt1 itself may also be a susceptibility gene for schizophrenia (Lang et al., 2007). Furthermore, Akt protein levels seem to be reduced in postmortem brains of patients with schizophrenia (Emamian et al., 2004). The Akt substrate GSK-β has also been implicated in the pathogenesis of schizophrenia (Jope and Roh, 2006). However, at this point conflicting data exist regarding the role of GSK-3β in schizophrenia. While some studies point towards increased GSK-3β protein activity in schizophrenia, others suggest the opposite (Jope and Roh, 2006). In any case, the PI3K/Akt/GSK-3β pathway seems to be an attractive target for future drug development as suggested by the fact that lithium is a GSK-3β inhibitor. Some of the main intracellular signal transduction cascades downstream of nicotinic cholinergic neurotransmission are depicted in Figure 8.2.

## 8.4
## Nicotine and Cognition

Nicotine modulates cognitive processes like learning/memory and attention in nicotine-addicted subjects as well as humans and laboratory animals that have never been exposed to nicotine before (Levin, McClernon, and Rezvani, 2006; Mansvelder et al., 2006). The majority of studies suggest an improvement of cognitive performance under nicotine. The neurophysiological correlate of learning and memory are modifications in the efficiency of synaptic transmission (LTP and LTD) in the hippocampus, a structure that is critically involved in the pathophysiology of schizophrenia, together with the prefrontal cortex and the striatum. nAChRs are located at pre- and postsynaptic sites on hippocampal pyramidal cells and interneurons. Depending on the experimental conditions, nicotine can induce LTP or LTD in the hippocampus – a fact that may explain the beneficial effects of nicotine on learning and memory (Ji, Lape, and Dani, 2001; Fujii et al., 1999; Mann and Greenfield, 2003; Ge and Dani, 2005). Furthermore, administration of nAChR antagonists in the hippocampus has a negative impact on cognitive performance (Ohno, Yamamoto, and Watanabe, 1993). Both, $\alpha_4\beta_2^*$ and $\alpha_7$ receptors seem to be involved in the plastic processes in the hippocampus that are thought to be neuronal correlates of learning and memory (Levin, McClernon, and Rezvani, 2006; Mansvelder et al., 2006).

Acute nicotine administration modulates the attentional network consisting of the prefrontal cortex, the anterior cingulate cortex (ACC) and parieto-temporal association areas, and may improve attention not only in nicotine addicted subjects during withdrawal, but also in nicotine naïve probands (Edwards, Austin, and Jarvic, 1995; Hahn, Shoaib, and Stolerman, 2002; Hahn et al., 2003; Harris et al., 2004; Houlihan et al., 1996; Houlihan, Pritchard, and Robinson, 1996; Rezvani and Levin, 2001; Sacco, Bannon, and George, 2004; Sherwood, Kerr, and Hindmarch, 1992;

Stolerman et al., 2000; Wesnes and Warburton, 1983). Interestingly, functional magnetic resonance imaging (fMRI) studies provided several possible explanations for this phenomenon: some results point towards an activation (as defined by a higher blood oxygen level-dependent (BOLD) signal) of the attentional network (Thiel, Zilles, and Fink, 2005; Lawrence, Ross, and Stein, 2002), but other effects may play a role. For instance, Thiel, Zilles, and Fink (2005) reported a nicotine-induced regional deactivation in the parietal cortex in response to stimuli outside the current focus of attention. Hahn et al. (2007) provided evidence that nicotine may induced a deactivation of the "default mode network" – a network of brain regions that is active at rest and becomes deactivated during the performance of specific tasks. This deactivation of the default mode network correlated with the improvement of attention upon nicotine exposure. The effects of nicotine on prefrontal functioning related to attention and working memory are less well studied on the cellular and network level. Until recently a nicotine-induced activation of the prefrontal cortex seemed to be the most likely mechanism associated with improved attention. However, an electrophysiological study published recently points in another direction. Couey et al. (2007) provided evidence that the key effect of nicotine on prefrontal functioning may not be "activation," but improvement of the signal-to-noise ratio in prefrontal microcircuits (Figure 8.3). In this study nicotine increased the activity of

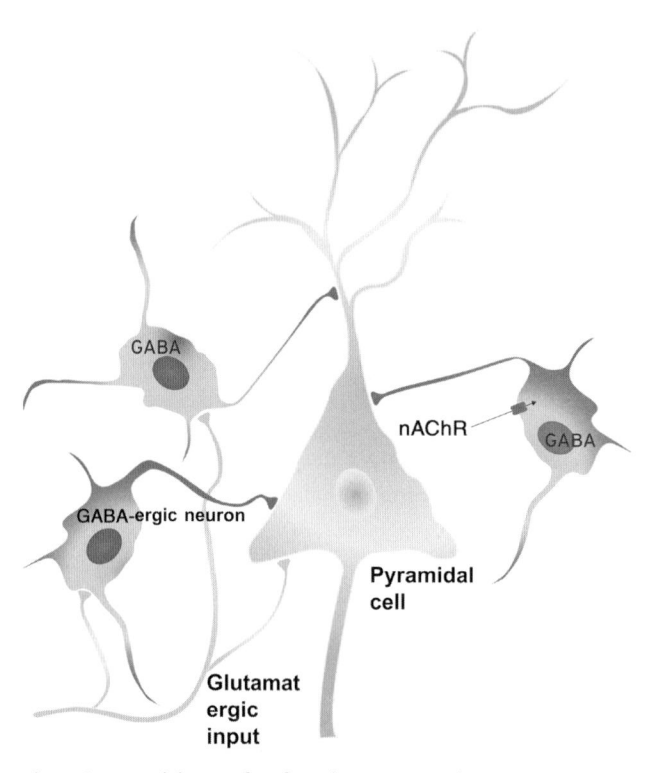

**Figure 8.3** Modulation of prefrontal microcircuits by nicotine.

inhibitory GABA-ergic interneurons, which lead to an elevated threshold for the occurrence of a specific type of synaptic plasticity (i.e., spike-timing-dependent-plasticity). The $\beta_2$-containing as well as $\alpha_7$ nAChRs on GABA-ergic interneurons in the prefrontal cortex seemed to be involved in this phenomenon (Couey *et al.*, 2007). In this study functional nAChRs were not found in prefrontal layer 5 pyramidal cells. However, other physiological processes may be involved in nicotine-induced improvement of attention. For instance, nicotine may modulate glutamateric or dopaminergic input from other brain areas. A number of studies from our group suggest that the signal-to-noise ratio in the prefrontal cortex of schizophrenics may be reduced (Winterer *et al.*, 2004, 2006b, 2006c; Rolls *et al.*, 2008). Thus, this group of people may smoke in order to optimize the signal-to-noise ratio in the prefrontal cortex, which could improve cognitive functioning.

Data from a number of clinical studies suggest that the beneficial effects of nicotine on attention may be pronounced in patients with impaired cognition. Patients with schizophrenia or attention deficit hyperactivity disorder (ADHD) smoke more frequently and also more heavily than the general population (de Leon *et al.*, 1995; Pomerleau *et al.*, 1995; Lasser *et al.*, 2000). It seems that that nicotine-induced improvement of cognitive performance may have some clinical/behavioral relevance, especially in this subset of smokers (Levine *et al.*, 1996, 2001; Adler *et al.*, 1993; Smith *et al.*, 2002; Harris *et al.*, 2004). Thus, smoking may serve the purpose of self-medication in smokers with ADHD or schizophrenia.

The effects of chronic nicotine on attention have hardly been studied in humans. Data from our group suggest that prefrontal activation may be diminished in smokers during cognitive task performance (Musso *et al.*, 2007).

## 8.5
## Nicotine and Reward

Most addictive drugs induce a release of dopamine in the NAcc – a part of the ventral striatum (Ikemoto, 2007) (Figure 8.4). Dopamine release in the NAcc correlates with the subjective experience of reward, and promotes addiction in humans and corresponding behavior in animal models such as self-administration or conditioned place preference. The NAcc receives dopaminergic input form a structure in midbrain, the ventral tegmental area (VTA). The dopaminergic projections from the VTA to the NAcc are the central structure of the mesolimbic reward system. Dopaminergic VTA neurons themselves receive excitatory glutamatergic input from the prefrontal cortex, and cholinergic input from brainstem and midbrain nuclei. Inhibitory input is provided by GABA-ergic neurons (Wonnacott, Sidhpura, and Balfour, 2005) (Figure 8.4). nAChRs are located on the cell bodies of dopaminergic VTA neurons, and in the presynaptic terminals of afferent GABA-ergic and glutamatergic neurons. Dopaminergic VTA neurons are activated by systemic administration as well as local application of nicotine and the release of dopamine in the NAcc is increased upon nicotinic stimulation (Wonnacott, Sidhpura, and Balfour, 2005; Pidoplichko *et al.*, 1997; Ferrari *et al.*, 2002; Balfour *et al.*, 2000). Nicotine modulates

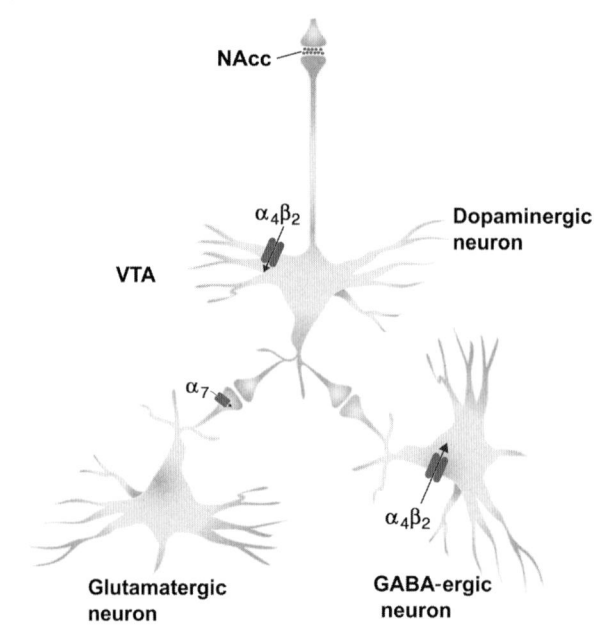

**Figure 8.4** Effects of nicotine on the reward system.

the activity of dopaminergic VTA neurons in at least three different ways: (i) direct activation via $\alpha_4\beta_2^*$ receptors on dopaminergic VTA neurons (Pidoplichko et al., 1997), (ii) increase of excitatory input via presynaptic $\alpha_7$ receptors on glutamatergic neurons (Mansvelder and McGee, 2000), and (iii) increase of inhibitory input via $\beta_2$-containing receptors on GABA-ergic neurons (Mansvelder, Keath, and McGehee, 2002). Mechanisms (i) and (ii) work in the same direction, and result in a net increase in activity of dopaminergic VTA neurons and subsequent dopamine release in the NAcc. In addition, the concerted pre- and postsynaptic stimulation promotes plastic changes in the system. The inhibitory input from GABA-ergic afferences is of short duration as the $\beta_2$-containing nAChRs on these neurons desensitize rapidly (Mansvelder, Keath, and McGehee, 2002).

The *in vivo* effects of chronic nicotine exposure on the reward system are less well understood (Besson et al., 2007). Most likely adaptive processes occur: receptor desensitization and upregulation of nAChRs on GABA-ergic VTA neurons, on one hand, and $\alpha_7$-receptor-mediated plastic changes, on the other hand, that balance one another in a way that causes little neurophysiological and behavioral abnormalities as long as nicotine is administered. Upon cessation of nicotine application, cholinergic understimulation occurs with the consequence of a hypodopaminergic state of the reward system that may cause withdrawal symptoms such as anhedonia and depression. There is a lot of circumstantial evidence that links nicotine addiction to mood regulation and affective disorders (Glassman et al., 1988; Keuthen et al., 2000;

Breslau, Kilbey, and Andreski, 1993; Breslau and Johnson, 2000). The reward system can be thought of as the anatomical link between nicotine (but also other) addiction(s) and mood regulation (Heinz, Schmidt, and Reischies, 1994). Therefore, it has been argued that nicotine consumption in depressed (or subclinically depressed smokers) may reflect an attempt to self-medicate depressive symptoms via dopaminergic stimulation of the reward system (Cardenas et al., 2002). The same may be true for smoking schizophrenics who frequently suffer from affective symptoms.

## 8.6
## Nicotine and Stress Response

Nicotine modulates the stress response system. Conversely, genetic variations (and alterations of the stress response system that have been acquired during life) impact on the effects of nicotine on the central nervous system and may thereby affect the risk to become a nicotine-addicted smoker.

The key structures of the stress response system are the hypothalamus, the pituitary and the adrenal gland, often called the hypothalamic–pituitary–adrenal (HPA) axis, which is controlled by autoregulation through steroids and by inputs from various brain regions, including monoaminergic nuclei, limbic structures, and the prefrontal cortex. It has been shown that nicotine causes a central, hypothalamic activation of the HPA axis and increases cortisol secretion (Al Abisi, 2006; Lovallo, 2006; Rohleder and Kirschbaum, 2006). However, the cellular effects of nicotine on the HPA axis are not well understood. There is evidence for a direct stimulation of the hypothalamus via hypothalamic nAChRs (Fuxe et al., 1989), but also for a nicotinic modulation of the monoaminerigic input in the hypothalamus (Matta et al., 1998). Upon chronic nicotine exposure, homeostatic adaptations in the system occur; however, the HPA axis can still be activated by nicotine.

The state of nicotine withdrawal is characterized by a blunted response of the HPA axis to stress. The alterations in stress response in the HPA axis seem to correlate with the risk of relapse in abstinent smokers (Al Abisi, 2006). The stress response system in smokers in early abstinence is dysregulated. Therefore, psychosocial stress causes an increase in nicotine consumption in smokers and is one of the most important predictors of relapse in abstinent smokers. There is converging evidence from human and animal studies that elevated stress sensitivity increases the risk and severity of nicotine dependence (Thorndike et al., 2006; Fagen et al., 2007). For instance, post-traumatic stress disorder is associated with an elevated risk for nicotine abuse (Thorndike et al., 2006). Fagen et al. (2007) recently reported that environmental or pharmacologically induced stress results in a state of increased susceptibility of the reward system to nicotinic stimulation. Thus, there may be a connection between stress response and reward system. Furthermore, it is suggested by data from animal studies that addictive behavior and stress response may be regulated by a common or at least overlapping set of genes (Fagen et al., 2007; Bilkei-Gorzo et al., 2008).

## 8.7
### Variation in nAChrR Genes and Smoking

Twin studies suggest that the heritability (the fraction of the total variance of a trait/phenotype/disorder in a population that is explained by genetic factors) of habitual cigarette smoking and nicotine addiction is high (Carmelli *et al.*, 1992; Edwards, Austin, and Jarvic, 1995; Sullivan and Kendler, 1999; Lessov *et al.*, 2004; Lessov-Schlagger *et al.*, 2008; Maes *et al.*, 2004). Considering the key role of nAChRs in nicotinic cholinergic neurotransmission, genes encoding for nAChR subunits are among the most likely risk genes for nicotine addiction. This view is also supported by experimental data obtained in mutant mice in which genes encoding for individual nAChR subunits have been knocked out. For instance, it has been shown that $CHRNB2^{-/-}$ mice do not self-administer nicotine (Epping-Jordan *et al.*, 1999).

Over the last view years some evidence has accumulated that genes encoding for nAChR subunits contribute to the genetic liability to nicotine addiction in humans. The most consistent data have been obtained for chromosomal region 15q24/15q25.1. This region contains the *CHRNA3–CHRNA5–CHRNB4* nAChR subunit gene cluster. An association of this gene cluster with smoking has been found in a number of candidate gene and genome-wide association studies (Berrettini *et al.*, 2008; Hung *et al.*, 2008; Thorgeirsson *et al.*, 2008; Chanock and Hunter, 2008; Bierut *et al.*, 2007, 2008; Saccone *et al.*, 2007). However, Breitling *et al.* (2009a) were not able to show that genetic variation in this region is also associated with smoking cessation.

Interestingly, in the three genome-wide association studies published by Hung *et al.*, Thorgeirsson *et al.* and Amos *et al.*, an association between single nucleotide polymorphisms (SNPs) in this gene cluster and lung cancer was also reported. However, it remains to be clarified if that association is mediated by nicotine addiction and tobacco abuse or independent of smoking behavior.

Certain aspects of smoking behavior may also be mediated by other nAChR subunit genes, namely *CHRNA6*, *CHRNB3* (Zeiger *et al.*, 2008), and *CHRNA4*. As the $\alpha_4\beta_2$ receptor subtype is the most abundant high-affinity nicotinic receptor in the brain and as transgenic mouse models suggest that it is critically involved in the effects of nicotine on brain function, it is no surprise that there is also accumulating evidence that polymorphisms in *CHRNA4* (the gene coding for the $\alpha_4$ subunit) impact on the genetic liability to nicotine addiction and smoking-related phenotypes (Voineskos *et al.*, 2007; Hutchinson *et al.*, 2007; Li *et al.*, 2005; Feng *et al.*, 2004; Breitling *et al.*, 2009b). Of particular interest is the paper by Hutchinson *et al.* (2007). Applying a translational (one may also say "systemic") approach the authors looked at the cellular effects of two SNPs in the 5'- and 3'-untranslated regions (UTRs) of the gene. SNPs in these regions of the gene may affect transcription or mRNA stability, respectively, and may thereby ultimately affect protein and receptor complex levels in the cell. In a series of *in vitro* experiments they show that the studied SNPs are functional (e.g., that they affect transcription factor binding or protein yield). They than moved on and studied the behavioral correlates of these SNPs (i.e., how they affect a subject's response to nicotine). Finally, they showed that their SNPs affect

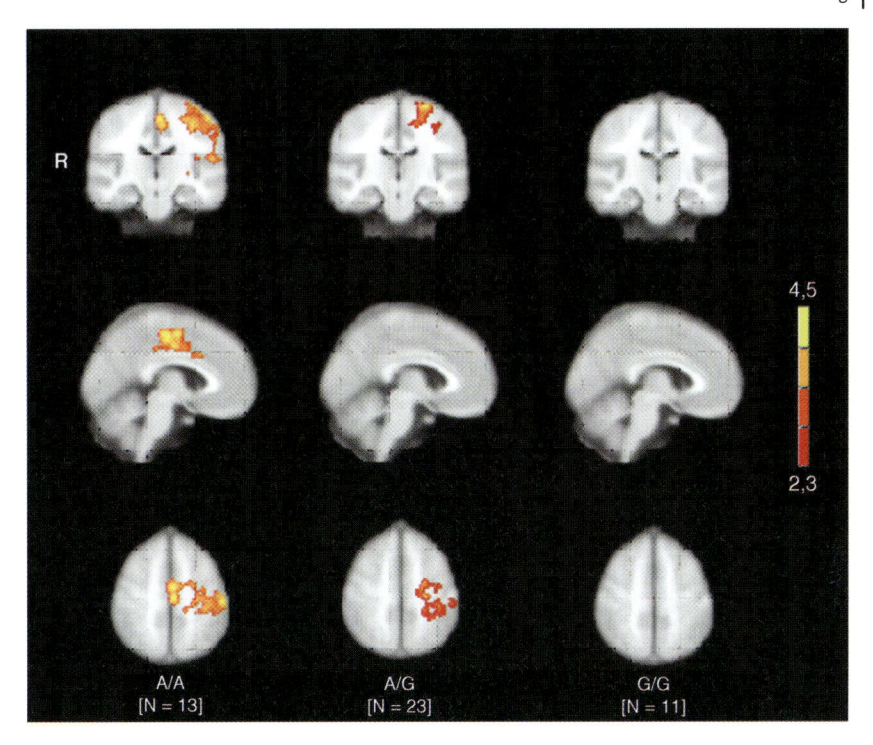

**Figure 8.5** Impact of a common genetic variant (SNP rs1044396) in *CHRNA4* – the gene coding for the $\alpha_4$ subunit of nAChRs – on brain activity during performance of a visual oddball task. Note the dose effect of the A-allele. (Reproduced form Winterer *et al.*, 2007, with kind permission of the publisher.)

treatment outcome in a cessation trial using different applications of nicotine (nasal spray or patch) as a cigarette substitute. This work exemplifies nicely how genetic data – when merged with behavioral data – may allow the optimization of treatment regimens based on genetic information. Furthermore, work from our own group suggests that genetic variation in *CHRNA4* affects the fMRI BOLD activation pattern during cognitive task performance (Winterer *et al.*, 2007) (see also Figure 8.5).

However, like in most other behavioral traits and neuropsychatric disorders, multiple genes contribute to the overall genetic liability to smoking and nicotine addiction. Other genes that may be involved include genes that encode for dopamine receptors and transporters, GABA receptors, and genes coding for enzymes involved in the metabolism of nicotine (Benowitz, 2008; see also Ho and Tyndale, 2007 for a detailed review of the literature). Finally, variations in *CHRNA7*, the gene encoding for the $\alpha_7$ subunit, have been implicated in the genetic liability of schizophrenia (Harrison and Weinberger, 2005; Leonard *et al.*, 2002; Adams and Stevens, 2007).

Except for the above-mentioned variation in *CHRNA7*, the genetic variations in nAChR subunits that are associated with nicotine addiction are SNPs. If a SNP alters the amino acid sequence of its gene product, protein function or protein stability may be altered, which could directly lead to a certain phenotype. For instance, the two alleles of SNP rs16969968 code for proteins that contain either aspartic acid or asparagine at position 398 of the amino acid chain of the $\alpha_5$ nAChR subunit (Bierut et al., 2008). However, most other SNPs that have been reported to be associated with smoking do not lead to a change in the amino acid sequence. In some cases they are synonymous, meaning that the SNP is located in the coding region of the gene, but the base triplet codes for the same amino acid. It is thought that SNPs of this kind are in linkage disequilibrium (meaning they are correlated) with other (known or unknown SNPs) which are the functional ones. Some functional SNPs are located in intronic regions of the gene and may affect mRNA splicing. Finally, SNPs in the promoter, 5'-UTR, and 3'-UTR of the gene may affect gene expression. For instance, Breitling et al. (2009b) showed that the synonymous SNP rs1044396 in the gene coding for the nAChR $\alpha_4$ subunit – a SNP that has been shown to be associated with fMRI BOLD activation during attentional task performance (Winterer et al., 2007) – is in strong linkage disequilibrium (meaning that it is highly correlated) with the functional 3'-UTR SNP rs2236196 described by Hutchinson et al. (2007). Further research is needed to understand how variation in nAChR genes contributes to the genetic liability to smoking and nicotine addiction.

As pointed out above, nicotine addiction – like most other chronic neuropsychiatric disorders – develops from multigenic disturbances where the contribution of each single gene is small. Therefore, only a limited number of putative genomic linkages have been replicated in independent studies (Li, 2008). This emphasizes the importance of the "biological context" when the relationship between genotype (e.g., a putative risk gene for nicotine addiction) and phenotype (habitual smoking/nicotine addiction) is studied. The biological context may be thought of as the sum of the "genetic background" and all nongenetic biological and environmental factors. Thus, gene–gene interactions of putative risk/protective genes could be crucial to further our understanding of how the phenotype is shaped from the genotype. With respect to nAChRs, this issue is largely unresolved. However, it has already been shown that genetic variation in the $\alpha_5$ subunit substantially alters the pharmacological properties of $\alpha_4\beta_2$-containing nAChRs (Bierut et al., 2008). It will be interesting to see if gene–gene interactions (e.g., *CHRNA4–CHRNA5*) can help to define risk constellations for nicotine addiction. In the past the major focus of neuropsychiatric genetics was genetic variation determined by SNP. However, recently it has been shown that another source of genetic variation may play an important role in the genetics of neuropsychiatric disorders – copy number variations (CNVs). CNVs arise from chromosomal microdeletions and microduplications that may involve one or more genes. CNVs have been shown to play a role in the genetic liability to autism (Sebat et al., 2007) and schizophrenia (Stefansson et al., 2008; Walsh et al., 2008). It remains to be seen how this aspect of genetics contribute to the genetic risk for nicotine addiction and smoking. Furthermore, the effects of genetic variation (SNPs and CNVs) are modified by other biological phenomena such as epigenetics (the regulation of transcriptional activity of certain genes by DNA methylation). DNA methylation has been shown to play a role in the etiology of neuropsychiatric disorders (Tsankova et al., 2007). Again, how this biological principle contributes to nicotine addiction is a question that remains to be answered.

To summarize, we have seen that nicotine affects various types of neurons that are involved in a variety of neuronal processes via activation of several nAChR subtypes. These receptors may be located on axonal terminals, dendrites, or cell bodies. Therefore, they affect neuronal activity in different ways (e.g., promoting transmitter release in the presynaptic neuron versus modulating excitability and firing rate of postsynaptic neurons). The net effect of nicotinic stimulation on the level of single neurons is generally an increased electrical/synaptic activity. However, dependent on the cell type that is predominantly stimulated within a neuronal network, the overall effect on the network may be inhibitory, as it is the case when GABA-ergic neurons are predominantly activated by nicotine. One may think of the effects of nicotine on neurons in two different ways. (i) immediate effects that are mediated by changes in neuronal electrical and synaptic activity. The increase in neuronal activity of dopaminergic VTA neurons is an example of such an effect and is supposed to be the neurophysiological correlate of the psychological concept of reward, and may be relevant to the addictive and mood-regulatory effects of nicotine. (ii) $Ca^{2+}$-dependent, "quasi-metabotropic" effects that are initiated by the CaMK, ERK/CREB and PI3K/Akt/GSK-3β pathways. These intracellular signal transduction cascades may lead to various types of synaptic plasticity. These effects on synaptic plasticity may explain some of the nicotine effects on cognition, but possibly also effects on other systems, including the reward system and stress regulation system.

People with certain psychiatric diseases (schizophrenia may serve as the most important example) may (ab)use nicotine to self-medicate several symptoms of the disease. (i) Nicotine may improve cognitive processes related to prefrontal functioning such as attention or working memory. It may do so by modulation of the signal-to-noise ratio and synchrony of neuronal activity in the prefrontal cortex as suggested by a recent electrophysiological study (Couey et al., 2007; McGehee, 2007) – a concept that is somewhat reminiscent of the effects of dopamine on prefrontal functioning (Winterer and Weinberger, 2004; Rolls et al., 2008). Imaging studies point in a similar direction, suggesting that in addition to "activation" of the attentional network, "deactivation" of the default mode network and "deactivation" in the context of stimuli that are outside the focus of the current cognitive task may be important aspects, too (Thiel, Zilles, and Fink, 2005; Lawrence, Ross, and Stein, 2002; Hahn et al., 2007). (ii) Nicotine may promote plastic processes in the hippocampus that could be beneficial with respects to other cognitive deficits in schizophrenia (i.e., deficits in learning and memory) (Ji, Lape, and Dani, 2001; Fujii et al., 1999; Mann and Greenfield, 2003; Ge and Dani, 2005). (iii) Nicotine may improve symptoms related to mood and stress by modulation of the reward and stress response systems of the brain (Glassman et al., 1988; Keuthen

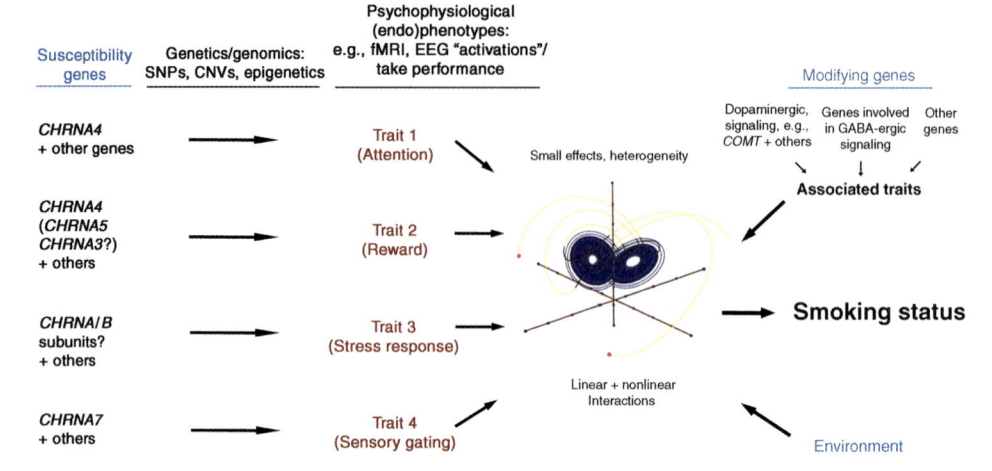

**Figure 8.6** Impact of multiple genes on human behavior and psychiatric disease. The nicotinic system. Note that the influence of potentially modifying factors such as CNVs and epigenetic variations on the genotype–phenotype relationship in the case of nicotine addiction is largely unclear at this point. Variants in susceptibility genes are over-represented in affected subjects (e.g., people with nicotine dependence), whereas modifying genes can be equally distributed in affected subjects and unaffected controls. They may exert their effect only in the presence of certain variants in susceptibility genes. However, a clear-cut distinction may not always be possible.

et al., 2000; Breslau and Johnson, 2000; Breslau, Kilbey, and Andreski, 1993; Thorndike et al., 2006).

It remains to be seen if nicotinic cholinergic neurotransmission, the intracellular signal transduction cascades downstream of nAChR stimulation, and the potential targets of the intracellular effects of nicotine (folding of nAChR subunits in the ER, assembly of nAChR complexes, and ER-associated proteasomal activity) can be used as targets for the development of drugs for the treatment of cognitive deficits that are associated with psychiatric diseases such as schizophrenia. Figure 8.6 illustrates the issue of the complex relationship between the genotype (variation in specific genes) and the ultimate phenotype (human behavior and psychiatric disease) with a focus on nicotinic cholinergic neurotransmission. It also illustrates how "endophenotypes" are thought to be closer linked to individual susceptibility genes.

## 8.8
### Future Perspectives

Much more research is needed in the field to clarify the specific (quantitative) contributions of the above-mentioned signaling pathways to the pathogenesis and symptomatology of nicotine addiction – and, by extension, to neuropsychiatric disorders that are characterized by frequent nicotine abuse such as schizophrenia.

It is important to state that future nicotinic drugs should not only be effective in the treatment of nicotine addiction, but also as specific as possible (i.e., they should not be addictive and should not promote uncontrolled cell survival and cell proliferation). The latter is an important issue as some of the intracellular signaling pathways downstream of nAChR stimulation such as the ERK/MAPK and PI3K/Akt pathways play important roles in the pathogenesis of cancer (Roberts and Der, 2007; LoPiccolo et al., 2007). Neuropsychiatric patients suffering, for instance, from schizophrenia may benefit in another way from future research in the field of the cellular effects of nicotine and nicotinic neurotransmission. While the fraction of smokers in the general population is decreasing in the developed countries – a phenomenon that is likely due to antismoking campaigns and cessation programs – people with schizophrenia usually do not benefit from these programs and keep smoking (Addington, 1998). In fact, it is estimated that roughly 45% of the cigarettes smoked in America are purchased by the mentally ill (Lasser et al., 2000). Thus, there is a need to develop more effective treatments for nicotine addiction in this population that is already prone to the development of medical problems such as cardiovascular disease (Hennekens, 2007).

## 8.9
### Role of Systems Neuroscience

Systems biology has been defined as the studies of how the phenotype is generated from the genotype with an emphasis on the role of the context/biological system in which a gene/protein of interest is functioning. The contextual/systemic forces that shape the phenotype may be genetic (also known as the "genetic background"), developmental, may depend on cell type, neuronal network, and finally the psychosocial interactions and learning history of a human being. Systems biology tries to define the molecular constituents that make up such a system and to understand how their interaction creates the phenotype. The ultimate goal of systems neuroscience may be a computer model of the brain or at least models of distinct neuronal systems. Such a model would allow us to study how a genetic manipulation of one component, such as a neurotransmitter receptor, would affect the neuronal system/brain function. Importantly, such an approach is not only promising for the study of genetic effects on human behavior and mental disease, it will also prove useful for the development of new drugs, because pharmacological manipulation of a system may be studied in the same way as genetic modulation. In that sense, systems neuroscience can contribute to translational drug development – a branch of pharmaceutical research in which during the process of drug development the research and development groups try to optimize the transition of test tube results to clinical trials – a process that takes on average 5–10 years and cost US$1 billion per compound. Though certain aspects of brain function are now studied with computer models of neuronal networks or transmitter systems such as nicotinic cholinergic or dopaminergic neurotransmission (Gutkin, Dehaene, and Changeux, 2006; Rolls et al., 2008; Durstewitz and Seamans, 2008), these current models are only beginning to integrate the different levels of complexity and the diversity of interactions within a neuronal (sub)system and in between these (sub)systems. A comprehensive computer model of the brain is not even on the horizon.

In the meantime, the "systemic" approach can be pursued by the application of experimental study designs that integrate molecular (e.g., genetic) or other neurophysiological information. Two upcoming fields are "genetic imaging" and "multimodal imaging." fMRI –a research tool developed in the 1990s – has made it possible to study the living human brain "at work" with unprecedented spatial resolution and much higher temporal resolution than positron emission tomography (PET), although with poorer temporal resolution that the electrophysiological methods of electroencephalography (EEG) and magnetoencephalography (MEG). In genetic imaging, the impact of a certain genotype (e.g., a polymorphism in a neurotransmitter receptor or in an enzyme involved in neurotransmitter metabolism) on brain structure (voxel-based morphometry), white matter connectivity (diffusion tensor imaging), or function (fMRI) is studied. Among the most studied molecules are the enzyme catechol-O-methyltransferase (COMT), which is involved in dopamine degradation in the cortex (Winterer *et al.*, 2006a, 2006c), and neuregulin, a growth-factor like cell–cell signaling molecule that has been implicated in the pathogenesis of schizophrenia (Harrison and Weinberger, 2005; Winterer *et al.*, 2008). However, it seems that genetic variation in the nicotinic system also has an impact on brain activity that can be captured with fMRI (Winterer *et al.*, 2007). Unfortunately, fMRI has its limitations. It does not have the temporal resolution of EEG or MEG of a few milliseconds, and does not have the specificity of ligand PET in terms of the transmitter system studied. Recent technical advances in EEG hardware development and artifact correction procedures have made it possible to study brain function with fMRI and EEG at the same time. The exciting field of simultaneous EEG/fMRI enables researches to ask the questions "where and when?" of the same dataset. However, the two signals obtained in fMRI and EEG (the BOLD signal or the scalp-recorded correlate of the local field potentials) are fundamentally different in nature and further research is needed to optimize integrated cross-modal data analysis (Debener *et al.*, 2006; Laufs *et al.*, 2008). Likewise, a new generation of "hybrid" MRI–PET scanners allows the analysis of brain structure and function using magnetic resonance techniques and PET simultaneously. However, issues related to the design of informative experiments that benefit from an integrated ((f)MRI-PET)) approach as well as issues related to optimal data analysis need to be resolved.

In the past, genetic imaging has mostly relied on defining groups of subjects based on the genotype of specific SNPs. That worked well for a few model SNPs (like the COMT Val/Met polymorphism) that proved the concept. However, the latest research in the fields of neuropsychiatric genetics and epidemiology suggest a major role of other biological factors such as CNVs and epigenetics in determining the liability to brain disorders. In the National Priority Program of the Deutsche Forschungsgemeinschaft, "Nicotine: Molecular and Physiological Mechanisms in the Central Nervous System," we use SNPs as a starting point to study the genotype–phenotype relationship with respect to nicotine addiction, but extend our analyses to CNVs, epigenetics, expression profiling, and "metabolomics" (www.nicotine-research.com). In the era of high-throughput "...omics" biological analyses (i.e., "genomics," "transcriptomics," "proteomics," and "metabolomics") it will be a challenge to the neuroscience field (be it genetic neuroimaging or a neighboring discipline) to integrate the multitude of epigenetic, post-translational, and other modifying factors that have an impact on the ultimate phenotype. This difficult job will be done by a systems neuroscientist.

### Acknowledgments

This work was supported within the National Priority Program SPP1226 "Nicotine: Molecular and Physiological Mechanisms in the Central Nervous System" (www.nicotine-research.com) funded by the Deutsche Forschungsgemeinschaft (Wi1316/7-1).

### References

Adams, C.E. and Stevens, K.E. (2007) Evidence for a role of nicotinic acetylcholine receptors in schizophrenia. *Frontiers in Bioscience*, **12**, 4755–4772.

Addington, J. (1998) Group treatment for smoking cessation among persons with schizophrenia. *Psychiatric Services*, **49**, 925–928.

Adler, L.E., Hoffer, L.D., Wiser, A., and Freedman, R. (1993) Normalization of auditory physiology by cigarette smoking in schizophrenic patients. *American Journal of Psychiatry*, **150**, 1856–1861.

Al Absi, M. (2006) Hypothalamic–pituitary–adrenocortical responses to psychological stress and risk for smoking relapse. *International Journal of Psychophysiology*, **59**, 218–227.

Andreas, S. and Loddenkemper, R. (2007) Tabakprävention. *Pneumologie*, **61**, 588–589.

Andreas, S., Herth, F.J., Rittmeyer, A., Kyriss, T., and Raupach, T. (2007) Tabakrauchen, chronisch obstruktive Lungenerkrankung und Lungenkarzinom. *Pneumologie*, **61**, 590–595.

Balfour, D.J.K., Wright, A.E., Benwell, M.E.M., and Birrell, C.E. (2000) The putative role of extra-synaptic mesolimbic dopamine in the neurobiology of nicotine dependence. *Behavioural Brain Research*, **113**, 73–83.

Batra, A. (2004) Leitlinie Tabakentwöhnung [Tobacco attributable diseases "guidelines tobacco cessation"]. Leitlinien der Deutschen Gesellschaft für Suchtforschung und Suchttherapie (DG-Sucht)/Deutschen Gesellschaft für Psychiatrie, Psychotherapie und Nervenheilkunde (DGPPN); http://www.uni-duesseldorf.de/AWMF/ll/076-006.htm.

Bierut, L.J., Madden, P.A., Breslau, N., Johnson, E.O., Hatsukami, D., and Pomerleau, O.F. *et al.* (2007) Novel genes identified in a high-density genome wide association study for nicotine dependence. *Human Molecular Genetics*, **16**, 24–35.

Bierut, L.J., Stitzel, J.A., Wang, J.C., Hinrichs, A.L., Grucza, R.A., Xuei, X., *et al.* (2008) Variants in nicotinic receptors and risk for nicotine dependence. *American Journal of Psychiatry*, **165**, 1163–1171.

Benowitz, N.L. (2008) Clinical pharmacology of nicotine: implications for understanding, preventing, and treating tobacco addiction. *Clinical Pharmacology and Therapeutics*, **83**, 531–541.

Berrettini, W., Yuan, X., Tozzi, F., Song, K., Francks, C., Chilcoat, *et al.* (2008) α-5/α-3 nicotinic receptor subunit alleles increase risk for heavy smoking. *Molecular Psychiatry*, **13**, 368–373.

Beaulieu, J.M., Tirotta, E., Sotnikova, T.D., Masri, B., Salahpour, A., Gainetdinov Raul, R., et al. (2007) Regulation of Akt signalling by $D_2$ and $D_3$ dopamine receptors in vivo. *Journal of Neuroscience*, **27**, 881–885.

Besson, M., Granon, S., Mameli-Engvall, M., Cloez-Tayarani, I., Maubourguet, N., Cormier, A., et al. (2007) Long-term effects of chronic nicotine exposure on brain nicotinic receptors. *Proceedings of the National Academy of Sciences of the United States of America*, **104**, 8155–8160.

Bilkei-Gorzo, A., Racz, I., Michel, K., Darvas, M., Maldonado, R., and Zimmer, A. (2008) A common genetic predisposition to stress sensitivity and stress-induced nicotine craving. *Biological Psychiatry*, **63**, 164–171.

Bitner, R.S., Bunnelle, W.H., Anderson, D.J., Briggs, C.A., Buccafusco, J., Curzon, P., et al. (2007) Broad-spectrum efficacy, across cognitive domains by α7 nicotinic acetylcholine receptor agonism correlates with activation of ERK 1/2 and CREB phosphorylation pathways. *Journal of Neuroscience*, **27**, 10578–10587.

Breitling, L.P., Dahmen, N., Mittelstraß, K., Illig, T., Rujescu, D., Raum, E., et al. (2009a) Smoking cessation and variations in nicotinic acetylcholine receptor subunits α-5, α-3, and β-4 genes. *Biological Psychiatry*, **65**, 691–695.

Breitling, L.P., Dahmen, N., Mittelstraß, K., Rujescu, D., Gallinat, J., Fehr, C., et al. (2009b) Association of nicotinic acetylcholine receptor subunit alpha-4 polymorphisms with nicotine dependence in 5500 Germans. *Pharmacogenomics Journal*, **9**, 219–224.

Breslau, N. and Johnson, E.O. (2000) Predicting smoking cessation and major depression in nicotine-dependent smokers. *American Journal of Public Health*, **90**, 1122–1127.

Breslau, N., Kilbey, M.M., and Andreski, P. (1993) Nicotine dependence and major depression. New evidence form a prospective investigation. *Archives of General Psychiatry*, **50**, 31–35.

Cardenas, L., Tremblay, L.K., Naranjo, C.A., Herrmann, N., Zack, M., and Busto, U.E. (2002) Brain reward system activity in major depression and comorbid nicotine dependence. *Journal of Pharmacology and Experimental Therapeutics*, **302**, 1265–1271.

Carmelli, D., Swan, G.E., Robinette, D., and Fabsitz, R. (1992) Genetic influence on smoking – a study of male twins. *New England Journal of Medicine*, **327**, 829–833.

Chanock, S.J. and Hunter, D.J. (2008) When the smoke clears.... *Nature*, **452**, 537–538.

Chen, L., Bohanick, J.D., Nishihara, M., Seamans, J.K., and Yang, C.R. (2007) Dopamine $D_1/D_5$ receptor-mediated long-term potentiation of intrinsic excitability in rat prefrontal cortical neurons: $Ca^{2+}$-dependent intracellular signalling. *Journal of Neurophysiology*, **97**, 2448–2464.

Couey, J.J., Meredith, R.M., Spijker, S., Poorthuis, R.B., Smit, A.B., Brussaard, A.B., et al. (2007) Distributed network actions by nicotine increase the threshold for spike-timing dependent plasticity in prefrontal cortex. *Neuron*, **54**, 73–87.

Dani, J.A. and De Biasi, M. (2001) Cellular mechanisms of nicotine addiction. *Pharmacology, Biochemistry and Behavior*, **70**, 439–446.

Debener, S., Ullsperger, M., Siegel, M., Fiehler, K., von Cramon, D.Y., and Engel, A.K. (2006) Single-trial EEG-fMRI reveals the dynamics of cognitive function. *Trends in Cognitive Sciences*, **10**, 558–563.

de Leon, J., Dadvand, M., Canuso, C., White, A.O., Stanilla, J.K., and Simpson, G.M. (1995) Schizophrenia and smoking: an epidemiological survey in a state hospital. *American Journal of Psychiatry*, **152**, 453–455.

Di Chiara, G., Bassareo, V., Fenu, S., De Luca, M.A., Spina, L., Cadoni, C., et al. (2004) Dopamine and drug addiction: the nucleus accumbens shell connection. *Neuropharmacology*, **47**, 227–241.

Durany, N., Zöchlinger, R., Boissl, K.W., Paulus, W., Ransmayr, G., Tatschner, T., et al. (2000) Human post-mortem striatal alpha4beta2 nicotinic acetylcholine receptor density in schizophrenia and Parkinson's syndrome. *Neuroscience Letters*, **287**, 109–112.

Durstewitz, D. and Seamans, J.K. (2008) The dual-state theory of prefrontal cortex dopamine function with relevance to catechol-*O*-methyltransferase genotypes and schizophrenia. *Biological Psychiatry*, **64**, 739–749.

Edwards, K.L., Austin, M.A., and Jarvic, G.P. (1995) Evidence for genetic influences on smoking in adult women twins. *Clinical Genetics*, 47, 236–244.

Emamian, E.S., Hall, D., Birnbaum, M.J., Karayiorgou, M., and Gogos, J.A. (2004) Convergent evidence for impaired Akt1–GSK3beta signaling in schizophrenia. *Nature Genetics*, 36, 131–137.

Epping-Jordan, M.P., Picciotto, M.R., Changeux, J.P., and Pich, E.M. (1999) Assessment of nicotinic acetylcholine receptor subunit contributions to nicotine self-administration in mutant mice. *Psychopharmacology*, 147, 25–26.

Fagen, Z.M., Mitchum, R., Vezina, P., and McGehee, D.S. (2007) Enhanced nicotinic receptor function and drug abuse vulnerability. *Journal of Neuroscience*, 27, 8771–8778.

Feng, Y., Niu, T., Xing, H., Xu, X., Chen, C., Peng, S., et al. (2004) A common haplotype of the nicotinic acetylcholine receptor $\alpha_4$ subunit gene is associated with vulnerability to nicotine addiction in men. *American Journal of Human Genetics*, 75, 112–121.

Ferrari, R., Le Novere, N., Picciotto, M.R., Changeux, J.P., and Zoli, M. (2002) Acute and long-term changes in the mesolimbic dopamine pathway after systemic or local single nicotine injections. *European Journal of Neuroscience*, 15, 1810–1818.

Freedman, R., Hall, M., Adler, L.E., and Leonard, S. (1995) Evidence in post-mortem brain tissue for decreased number of hippocampal nicotinic receptors in schizophrenia. *Biological Psychiatry*, 38, 22–33.

Freedman, R., Adams, C.E., and Leonard, S. (2000) The alpha7-nicotinic acetylcholine receptor and the pathology of hippocampal interneurons in schizophrenia. *Journal of Chemical Neuroanatomy*, 20, 299–306.

Fujii, S., Ji, Z., Morita, N., and Sumikawa, K. (1999) Acute and chronic nicotine exposure differentially facilitate the induction of LTP. *Brain Research*, 846, 137–143.

Fuxe, K., Andersson, K., Eneroth, P., Harfstrand, A., and Agnati, L.F. (1989) Neuroendocrine actions of nicotine and of exposure to cigarette smoke: medical implications. *Psychoneuroendocrinology*, 14, 19–41.

Ge, S. and Dani, J.A. (2005) Nicotinic acetylcholine receptors at glutamate synapses facilitate long-term depression or potentiation. *Journal of Neuroscience*, 25, 6084–6091.

Glassman, A.H., Stetner, F., Walsh, B.T., Raizman, P.S., Fleiss, J.L., Cooper, T.B., et al. (1988) Heavy smokers, smoking cessation, and clonidine. Results of a double-blind, randomized trial. *Journal of the American Medical Association*, 259, 2863–2866.

Gotti, C., Zoli, M., and Clementi, F. (2006) Brain nicotinic acetylcholine receptors: native subtypes and their relevance. *Trends in Pharmacological Sciences*, 27, 482–491.

Gutkin, B.S., Dehaene, S., and Changeux, J.P. (2006) A neurocomputational hypothesis for nicotine addiction. *Proceedings of the National Academy of Sciences of the United States of America*, 103, 1106–1111.

Hahn, B., Shoaib, M., and Stolerman, I.P. (2002) Nicotine-induced enhancement of attention in the five-choice serial reaction time task. The influence of task demands. *Psychopharmacology*, 162, 129–137.

Hahn, B., Sharples, C.G., Wonnacott, S., Shoaib, M., and Stolerman, I.P. (2003) Attentional effects of nicotinic agonists in rats. *Neuropharmacology*, 44, 1054–1067.

Hahn, B., Ross, T.J., Yang, Y., Kim, I., Huestis, M.A., and Stein, E.A. (2007) Nicotine enhances visuospatial attention by deactivating areas of the resting brain default network. *Journal of Neuroscience*, 27, 3477–3489.

Harris, J.G., Kongs, S., Allensworth, D., Martin, L., Tregellas, J., Sullivan, B., et al. (2004) Effects of nicotine on cognitive deficits in schizophrenia. *Neuropsychopharmacology*, 29, 1378–1385.

Harrison, P.J. and Weinberger, D.R. (2005) Schizophrenia genes, gene expression and neuropathology: on the matter of their convergence. *Molecular Psychiatry*, 10, 40–68.

Heath, A.C., Cates, R., Martin, N.G., Meyer, J., Hewitt, J.K., Neale, M.C. et al. (1993) Genetic contribution to risk of smoking initiation: comparison across birth cohorts and across cultures. *Journal of Substance Abuse*, 5, 221–246.

Heath, A.C. and Martin, N.G. (1993) Genetic models for the natural history of smoking:

evidence for a genetic influence on smoking persistence. *Addictive Behaviors,* **18,** 19–34.

Heinz, A., Schmidt, L.G., and Reischies, F.M. (1994) Anhedonia in schizophrenic, depressed and alcohol-dependent patients – neurobiological correlates. *Pharmacopsychiatry,* **27** (Suppl. 1), 7–10.

Hennekens, C.H. (2007) Increasing global burden of cardiovascular in general populations and patients with schizophrenia. *Journal of Clinical Psychiatry,* **68** (Suppl. 4), 4–7.

Ho, M.K. and Tyndale, R.F. (2007) Overview of the pharmacogenomics of cigarette smoking. *Pharmacogenomics Journal,* **7,** 81–98.

Houlihan, M.E., Pritchard, W.S., Krieble, K.K., Robinson, J.H., and Duke, D.W. (1996) Effects of cigarette smoking on EEG spectral-band power, dimensional complexity, and nonlinearity during reaction time task performance. *Psychophysiology,* **33,** 740–746.

Houlihan, M.E., Pritchard, W.S., and Robinson, J.H. (1996) Faster P300 latency after smoking in visual but not auditory oddball tasks. *Psychopharmacology,* **123,** 231–238.

Hung, R.J., McKay, J.D., Gaborieau, V., Boffetta, P., Hashibe, M., Zaridze, D., et al. (2008) A susceptibility locus for lung cancer maps to nicotinic acetylcholine receptor subunit genes on 15q25. *Nature,* **452,** 633–637.

Hutchinson, K.E., Allen, D.L., Filbey, F.M., Jepson, C., Lerman, C., Benowitz, N.L., et al. (2007) *CHRNA4* and tobacco dependence – from gene regulation to treatment outcome. *Archives of General Psychiatry,* **64,** 1078–1086.

Ikemoto, S. (2007) Dopamine reward circuitry: two projection systems from the ventral midbrain to the nucleus accumbens – olfactory tubercle complex. *Brain Research Rev.,* **56,** 27–78.

Inoue, Y., Yao, L., Hopf, W., Fan, P., Jiang, Z., Bronci, A., and Diamond, I. (2007) Nicotine and ethanol activate protein kinase A synergistically via $G_i\beta\gamma$ subunits in nucleus acumbens/ventral tegmental cocultures: the role of dopamine $D_1/D_2$ and adenosine $A_{2A}$ receptors. *Journal of Pharmacology and Experimental Therapeutics,* **322,** 23–29.

Ji, D., Lape, R., and Dani, J.A. (2001) Timing and location of nicotinic activity enhances or depresses hippocampal synaptic plasticity. *Neuron,* **31,** 131–141.

Jope, R.S. and Roh, M.S. (2006) Glycogen synthase kinase-3 (GSK3) in psychiatric disease and therapeutic interventions. *Current Drug Targets,* **7,** 1421–1434.

Kanakry, C.G., Li, Z., Nakai, Y., and Weinberger, D.R. (2007) Neuregulin-1 regulates cell adhesion via an ErbB2/phosphoinositide-3 kinase/Akt-dependent pathway: potential implications for schizophrenia and cancer. *PLoS One,* **2,** e1369.

Kelleher, R.J., Govindarajan, A., and Tonegawa, S. (2004) Translational regulatory mechanisms in persistent forms of synaptic plasticity. *Neuron,* **44,** 59–73.

Keuthen, N.J., Niaura, R.S., Borrelli, B., Goldstein, M., DePue, J., Murphy, C., et al. (2000) Comorbidity, smoking behavior and treatment outcome. *Psychotherapy and Psychosomatics,* **69,** 244–250.

Kihara, T., Shimohama, S., Sawada, H., Honda, K., Nakamizo, T., Shibasaki, H., et al. (2001) $\alpha_7$ nicotinic receptor transduces signals to phosphatidylinositol 3-kinase to block A$\beta$-amyloid-induced neurotoxicity. *Journal of Biological Chemistry,* **276,** 13541–13546.

Kirschner, M.W. (2005) The meaning of systems biology. *Cell,* **121,** 503–504.

Lang, U.E., Puls, I., Müller, D.J., and Strutz-Seebohm Gallinat, J. (2007) Molecular mechanisms of schizophrenia. *Cellular Physiology and Biochemistry,* **20,** 687–702.

Lasser, K., Boyd, J.W., Woolhandler, S., Himmelstein, D.U., McCormick, D., and Bor, D.H. (2000) Smoking and mental illness: a population-based prevalence study. *Journal of the American Medical Association,* **284,** 2606–2610.

Laufs, H., Daunizeau, J., Carmichael, D.W., and Kleinschidt, A. (2008) Recent advances in recording electrophysiological data simultaneously with magnetic resonance imaging. *NeuroImage,* **40,** 515–528.

Lawrence, N.S., Ross, T.J., and Stein, E.A. (2002) Cognitive mechanisms of nicotine on visual attention. *Neuron,* **36,** 539–548.

Leonard, S., Gault, J., Hopkins, J., Logel, J., Vianzon, R., Short, M., et al. (2002)

Association of promoter variants in the alpha 7 nicotinic acetylcholine receptor subunit gene with an inhibitory deficit found in schizophrenia. *Archives of General Psychiatry*, **59**, 1085–1096.

Lessov, C.N., Martin, N.G., Statham, D.J., Todorov, A.A., Slutske, W.S., Bucholz, K.K., et al. (2004) Defining nicotine dependence for genetic research: evidence from Australian twins. *Psychological Medicine*, **34**, 865–879.

Lessov-Schlagger, C.N., Pergadia, M.L., Khroyan, T.V., and Swan, G.E. (2008) Genetics of nicotine dependence and pharmacotherapy. *Biochemical Pharmacology*, **75**, 178–195.

Levine, E.D., Conners, C.K., Sparrow, E., Hinton, S.C., Erhardt, D., Meck, W.H., et al. (1996) Nicotine effects on adults with attention-deficit/hyperactivity disorder. *Psychopharmacology*, **123**, 55–63.

Levine, E.D., Conners, C.K., Silva, D., Canu, W., and March, J. (2001) Effects of chronic nicotine and methylphenidate in adults with ADHD. *Experimental and Clinical Pharmacology*, **9**, 83–90.

Levin, E.D., McClernon, F.J., and Rezvani, A.H. (2006) Nicotinic effects on cognitive function: behavioral characterization, pharmacological specification and anatomical localization. *Psychopharmacology*, **184**, 523–539.

Li, M.D., Beuten, J., Ma, J.Z., Payne, T.J., Lou, X.Y., Garcia, V., et al. (2005) Ethnic- and gender-specific association of the nicotinic acetylcholine receptor alpha4 subunit gene (*CHRNA4*) with nicotine dependence. *Human Molecular Genetics*, **14**, 1211–1219.

Li, M.D. (2008) Identifying susceptibility loci for nicotine dependence: 2008 update based on recent genome-wide linkage analyses. *Human Genetics*, **123**, 119–131.

LoPiccolo, J., Granville, C.A., Gills, J.J., and Dennis, P.A. (2007) Targeting Akt in cancer therapy. *Anti-Cancer Drugs*, **18**, 861–874.

Lovallo, W.R. (2006) Cortisol secretion patterns in addiction and addiction risk. *International Journal of Pharmacology*, **59**, 195–202.

Mann, E.O. and Greenfield, S.A. (2003) Novel modulatory mechanisms revealed by the sustained application of nicotine in the guinea-pig hippocampus in vitro. *Journal of Physiology*, **551**, 539–550.

Manning, B.D. and Cantley, L.C. (2007) Akt/PKB signalling: navigating downstream. *Cell*, **129**, 1261–1274.

Mansvelder, H.D. and McGee, D.S. (2000) Long-term potentiation of excitatory inputs to brain reward areas by nicotine. *Neuron*, **27**, 349–357.

Mansvelder, H., Keath, J.R., and McGehee, D.S. (2002) Synaptic mechanisms underlie nicotine-induced excitability of brain reward areas. *Neuron*, **33**, 905–919.

Mansvelder, H.D., van Aerde, K.I., Couey, J.J., and Brussaard, A.B. (2006) Nicotinic modulation of neuronal networks: from receptors to cognition. *Psychopharmacology*, **184**, 292–305.

Matta, S.G., Fu, Y., Valentine, J.D., and Sharp, B.M. (1998) Response of the hypothalamo-pituitary-adrenalo axis to nicotine. *Psychoneuroendocrinology*, **23**, 103–113.

McGehee, D.S. (2007) Nicotine and synaptic plasticity in the prefrontal cortex. *Science's STKE: Signal Transduction Knowledge Environment*, **399**, e44.

Maes, H.H., Sullivan, P.F., Bulik, C.M., Neale, M.C., Prescott, C.A., Eaves, L.J., et al. (2004) A twin study of genetic and environmental influences on tobacco initiation, regular tobacco use and nicotine dependence. *Psychological Medicine*, **34**, 1251–1261.

Mineur, Y.S. and Picciotto, M.R. (2008) Genetics of nicotinic acetylcholine receptors: relevance to nicotine addiction. *Biochemical Pharmacology*, **75**, 323–333.

Musso, F., Bettermann, F., Vucurevic, G., Stoeter, P., Konrad, A., and Winterer, G. (2007) Smoking impacts on prefrontal attention network function in young adult brains. *Psychopharmacology*, **191**, 159–169.

Nakayama, H., Numakawa, T., Ikeuchi, T., and Hatanaka, H. (2001) Nicotine-induced phosphorylation of extracellular signal-regulated protein kinase and CREB in PC12h cells. *Journal of Neurochemistry*, **79**, 489–498.

Ohno, M., Yamamoto, T., and Watanabe, S. (1993) Blockade of hippocampal nicotinic receptors impairs working memory but not reference memory in rats. *Pharmacology, Biochemistry, and Behavior*, **45**, 89–93.

Peineau, S., Taghibiglou, C., Bradley, C., Wong, T.P., Liu, L., Lu, J., et al. (2007) LTP inhibits LTD in the hippocampus via regulation of GSK3β. *Neuron*, **53**, 703–717.

Peng, X., Gerzanich, V., Anand, R., Whiting, P.J., and Lindstrom, J. (1994) Nicotine-induced increase in neuronal nicotinic receptors results from a decrease in the rate of receptor turnover. *Molecular Pharmacology*, **46**, 523–530.

Pidoplichko, V.I., DeBiasi, M., Williams, J.T., and Dani, J.A. (1997) Nicotine activates and desensitizes midbrain dopamine neurons. *Nature*, **390**, 401–404.

Pomerleau, O.F., Downey, K.K., Stelson, F.W., and Pomerleau, C.S. (1995) Cigarette smoking in adult patients diagnosed with attention deficit hyperactivity disorder. *Journal of Substance Abuse*, **7**, 373–378.

Raupach, T., Nowak, D., Hering, T., Batra, A., and Andreas, S. (2007) Rauchen und pneumologische Erkrankungen, positive Effekte der Tabakentwöhnung. *Pneumologie*, **61**, 11–14.

Raymond, C.R. (2007) LTP forms 1, 2 and 3: different mechanism for the "long" in long-term potentiation. *Trends in Neurosciences*, **30**, 167–175.

Rezvani, A.H. and Levin, E.D. (2001) Cognitive effects of nicotine. *Biological Psychiatry*, **49**, 258–267.

Rezvani, K., Teng, Y., Shim, D., and De Biasi, M. (2007) Nicotine regulates multiple synaptic proteins by inhibiting proteasomal activity. *Journal of Neuroscience*, **27**, 10508–10519.

Roberts, P.J. and Der, C.J. (2007) Targeting the Raf–MEK–ERK mitogen-activated protein kinase cascade for the treatment of cancer. *Oncogene*, **26**, 3291–3310.

Robinson, S.E., Vann, R.E., Britton, A.F., O'Connell, M.M., James, J.R., and Rosecrans, J.A. (2007) Cellular nicotinic receptor desensitization correlates with nicotine-induced acute behavioural tolerance in rats. *Psychopharmacology*, **192**, 71–78.

Rohleder, N. and Kirschbaum, C. (2006) The hypothalamic–pituitary–adrenal (HPA) axis in habitual smokers. *International Journal of Psychophysiology*, **59**, 236–243.

Rolls, E.T., Loh, M., Deco, G., and Winterer, G. (2008) Computational models of schizophrenia and dopamine modulation in the prefrontal cortex. *Nature Reviews. Neuroscience*, **9**, 696–709.

Sacco, K.A., Bannon, K.L., and George, T.P. (2004) Nicotinic receptor mechanisms and cognition in normal states and neuropsychiatric disorders. *Journal of Psychopharmacology*, **18**, 457–474.

Saccone, S.F., Hinrichs, A.L., Saccone, N.L., Chase, G.A., Konvicka, K., Madden, P.A. et al. (2007) Cholinergic nicotinic receptor genes implicated in a nicotine dependence association study targeting 348 candidate genes with 3713 SNPs. *Human Molecular Genetics*, **16**, 36–49.

Sallette, J., Pons, S., Devillers-Thiery, A., Soudant, M., Prado de Carvalho, L., Changeux, J.P., et al. (2005) Nicotine upregulates its own receptors through enhanced intracellular maturation. *Neuron*, **46**, 595–607.

Schmitt, J.M., Guire, E.S., Saneyoshi, T., and Soderling, T.R. (2005) Calmodulin-dependent kinase kinase/calmodulin kinase I activity gates extracellular-regulated kinase-dependent long-term potentiation. *Journal of Neuroscience*, **25**, 1281–1290.

Sebat, J., Lakshmi, B., Malhotra, D., Troge, J., Lese-Martin, C., Walsh, T., et al. (2007) Strong association of de novo copy number mutations with autism. *Science*, **316**, 445–449.

Sherwood, N., Kerr, J.S., and Hindmarch, I. (1992) Psychomotor performance in smokers following single and repeated doses of nicotine gum. *Psychopharmacology*, **108**, 432–436.

Smith, R.C., Singh, A., Infante, M., Khandat, A., and Kloos, A. (2002) Effects of cigarette smoking and nicotine nasal spray on psychiatric symptoms and cognition in schizophrenia. *Neuropsychopharmacology*, **27**, 479–497.

Staley, J.K., Krishnan-Sarin, S., Kelly, P., Cosgrove, K.P., Krantzler, E., Frohlich, E., et al. (2006) Human tobacco smokers in early abstinence have higher levels of $\beta_2^*$ nicotinic acetylcholine receptors than nonsmokers. *Journal of Neuroscience*, **26**, 8707–8714.

Stefansson, H., Rujescu, D., Cichon, S., Pietiläinen, O.P., Ingason, A., Steinberg, S., et al. (2008) Large recurrent microdeletions associated with schizophrenia. *Nature*, **455**, 232–236.

Steiner, R.C., Heath, C.J., and Picciotto, M.R. (2007) Nicotine-induced phosphorylation of ERK in mouse primary cortical neurons: evidence for involvement of glutamatergic signalling and CaMKII. *Journal of Neurochemistry*, **103**, 666–678.

Stolerman, I.P., Mirza, N.R., Hahn, B., and Shoaib, M. (2000) Nicotine in an animal model of attention. *European Journal of Pharmacology*, **393**, 147–154.

Sugano, T., Yanagita, T., Yokoo, H., Satoh, S., Kobayashi, H., and Wada, A. (2006) Enhancement of insulin-induced PI3K/Akt/GSK-3β and ERK signalling by neuronal nicotinic receptor/PKC-α/ERK pathway: upregulation of IRS-1/-2 mRNA and protein in adrenal chromaffin cells. *Journal of Neurochemistry*, **98**, 20–33.

Sullivan, P.F. and Kendler, K.S. (1999) The genetic epidemiology of smoking. *Nicotine & Tobacco Research*, **1** (Suppl. 2), 51–57.

Thiel, C.M., Zilles, K., and Fink, G.R. (2005) Nicotine modulates reorienting of visuospatial attention and neural activity in human parietal cortex. *Neuropsychopharmacology*, **30**, 810–820.

Thorgeirsson, T.E., Geller, F., Sulem, P., Rafnar, T., Wiste, A., Magnusson, K., et al. (2008) A variant associated with nicotine dependence, lung cancer and peripheral arterial disease. *Nature*, **452**, 638–641.

Thorndike, F.F., Wernicke, R., Pearlman, M.Y., and Haaga, D.A.F. (2006) Nicotine dependence. PTSD symptoms and depression proneness among male and female smokers. *Addictive Behaviors*, **31**, 223–231.

Tsankova, N., Renthal, W., Kumar, A., and Nestler, E.J. (2007) Epigenetic regulation in psychiatric disorders. *Nature Reviews Neuroscience*, **8**, 355–367.

van der Heide, L.P., Ramakers, G.M.J., and Smidt, M.P. (2006) Insulin signaling in the central nervous system: learning to survive. *Progress in Neurobiology*, **79**, 205–221.

Voineskos, S., De Luca, V., Mensah, A., Vincent, J.B., Potapova, N., and Kennedy, J.L. (2007) Association of α4β2 nicotinic receptor and heavy smoking in schizophrenia. *Journal of Psychiatry and Neuroscience*, **32**, 412–416.

Walsh, T., McClellan, J.M., McCarthy, S.E., Addington, A.M., Pierce, S.B., Cooper, G.M., et al. (2008) Rare structural variants disrupt multiple genes in neurodevelopmental pathways in schizophrenia. *Science*, **320**, 539–543.

Wang, Q., Liu, L., Pei, L., Ju, W., Ahmadian, G., Lu, J., et al. (2003) Control of synaptic strength, a novel function of Akt. *Neuron*, **38**, 915–928.

Weeber, E.J. and Sweatt, J.D. (2002) Molecular neurobiology of human cognition. *Neuron*, **33**, 845–848.

Welte, R., König, H.H., and Leidl, R. (2000) Cost of health damage and productivity losses attributable to cigarette smoking in Germany. *European Journal of Public Health*, **10**, 31–38.

Wesnes, K. and Warburton, D.M. (1983) Effects of smoking on rapid information processing performance. *Neuropsychobiology*, **9**, 223–229.

Winterer, G. and Weinberger, D.R. (2004) Genes, dopamine and cortical signal to noise ratio in schizophrenia. *Trends in Neurosciences*, **27**, 683–690.

Winterer, G., Coppola, R., Goldberg, T., Egan, M., Jones, D., Sanchez, C.E., et al. (2004) Prefrontal broadband noise, working memory and genetic risk for schizophrenia. *American Journal of Psychiatry*, **161**, 490–500.

Winterer, G., Musso, F., Vucurevic, G., Stoeter, P., Konrad, A., Seker, B., et al. (2006a) COMT genotype predicts BOLD signal and noise characteristics in prefrontal circuits. *NeuroImage*, **32**, 1722–1732.

Winterer, G., Musso, F., Beckmann, C., Mattay, V., Egan, M.F., Jones, D.W., et al. (2006b) Instability of prefrontal signal processing in schizophrenia. *American Journal of Psychiatry*, **163**, 1960–1968.

Winterer, G., Egan, M.F., Kolachana, B.S., Goldberg, T.E., Coppola, R., and Weinberger, D.R. (2006c) Prefrontal electrophysiologic "noise" and catechol-O-methyltransferase genotype in schizophrenia. *Biological Psychiatry*, **60**, 578–584.

Winterer, G., Musso, F., Konrad, A., Vucurevic, G., Stoeter, P., Sander, T., et al. (2007) Association of attentional network function with exon 5 variations of the

CHRNA4 gene. *Human Molecular Genetics*, **16**, 2165–2174.

Winterer, G., Konrad, A., Vucurevic, G., Musso, F., Stoeter, P., and Dahmen, N. (2008) Association of 5′ end neuregulin-1 (*NRG1*) gene variation with subcortical medial frontal microstructure in humans. *NeuroImage*, **40**, 712–718.

Wonnacott, S., Sidhpura, N., and Balfour, D.J.K. (2005) Nicotine: from molecular mechanisms to behaviour. *Current Opinion in Pharmacology*, **5**, 53–59.

Yoshii, A. and Constantine-Paton, M. (2007) BNDF induces transport of PSD-95 to dendrites through PI3K–Akt signalling after NMDA receptor activation. *Nature Neuroscience*, **10**, 702–711.

Zeiger, J.S., Haberstick, B.C., Schlaepfer, I., Collins, A.C., Corley, R.P., Crowley, T.J., et al. (2008) The neuronal nicotinic receptor subunit genes (*CHRNA6* and *CHRNB3*) are associated with the subjective response to tobacco. *Human Molecular Genetics*, **17**, 724–734.

# 9
# Progress in Psychopharmacotherapy though Molecular Imaging
*Ingo Vernaleken, Gerhard Gruender, and Paul Cumming*

## 9.1
Optimizing Psychopharmacotherapy through Molecular Imaging

### 9.1.1
Techniques for Molecular Imaging in Living Brain

Molecular imaging is an extension of classic methods employing radiopharmaceuticals for visualizing receptors and assaying enzyme activities in biological samples, now applied to the study of the living organism. This chapter considers the most widely employed forms of molecular imaging for brain research: positron emission tomography (PET) and the allied method of single photon emission computed tomography (SPECT). Countless enzyme substrates and receptor ligands have been prepared in radioactive forms, so that their binding and metabolic transformation can be monitored using laboratory techniques. Certain radionuclides have emissions of sufficient energy that they can be detected from some distance. If these radionuclides can be incorporated into a biological molecule or drug, then the distribution of the drug in the living organism can be monitored using noninvasive methods. As such, contemporary PET and SPECT procedures are derived from scintigraphy – a radiological procedure that was first employed clinically for the detection of radioactive iodine accumulation in the thyroid gland. However, instrumentation and image reconstruction procedures have developed enormously in the past five decades, such that it is now possible to monitor and quantify molecular events in brain structures as small as a pea.

The advent of molecular brain imaging has brought about a revolution in neurobiology, most specifically psychopharmacotherapy. However, PET and SPECT methods suffer from certain inherent limitations, summarized in Table 9.1.

#### 9.1.1.1 Methodologic Background
Process specificity of molecular imaging is imparted by the pharmacologic specificity of the tracer molecule, which must first be substantiated by testing *in vitro* and *in vivo*. Next, the molecule must be labeled with an appropriate radioactive isotope, depending on the requirements of PET or SPECT methods.

*Systems Biology in Psychiatric Research.*
Edited by F. Tretter, P.J. Gebicke-Haerter, E.R. Mendoza, and G. Winterer
Copyright © 2010 WILEY-VCH Verlag GmbH & Co. KGaA, Weinheim
ISBN: 978-3-527-32503-0

**Table 9.1** Advantages and limitations of PET and SPECT.

| Advantages | Limitations |
| --- | --- |
| High molecular specificity | Temporal resolution approximately 15 min |
| High process sensitivity | Spatial resolution in the range 1–10 mm |
| Can provide three-dimensional spatial information | Complex procedures for analysis of images |
| Amenable for pharmacological challenge paradigms | Radiation exposure to subjects and experimenters |
| Kinetically defined outcome parameters ($B_{max}$, occupancy, etc.) | Costs approximately US$1000 per investigation |

**PET** The nuclei of positron-emitting isotopes are deficient in neutrons. These isotopes undergo physical decay, with the emission of a positron ($\beta^+$), and the reconfiguration of the nucleus into a stable form, with atomic number reduced by one. The tracer molecule carries within its chemical structure the unstable isotope for some time, until the inevitable decay. This event releases the positron particle from wherever in the living organism the tracer molecule happens to be at the time of decay. The positron then travels a relatively brief distance (in spite of its high kinetic energy) before encountering its counterpart, an electron ($\beta^-$). The star-crossed encounter of the two particles leads to their mutual annihilation, with the rest mass-energy ($E = mc^2$) of the matter and antimatter pair released as two high-energy (511-keV) $\gamma$-rays, which diverge at very nearly 180° from the point of annihilation.

Positron-emitting isotopes characteristically have very brief physical half-lives and must be prepared at a cyclotron facility. Given its 2-h half-life, $^{18}$F can be distributed from a central production facility, but the briefer half-life of the other common isotopes imposes the requirement of an on-site cyclotron facility (Table 9.2). This is an important financial consideration, since radiochemistry laboratories are much less expensive than cyclotron facilities.

Positrons have a characteristic mean path in brain tissue (Table 9.2). This imposes a limitation on the maximum spatial resolution to be obtained with a given isotope, since the free path prior to decay is a stochastic process, resulting in uncertainty about the point of origin of a given positron. In this regard, $^{18}$F is distinctly superior to

**Table 9.2** PET isotopes: half-lifes and spatial resolution.

| Isotope | Half-life (min) | Spatial resolution loss due to mean free path (mm)[a] |
| --- | --- | --- |
| $^{11}$C | 20.4 | 0.96 |
| $^{13}$N | 10 | 1.26 |
| $^{15}$O | 2.2 | 1.87 |
| $^{18}$F | 109.8 | 0.56 |

a) According to Sanchez-Crespo, Andreo, and Larsson (2004); soft-tissue data.

$^{15}$O. However, it is only in recent years that the spatial resolution of tomographs has approached the limits imposed by the positron free path.

PET imaging is based upon the synchronous detection of two γ-rays, presumed to originate from a single decay event lying somewhere between the two detectors, in a vector known as the line of response (LOR). The very first positron emission detectors consisted of a pair of detectors; however, tomographic imaging required the innovation of a circular ring or polygon of detectors, such that spatial information about the radiation field could be obtained (Figures 9.1 and 9.2). This basic design principle has not changed since the mid-1960s, but current instruments consist of dozens of detector rings, each containing hundreds of detector elements. These instruments can operate in two-dimensional mode, in which each ring is used to produce a single slice, or three-dimensional mode, in which coincidence events from different rings are registered. The three-dimensional process improves sensitivity, but with a penalty due to increase registering of random events and much greater computational requirements for the image reconstruction. A raw PET recording obtained for a given interval consists of a sinogram, showing the frequency of decay events observed for each LOR. Using the computational procedure of filtered back projection, an image of radioactivity concentration is calculated for each slice.

**SPECT** As in PET, single-photon-emitting isotopes are chemically bound to the molecule of interest. However, the SPECT isotopes decay by direct emission of a single γ-ray, with energy of less than 200 keV. Consequently, the localization of these decay events cannot be defined by a LOR, but relies upon collamination: the SPECT detectors are shielded with lead so as to admit only γ-rays with a narrow range of trajectories. In modern SPECT instruments, two or three γ-cameras rotate around the subject, pausing at intervals of 3–6°, so as to obtain spatial information about the radiation field. With the exception of pinhole SPECT, in which radioactivity is

**Figure 9.1** A polygonal array of photoelectron detector crystals and photoelectron multipliers along with associated circuitry within a PET scanner. (Reproduced with kind permission from Dr. Kieran Maher)

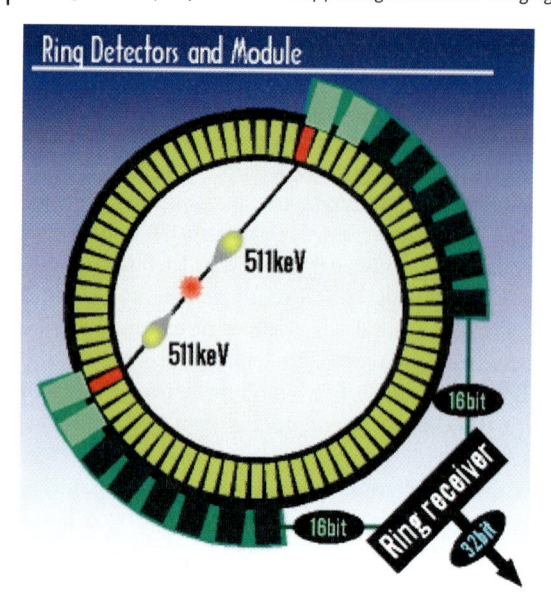

**Figure 9.2** Basic principal of annihilation, LOR, and temporal coincidence. (With kind permission from the University of British Columbia, TRIUMF.)

measured from a single "point source," SPECT has inherently lower spatial resolution than does PET; the best modern instruments have obtained a resolution of 7 mm. In the context of biological psychiatry, this resolution may be just adequate to resolve the caudate nucleus and the putamen.

SPECT has a considerable advantage with respect to procurement of the radio-isotopes. The widely used isotope $^{99m}$Tc is generated on site using $^{99m}$Tc generators that contain $^{99}$Mo as parent compound and decays with a half-life of 6 h. As such, a single radiosynthesis can last a day or more, which allows for long scanning protocols and easier management of scanning schedules. $^{123}$I, another widely used SPECT isotope, is generated at a nuclear reactor and can be widely distributed, given its half-life of 13 h.

**Radiotracers** Radiotracers for PET and SPECT are obtained by syntheses in which the radioactive atom is introduced into a precursor molecule through conventional chemical reactions. The product is then prepared in pure and sterile form, suitable for intravenous injection into a living subject. Radiotracer synthesis is subject to the limitations of what is chemically possible and convenient; simple and rapid synthesis is especially important in the case of short-lived PET isotopes. Many pharmaceuticals contain in their structure *N*-alkyl groups: [$^{11}$C]methyl-iodide is easy to prepare and reacts readily with desmethylated precursor substances. [$^{18}$F]Fluoroethyl substitution presents another route to synthesis of PET tracers. Radiosynthesis involving several steps can suffer from low yield, due to extensive radioactive decay. In the case of $^{11}$C chemistry, specific activity can be a particular problem; the radiochemical reaction can be poisoned by ambient carbon dioxide, resulting is inadequate specific activity of the tracer, such that an excessive mass must be injected to the subject. Inadequate monitoring of the specific activity of $^{11}$C ligands can result in undesirable pharmacological effects; an ideal PET tracer is of such high specific activity that the amount of mass injected is practically zero (in this respect, $^{18}$F presents a distinct advantage over $^{11}$C). Finally, it must be considered that radiochemical modification of a drug may result in loss of biological activity or specificity of the resultant tracer. The ideal molecular imaging agent (and there are few) satisfies the following conditions:

- High radiochemical yield and specific activity at the end of synthesis.
- High affinity for the target receptor.
- High specificity for the target receptor or process.
- Ready transfer across the blood–brain barrier.
- Low nonspecific binding.
- Absence of brain-penetrating metabolites.
- Rapid kinetics, relative to the time limits imposed by physical decay of the tracer.

**Data Analysis** Kinetic analysis is central to molecular imaging. PET and SPECT measure simply the amount of radioactivity in a given brain region as a function of time after tracer injection. For the case of receptor ligands, the initial phase of a PET or SPECT recording reveals the partitioning of the tracer across the blood–brain barrier, which is defined by the clearance to brain ($K_1$, which has units of cerebral blood flow) and the fractional rate constant for diffusion back to blood ($k_2$). The time–activity curve (TAC) measured in a brain region devoid of specific binding can be interpreted relative to the TAC measured in serial arterial blood samples (known as the arterial input) in terms of these two kinetic parameters. In brain regions containing the binding site of interest, the ligand may associate with its receptor at a rate constant known as $k_3$ (which is proportional to the local abundance or $B_{max}$ of the binding site) and dissociate at rate constant $k_4$. Given sufficient time, an equilibrium state will arise, in which the proportions of tracer in bound and unbound state are fairly constant. This dynamic process can be seen qualitatively in Figure 9.3, where the

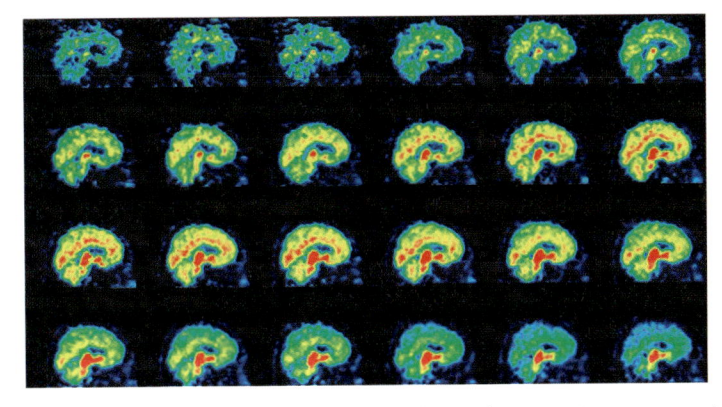

**Figure 9.3** Time-series of PET recordings in the initial minutes (upper two rows) after injection of the serotonin uptake site ligand [$^{11}$C]DASB. (Reproduced with kind permission from Dr. Jan Kalbitzer and Professor Gitte Moos Knudsen, Neurobiology Research Unit, Copenhagen University.)

tracer initially distributes homogeneously throughout brain, and then washes out of nonbinding regions, while "sticking" in brain regions richly endowed with the binding site for the tracer.

For many PET and SPECT tracers of receptors, the end-point is this equilibrium condition, defined by the ratio ($k_3/k_4$), which is commonly known as the binding potential ($BP_{ND}$). This index of specific binding is calculated relative to the radioactivity in a nonbinding reference region, such as cerebellum, as indicated by the subscript ND ("nondisplaceable"). A more rigorous definition of $BP_{ND}$ in terms of Michaelis–Menten kinetics is:

$$BP_{ND} = \frac{(B_{AVAIL} \times f_{ND})}{K_D},$$

where $B_{AVAIL}$ is the total number of binding sites ($B_{max}$), $f_{ND}$ is the free fraction of tracer and $K_D$ is the half-saturation concentration of the ligand.

## 9.2
### Characterization of Neurotransmitter Systems with Molecular Imaging

PET and SPECT studies of neurotransmission are an extension of classical methods employing neuroreceptor ligands and substrates for enzymatic assays, but applied to the case of the living brain. As such, molecular imaging methods can suffer from the same limitations as classical pharmacology, namely the incomplete specificity of ligands.

The dopamine $D_2$-like receptor was the first neuroreceptor to be examined by PET in humans using the butyrophenone ligand [$^{11}$C]N-methylspiperone ([$^{11}$C]NMSP) (Wagner et al., 1983; Wong et al., 1984). In the presence of a nonradioactive competitor, [$^{11}$C]NMSP has been used to estimate the saturation binding parameters ($B_{max}$ and $K_D$) for $D_2$-like receptors in living human brain (Wong et al., 1986a), with the caveat that [$^{11}$C]NMSP-PET reveals the composite of dopamine $D_2$, $D_3$, and $D_4$, as well as serotonin (5-hydroxytryptamine) 5-HT$_{2A}$ receptors. Greater specificity is afforded by the benzamide [$^{11}$C]raclopride, which binds with equal affinity to dopamine $D_2$ and $D_3$ receptors (Ehrin et al., 1985). The development of [$^{123}$I]iodobenzamide ([$^{123}$I]IBZM) made SPECT studies of dopamine $D_{2/3}$ receptors possible (Kung et al., 1989). Whereas all of the aforementioned ligands are suited for detecting $D_2$ receptors where they are abundant (in striatum), detection of the cortical binding sites required development of ligands of considerably higher affinity, such as [$^{18}$F]fallypride and [$^{11}$C]FLB 457 (de Paulis, 2003) (see Figure 9.4).

PET studies with the diverse ligands have confirmed that the abundance in striatum of dopamine $D_2$-like receptors declines by almost 7% with each decade of healthy aging (Antonini et al., 1993; Mukherjee et al., 2002; Volkow et al., 1998). Physiological asymmetry in $D_2$ abundance of some 10% is evident in the caudate nucleus of healthy young subjects, but declines as a function of healthy aging (Vernaleken et al., 2007a).

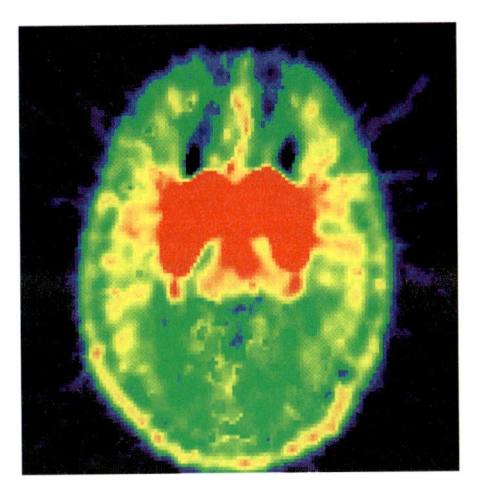

**Figure 9.4** Parametric [$^{18}$F]fallypride $BP_{ND}$ image of an unmedicated healthy subject. Beside striatal structures, cortical and thalamical regions also show significant ligand binding.

[$^{11}$C]Raclopride and other benzamide ligands bind to dopamine receptor in a manner sensitive to competition from endogenous dopamine in the interstitial milieu: depletion of dopamine with reserpine decreased the apparent $K_D$ of [$^{3}$H]raclopride in mouse striatum, whereas amphetamine-evoked dopamine release increased the apparent $K_D$ (Ross and Jackson, 1989). Once this was appreciated, it became possible to conduct an array of pharmacological challenges of the nigrostriatal dopamine system using PET and SPECT competition protocols. For example, the binding of [$^{11}$C]raclopride in striatum of living pig was reduced by treatment with the designer amphetamine MDMA (3,4-methylenedioxymethamphetamine; "Ecstasy"), revealing the MDMA-evoked dopamine release (Rosa-Neto et al., 2004). The competition paradigm can be used to test interactions between neurotransmitter systems. For example, serotonin release evoked by fenfluramine decreases the binding of [$^{11}$C]raclopride in human striatum (Smith et al., 1997); similar effects could be seen in striatum of pigs challenged with a direct serotonin agonist, the potent hallucinogen LSD (Minuzzi et al., 2005), and in a microPET study of rat striatum after challenge with a selective 5-HT$_{2C}$ antagonist (Egerton et al., 2008). Altered [$^{11}$C]raclopride binding can also be obtained in cognitive challenge paradigms. In an influential study, performance of a video game for monetary reward evoked a decline in striatal [$^{11}$C]raclopride binding, presumably due to dopamine release (Koepp et al., 1998).

Although great emphasis has been placed on PET and SPECT studies of dopamine $D_2$ receptors, $D_1$-like receptors are also abundant in striatum and also in cerebral cortex, where they are implicated in aspects of cognition. PET studies of $D_1$ receptors became possible with the advent of [$^{11}$C]SCH 23390 (Farde et al., 1987a) and [$^{11}$C]NNC 112. However, [$^{11}$C]NNC 112 is now understand to bind extensively to serotonin 5-HT$_2$ receptors in cortex (Slifstein et al., 2007). The binding of $D_1$ receptor ligands in monkey striatum was unaffected by pharmacological challenge with reserpine or amphetamine (Chou et al., 1999). So far, the competition model has been most clearly established for the case of dopamine $D_2$ receptors and has not been generalized to other neurotransmitter receptor types.

Almost all dopamine receptors in striatum are expressed on the medium spiny neurons, which comprise an important element of the cortico-striatal-thalamic loop. The integrity of presynaptic dopamine terminals can be assessed in molecular imaging studies with ligands for the dopamine transporter (DAT). Successful DAT ligands include [$^{11}$C]cocaine (Logan et al., 1990), [$^{11}$C]methylphenidate (Doudet et al., 2006), and [$^{99m}$Tc]TRODAT-1 – a SPECT tracer (Kung et al., 1996) that has found wide use for diagnosis of nigrostriatal degeneration.

The DOPA (3,4-dihydroxyphenylalanine) decarboxylase substrate 6-[$^{18}$F]fluoro-DOPA ([$^{18}$F]6-[$^{18}$F]fluoro-3,4-dihydroxyphenylalanine[$^{18}$F]FDOPA) occupies a special position in the history of molecular brain imaging, as one of the very first successful PET tracers of the dopamine system (Garnett, Firnau, and Nahmias, 1983), and by virtue of its almost unique mechanism. Whereas all the above-mentioned ligands bind to a receptor or transporter, [$^{18}$F]FDOPA is metabolically trapped as a tracer in the pathway for dopamine synthesis. As such, [$^{18}$F]FDOPA modeling is based upon the [$^{18}$F]fluorodeoxyglucose ([$^{18}$F]FDG) method for the analysis of cerebral glucose consumption. However, the analysis of [$^{18}$F]FDOPA uptake to striatum is notoriously complex and requires extensive simplification of the kinetic model (Cumming and Gjedde, 1998). Among the alternate tracers for dopamine synthesis, [$^{11}$C]levodopa presents some advantages with respect to simplicity of analysis (Hartvig et al., 1992), but has not become widely used due to its difficult radiosynthesis.

## 9.3
### Focus Schizophrenia

Schizophrenia is a difficult disease for its sufferers and is only partially responsive to pharmacotherapy. By its nature, schizophrenia has high intersubject variability and a multifactorial origin, unlikely to be attributable to a unique or focal pathophysiological mechanism. In addition to the well-known positive symptoms, patients also suffer from more subtle disturbances of cognition, particularly selective attention, speech production, working memory, and set shifting. Furthermore, patients are impaired in their social interactions, mood regulation, and incentives. Finally, schizophrenia patients suffer from disturbances in several somatic systems, manifesting in immunological changes and generally shorter life expectancies, independent of antipsychotic treatment. No uniform time-course of the disease progress can be defined and in most cases the disease has gradual onset, preceded by prodromal states lasting several years. Although schizophrenia has approximately 50% heritability, several generations of genetic investigations have identified only a very few susceptibility genes, with weak odds ratios. Whereas the symptoms of Parkinson's disease can, to the first approximation, be attributed to a simple dopamine deficiency, the protean nature of schizophrenia suggests a disturbance in brain networks, rather than a focal lesion.

According to the dopamine hypothesis of schizophrenia, the positive schizophrenic symptoms such as hallucinations and delusions result from a functional excess of dopaminergic transmission. The claim was based upon clinical observations that typical antipsychotic drugs block dopamine $D_2$-like receptors and that indirect dopamine agonists such as cocaine or amphetamine can induce an acute psychosis similar to symptoms of the paranoid subtype of schizophrenia (Carlsson, 1988). Based on the dopamine hypothesis, molecular imaging has been used extensively for studying the pathophysiology of schizophrenia and for optimizing pharmacotherapy with dopamine $D_2$-like antagonists. The early report of increased [$^{11}$C]NMSP binding in patients with schizophrenia (Wong et al., 1986b) was not replicated in a subsequent study using [$^{11}$C]raclopride (Farde et al., 1987b). In a subsequent [$^{11}$C]NMSP study, it was reported that the age-related decline in $D_2$-like binding was accelerated in patients (Wong et al., 1997), such that the detection of disease-related differences was critically sensitive to the age composition of the study groups. A meta-analysis suggests that $D_2$ receptor binding is increased in patients by 5–10% (Weinberger and Laruelle, 2001). However, the intersubject variability is markedly high; again, suggesting profound heterogeneity of the disorder. The initial finding of increased striatal $D_{2/3}$ binding in early schizophrenia was supported by some congruent results in unaffected twins of patients (Frankle, 2007) as well as in of postmortem studies of schizophrenia patients (Soares and Innis, 1999).

As noted above, high-affinity ligands have made possible the detection of dopamine $D_{2/3}$ receptors in extrastriatal regions, notably in cerebral cortex and in thalamus. Binding of [$^{11}$C]FLB 457 to dopamine $D_{2/3}$ receptors was reduced in the thalamus of patients with schizophrenia, notably in the dorsomedial division that is functionally and anatomically associated with the nucleus accumbens and frontal cortex (Suhara et al., 2002). Likewise, reduced [$^{18}$F]fallypride binding was seen in the thalamus of patients with schizophrenia, in whom binding was also reduced in the cingulate and temporal cortices (Buchsbaum et al., 2006). Others found [$^{18}$F]fallypride binding to be intact or elevated in cerebral cortex of patients, but significantly increased in the substantia nigra (Kessler et al., 2009; Vernaleken et al., 2008a). In all of these studies, the relative standard deviation of $D_{2/3}$ receptor availability was greater in the patients; as always, it must be considered that disease duration and heterogeneity of disease symptoms complicate the interpretation of these data. If schizophrenia is acknowledged to be a heterogeneous condition, it may be that PET studies in very large cohorts could provide the basis for stratification of patients into distinct categories.

The $D_2$ receptor competition model described above has been employed to great effect for the study of schizophrenia. Dopamine depletion brought about with the tyrosine hydroxylase inhibitor α-methyl-p-tyrosine (AMPT) increased the availability of dopamine receptors measured by [$^{123}$I]IBZM-SPECT to a greater extent in patients with schizophrenia (+19%) than in healthy controls subjects (+9%), indicating higher basal occupancy at the $D_2$ receptors by endogenous dopamine (Abi-Dargham et al., 2000). Conversely, amphetamine-induced dopamine release provoked a greater decrease [$^{123}$I]IBZM binding in patients, indicative of the presence of a larger amphetamine-releasable dopamine pool than in healthy control subjects (Abi-Dargham et al., 1998). The same research group has shown that

that amphetamine-induced decrease correlated with AMPT-induced increases in [$^{123}$I]IBZM binding in patients, but not in control subjects (Abi-Dargham et al., 2009). This phenomenon may be due to a Type II error, given that the dynamic range of the competition is manifestly greater in patients, who might be characterized as possessing a more pharmacologically labile (and variable) dopamine system.

There have been relatively few studies of dopamine $D_1$ receptors in schizophrenia. Healthy carriers of the Val/Val allele of catechol-O-methyltransferase – a genotype that is commonly associated with higher risk of schizophrenia – had higher [$^{11}$C]NNC 112 binding in striatum (Slifstein et al., 2008). A [$^{11}$C]SCH 23390 PET study in 10 drug-naïve patients did not reveal any difference in striatal binding, as compared to healthy controls (Karlsson et al., 2002). High [$^{11}$C]NNC 112 binding in dorsolateral prefrontal cortex (PFC) correlated with poor performance in working memory in patients with schizophrenia (Abi-Dargham et al., 2002). Elevated [$^{11}$C]SCH 23390 binding in the medial PFC and other cortical areas was associated with greater risk for schizophrenia in a study of twins discordant for the trait; this study furthermore showed that binding was lower in proportion to the dose of antipsychotic medication (Abi-Dargham et al., 2002; Hirvonen et al., 2006).

The dopamine hypothesis of schizophrenia might predict an elevated density of dopamine innervation in the basal ganglia of patients. However, the binding of the SPECT ligand [$^{123}$I]N-ω-fluoropropyl-2β-carbomethoxy-3β-(4-iodophenyl)-nortropane to DAT was entirely normal in striatum of patients with schizophrenia, irrespective of medication status (Lavalaye et al., 2001).

An early [$^{18}$F]FDOPA study showed that dopamine synthesis capacity was elevated in striatum of untreated patients with schizophrenia and also among epileptic patients suffering from interictal psychosis (Reith et al., 1994). This finding has been replicated in a number of independent PET studies, is particularly associated with the positive symptoms, and seems to be most pronounced in the (limbic) ventral striatum (Hietala et al., 1995; McGowan et al., 2004). Using [$^{11}$C]levodopa, others report increased dopamine synthesis throughout striatum and in the medial PFC (Lindstrom et al., 1999) or only in the left caudate nucleus (Nozaki et al., 2009) of patients with schizophrenia. Recently, [$^{18}$F]FDOPA influx was found to be elevated in striatum of first-degree relative of patients (Huttunen et al., 2008) and in patients with prodromal symptoms of schizophrenia (Howes et al., 2009). Based upon very advanced FDOPA kinetic modeling, we have recently shown that [$^{18}$F]FDOPA utilization is indeed elevated in striatum of patients and that the rate of consumption of [$^{18}$F]fluorodopamine is nearly doubled (Kumakura et al., 2007). The net effect of this "churning" of the dopamine pathway was reduced steady-state dopamine storage – a circumstance which we described as "poverty in the midst of plenty."

According to textbooks in neurochemistry, the rate-limiting step in the pathway for dopamine synthesize is mediated by tyrosine hydroxylase; we have shown that the activity of DOPA decarboxylase can also influence dopamine synthesis in brain, due to the presence of alternate pathways for levodopa in brain (Cumming and Gjedde, 1998). Consistent with this model, acute treatment with a dopamine receptor antagonist increased the rate of [$^3$H]levodopa decarboxylation in striatum of living rats, presumably due to blockade of autoreceptors tonically inhibiting dopamine synthesis (Cumming et al., 1997). Using FDOPA-PET, we have reported similar regulation of DOPA decarboxylase in anesthetized pigs (Danielsen et al., 2001), and in healthy humans (Vernaleken et al., 2006), in whom baseline [$^{18}$F]FDOPA kinetics correlated with performance of executive functions of the frontal cortex (Vernaleken et al., 2007b) and also predicted the cognitive effects of challenge with haloperidol (Vernaleken et al., 2008b). In contrast to the findings in healthy subjects, others report that [$^{18}$F]FDOPA utilization correlated inversely with frontal cortex function of patients with schizophrenia, as measured by a functional magnetic resonance imaging paradigm (Meyer-Lindenberg et al., 2002). Together, these findings link striatal dopamine transmission and frontal cortical function. We have found that the utilization of [$^{18}$F]FDOPA in striatum of patients with schizophrenia declined after a subacute treatment with haloperidol (Grunder et al., 2003); we attributed this phenomenon to the onset of depolarization block – a decline in the firing rate of dopamine neurons seen in rats after prolonged antipsychotic treatment.

Cortical glutamatergic neurons densely innervate the striatum, where signaling is mediated by N-methyl-D-aspartate (NMDA) receptors. Pharmacologic blockade of these receptors can transiently provoke in healthy subjects schizophrenia-like symptoms, both negative and positive symptoms. In one [$^{11}$C]raclopride PET investigation, NMDA antagonism with ketamine reduced the availability of dopamine $D_{2/3}$ receptors in striatum of healthy control subjects, suggesting a mechanism whereby a primary dysfunction of frontal cortex could propagate to dysregulation of striatal dopamine signaling (Vollenweider et al., 2000). Similarly, challenge with a weak NMDA receptor antagonist, amantadine, stimulated [$^{18}$F]FDOPA uptake in ventral striatum of healthy subjects (Deep et al., 1999). Cortico-striatal as well as striato-thalamical pathways are mediated by inhibitory γ-aminobutyric acid (GABA)-ergic interneurons and projection neurons, with direct and indirect influence on the brainstem, striatum, and thalamus. Depending on the number of synapses involved, the net influence may be excitatory or inhibitory, but inhibitory processes seem to predominate in the healthy brain. Dopaminergic transmission in cortex and striatum can modulate glutamatergic and GABA-ergic neuron activity. In general, $D_1$ and $D_2$ receptors frequently exert opposite effects with respect to effects on membrane potential, target pathways, and the time-course of action (Vernaleken, Cumming, and Grunder, 2008). Computational models of these circuits suggest that dopamine serves a tuning function for "signal-to-noise" optimization (Winterer, 2007). In this hypothesis, $D_1$ transmission, especially in medial PFC, facilitates the response to salient stimuli, while reducing sensitivity to signals that are to be ignored. Excessive signaling mediated by $D_2$-like receptors opposes this effect, leading to inappropriate emotional processing, and inability to filter extraneous stimuli.

## 9.4
### Action of Psychopharmaceuticals in Schizophrenia

The protean clinical nature and unpredictable course of schizophrenia, and the lack of clearly defined pathology, suggest that the disease may more properly be called a syndrome, much as the descriptive term "fever" referred in former times to diverse clinical entities, requiring quite different treatment. Similarly, unipolar depression seems to arise from diverse causes, and can be treated with unrelated classes of medications targeting different neurotransmitter systems (noradrenaline, serotonin), mood stabilizers of uncertain mechanism, entirely nonselective treatments such as electroconvulsant shock therapy, and indeed, nonpharmacological interventions such as psychotherapy and aerobic exercise. In contrast, all approved antipsychotic drugs exert antagonism at $D_2$ receptors or otherwise impair dopamine transmission (i.e., dopamine depletion with reserpine). This common treatment strategy did not arise as a rational consequence of neuropathological findings, as in the case of levodopa for the treatment of Parkinson's disease. Rather, the first effective antipsychotic drugs were discovered accidently, in the course of developing new classes of sedatives (Delay, Deniker, and Harl, 1952). Some decades later, their antipsychotic action was linked to effects on the dopamine system (Carlsson and Lindqvist, 1963). Still later, once dopamine receptor pharmacology had been characterized, it was reported that average daily doses of diverse antipsychotic medications used in clinical practice correlated with their affinity for blocking dopamine $D_2$ receptors, which was interpreted to suggest a causal mechanism (Seeman and Lee, 1975).

Clinical effective doses of any of the early antipsychotic medications (chlorpromazine, haloperidol) can cause a severe syndrome of parkinsonism, known as extrapyramidal side-effects (EPMSs). Indeed, this is to be expected if dopamine transmission is substantially blocked. In the late 1970s, new antipsychotic drugs were discovered through behavioral screening in animals, rather than rational pharmacological strategy. The new drugs, the prototype of which is clozapine, had little or no propensity to evoke EPMS, but nonetheless proved to be effective antipsychotics and came to be known in clinical practice as atypical antipsychotics. The term "atypicality" seems to refer to any drug for which the antipsychotic properties cannot be solely linked to dopamine receptor antagonism, but also to some other mechanism. Molecular imaging results have been instrumental in the more precise pharmacological definition of atypicality, as will be reviewed below.

Insofar as dopamine $D_2$ receptor antagonism has been linked to antipsychotic efficacy, it might be asked what extent of receptor blockade should be required in order to produce antipsychotic effects, without crossing the threshold to EPMSs. Competitive binding paradigms with PET or SPECT ligands selective for dopamine $D_{2/3}$ receptors have been brought to bear on this topic; by comparing the receptor availability in control subjects or untreated patients to that seen in patients under treatment with receptor antagonists, it has been shown that clinical response occurs when dopamine $D_{2/3}$ receptor occupancies exceed 60%, whereas EPMS symptoms emerge when striatal occupancies exceed 80% (Farde et al., 1992; Nordstrom et al., 1993). It is still debated whether 60% occupancy is merely a threshold with independent factors influencing treatment (non)response or if it arises from a correlation between $D_{2/3}$ receptor occupancy and treatment effect. As the early receptor occupancy PET studies used low-affinity ligands and first-generation antipsychotics, all of these thresholds were defined for striatal $D_2$ receptor binding only. It is of course possible that the locus of therapeutic action of antipsychotics lies outside the striatum; it might still be the case that it is the cortical blockade that is critical for antipsychotic action, whereas the striatum blockade might be an unfortunate epiphenomenon. Furthermore, it might be that antipsychotic drugs establish differential levels of $D_{2/3}$ receptor occupancy in striatal and extrastriatal regions. Recent PET studies with the high-affinity ligands [$^{18}$F]fallypride or [$^{11}$C]FLB 457 have been used to test for differential occupancies in cortex and striatum, revealing several discrepancies. For many "atypicals" (more appropriately called second-generation antipsychotics) with moderate affinity to $D_{2/3}$ receptors (i.e., ziprasidone, amisulpride, risperidone, and olanzapine), the striatal occupancy was 10–20% lower than that seen in cerebral cortex, over a broad range of clinically relevant plasma concentrations (Vernaleken, Cumming, and Grunder, 2008). Clozapine, which evokes absolutely no EPMSs, has rather low affinity for $D_{2/3}$ receptors, such that striatal occupancies greater than 60% cannot be achieved under clinical conditions. In extrastriatal regions, however, clozapine occupancy exceeded the 60% threshold (Grunder et al., 2006). In contrast, preferential extrastriatal binding has never been demonstrated for first-generation antipsychotics, which are most apt to produce EPMSs. However, it must be mentioned that several studies using [$^{11}$C]FLB 457 and also a low-affinity ligand (dual-tracer approach) found uniform occupancy by second-generation antipsychotics (for review, see Frankle, 2007). However, a meta-analysis of $D_{2/3}$ receptor occupancy studies supports claims of preferential extrastriatal binding by second-generation antipsychotics and confirms the validity of the occupancy window mainly for the cerebral cortex (first-generation antipsychotics: 77%; second-generation antipsychotics 67% receptor occupancy). Corresponding clinically effective occupancies in striatum were 74% for typical versus only 49% for second-generation antipsychotics (Stone et al., 2009).

In spite of this evidence suggesting antipsychotic efficacy at cortical dopamine receptors, there is reported to be a significant correlation specifically with $D_{2/3}$ receptor occupancy (measured by [$^{11}$C]raclopride and [$^{11}$C]FLB 457) and clinical improvement of positive symptoms (Agid et al., 2007). However, low-affinity antipsychotics such as clozapine and quetiapine were not included in this study. It seems questionable whether such a correlation would remain if drugs imposing less than 60% receptor occupancy under clinical effective doses were also considered.

In an [$^{18}$F]FDG-PET study, the typical antipsychotic haloperidol increased energy metabolism in striatum in proportion to the clinical outcome (Bartlett et al., 1998). However, the argument still applies that processes in the striatum may be a surrogate parameter for other cerebral regions. The relatively low abundance of $D_{2/3}$ receptors in cortex does not necessarily mean that they are unimportant; animal studies have shown important cognitive effects to be mediated by $D_{2/3}$ receptors, especially in the PFC (Rinaldi et al., 2007). Unfortunately, even high-affinity PET ligands are unsuited for quantitation in PFC, so occupancy measurements in temporal cortex, for example, may surrogates for the unmeasurable occupancies in key brain regions subserving aspects of disturbed cognition in schizophrenia.

$D_1$ receptor agonists (e.g., DAR 0100) potentially may have beneficial effects on cognitive functioning levels in schizophrenia (Mu et al., 2007). As noted above, $D_1$

receptor-mediated transmission in the PFC may serve to identify relevant stimuli and modulate selective attention, but $D_{2/3}$ receptors carry in some sense more weight, by decreasing the threshold for neuronal activation and perception. Thus, informational processing is less focused but highly sensitive. In this theoretical model, with opposing effects of $D_1$- and $D_2$-like signaling in the cerebral cortex (the same arguments apply for striatum), $D_2$ receptor antagonism have a key position for rectifying the cognitive dysfunction of schizophrenia. Given the disruptive effect of agonism at thalamic $D_2$ receptors on sensory gating (Young, Randall, and Wilcox, 1995) and the reports of altered thalamic binding cited above, an integrated model of the mechanism of antipsychotic action must consider not just striatum and PFC, but thalamus also.

This chapter, like most psychopharmacology research of schizophrenia, has placed almost entire emphasis on the dopamine system. Serotonin is among the many other neurotransmitter systems that might also be involved in the pathology and therapeutics of schizophrenia. Touching briefly on this topic, PET occupancy studies with selective serotonin ligands have been used to investigate the nature of atypicality. Treatment of with clozapine evoked 85% occupancy at serotonin 5-$HT_2$ receptors in cerebral cortex of patients with schizophrenia (Nordstrom, Farde, and Halldin, 1993). In a triple-tracer PET study, flupentixol at a therapeutic dose evoked 20% occupancy at cortical serotonin 5-$HT_2$ receptors, 20% occupancy at striatal dopamine $D_1$ receptors, and 60% occupancy at striatal $D_2$ receptors (Reimold et al., 2007). In another triple-tracer PET study, the second-generation antipsychotic aripiprazole (which uniquely exerts agonism at $D_2$ receptors) evoked 90% occupancy at striatal dopamine $D_{2/3}$ receptors, 55% occupancy at cortical serotonin 5-$HT_2$ receptors, and 16% occupancy at cortical serotonin 5-$HT_{1A}$ receptors (Mamo et al., 2007). Many, but not all, second-generation antipsychotics exert considerable levels of 5-$HT_{2A}$ receptor occupancies. These observations culminated in the – finally unsuccessful – attempt to use a selective 5-$HT_{2A}$ antagonist (M 100907) for treatment of psychotic symptoms. In fact, molecular imaging methods have given only scanty evidence for any abnormality in serotonin systems of patients with schizophrenia. The density of serotonin transporters labeled with [$^{11}$C]-3-amino-4-(2-dimethylaminomethylphenylsulfanyl)benzonitrile ([$^{11}$C]DASB) was entirely normal in patients with schizophrenia (Frankle et al., 2005). PET studies with antagonists for 5-$HT_{1A}$ receptors have no differences in patients with schizophrenia (Frankle et al., 2006) or an increase in cortical binding (Tauscher et al., 2002). However, the abundance of serotonin 5-$HT_{2A}$ receptors labeled with [$^{18}$F]altanserin was low in frontal cortex of a small group of prodromal patients at risk for psychosis (Hurlemann et al., 2005) Others report no change at all in 5-$HT_{2A}$ binding (Verhoeff et al., 2000) or no cortical change, but an increase in striatal binding among unmediated first episode patients (Erritzoe et al., 2008). As is frequently the case in molecular imaging studies of schizophrenia, factors such as stage of disease, medication status, and clinical heterogeneity may all tend to obscure possible abnormalities of serotonin in schizophrenia.

Occupancies by antipsychotic drugs at other classes of serotonin receptors, and many other types of neurotransmitter receptors (i.e., opioid, adrenergic, histamine), remain entirely un-investigated by molecular imaging techniques.

# References

Abi-Dargham, A. et al. (1998) Increased striatal dopamine transmission in schizophrenia: confirmation in a second cohort. *American Journal of Psychiatry*, **155**, 761–767.

Abi-Dargham, A. et al. (2000) Increased baseline occupancy of $D_2$ receptors by dopamine in schizophrenia. *Proceedings of the National Academy of Sciences of the United States of America*, **97**, 8104–8109.

Abi-Dargham, A. et al. (2002) Prefrontal dopamine $D_1$ receptors and working memory in schizophrenia. *Journal of Neuroscience*, **22**, 3708–3719.

Abi-Dargham, A., Giessen, E.V., Slifstein, M., Kegeles, L.S., and Laruelle, M. (2009) Baseline and amphetamine-stimulated dopamine activity are related in drug-naive schizophrenic subjects. *Biological Psychiatry*.

Agid, O. et al. (2007) Striatal vs extrastriatal dopamine $D_2$ receptors in antipsychotic response – a double-blind PET study in schizophrenia. *Neuropsychopharmacology*, **32**, 1209–1215.

Antonini, A. et al. (1993) Effect of age on $D_2$ dopamine receptors in normal human brain measured by positron emission tomography and $^{11}$C-raclopride. *Archives of Neurology*, **50**, 474–480.

Bartlett, E.J. et al. (1998) Effect of a haloperidol challenge on regional brain metabolism in neuroleptic-responsive and nonresponsive schizophrenic patients. *American Journal of Psychiatry*, **155**, 337–343.

Buchsbaum, M.S. et al. (2006) $D_2/D_3$ dopamine receptor binding with [F-18] fallypride in thalamus and cortex of patients with schizophrenia. *Schizophrenia Research*, **85**, 232–244.

Carlsson, A. (1988) The current status of the dopamine hypothesis of schizophrenia. *Neuropsychopharmacology*, **1**, 179–186.

Carlsson, A. and Lindqvist, M. (1963) Effect of chlorpromazine or haloperidol on formation of 3methoxytyramine and normetanephrine in mouse brain. *Acta Pharmacologica et Toxicologica*, **20**, 140–144.

Chou, Y.H., Karlsson, P., Halldin, C., Olsson, H., and Farde, L. (1999) A PET study of $D_1$-like dopamine receptor ligand binding during altered endogenous dopamine levels in the primate brain. *Psychopharmacology*, **146**, 220–227.

Cumming, P. and Gjedde, A. (1998) Compartmental analysis of dopa decarboxylation in living brain from dynamic positron emission tomograms. *Synapse*, **29**, 37–61.

Cumming, P., Ase, A., Laliberte, C., Kuwabara, H., and Gjedde, A. (1997) *In vivo* regulation of DOPA decarboxylase by dopamine receptors in rat brain. *Journal of Cerebral Blood Flow and Metabolism*, **17**, 1254–1260.

Danielsen, E.H., Smith, D., Hermansen, F., Gjedde, A., and Cumming, P. (2001) Acute neuroleptic stimulates DOPA decarboxylase in porcine brain *in vivo*. *Synapse*, **41**, 172–175.

de Paulis, T. (2003) The discovery of epidepride and its analogs as high-affinity radioligands for imaging extrastriatal dopamine $D_2$ receptors in human brain. *Current Pharmaceutical Design*, **9**, 673–696.

Deep, P., Dagher, A., Sadikot, A., Gjedde, A., and Cumming, P. (1999) Stimulation of dopa decarboxylase activity in striatum of healthy human brain secondary to NMDA receptor antagonism with a low dose of amantadine. *Synapse*, **34**, 313–318.

Delay, J.P., Deniker, P., and Harl, J.M. (1952) Utilisation en thérapeutique d'une phénothiazine d'action centrale selective. *Annales Médico-psychologiques*, **110**, 112–117.

Doudet, D.J. et al. (2006) Effect of age on markers for monoaminergic neurons of normal and MPTP-lesioned rhesus monkeys: a multi-tracer PET study. *NeuroImage*, **30**, 26–35.

Egerton, A., Ahmad, R., Hirani, E., and Grasby, P.M. (2008) Modulation of striatal dopamine release by 5-HT$_{2A}$ and 5-HT$_{2C}$ receptor antagonists: [$^{11}$C]raclopride PET studies in the rat. *Psychopharmacology*, **200**, 487–496.

Ehrin, E. et al. (1985) Preparation of $^{11}$C-labelled raclopride, a new potent dopamine receptor antagonist: preliminary PET studies of cerebral dopamine receptors in the monkey. *International Journal of Applied Radiation and Isotopes*, **36**, 269–273.

Erritzoe, D. et al. (2008) Cortical and subcortical 5-HT$_{2A}$ receptor binding in neuroleptic-naive first-episode schizophrenic patients. *Neuropsychopharmacology*, **33**, 2435–2441.

Farde, L., Halldin, C., Stone-Elander, S., and Sedvall, G. (1987) PET analysis of human dopamine receptor subtypes using $^{11}$C-SCH 23390 and $^{11}$C-raclopride. *Psychopharmacology*, **92**, 278–284.

Farde, L. et al. (1987) No D$_2$ receptor increase in PET study of schizophrenia. *Archives of General Psychiatry*, **44**, 671–672.

Farde, L. et al. (1992) Positron emission tomographic analysis of central D$_1$ and D$_2$ dopamine receptor occupancy in patients treated with classical neuroleptics and clozapine. Relation to extrapyramidal side effects. *Archives of General Psychiatry*, **49**, 538–544.

Frankle, W.G. (2007) Neuroreceptor imaging studies in schizophrenia. *Harvard Review of Psychiatry*, **15**, 212–232.

Frankle, W.G. et al. (2005) Serotonin transporter availability in patients with schizophrenia: a positron emission tomography imaging study with [$^{11}$C]DASB. *Biological Psychiatry*, **57**, 1510–1516.

Frankle, W.G. et al. (2006) Serotonin 1A receptor availability in patients with schizophrenia and schizo-affective disorder: a positron emission tomography imaging study with [$^{11}$C]WAY 100635. *Psychopharmacology*, **189**, 155–164.

Garnett, E.S., Firnau, G., and Nahmias, C. (1983) Dopamine visualized in the basal ganglia of living man. *Nature*, **305**, 137–138.

Grunder, G. et al. (2003) Subchronic haloperidol downregulates dopamine synthesis capacity in the brain of schizophrenic patients *in vivo*. *Neuropsychopharmacology*, **28**, 787–794.

Grunder, G. et al. (2006) The striatal and extrastriatal D$_2$/D$_3$ receptor-binding profile of clozapine in patients with schizophrenia. *Neuropsychopharmacology*, **31**, 1027–1035.

Hartvig, P. et al. (1992) Regional brain kinetics of 6-fluoro-(β-$^{11}$C)-L-dopa and (β-$^{11}$C)-L-dopa following COMT inhibition. A study *in vivo* using positron emission tomography. *Journal of Neural Transmission: General Section*, **87**, 15–22.

Hietala, J. et al. (1995) Presynaptic dopamine function in striatum of neuroleptic-naive schizophrenic patients. *Lancet*, **346**, 1130–1131.

Hirvonen, J. et al. (2006) Brain dopamine D$_1$ receptors in twins discordant for schizophrenia. *American Journal of Psychiatry*, **163**, 1747–1753.

Howes, O.D. et al. (2009) Elevated striatal dopamine function linked to prodromal signs of schizophrenia. *Archives of General Psychiatry*, **66**, 13–20.

Hurlemann, R. et al. (2005) Decreased prefrontal 5-HT$_{2A}$ receptor binding in subjects at enhanced risk for schizophrenia. *Anatomy and Embryology*, **210**, 519–523.

Huttunen, J. et al. (2008) Striatal dopamine synthesis in first-degree relatives of patients with schizophrenia. *Biological Psychiatry*, **63**, 114–117.

Karlsson, P., Farde, L., Halldin, C., and Sedvall, G. (2002) PET study of D$_1$ dopamine receptor binding in neuroleptic-naive patients with schizophrenia. *American Journal of Psychiatry*, **159**, 761–767.

Kessler, R.M. et al. (2009) Dopamine D$_2$ receptor levels in striatum, thalamus, substantia nigra, limbic regions, and cortex in schizophrenic subjects. *Biological Psychiatry*, **65**, 1024–1031.

Koepp, M.J. et al. (1998) Evidence for striatal dopamine release during a video game. *Nature*, **393**, 266–268.

Kumakura, Y. et al. (2007) Elevated [$^{18}$F]fluorodopamine turnover in brain of patients with schizophrenia: an [$^{18}$F]fluorodopa/positron emission tomography study. *Journal of Neuroscience*, **27**, 8080–8087.

Kung, H.F. et al. (1989) *In vitro* and *in vivo* evaluation of [$^{123}$I]IBZM: a potential CNS D-2 dopamine receptor imaging agent. *Journal of Nuclear Medicine*, **30**, 88–92.

Kung, H.F. et al. (1996) Imaging of dopamine transporters in humans with technetium-99m TRODAT-1. *European Journal of Nuclear Medicine*, **23**, 1527–1530.

Lavalaye, J. et al. (2001) Dopamine transporter density in young patients with schizophrenia assessed with [$^{123}$I]FP-CIT SPECT. *Schizophrenia Research*, **47**, 59–67.

Lindstrom, L.H. et al. (1999) Increased dopamine synthesis rate in medial prefrontal cortex and striatum in schizophrenia indicated by L-(beta-$^{11}$C)DOPA and PET. *Biological Psychiatry*, 46, 681–688.

Logan, J. et al. (1990) Graphical analysis of reversible radioligand binding from time-activity measurements applied to [$N$-$^{11}$C-methyl]-(−)-cocaine PET studies in human subjects. *Journal of Cerebral Blood Flow and Metabolism*, 10, 740–747.

Mamo, D. et al. (2007) Differential effects of aripiprazole on $D_2$, 5-$HT_2$, and 5-$HT_{1A}$ receptor occupancy in patients with schizophrenia: a triple tracer PET study. *American Journal of Psychiatry*, 164, 1411–1417.

McGowan, S., Lawrence, A.D., Sales, T., Quested, D., and Grasby, P. (2004) Presynaptic dopaminergic dysfunction in schizophrenia: a positron emission tomographic [$^{18}$F]fluorodopa study. *Archives of General Psychiatry*, 61, 134–142.

Meyer-Lindenberg, A. et al. (2002) Reduced prefrontal activity predicts exaggerated striatal dopaminergic function in schizophrenia. *Nature Neuroscience*, 5, 267–271.

Minuzzi, L. et al. (2005) Interaction between LSD and dopamine $D_{2/3}$ binding sites in pig brain. *Synapse*, 56, 198–204.

Mu, Q. et al. (2007) A single 20 mg dose of the full $D_1$ dopamine agonist dihydrexidine (DAR-0100) increases prefrontal perfusion in schizophrenia. *Schizophrenia Research*, 94, 332–341.

Mukherjee, J. et al. (2002) Brain imaging of $^{18}$F-fallypride in normal volunteers: blood analysis, distribution, test–retest studies, and preliminary assessment of sensitivity to aging effects on dopamine D-2/D-3 receptors. *Synapse*, 46, 170–188.

Nordstrom, A.L. et al. (1993) Central $D_2$-dopamine receptor occupancy in relation to antipsychotic drug effects: a double-blind PET study of schizophrenic patients. *Biological Psychiatry*, 33, 227–235.

Nordstrom, A.L., Farde, L., and Halldin, C. (1993) High 5-$HT_2$ receptor occupancy in clozapine treated patients demonstrated by PET. *Psychopharmacology*, 110, 365–367.

Nozaki, S. et al. (2009) Regional dopamine synthesis in patients with schizophrenia using L-[beta-$^{11}$C]DOPA PET. *Schizophrenia Research*, 108, 78–84.

Reimold, M. et al. (2007) Occupancy of dopamine $D_1$, $D_2$ and serotonin$_{2A}$ receptors in schizophrenic patients treated with flupentixol in comparison with risperidone and haloperidol. *Psychopharmacology*, 190, 241–249.

Reith, J. et al. (1994) Elevated dopa decarboxylase activity in living brain of patients with psychosis. *Proceedings of the National Academy of Sciences of the United States of America*, 91, 11651–11654.

Rinaldi, A., Mandillo, S., Oliverio, A., and Mele, A. (2007) $D_1$ and $D_2$ receptor antagonist injections in the prefrontal cortex selectively impair spatial learning in mice. *Neuropsychopharmacology*, 32, 309–319.

Rosa-Neto, P. et al. (2004) MDMA-evoked changes in [$^{11}$C]raclopride and [$^{11}$C]NMSP binding in living pig brain. *Synapse*, 53, 222–233.

Ross, S.B. and Jackson, D.M. (1989) Kinetic properties of the accumulation of $^3$H-raclopride in the mouse brain *in vivo*. *Naunyn-Schmiedeberg's Archives of Pharmacology*, 340, 6–12.

Sanchez-Crespo, A., Andreo, P., and Larsson, S.A. (2004) Positron flight in human tissues and its influence on PET image spatial resolution. *European Journal of Nuclear Medicine and Molecular Imaging*, 31, 44–51.

Seeman, P., and Lee, T. (1975) Antipsychotic drugs: direct correlation between clinical potency and presynaptic action on dopamine neurons. *Science*, 188, 1217–1219.

Slifstein, M. et al. (2007) [$^{11}$C]NNC 112 selectivity for dopamine $D_1$ and serotonin 5-$HT_{2A}$ receptors: a PET study in healthy human subjects. *Journal of Cerebral Blood Flow and Metabolism*, 27, 1733–1741.

Slifstein, M. et al. (2008) COMT genotype predicts cortical-limbic $D_1$ receptor availability measured with [$^{11}$C]NNC112 and PET. *Molecular Psychiatry*, 13, 821–827.

Smith, G.S. et al. (1997) Serotonergic modulation of dopamine measured with [$^{11}$C]raclopride and PET in normal human subjects. *American Journal of Psychiatry*, 154, 490–496.

Soares, J.C. and Innis, R.B. (1999) Neurochemical brain imaging investigations of schizophrenia. *Biological Psychiatry*, **46**, 600–615.

Stone, J.M., Davis, J.M., Leucht, S., and Pilowsky, L.S. (2009) Cortical dopamine $D_2/D_3$ receptors are a common site of action for antipsychotic drugs – an original patient data meta-analysis of the SPECT and PET *in vivo* receptor imaging literature. *Schizophrenia Bulletin*, **35**, 789–797.

Suhara, T. *et al.* (2002) Decreased dopamine $D_2$ receptor binding in the anterior cingulate cortex in schizophrenia. *Archives of General Psychiatry*, **59**, 25–30.

Tauscher, J. *et al.* (2002) Brain serotonin 5-HT$_{1A}$ receptor binding in schizophrenia measured by positron emission tomography and [$^{11}$C]WAY-100635. *Archives of General Psychiatry*, **59**, 514–520.

Verhoeff, N.P. *et al.* (2000) A voxel-by-voxel analysis of [$^{18}$F]setoperone PET data shows no substantial serotonin 5-HT$_{2A}$ receptor changes in schizophrenia. *Psychiatry Research*, **99**, 123–135.

Vernaleken, I. *et al.* (2006) Modulation of [$^{18}$F] fluorodopa (FDOPA) kinetics in the brain of healthy volunteers after acute haloperidol challenge. *NeuroImage*, **30**, 1332–1339.

Vernaleken, I. *et al.* (2007a) Asymmetry in dopamine $D_{2/3}$ receptors of caudate nucleus is lost with age. *NeuroImage*, **34**, 870–878.

Vernaleken, I. *et al.* (2007b) "Prefrontal" cognitive performance of healthy subjects positively correlates with cerebral FDOPA influx: an exploratory [$^{18}$F]-fluoro-L-DOPA-PET investigation. *Human Brain Mapping*, **28**, 931–939.

Vernaleken, I. *et al.* (2008a) Elevated $D_{2/3}$-receptor availability in schizophrenia: a [$^{18}$F]fallypride study. *NeuroImage*, **41** (Suppl 2), T145.

Vernaleken, I. *et al.* (2008b) Baseline [$^{18}$F]-FDOPA kinetics are predictive of haloperidol-induced changes in dopamine turnover and cognitive performance: a positron emission tomography study in healthy subjects. *NeuroImage*, **40**, 1222–1231.

Vernaleken, I., Cumming, P., and Grunder, G. (2008) Imaging studies – differential action of typical and atypical antipsychotics in a network perspective. *Pharmacopsychiatry*, **41** (Suppl. 1), S60–S69.

Volkow, N.D. *et al.* (1998) Parallel loss of presynaptic and postsynaptic dopamine markers in normal aging. *Annals of Neurology*, **44**, 143–147.

Vollenweider, F.X., Vontobel, P., Oye, I., Hell, D., and Leenders, K.L. (2000) Effects of (S)-ketamine on striatal dopamine: a [$^{11}$C] raclopride PET study of a model psychosis in humans. *Journal of Psychiatric Research*, **34**, 35–43.

Wagner, H.N. Jr. *et al.* (1983) Imaging dopamine receptors in the human brain by positron tomography. *Science*, **221**, 1264–1266.

Weinberger, D.R. and Laruelle, M. (2001) Neurochemical and neuropharmacological imaging in schizophrenia, in *Neuropharmacology – The Fifth Generation of Progress* (eds. K.L. Davis, D.S. Charney, J.T. Coyle, and C. Nemeroff), Lippincott, Williams, & Wilkins, Philadelphia, PA, pp. 883–885.

Winterer, G. (2007) Prefrontal dopamine signaling in schizophrenia – the corticocentric model. *Pharmacopsychiatry*, **40**, S45–S53.

Wong, D.F. *et al.* (1984) Effects of age on dopamine and serotonin receptors measured by positron tomography in the living human brain. *Science*, **226**, 1393–1396.

Wong, D.F. *et al.* (1986a) Quantification of neuroreceptors in the living human brain. II. Inhibition studies of receptor density and affinity. *Journal of Cerebral Blood Flow and Metabolism*, **6**, 147–153.

Wong, D.F. *et al.* (1986b) Positron emission tomography reveals elevated $D_2$ dopamine receptors in drug-naive schizophrenics. *Science*, **234**, 1558–1563.

Wong, D.F. *et al.* (1997) Quantification of neuroreceptors in the living human brain: IV. Effect of aging and elevations of $D_2$-like receptors in schizophrenia and bipolar illness. *Journal of Cerebral Blood Flow and Metabolism*, **17**, 331–342.

Young, K.A., Randall, P.K., and Wilcox, R.E. (1995) Startle and sensorimotor correlates of ventral thalamic dopamine and GABA in rodents. *Neuroreport*, **6**, 2495–2499.

# 10
# The Marriage of Phenomics and Genetical Genomics: A Systems Approach to Complex Trait Analysis

*Laura M. Saba, Paula L. Hoffman, Lawrence E. Hunter, and Boris Tabakoff*

## 10.1
### Introduction and Brief History of Genetical Genomics

In 1970, Francis Crick clarified his version of the central dogma of molecular biology that he originally articulated in 1958. In the central dogma, Crick outlined the general direction of information flow from DNA to protein along with a few paths of minor or "special" flow. This information flow can be extended to include phenotype as the final step after protein synthesis. In a few cases, biology is so simple that a difference in DNA results in a direct and obvious difference in phenotype (Kerem *et al.*, 1989; The Huntington's Disease Collaborative Research Group, 1993). However, most phenotypes are not manifested as an all-or-none phenomenon, but are exhibited as quantitative traits. More often than not, variations throughout the entire pathway from DNA to protein to phenotype are responsible for quantitative variation in a phenotype.

In the early years of genetic research, investigators focused on the variation in DNA sequence with the assumption that trait variation was a direct result of DNA polymorphisms that caused functional aberrations in resultant proteins. As technology evolved, researchers began to see the potential of quantitative variation within each step along the pathway, as well as the role that environment played throughout. Before the sequencing of the human genome, researchers hypothesized that there were over 100 000 human genes. After sequencing, the estimated number of genes was reduced significantly and the concept that phenotypic variation is a direct result of functional polymorphisms in the coding sequence for gene products lost some of its appeal. The development of newer technologies for proteomics and transcriptomics allowed investigators to make the assertion that multiple transcripts can arise from a single gene due to alternative splicing, and that phenotypic differences are more likely to arise from quantitative differences at the transcript or protein level.

An important advance occurred in the study of genetics of non-Mendelian traits when large panels of molecular genetic markers that could be assigned to specific locations within the genome of experimental animals, including humans, became available (International Human Genome Sequencing Consortium, 2004), and statistical analyses that allow associations to be established between regions of a genome and the phenotype of a complex trait were developed (Thoday, 1961; Lander and Botstein, 1989). The process of defining genomic regions associated with a trait that is inherited in a non-Mendelian manner (quantitative trait) is referred to as quantitative trait locus (QTL) analysis. The techniques of QTL analysis have been refined since their inception by the development of more detailed genetic maps, and by the development of techniques such as interval mapping (Lander and Botstein, 1989), composite interval mapping (Zeng, 1993), and multiple QTL mapping (Jansen, 1993).

Another breakthrough in the study of the genetics of quantitative traits occurred with the advent of whole-genome mRNA expression analyses. Microarray and high-throughput cDNA sequencing technology allows for the simultaneous measurement of tens of thousands of transcripts. Researchers can use these data for a variety of analyses, including expression differences between cases and controls, correlation of mRNA expression with a continuous phenotype, the clustering of genes based on transcript level across time, or the identification of disease subtypes by transcription profiling, to name but a few. Technology and statistical methodology have been evolving quite rapidly in this relatively young field.

Analysis methodologies for associating DNA polymorphisms and mRNA levels with phenotypes have flourished, but until the beginning of the new millennium, they did so in a parallel not in an integrated fashion. In 2001, Jansen and Nap coined the phrase "genetical genomics," and urged the combination of DNA sequence and transcriptome studies. In their seminal paper, Jansen and Nap articulated the need for a field of research that investigated the genetic determinants of levels of mRNA expression. Using mRNA expression as a "molecular phenotype" in a typical QTL analysis, one can begin to unravel the different mechanisms of transcriptional control. Since the publication of the paper by Jansen and Nap (2001), the locations within the genome that control gene expression have been referred to as expression QTLs (expression quantitative trait locus eQTLs).

After the Jansen and Nap paper, several different groups began applying this concept to different species. In 2002, Brem *et al.* studied genetical genomics in the yeast *Saccharomyces cerevisiae*. This was followed quickly by a genetical genomic survey across maize, mice, and humans (Schadt *et al.*, 2003). In further work with humans, Morley *et al.* (2004) and Cheung *et al.* (2005) published related papers on eQTLs derived from studies of members of the Centre d'Etude du Polymorphisme Humain Utah pedigrees. Then, in 2005, three significant papers were published in a single edition of *Nature Genetics* (Bystrykh *et al.*, 2005; Chesler *et al.*, 2005; Hubner *et al.*, 2005), describing genetical genomics in mice and rats and using this information to help identify disease candidate genes.

By 2005, genetical genomics was quickly establishing its place in systems biology. Researchers moved away from the simplistic view that the pathway from DNA to phenotype is straightforward, but instead began to view the process as a system with many inputs, influences, and variations in outcome. In Li and Burmeister's (2005) review of genetical genomics, they state that:

> Although this novel field is still developing, understanding the genetic basis of molecular phenotypes such as gene expression is expected to shed light on the intermediate processes that connect genotype to cellular and organismal traits and represents a critical step towards true systems biology.

### 10.1.1
### Characteristics of eQTLs

There have been several well written overviews of eQTL analyses published in the last few years, particularly the papers by Rockman and Kruglyak (2006) and Gilad, Rifkin, and Pritchard (2008). Many of these reviews touched on three key points: genetic heritability of gene expression, characterization of cis- and trans-regulating eQTLs, and evidence for multiple loci involved in the control of transcriptional events.

It has been stated that the quantitative variability in gene expression is a heritable trait and may be one of the major contributors to species differentiation (King and Wilson, 1975). The global exploration of the genetic heritability of gene expression was made possible by the advent of microarray technology in the mid-1990s. Although, prior to that time, scientists were able to quantitate the expression of individual genes, microarrays allowed investigators to survey the heritability of gene expression across all genes in a genus or species genome without bias. Investigators have been able to show the varying amount of genetic heritability across genes in many species (see references in Rockman and Kruglyak, 2006). One also needs to note that mRNA levels do not strictly reflect the rate of transcription, but also reflect several other mechanisms that control the amount of mRNA in a cell. Many of these mechanisms (RNA degradation, etc.) could also be genetically controlled.

One of the characteristics of eQTLs that gives insight into control of transcription is the classification of eQTLs as local or distal. An eQTL that is located "near" the physical location of the gene is considered to be local or cis-regulating, whereas an eQTL that is distal to the physical location of the gene or is not even on the same chromosome as the gene is considered to be distal or trans-regulating. Rockman and Kruglyak (2006) urge the use local and distal versus cis and trans, since these mapping studies do not reveal the particular molecular mechanism that controls transcription nor the exact location of this control, which constitute knowledge implied by the terms cis and trans. There are many potential mechanisms for control of transcriptional events by both local and distal eQTL. Local eQTLs could potentially reflect a polymorphism in a transcription factor binding site in the promoter region of the gene that modifies the initiation of transcription, a polymorphism within the gene that affects the stability of mRNA produced, or a polymorphism in a neighboring gene that is needed for transcription initiation, to name a few. Likewise, a distal eQTL could be a functional polymorphism within a coding sequence of a transcription factor, which alters its interaction properties, or the stability of the transcription factor mRNA, which alters the transcription of a gene of interest. A distal eQTL can even be related to production of factors necessary for transport of the gene of interest's mRNA for translation. As Jansen and Nap (2001) pointed out, a distal eQTL can represent a polymorphism in a gene whose influence occurred at a time prior to analysis of mRNA levels. In general, most studies have shown that local eQTLs tend to have lower *p*-values (higher statistical significance) than distal eQTLs, making it more probable to detect local versus distal eQTLs in such a QTL analysis. Hypothetically, this could happen for several reasons. A polymorphism in the promoter sequence of the gene could have a more direct effect than a polymorphism elsewhere that must interact with several factors before affecting the transcription of the gene. Also, the local polymorphisms (eQTLs) may be less likely to be affected by environmental influences. The statistical robustness of local eQTLs also makes these eQTLs more likely to be detected in different tissues, compared to their distal counterparts.

As with most quantitative traits, a gene expression trait is likely to exhibit control by multiple loci working additively and/or epistatically. Brem and Kruglyak (2005) performed a detailed analysis of the genetic complexity within transcript expression levels from a cross between two strains of *S. cerevisiae*. They limited their analysis to highly heritable transcription traits to help eliminate transcripts with large environmental effects. Within this set of highly heritable transcription events, only 3% were consistent with monogenic inheritance, whereas almost half of the highly heritable transcription events had linkage statistics that indicated that at least six loci influenced the level of the transcript. They also estimated that at least 16% of the highly heritable transcripts were influenced by epistatic effects between loci. That same year, Brem *et al.* (2005) published another report that looked directly for epistatic effects in the transcription of yeast genes. In this paper, they estimated that 57% of transcripts were affected by interacting loci. Multiple loci and epistatic effects are hard to demonstrate with strict statistical criteria, and particularly in the case of eQTLs, due to the large number of comparisons involved.

### 10.1.2
### Recent Developments

Several recent developments have furthered the field of genetic genomics. The sharing of genetic marker datasets and microarray data from inbred and recombinant inbred (RI) rodent panels has been particularly helpful in reducing the workload for an eQTL study. Recent developments and advances in microarray and genotype technology such as exon arrays and whole-genome genotyping improve the quality and quantity of the available data. Also, as the field gathers momentum, statistical methodology is being developed for analysis of the massive datasets. New methodology also includes moving beyond a focus on individual genes and QTLs, toward creating gene networks from the abundance of generated data.

The Wellcome Trust and Illumina have created a single nucleotide polymorphism (SNP) dataset for 480 strains of mice utilizing 13 370 SNPs (Shifman *et al.*, 2006). They chose many of the common inbred strains and several RI panels, including the commonly used BXD (cross between C57BL/6J and DBA/2J) RIs and the large set of LXS (cross between ILS (inbred long sleep) and ISS (inbred short sleep)) RI mice. The data are easily accessible through their web interface at http://www.well.ox.ac.uk/mouse/INBREDS. They have taken their dataset a step further and analyzed haplotype block structure within the eight RI panels. This effort was one of the first of its kind and it enabled many of the original influential eQTL analyses.

More recently a group from the Jackson Laboratory has created a SNP panel of 7.87 million loci across 74 common inbred strains of mice (Szatkiewicz *et al.*, 2008). They created this panel by gathering data from multiple public databases and then using a hidden Markov model to impute any missing information. They provided genotype calls for each of the SNPs and strains, and then also assigned a confidence score to each call based on its posterior probability and the actual genotype probability. This SNP dataset is also publicly available on the web at http://cgd.jax.org/ImputedSNPData/v1.1/.

Likewise, for several common inbred strains and a RI panel of rats, the STAR consortium has developed a SNP dataset that contains 20 238 SNPs across 167 inbred strains and 64 RI strains (The STAR Consortium, 2008). This SNP dataset is available to the public through the Ensembl (www.ensembl.org) and BioMart (www.biomart.org) web sites.

A recent technological advancement in microarrays that has a potential to have a large impact on genetical genomics is the introduction of exon arrays to measure alternatively spliced transcripts. Instead of probing the 3′-untranslated region of a gene as in earlier arrays, these exon arrays have several probes for each exon. The resulting data can then either be combined to create a general expression estimate for a gene, an expression estimate for a specific exon, or to determine if differences in alternative splicing are occurring between samples. These exon arrays have been shown to be accurate in their assessment of gene expression (Kapur *et al.*, 2007). In addition, they have been used in at least one genetical genomics study thus far (Zhang *et al.*, 2008).

As with most large-scale genetic methodological breakthroughs, the statistics methodology is racing to keep up. In the years since the term genetical genomics was first introduced, there have been many statistical advances to help appropriately analyze the data. The statistical fields that represent microarray and QTL analysis have undergone tremendous growth in the past two decades. The new challenge that genetical genomics presented was the massive amount of multiple comparisons and the demand on computational systems. In a typical whole-genome QTL analysis, a researcher may test the association of a phenotype with thousands of markers. With eQTL analysis, not only are thousands of markers being tested, but also thousands of transcripts at each marker, resulting in millions of comparisons in one analysis. That estimate does not even consider joint or epistatic effects between markers that are likely to occur. When one adds the complication that in many eQTL datasets the subjects are related in some fashion, whether they are multiple members of the same human family or they are multiple rodents from the same strain, traditional QTL models are inefficient. The most accurate eQTL analysis would thus have to account for the correlation between subjects not to mention accounting for technical replicates.

Two reports have well addressed the problem of multiple comparisons and more specifically the multilocus analysis (Storey, Akey, and Kruglyak, 2005; Chen and Kendziorski, 2007). Storey, Akey, and Kruglyak (2005) suggest a stepwise approach to selecting multiple loci that influence expression of a particular transcript, as opposed to an exhaustive two-dimensional scan of all SNP pairs. They calculate a *p*-value based on the premise that either of the loci can be false positives rather than the null hypothesis that both of the loci are false positives as in the more traditional approach. They were able to show that their method was more powerful in a yeast dataset. Chen and Kendziorski adapt the Sen–Churchill framework (Sen and Churchill, 2001) to allow more phenotypes or in this case, gene expression traits, than the number of subjects. This model, called pseudo-marker mixture-over-marker, allows the user to share information across markers and across transcripts in a Bayesian framework.

The problem with related (correlated) subjects is a common occurrence in the rodent studies of eQTLs. For inbred and RI strains, the most common approach is to simply calculate eQTLs based on the strain means. This can be problematic when there are not equal numbers of samples within each strain. It is always with good intentions that researchers begin a microarray project with a balanced study design, the same number of samples in each group, but inevitably a few arrays do not meet quality control standards for various reasons. A mixed model approach where a random effect of strain is accounted for in the analyses is a logical solution to the problem of correlation between subjects. However, because of the estimation procedures involved in such a model, computation time becomes a major and nearly insurmountable concern. In 2005, Carlborg *et al.* proposed a weighted analysis approach. Their method gave more weight to strain means that were calculated using more subjects and the amount of weight was dependent on the heritability of the particular trait. For example, if a trait was highly heritably then there would be little difference in weight given to strain means with differing number of samples. However, if the trait had a low heritability, the number of samples per strain would have a larger effect on the weights.

One of the statistical areas of research that has been developing and is closely related to systems biology is gene networks. Zhu *et al.* (2004) was one of the first papers to address gene networks with respect to eQTL analyses. They implemented a Bayesian network that not only linked genes, but also attempted to determine the causal relationship with regard to expression between any two genes (i.e., gene A controls gene B). More recently Liu, de la Fuente, and Hoeschele (2008) proposed a three-step approach to generating a gene network from genotype and gene expression data. The first step was to identify eQTLs within their dataset and categorize genes as cis-regulated (local regulation), cis–trans-regulated, or trans-regulated (distal regulation). Then regulator–target pairs were identified and combined into an encompassing directed network (EDN) using local structure equation modeling. Finally, structural equation modeling was used to reduce the EDN from the previous step. Networking models like these are essential to a systems biology approach. They help to describe the feedback and interactive processes that control gene transcription, and the functional consequences of alterations in the transcription process.

## 10.1.3
### Extension of Genetical Genomics to Include a Systems Approach to Phenomics

A logical extension of the systems approach to generating networks for genetical genomics is to include downstream events into the analysis. Complex/quantitative/physiologic/behavioral phenotypes that are heritable can be profitably added to a genetical genomic analysis. The three essential datasets needed for such work and their relationships are outlined in Figure 10.1. When extending the network model to include physiologic/behavioral complex traits, mRNA levels become an intermediate molecular phenotype that can help bridge the biological connection between DNA sequences and physiology/behavior.

The simplest approach to incorporating the three types of data is to require that candidate genes are not only associated with the physiologic/behavioral phenotype of interest, but also that the candidate gene has a significant eQTL that overlaps with a physiologic/behavioral phenotypic QTL for the quantitative trait. The general reasoning behind this criterion is that if variation in a gene's expression is truly associated with the physiologic/behavioral trait, then gene expression should be controlled in the same region of the genome that controls the physiological/behavioral trait. This approach has now been applied by many investigators, such as Schadt *et al.* (2005), Bystrykh *et al.* (2005), Chesler *et al.* (2005), Saba *et al.* (2006), Hu *et al.* (2008), and Tabakoff *et al.* (2008).

Schadt *et al.* (2005) took the integration of genetical genomics and "complex" physiologic/behavioral traits a step further and developed a likelihood-based causality model selection method to test relationships between DNA loci, mRNA levels, and complex traits as to ascertain whether the traits are causal, reactive, or independent. With this method, they were able to develop a comprehensive picture of genetic determinants of obesity.

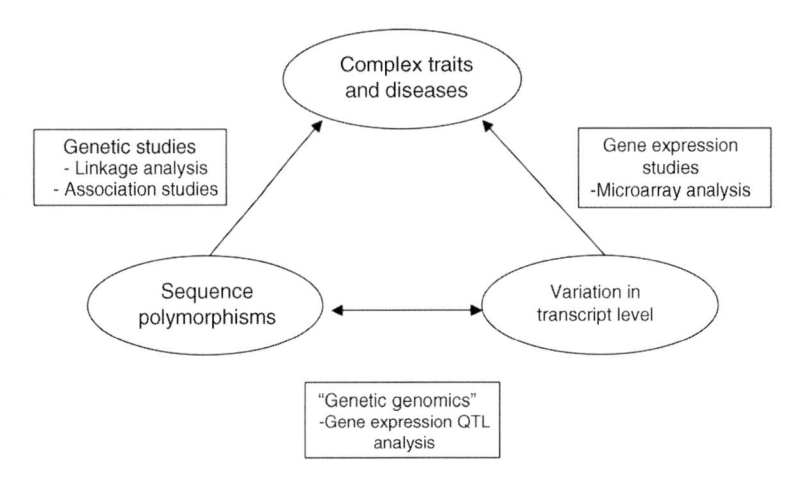

**Figure 10.1** "Genetical genomics" brings together traditional genetic analyses and gene expression studies by directly characterizing the genetic influence of gene expression. (Adapted from Li and Burmeister, 2005.)

The methods that combine genetical genomics and phenomics do not provide the details of molecular function, such as finding the causative SNP and understanding the mechanism involved in changing the intermediate phenotype, mRNA, or how that translates to variation in the psysiologic/behavioral phenotype. Instead, this global approach looks at networks as determinants of causality and purposefully leaves the direct molecular mechanisms ambiguous.

Mendelian approaches to genetic causality of phenotype have predisposed many to desire the identification of "a gene" that is responsible for any trait and particularly complex traits (e.g., the "gene" for alcoholism or the "gene" for schizophrenia). Aside from the problem of the definition of the phenotypes of alcoholism or schizophrenia, and so on, one needs to realize that most "complex" quantitative traits are polygenic and a fruitful endeavor may be to search for candidate pathways as well as candidate genes.

## 10.2
### Potential Pitfalls in Phenotype Selection and Current Technology

Most researchers define a "complex" quantitative phenotype or trait as a phenotype/trait that is caused by the interaction of multiple genetic factors and possibly multiple environmental factors (Malats and Calafell, 2003). A critical issue regarding a phenotype is the robustness of its definition and replicability across individuals. For example, many psychiatric disorders are defined by the presence or absence of some minimal number of identified symptoms from a longer list of possibilities. Such heterogeneous definitions of disorders and the sometimes subjective evaluation of disease may lead to inconsistent results in genetic studies. To help alleviate some of these problems, genetic researchers have turned to the use of more tightly defined endophenotypes. Endophenotypes, also referred to as intermediate phenotypes or subclinical traits, represent a simpler, more definitive trait that is in the direct causal pathway between the genetic/biological source of the disease and the clinical manifestation of that disease. An ideal endophenotype for genetic study is a precisely and objectively measured trait that may manifest itself within the clinical spectrum of the disease. To be precise, Gottesman and Gould (2003) proposed the four following criteria plus the optional fifth criterion for identifying endophenotypes for genetic studies:

i) The endophenotype is associated with illness in the population.
ii) The endophenotype is heritable.
iii) The endophenotype is primarily state-independent (manifests in an individual whether or not illness is active).
iv) Within families, endophenotype and illness cosegregate.
v) [*Optional*] The endophenotype found in affected family members is found in nonaffected family members at a higher rate than in the general population.

Endophenotypes are particularly useful when studying complex traits in model organisms. For example, alcoholism, as defined in the *Diagnostic and Statistical*

*Manual of Mental Disorders*, text revision, 4th edn (DSM-IV), is difficult to measure in humans, not to mention rodent populations. Instead, alcoholism researchers often use endophenotypes such as alcohol preference in a two-bottle choice paradigm or tolerance to the discoordinating effects of alcohol intoxication as a surrogate endophenotype for studies of "alcoholism" in rodents. Both traits have been shown to be associated with alcoholism in humans and both traits show evidence of genetic heritability in rodents.

There are several potential pitfalls in selecting phenotypes and endophenotypes for genetical genomic/phenomic studies. Finding endophenotypes that adequately represent the human phenotype of interest is not a trivial task and when one confronts a study involving neurobiology, one is also confronted with the need to address the function of a complex, heterogeneous tissue – the brain.

In the field of neurobiology, much has been intimated regarding the use of dissected areas of brain versus whole-brain transcriptome analyses. The premise used for justifying and, in fact preferring, the use of anatomically defined brain regions for analysis of mRNA levels, versus measuring the whole-brain transcriptome, is that this procedure brings one closer to the source of possible variation responsible for functional processes, which lead to normal behavior or psychiatric/neurologic disease states. Unless one is confident that they totally understand the anatomical paths and integrative centers (nuclei) that control a physiologic function or behavior, focusing attention on one or a few areas of brain for studies of mRNA levels is fraught with caveats. mRNA synthesis (transcription) related to maintaining neuronal function is an activity localized (primarily) to the cell nucleus within the cell body. The location of the cell body could well be in an area of brain that is at a distance from where the functional consequences of neuronal function/malfunction may be evidenced (i.e., the synapse). A very well known example would be in the realm of movement disorders, such as Parkinsonism. The striatal (caudate nucleus) control of coordinated movement may lead one to examine the caudate mRNA levels as a cause of tremor in Parkinsonism. This would provide little information on the primary degenerative process taking place in the dopaminergic neuron cell bodies that reside in the substantia nigra and send their axonal projections to the caudate.

The brain is arguably the most complex organ in the body, and not only is the brain made up of several parts, but many of the parts have a collection of heterogeneous cells that are quite different from each other and form anatomical networks through intercellular communication between and among brain regions. Hippocampus, for example, contains morphologically distinct granule cells, basket cells, glial cells, and pyramidal cells. These cells generate and respond to numerous neurotransmitters, and the neurons project from hippocampus to a number of brain regions.

The concern over using whole-brain transcriptome data versus data from individual brain parts, with specific reference to eQTL data, was highlighted in an article by Hovatta *et al.* (2007). In their study, expression data were compared across five brain regions and six inbred strains. Although they asserted that their study indicated that eQTL analyzes need to be carried out on data from brain parts, rather than whole-brain data, they did not have enough statistical power (too few strains), and no data from whole brain, to actually compare eQTL results between whole-brain data and individual brain parts. If an investigator knows that a single brain region is relevant for his/her study, multiple brain regions would not be necessary to include in the experiment, allowing for a more optimal design solution. However, it is probably never the case that a single brain region can account for a complex behavior. As noted, mRNA is localized in cell bodies and generates protein products that are transported to neuron terminals that may be localized in different brain areas or in different nuclei within the same general brain region. These transported proteins may influence mRNA synthesis and function at a distance from the anatomical area chosen for initial analysis. Therefore, to understand the implications of differences in gene expression, one would necessarily need to assess this expression in all of the interacting brain regions, nuclei, or cells and would have to understand these systems with respect to the behavior of interest.

Philosophically, an eQTL "phenotype" reflects a variation at a genetic locus that interacts with cellular environment to regulate gene expression. When an eQTL is present in one brain region or organ, but not another, it is implied that the environmental interaction with the locus causes that particular gene to be expressed differently in the particular brain region. The question then arises, why a researcher would not get more specific than studying whole-brain or brain-regional gene expression, and would instead perform investigations at the cellular level? This would enhance control of the exact "environment." The reality is that one needs to study gene expression at all levels of anatomy to arrive at information useful for linking genetics to phenotype. By looking at whole-brain data, we may miss eQTLs that are conditional on transcription factors expressed particularly in one or another brain area, but the eQTLs that we detect are more likely to be generally applicable to many cell types or brain regions. Therefore, the whole-brain results could be relevant to the researcher studying a heterogeneous brain region such as hypothalamus or to those not having complete knowledge of the anatomy contributing to a phenotype. An example of a complicated scenario would be in studies related to the reinforcing effects of certain stimuli, including addictive drugs. Attempts have been made to examine mRNA levels in several areas of brain of rats in an attempt to generate insights into the determinants of the reinforcing effects of ethanol (Rodd *et al.*, 2008). The anatomy of a process such as drug reinforcement is complicated (Kalivas and O'Brien, 2008), and trying to define the determinants at any molecular level without clear understanding of circuit anatomy and the stoichiometry of interactions leads to confusion (Rodd *et al.*, 2008). Whole-brain transcriptome analysis may, at times, not provide the sensitivity to monitor small changes occurring in a few neurons, but if reliable changes are evidenced, one can try to construct relationships between the witnessed changes, including attempts at defining anatomical pathways (Tabakoff *et al.*, 2008). The Allen Brain Atlas (www.brain-map.org) is an important aid in understanding the anatomy of gene expression and assisting in inferring anatomical relationships that contribute to complex physiologic/behavioral events (see below).

A recent article (Oldham *et al.*, 2008) may help alleviate some of the concern over the use of whole brain and the intimated necessity to only use single cells within the brain for microarray analyses. Oldham *et al.* (2008) were able to identify clusters of genes that were transcribed in either neurons, oligodendrocytes, astrocytes, or microglia from cerebral cortex, caudate nucleus, and cerebellum in human brain. These four cell-specific clusters were conserved across the three brain regions, and across assay platforms and over a significant number of samples. In their conclusion they state, "cellular heterogeneity contributes in *measurable* and *predictable* ways to expression levels quantified by microarrays using messenger RNA extracted from whole-brain tissue." The application of the analytical techniques described by Oldham *et al.* (2008) can allow one to segregate cell-type-specific information on gene expression from data derived from whole brain.

In addition to potential pitfalls in choice of phenotype and tissue, the advent of new technology inevitably produces a new set of potential pitfalls. The new technologies involved in genetical genomics are no exception. With microarrays, researchers are constantly working toward more accurate chip design and making adjustments to accommodate the ever-changing genome landscape. Since microarrays probe so many of the products of gene transcription simultaneously, even slight changes in the genome build (www.ensembl.org) can have detrimental effects on construction of probes. Statisticians are also working diligently to develop new and improved methods for normalization (Ni *et al.*, 2008). Likewise with genetic maps, new technology constantly emerges that allows for the genotyping of more markers at less cost. What is unsettling to many researchers involved in genetical genomics is the lack of confirmation of strong eQTLs using such techniques as quantitative reverse transcription polymerase chain reaction or allele-specific transcript quantification.

One of the common pitfalls that many of the initial eQTL studies experienced is the identification of strong cis eQTLs that were due to differential hybridization because of a SNP within the probe's target sequence and not a true difference in the amount of mRNA present between subjects. A microarray probe is designed to quantify the amount of mRNA that mirrors the probe sequence. Currently, most microarray probes are created using a sequence of 25–60 nucleotide bases. When a SNP is located within the nucleotide sequence of mRNA, it influences the mRNA's ability to hybridize to the probe. It has been shown that the length of the probe, the position of the SNP within the probe, and even the specific nucleotide can affect hybridization (Hughes *et al.*, 2001).

For example, probe sequences for the Affymetrix (Santa Clara, CA, USA) mouse microarray were designed based on the genome sequence of the C57BL/6J mouse. Many of the current eQTL studies in mice have involved the BXD RI panel, which consists of inbred mice derived from brother/sister mating initiated from the $F_2$ offspring of a C57BL/6J (B6) cross with DBA/2J (D2). For 100 strongest cis eQTLs in their study, Alberts *et al.* (2007) determined whether the B6 allele was associated with the higher transcript frequency or if the D2 allele was associated with the higher frequency. Without the influence of SNPs on hybridization with probes, it is expected that 50% of the cis eQTLs will show higher transcript abundance with the B6 allele and 50% will show higher transcript abundance with the D2 allele. Instead, Alberts *et al.* (2007) reported that 70% of the cis eQTLs were associated with a higher abundance in the B6 allele. This indicates that over half of the top cis eQTLs are potentially false positives due to SNPs that influence mRNA probe interactions. It should be noted that most microarray platforms utilize cDNA generated from mRNA for hybridization to probes.

Many researchers were quick to acknowledge this problem. One of the most common solutions to the problem is to eliminate probes that differentially bind mRNA with SNPs that differentiate subjects within the sample population. In the Affymetrix array design, several probes are available to get estimates of transcript levels, so the elimination of probes has limited effect on the coverage of the chip. Other platforms, such as CodeLink Bioarrays (Applied Microarrays, Tempe, AZ, USA), have only one or two probes per transcript, so the elimination of probes can have a dramatic effect on the coverage of the chip. Several researchers have created probe masks, which eliminate problematic probes, based on the currently known SNPs (Walter *et al.*, 2007). This approach dramatically reduces the number of false-positive eQTLs even though it does not eliminate all false positives associated with this specific problem.

Still others have attacked the problem in the opposite direction (Alberts *et al.*, 2007). Instead of relying on the localization of known SNPs, Alberts *et al.* (2007) built a model for use with the Affymetrix platform to systematically eliminate probes if that probe's hybridization characteristics were different from the majority of the probes within the probe set. They applied their method to the 100 cis eQTLs mentioned above and reduced the number of cis eQTLs associated with a higher transcript abundance in the B6 allele to 45 and the number of cis eQTLs associated with a higher transcript abundance in the D2 allele to 29. Their B6 cis eQTLs were then no longer in significant excess.

## 10.3
### General Strategy for Identifying Candidate Pathways

One general strategy for identifying candidate pathways, as mentioned briefly above, is to incorporate genetical genomics into a physiologic/behavioral QTL analysis and microarray analysis through a series of biologically relevant filters. Many microarray studies simply look to identify genes that are differentially expressed between two groups of subjects or genes whose expression is correlated with a phenotype of interest. Often, the list can contain hundreds or even thousands of genes. Assessing the biological relevance of transcripts that are differentially expressed between two or several groups of subjects that differ in a phenotype is many times frustrating because one usually does not know every difference that exists, in addition to the measured phenotype, between groups. To demonstrate a more encompassing approach, we included genetical genomics with standard behavioral QTL analysis and transcriptome analysis to identify candidate pathways.

We will briefly describe the analysis in Tabakoff *et al.* (2008) that led to the identification of a list of candidate genes and a candidate pathway, including anatomical aspects, for alcohol preference in mice. For this analysis, we used microarray data and phenotype (alcohol preference) data for from panels of inbred and RI mice, as well as selectively bred mice. In addition, we obtained eQTL data from a panel of BXD RI mice (all of our microarray data can be accessed through Phenogen.uchsc.edu), and we took advantage of the identification of behavioral QTLs (behavioral quantitative trait locus bQTLs) for alcohol preference from the literature (also accessible through our website). The first step was to eliminate transcripts that did not have a significant eQTL in the BXD panel. Next, transcripts were eliminated if their eQTL did not overlap any of the published bQTLs for alcohol preference (based on their 95% confidence interval). This eQTL/bQTL filter was executed as the first step in order to reduce the burden of multiple testing correction. This criterion was also used to restrict candidate transcripts to those whose expression is measurably influenced by a polymorphism in the DNA sequence that also contributes to the alcohol preference phenotype. For the transcripts which were retained, the microarray experiments were combined in a meta-analysis that tested for association between transcript abundance and a quantitative measure of alcohol preference. Transcripts were retained if their levels were significantly (false discovery rate below 0.001) correlated with alcohol preference. These transcripts were further scrutinized for heritability and probe specificity to reduce the possibility of false positives. The final list of candidate transcripts contained eight unique genes. With only eight genes to deal with, we were able to do a full literature search for each gene and we were able to perform multivariate analyses that would have been impossible with hundreds of genes. The caveat to this type of analysis is that the focus is on eliminating false positives, rather than reducing the number of false negatives. We concentrated on finding candidate genes that we were "confident" were involved in some capacity with a predisposition to alcohol preference. This came at the expense of being so stringent that other genes involved in influencing alcohol preference were likely to be missed. However, our strategy is worthwhile with microarray analysis, where the purpose would be to produce a limited number of the most probable candidates which influence a trait of interest. Many times the follow-up studies for candidate genes and candidate pathways are expensive and time-consuming, so it is important that the gene being investigated has a high probability of being relevant. Generating a list of candidate genes is just the beginning of a "systems" approach to understanding the genetic underpinnings of neurobiology (or any other biology).

There are several methods for constructing candidate pathways from a list of candidate genes. One of the most common approaches when one has a longer list of candidate genes (more than 15) is to group genes based on common Gene Ontology (GO) (www.geneontology.org; Ashburner *et al.*, 2000). GO is a hierarchical set of controlled vocabulary terms to describe genes/gene products and their attributes. The three major domains that are covered by GO include the cellular component that contains the gene product, the biological processes the gene product is associated with, and the molecular function of the gene product. With respect to candidate gene lists, tools have been developed (Draghici *et al.*, 2003) to test for over-representation of a GO category in a list of candidate genes. For instance, if one has generated a list of candidate genes from a microarray experiment, one can test to see if the number of genes within the list that are associated with cell proliferation is larger than would be expected in a completely random sample of all the genes on the microarray.

Another database for finding commonality among candidate genes is the Kyoto Encyclopedia of Genes and Genomes (KEGG) (http://www.genome.jp/kegg/; Kanehisa *et al.*, 2008). The KEGG system is a collection of databases used to describe biological systems through genomic, chemical, and systemic functional information. The KEGG PATHWAY database currently has 120 pathway maps that can be used to find biological connections between candidate genes. Pathways in this system have been manually constructed and curated from published materials.

Yet another way for finding commonality among candidate genes is to examine their regional expression. For instance, many of our microarray studies on mice (Tabakoff *et al.*, 2008) involved data on whole-brain gene expression. The Allen Brain Atlas (as mentioned previously) is a database with web-enabled tools that allow for the examination of regional brain expression of over 21 000 genes. This image database can be used to determine if there is a specific region or set of regions of the brain where the candidate genes are expressed. Localization to a particular area of the brain can give some insight into biological processes that the genes have in common. This particular data has been used for such tasks as identifying genes that show regionally enriched expression (D'Souza *et al.*, 2008). In our studies (Tabakoff *et al.*, 2008), we were able to localize the expression of many of our candidate genes to olfactory systems and the limbic and forebrain structures of brain. This allowed us to postulate that orosensory systems in brain are particularly important for alcohol preference in mice.

Besides simply looking for similar GO or KEGG function or regional expression of candidate genes, many researchers have looked for pathways that are less dependent on common functionality of genes and more on how the candidate genes are related in a transcriptional regulation network. One of the areas in which GO and other ontology solutions fall short is that they do not take into account that most genes and gene products have multiple context-dependent functions. Pathway analysis tools such as Pathway Studio use scientific literature to characterize relationships between and among genes and gene products from reports that have appeared in PubMed or other databases. Sivachenko *et al.* (2007) have published an overview of several methods that incorporate both the microarray-based association between transcripts and the literature-based association between transcripts and proteins. We have used an earlier version of Pathway Studio, Pathway Assist, to generate pathways that illustrate the relationships among candidate genes and gene products. An example of such a pathway, based on an artist's rendering of the results from Pathway Assist, is shown in Figure 10.2.

A further promising method of integrating literature and microarray results from gene networks was developed by Lawrence Hunter and his group at the University of Colorado, Denver. This group has developed a novel computational approach for integrating genome-scale data and current existing knowledge (literature) that they refer to as Hanalyzer (high-throughput analyzer). This approach can be classified as a "3R" system because it is based on three classes of algorithms: reading, reasoning, and reporting. In general, the goal of 3R systems is to help researchers interpret events or trends in genome-scale data and to enable the formation of relevant hypotheses for future research. The basis of approach is the weighted comparison

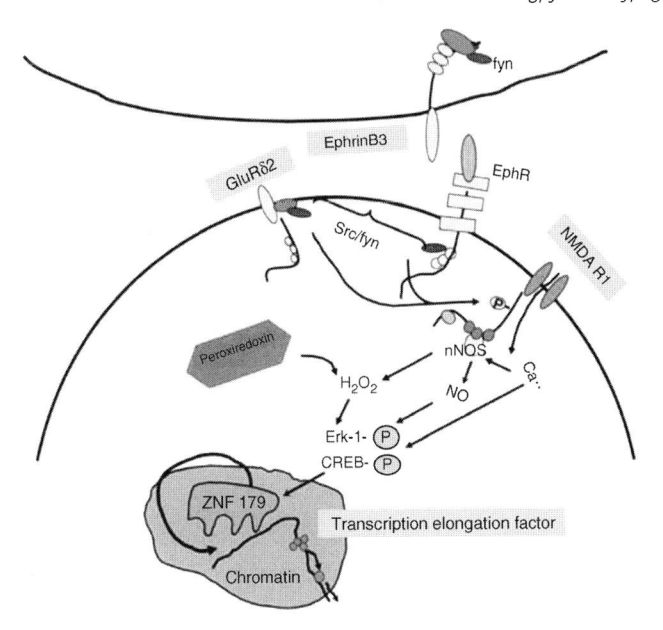

**Figure 10.2** Putative pathways mediating acute functional tolerance to ethanol. Artist's rendering of a pathway created using Pathway Assist and transcripts that were differentially expressed between high acute functional tolerance and low acute functional tolerance mice and localized to bQTLs for acute functional tolerance to a single exposure to ethanol. (The pathway was supplemental information to Tabakoff, Bhave, and Hoffman, 2003.)

of a knowledge network and a data network. The knowledge network consists of the existing knowledge of gene products and how they interact with one another. Existing knowledge is extracted from a variety of public sources. The data network is based solely on the results of a specific high-throughput experiment.

The system diagram in Figure 10.3 breaks down the three classes of algorithms. In the reading algorithm, data is extracted from a multitude of public resources including ontology annotation, publication abstracts, and protein–protein interaction databases. The reasoning algorithm integrates this information to characterize relationships among genes and gene products by estimating the likelihood of a biological relationship. The reporting algorithm integrates the knowledge and data networks, and creates a visual network that can be easily explored.

The 3R approach differs from most other approaches mentioned previously in three critical areas. First, prior knowledge about gene products and data from a high-throughput experiment are integrated before generating biological conclusions rather than first finding differentially expressed genes and then interpreting the relationship of the genes based on prior knowledge. Next, nodes in this system are meant to represent integration centers. This has critical implications in the types of knowledge graphs that can be created and utilized. Finally, this system separates the tasks of reading, reasoning, and reporting. By making these tasks mutually exclusive many different types of algorithms can be applied and optimized for each task.

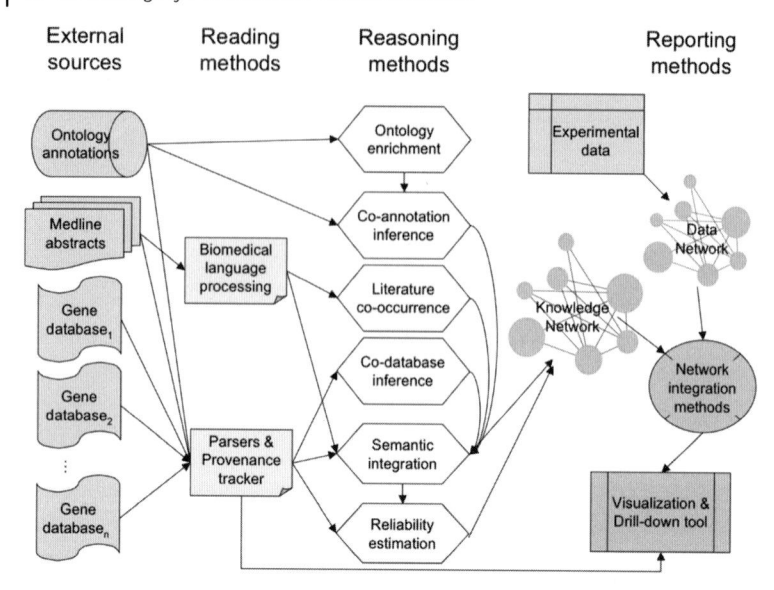

**Figure 10.3** System diagram describing the modules of the Hanalyzer. Reading methods (green) take external sources of knowledge (blue) and extract information from them, either by parsing structured data or biomedical language processing to extract information from unstructured data. Reading modules are responsible for tracking the provenance of all knowledge. Reasoning methods (yellow) enrich the knowledge that results from reading by, for example, noting two genes that are annotated to the same ontology term or database entry. All knowledge sources, read or reasoned, are assigned a reliability score, and all are combined using that score in a knowledge network (orange) that represents the integration of all sorts of relationship between a pair of genes and a combined reliability score. A data network (also orange) is created from experimental results to be analyzed. The reporting modules (pink) integrate the data and knowledge networks, producing visualizations that can be queried with the associated drill-down tool. (Figure and caption taken from Leach et al., 2009.)

More information on the construction and use of this system can be obtained from Bada and Hunter (2007), Leach et al. (2007), Hunter et al. (2008), and Tipney et al. (2009).

## 10.4
### Conclusions: Contributions of Genetical Genomic Phenomics to Systems Biology and Medicine

The integration of the large amounts of data obtained from transcriptional and genetic analyses, along with phenotypic data, into an interpretable framework requires not only an *a priori* focus on experimental design, but also the application of rigorous statistical approaches and bioinformatic technologies as described above. Even with these analyses, one has to keep in mind that transcriptional differences do not represent the only means by which genes influence behavior – some polymorphisms in the coding regions of genes do affect the function of the resulting proteins and these candidate genes are also valid modifiers of physiological phenotypes. In addition, while regulation of transcriptional networks can play a key role in the generation of phenotypic differences, the steps of translation to proteins and post-translational modifications (proteomics) also have to be considered in any discussion of systems biology. Nevertheless, as noted in Section 10.1, with the currently available and developing tools, researchers are continuing to unravel the regulatory mechanisms that control quantitative variation in transcript levels. These studies elucidate the transcriptional networks that underpin normal and disease phenotypes, and in the process illuminate the complex connections between genotype and physiological or behavioral traits that are the basis of systems biology.

The systems biology approach is already making an impact on the understanding of pathophysiologic process and enlightening research in the area of medication development. Work by Eric Shadt and his colleagues (Schadt et al., 2005; Emilsson et al., 2008) has produced a substantial understanding of the biochemical pathways and networks involved in obesity and type 1 diabetes. The experimental design of Schadt's studies (Drake, Schadt and Lusis, 2006) utilizes on many of the elements described in this chapter. The current reality in medication development, which is scientifically obsolete, is the scientific insistence that a single target can, and needs to be found and specifically modulated, to treat a complex phenotype that we define as a disease (e.g., hypertension, chronic pain, etc.). The scientific advancements in the understanding of systems biology and, in particular, the integration of genetic/functional genomic/phenomic approaches to identify the polygenic pathways that predispose disease, contribute to disease progression, and even determine an organisms response to medication are highly under-utilized in establishing remedial actions. The demonstration that networks, and not simply singular targets, are responsible for the most common human maladies, and that even many classic monogenic diseases are more adequately described as oligogenic due to other genes that modify disease intensity (Chial, 2008), is only recently penetrating the medication development planning process. The enlightenment can be seen in several recent publications that discuss the benefits of "promiscuous pharmacology" in the development of novel medications (Hopkins, Mason, and Overington, 2006) for treatment of conditions such as chronic pain, cancer, or neurodegenerative disease syndromes (Snell et al., 2000; Sivachenko, Kalinin, and Yuryev, 2006; Sivachenko and Yuryev, 2007; Schrattenholz and Soskic, 2008). Promiscuous pharmacology refers to the development of medications that can simultaneously modulate the function of more than one target in a biological network contributing to disease, while maintaining an acceptable safety profile (Harrill and Rusyn, 2008). One, therefore, would hope that systems biology will not only be an interesting (and at this point somewhat esoteric) scientific avocation, but will soon generally instruct all attempts to treat and prevent disease.

### Acknowledgments

The author's work has been supported by the National Institute on Alcohol Abuse and Alcoholism, the National Library of Medicine, and the Banbury Fund of the USA.

## References

Alberts, R., Terpstra, P., Li, Y., Breitling, R., Nap, J.P., and Jansen, R.C. (2007) Sequence polymorphisms cause many false cis eQTLs. PLoS ONE, 2, e622.

Ashburner, M., Ball, C.A., Blake, J.A., Botstein, D., Butler, H., Cherry, J.M., Davis, A.P., Dolinski, K., Dwight, S.S., Eppig, J.T., Harris, M.A., Hill, D.P., Issel-Tarver, L., Kasarskis, A., Lewis, S., Matese, J.C., Richardson, J.E., Ringwald, M., Rubin, G.M., and Sherlock, G. (2000) Gene Ontology: tool for the unification of biology. The Gene Ontology Consortium. Nature Genetics, 25, 25–29.

Bada, M. and Hunter, L. (2007) Enrichment of OBO ontologies. Journal of Biomedical Informatics, 40, 300–315.

Brem, R.B. and Kruglyak, L. (2005) The landscape of genetic complexity across 5700 gene expression traits in yeast. Proceedings of the National Academy of Sciences of the United States of America, 102, 1572–1577.

Brem, R.B., Storey, J.D., Whittle, J., and Kruglyak, L. (2005) Genetic interactions between polymorphisms that affect gene expression in yeast. Nature, 436, 701–703.

Brem R.B., Yvert, G., Clinton, R., and Kruglyak, L. (2002) Genetic dissection of transcriptional regulation in budding yeast. Science, 296, 752–755

Bystrykh, L., Weersing, E., Dontje, B., Sutton, S., Pletcher, M.T., Wiltshire, T., Su, A.I., Vellenga, E., Wang, J., Manly, K.F., Lu, L., Chesler, E.J., Alberts, R., Jansen, R.C., Williams, R.W., Cooke, M.P., and de Haan, G. (2005) Uncovering regulatory pathways that affect hematopoietic stem cell function using "genetical genomics". Nature Genetics, 37, 225–232.

Carlborg, Ö., De Koning, D.J., Manly, K.F., Chesler, E., Williams, R.W., and Haley, C.S. (2005) methodological aspects of the genetic dissection of gene expression. Bioinformatics, 21, 2383–2393.

Chen, M. and Kendziorski, C. (2007) A statistical framework for expression quantitative trait loci mapping. Genetics, 177, 761–771.

Chesler, E.J., Lu, L., Shou, S., Qu, Y., Gu, J., Wang, J., Hsu, H.C., Mountz, J.D., Baldwin, N.E., Langston, M.A., Threadgill, D.W., Manly, K.F., and Williams, R.W. (2005) Complex trait analysis of gene expression uncovers polygenic and pleiotropic networks that modulate nervous system function. Nature Genetics, 37, 233–242.

Cheung, V.G., Spielman, R.S., Ewens, K.G., Weber, T.M., Morley, M., and Burdick, J.T. (2005) Mapping determinants of human gene expression by regional and genome-wide association. Nature, 437, 1365–1369.

Chial, H. (2008) Mendelian genetics: patterns of inheritance and single-gene disorders. Nature Education, 1 (1).

D'Souza, C.A., Chopra, V., Varhol, R., Xie, Y.Y., Bohacec, S., Zhao, Y., Lee, L.L., Bilenky, M., Portales-Casamar, E., He, A., Wasserman, W.W., Goldowitz, D., Marra, M.A., Holt, R.A., Simpson, E.M., and Jones, S.J. (2008) Identification of a set of genes showing regionally enriched expression in the mouse brain. BMC Neuroscience, 9, 66.

Draghici, S., Khatri, P., Bhavsar, P., Shah, A., Krawetz, S.A., and Tainsky, M.A. (2003) Onto-tools, the toolkit of the modern biologist: Onto-express, Onto-compare, Onto-design and Onto-translate. Nucleic Acids Research, 31, 3775–3781.

Drake, T.A., Schadt, E.E., and Lusis, A.J. (2006) Integrating genetic and gene expression data: application to cardiovascular and metabolic traits in mice. Mammalian Genome, 17, 466–479.

Emilsson, V. et al. (2008) Genetics of gene expression and its effect on disease. Nature, 452, 423–428.

Gilad, Y., Rifkin, S.A., and Pritchard, J.K. (2008) Revealing the architecture of gene regulation: the promise of eQTL studies. Trends in Genetics, 24, 408–415.

Gottesman, I.I. and Gould, T.D. (2003) The endophenotype concept in psychiatry: etymology and strategic intentions. American Journal of Psychiatry, 160, 636–645.

Harrill, A.H. and Rusyn, I. (2008) Systems biology and functional genomics approaches for the identification of cellular responses to drug toxicity. Expert Opinion on Drug Metabolism and Toxicology, 4, 1379–1389.

Hopkins, A.L., Mason, J.S., and Overington, J.P. (2006) Can we rationally design promiscuous drugs? *Current Opinion in Structural Biology*, **16**, 127–136.

Hovatta, I., Zapala, M.A., Broide, R.S., Schadt, E.E., Libiger, O., Schork, N.J., Lockhart, D.J., and Barlow, C. (2007) DNA variation and brain region-specific expression profiles exhibit different relationships between inbred mouse strains: implications for eQTL mapping studies. *Genome Biology*, **8**, R25.

Hu, W., Saba, L., Kechris, K., Bhave, S.V., Hoffman, P.L., and Tabakoff, B. (2008) Genomic insights into acute alcohol tolerance. *Journal of Pharmacology and Experimental Therapeutics*, **326**, 792–800.

Hubner, N. et al. (2005) Integrated transcriptional profiling and linkage analysis for identification of genes underlying disease. *Nature Genetics*, **37**, 243–253.

Hughes, T.R. et al. (2001) Expression profiling using microarrays fabricated by an ink-jet oligonucleotide synthesizer. *Nature Biotechnology*, **19**, 342–347.

Hunter, L., Lu, Z., Firby, J., Baumgartner, W.A. Jr., Johnson, H.L., Ogren, P.V., and Cohen, K.B. (2008) OpenDMAP: an open source, ontology-driven concept analysis engine, with applications to capturing knowledge regarding protein transport, protein interactions and cell-type-specific gene expression. *BMC Bioinformatics*, **9**, 78.

International Human Genome Sequencing Consortium (2004) Finishing the euchromatic sequence of the human genome. *Nature*, **431**, 931–945.

Jansen, R.C. (1993) Interval mapping of multiple quantitative trait loci. *Genetics*, **135**, 205–211.

Jansen, R.C. and Nap, J.P. (2001) Genetical genomics: the added value from segregation. *Trends in Genetics*, **17**, 388–391.

Kalivas, P.W. and O'Brien, C. (2008) Drug addiction as a pathology of staged neuroplasticity. *Neuropsychopharmacology*, **33**, 166–180.

Kanehisa, M., Araki, M., Goto, S., Hattori, M., Hirakawa, M., Itoh, M., Katayama, T., Kawashima, S., Okuda, S., Tokimatsu, T., and Yamanishi, Y. (2008) KEGG for linking genomes to life and the environment. *Nucleic Acids Research*, **36**, D480–D484.

Kapur, K., Xing, Y., Ouyang, Z., and Wong, W.H. (2007) Exon arrays provide accurate assessments of gene expression. *Genome Biology*, **8**, R82.

Kerem, B., Rommens, J.M., Buchanan, J.A., Markiewicz, D., Cox, T.K., Chakravarti, A., Buchwald, M., and Tsui, L.C. (1989) Identification of the cystic fibrosis gene: genetic analysis. *Science*, **245**, 1073–1080.

King, M.C. and Wilson, A.C. (1975) Evolution at two levels in humans and chimpanzees. *Science*, **188**, 107–116.

Lander, E.S. and Botstein, D. (1989) Mapping Mendelian factors underlying quantitative traits using RFLP linkage maps. *Genetics*, **121**, 185–199.

Leach, S., Gabow, A., Hunter, L., and Goldberg, D.S. (2007) Assessing and combining reliability of protein interaction sources, in *Proceedings of Pacific Symposium on Biocomputing* (eds. R.B. Altman, A.K. Dunker, L. Hunter, T. Murray, and T.E. Klein) World Scientific, Singapore, pp. 433–444.

Leach, S.M., Tipney, H., Feng, W., Baumgartner, W.A., Kasliwal, P., Schuyler, R.P., Williams, T., Spritz, R.A., and Hunter, L. (2009) Biomedical discovery acceleration, with applications to craniofacial development. *PLoS Computational Biology*, **5**, e1000215.

Li, J. and Burmeister, M. (2005) Genetical genomics: combining genetics with gene expression analysis. *Human Molecular Genetics*, **14** (Spec. No. 2), R163–R169.

Liu, B., de la Fuente, A., and Hoeschele, I. (2008) Gene network inference via structural equation modeling in genetical genomics experiments. *Genetics*, **178**, 1763–1776.

Malats, N. and Calafell, F. (2003) Basic glossary on genetic epidemiology. *Journal of Epidemiology and Community Health*, **57**, 480–482.

Morley, M., Molony, C.M., Weber, T.M., Devlin, J.L., Ewens, K.G., Spielman, R.S., and Cheung, V.G. (2004) Genetic analysis of genome-wide variation in human gene expression. *Nature*, **430**, 743–747.

Ni, T.T., Lemon, W.J., Shyr, Y., and Zhong, T.P. (2008) Use of normalization methods for analysis of microarrays containing a high degree of gene effects. *BMC Bioinformatics*, **9**, 505.

Oldham, M.C., Konopka, G., Iwamoto, K., Langfelder, P., Kato, T., Horvath, S., and

Geschwind, D.H. (2008) Functional organization of the transcriptome in human brain. *Nature Neuroscience*, **11**, 1271–1282.

Rockman, M.V. and Kruglyak, L. (2006) Genetics of global gene expression. *Nature Reviews Genetics*, **7**, 862–872.

Rodd, Z.A., Kimpel, M.W., Edenberg, H.J., Bell, R.L., Strother, W.N., McClintick, J.N., Carr, L.G., Liang, T., and McBride, W.J. (2008) Differential gene expression in the nucleus accumbens with ethanol self-administration in inbred alcohol-preferring rats. *Pharmacology, Biochemistry, and Behavior*, **89**, 481–498.

Saba, L., Bhave, S.V., Grahame, N., Bice, P., Lapadat, R., Belknap, J., Hoffman, P.L., and Tabakoff, B. (2006) Candidate genes and their regulatory elements: alcohol preference and tolerance. *Mammalian Genome*, **17**, 669–688.

Schadt, E.E., Monks, S.A., Drake, T.A., Lusis, A.J., Che, N., Colinayo, V., Ruff, T.G., Milligan, S.B., Lamb, J.R., Cavet, G., Linsley, P.S., Mao, M., Stoughton, R.B., and Friend, S.H. (2003) Genetics of gene expression surveyed in maize, mouse and man. *Nature*, **422**, 297–302.

Schadt, E.E. et al. (2005) An integrative genomics approach to infer causal associations between gene expression and disease. *Nature Genetics*, **37**, 710–717.

Schrattenholz, A. and Soskic, V. (2008) What does systems biology mean for drug development? *Current Medicinal Chemistry*, **15**, 1520–1528.

Sen, S. and Churchill, G.A. (2001) A statistical framework for quantitative trait mapping. *Genetics*, **159**, 371–387.

Shifman, S., Bell, J.T., Copley, R.R., Taylor, M.S., Williams, R.W., Mott, R., and Flint, J. (2006) A high-resolution single nucleotide polymorphism genetic map of the mouse genome. *PLoS Biology*, **4**, e395.

Sivachenko, A.Y. and Yuryev, A. (2007) Pathway analysis software as a tool for drug target selection, prioritization and validation of drug mechanism. *Expert Opinion on Therapeutic Targets*, **11**, 411–421.

Sivachenko, A., Kalinin, A., and Yuryev, A. (2006) Pathway analysis for design of promiscuous drugs and selective drug mixtures. *Current Drug Discovery Technologies*, **3**, 269–277.

Sivachenko, A.Y., Yuryev, A., Daraselia, N., and Mazo, I. (2007) Molecular networks in microarray analysis. *Journal of Bioinformatics and Computational Biology*, **5**, 429–456.

Snell, L.D., Claffey, D.J., Ruth, J.A., Valenzuela, C.F., Cardoso, R., Wang, Z., Levinson, S.R., Sather, W.A., Williamson, A.V., Ingersoll, N.C., Ovchinnikova, L., Bhave, S.V., Hoffman, P.L., and Tabakoff, B. (2000) Novel structure having antagonist actions at both the glycine site of the N-methyl-D-aspartate receptor and neuronal voltage-sensitive sodium channels: biochemical, electrophysiological, and behavioral characterization. *Journal of Pharmacology and Experimental Therapeutics*, **292**, 215–227.

Storey, J.D., Akey, J.M., and Kruglyak, L. (2005) Multiple locus linkage analysis of genomewide expression in yeast. *PLoS Biology*, **3**, e267.

Szatkiewicz, J.P., Beane, G.L., Ding, Y., Hutchins, L., Pardo-Manuel de Villena, F., and Churchill, G.A. (2008) An imputed genotype resource for the laboratory mouse. *Mammalian Genome*, **19**, 199–208.

Tabakoff, B., Bhave, S.V., and Hoffman, P.L. (2003) Selective breeding, quantitative trait locus analysis, and gene arrays identify candidate genes for complex drug-related behaviors. *Journal of Neuroscience*, **23**, 4491–4498.

Tabakoff, B., Saba, L., Kechris, K., Hu, W., Bhave, S.V., Finn, D.A., Grahame, N.J., and Hoffman, P.L. (2008) The genomic determinants of alcohol preference in mice. *Mammalian Genome*, **19**, 352–365.

The Huntington's Disease Collaborative Research Group (1993) A novel gene containing a trinucleotide repeat that is expanded and unstable on Huntington's disease chromosome. *Cell*, **72**, 971–983.

The STAR Consortium (2008) SNP and haplotype mapping for genetic analysis in the rat. *Nature Genetics*, **40**, 560–566.

Thoday, J.M. (1961) Location of polygenes. *Nature*, **191**, 368–370.

Tipney, H., Leach, S., Feng, W., Spritz, R., and Williams, T. (2009) Leveraging existing biological knowledge in the identification of candidate genes for facial dysmorphology. *BMC Bioinformatics*, **10** (Suppl. 2), S12.

Walter, N.A., McWeeney, S.K., Peters, S.T., Belknap, J.K., Hitzemann, R., and Buck, K.J. (2007) SNPs matter: impact on detection of differential expression. *Nature Methods*, **4**, 679–680.

Zeng, Z.B. (1993) Theoretical basis for separation of multiple linked gene effects in mapping quantitative trait loci. *Proceedings of the National Academy of Sciences of the United States of America*, **90**, 10972–10976.

Zhang, W., Duan, S., Kistner, E.O., Bleibel, W.K., Huang, R.S., Clark, T.A., Chen, T.X., Schweitzer, A.C., Blume, J.E., Cox, N.J., and Dolan, M.E. (2008) Evaluation of genetic variation contributing to differences in gene expression between populations. *American Journal of Human Genetics*, **82**, 631–640.

Zhu, J., Lum, P.Y., Lamb, J., GuhaThakurta, D., Edwards, S.W., *et al.* (2004) An integrative genomics approach to the reconstruction of gene networks in segregating populations. *Cytogenetic and Genome Research*, **105**, 363–374.

# Part Four
# Data Mining and Modeling

Chapter 11 by Eduardo Mendoza is devoted to the methodology of modeling, which in our view is the important section in the field of theoretical psychiatry that soon should be developed. It is important to know that modeling starts before the application of mathematical formulation of equations – the conceptual model is the first stage of modeling when both experimentalists and theoreticians can begin to talk to each other.

Chapter 12 by James Smith and Marc-Thorsten Hütt demonstrates the new approach of graph-theoretical analysis of laboratory data. The aim of the application of this tool is to determine the pattern of interactions of genes and proteins in order to identify key players in complex molecular networks.

In Chapter 13, Peter J. Gebicke-Haerter and his working group present mathematical methods that can help to identify patterns in complex datasets generated in the molecular biological laboratory. Various methods that start with multivariate statistics such as principal component analysis are demonstrated and compared with newer approaches like the approach by Bayes based on conditional probabilities. These methods may be useful to resolve spatio-temporal patterns of on/off states of genes and interactions of proteins.

Mia Linskog *et al.* outline options of various computational modeling strategies applicable to the molecular network of the intracellular dopamine signal transduction network in Chapter 14. The effects of integrating noise terms into deterministic equations as shown in this chapter offers interesting perspectives in understanding molecular networks.

In Chapter 15, Marco Loh *et al.* develop a conceptual framework and model that allows the study of the dynamic properties of cortical networks. They relate short-term memory, attention, and decision making to theoretical concepts such as attractors, and show how changes in the $\alpha$-amino-3-hydroxy-5-methyl-4-isoxazol-propionacid (AMPA), *N*-methyl-D-aspartate (NMDA), and $\gamma$-aminobutyric acid (GABA) synaptic conductances could arise as a result of instability of attractor networks. Potential landscapes and the heuristic value of these concepts are discussed.

Finally, in an Epilogue in Chapter 16, Peter J. Gebicke-Haerter tries to summarize the essentials of a view of systems biology on psychiatric issues.

# 11
# From Communicational to Computational: Systems Modeling Approaches for Psychiatric Research

*Eduardo R. Mendoza*

## 11.1
### Introduction

In a recent review of models of circadian rhythms, we emphasized that modeling is much more than formulating differential equations that are run on a computer (Roenneberg *et al.*, 2008). Indeed, even in the language that we use to describe a phenomenon, to formulate a scientific question, or to explain methods and results, *implicit* models and concepts are prevalent and often influence the outcome of the activity. The fact that the circadian system was labeled a "clock" early on in the eighteenth century is an example of a model/concept of the circadian system that has profoundly influenced subsequent research. Conceptual models are made *explicit* in a variety of forms, which range from simple diagrams to complex systems of equations or rules.

A taxonomy tree (Figure 11.1) characterizes the different forms of explicit models as "selection points" in the modeling process. The "levels of modeling" introduced in Chapter 3 by Schulz and Klipp are subsumed as particular computational forms: qualitative (e.g., Boolean networks) or quantitative, which in turn may be deterministic or stochastic. The usefulness of qualitative models is shown in Chapter 12 by Smith and Hütt, where various techniques are described in the context of graph theory. Quantitative models in turn may be deterministic or stochastic. Ordinary differential equation(ODE) models are discussed in Chapter 3 in detail since they comprise the majority of computational models of biological systems today. Chapter 14 by Lindskog *et al.* provides illustrative examples of ODE models of the synapse in the context of psychiatric disease. The phenomenon of noise in neural networks discussed in Chapter 15 by Loh, Bolls and Deco provides a glimpse into stochastic aspects of modeling biological systems.

As always, the distinction between classes can blur since components of a model sometimes can be described in more than one category. Some elements may be continuous (concentrations), others discrete (phosphorylation or nuclear entry). Some, predominantly deterministic models may include stochastic processes and/or noise. Models that combine different methods are usually called "hybrid"

```
                                          ┌─ Diagrammatic/
                          ┌───────────────┤  Communicational
             Conceptual ──┤
                          │                ┌─ Discrete
                          │   ┌─ Qualitative ──┤
                          │   │            └─ Continuous
                          └─ Computational ──┤
                              │             ┌─ Stochastic
                              ├─ Quantitative ──┤
                              │             └─ Deterministic
                              └─ Hybrid
```

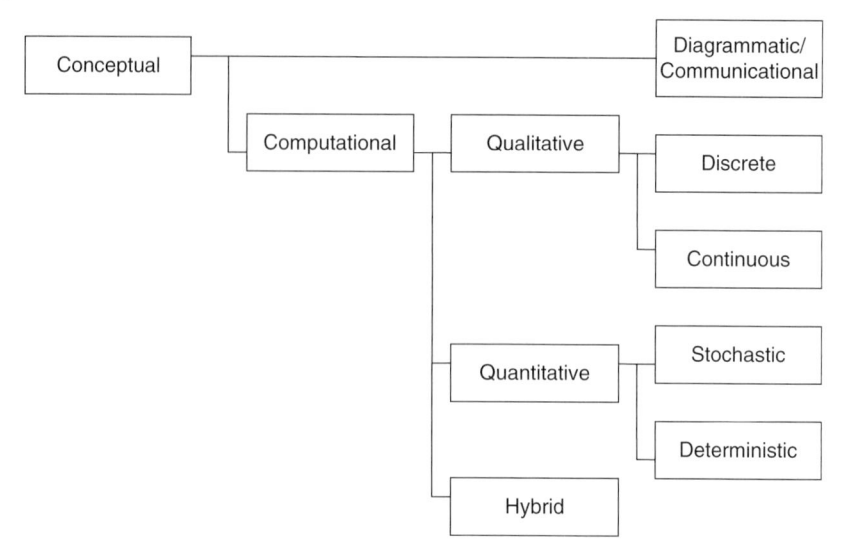

**Figure 11.1** Taxonomy of models for circadian clocks showing only the conceptual scheme. The full taxonomy in Roenneberg et al. (2008) includes both conceptual and contextual schemes wherein the classification of the contextual scheme is the same as the conceptual scheme. (Adapted from Marin-Sanguino, del Rosario, and Mendoza, 2009.)

in the modeling community and are quite popular in engineering. In a recent paper (Marin-Sanguino and Mendoza, 2008), we recommended using a combination of two well-established methods: Petri Nets on the qualitative side and biochemical systems theory (BST) models on the quantitative side for hybrid modeling in computational neuropsychiatry.

An important feature of the tree is that it also encompasses not only computational models, but also less formal ones, which are called "diagrammatic." These are the often colorful and imaginative figures found in most papers, fondly called "cartoons" by the biological community. Their main purpose is to communicate aspects of the system important for the study. Hence, the term "communicational" is probably better, as it better contrasts with "computational," indicating the different goals that the model creators have in mind. Diagrams are simply employed as one of the most effective means for this. The term "communicational" also better explains the great variety of such drawings – much to the chagrin of the computationally minded reader. While the renowned mathematician Henri Poincare once said "Mathematics is the art of giving the same name to different things," one is sometimes tempted to quip "Biology is the science of drawing many pictures of the same thing." However, once one understands that this sometimes bewildering variety is more due to different communication needs than lack of insight, one sees the importance of developing approaches to transform such communicational models to computational ones through a systematic dialogue with the experimentalists.

In this chapter, we first review – independent of particular methods – the various steps in modeling complex biological systems, following a recent paper by Voit, Qi, and Miller (2008). We then present a general approach using the theory of Petri nets for carrying out these steps in constructing qualitative and quantitative models, and illustrate it with various biological examples. In conclusion, we review a further method, the "concept map approach," which enables a direct transition from communicational models to a particular kind of ODE models, called canonical models. This approach has recently been successfully used for modeling dopamine metabolism in presynapses.

## 11.2
## Steps of the Modeling Process

For newcomers to the field of modeling, a statement by the statistician G. Box is very instructive: "All models are wrong, but some are useful." This witty remark (sometimes referred to as "Box's Law of Modeling") succinctly emphasizes one essential characteristic of modeling activities – the need to decide (beforehand) which aspects of the system are important (i.e., "useful") and then to leave out those details not relevant for those aspects (i.e., the model is necessarily incomplete or "wrong"). The second essential aspect concerns the data available with which the model can be constructed and, in a later step, validated. A well-known biochemist recently told me that, in writing a paper, he first asks his students to draw the figures (including the diagrams or "communicational models"), and then write the sections to detail the context (introduction), the approaches (methods), and evidence (results and discussion) for these – this can indeed be viewed as an effective process for communication and validation of such informal "diagrammatic" models.

For computational models, one needs a more structured and formal process. Voit, Qi and Miller (2008) provide a comprehensive and very readable description which is summarized in Figure 11.2. The phases (which are also largely valid for nonbiological systems) can be summarized as follows:

i) *Model selection.* Model selection for computational models essentially traverses the taxonomy tree, based on the two essential criteria of model purpose and available data. For biological systems, Voit *et al.* introduce a first decision point regarding the model purpose: should the model be explanatory or correlative. An explanatory model attempts to connect inputs and outputs of a system through a description of the actual biological mechanisms, while a correlative approach tries to relate them in a direct, global fashion (e.g., in a regression model).

ii) *Model design.* In the model design phase, the processes, the variables, and their known relationships need to be specified first in symbolic equations and, with the help of various parameter estimation techniques, in numerical equations.

iii) *Model diagnosis (or validation).* Once the parameters are specified, various steps are taken to test its validity, the first of which is to ensure that the logic of the model is correct. Other considerations include the (local) stability of the model, which would imply that it can recover from small perturbations as well as the sensitivity of parameters, which characterizes the effect of small changes in the parameters on aspects of the system behavior.

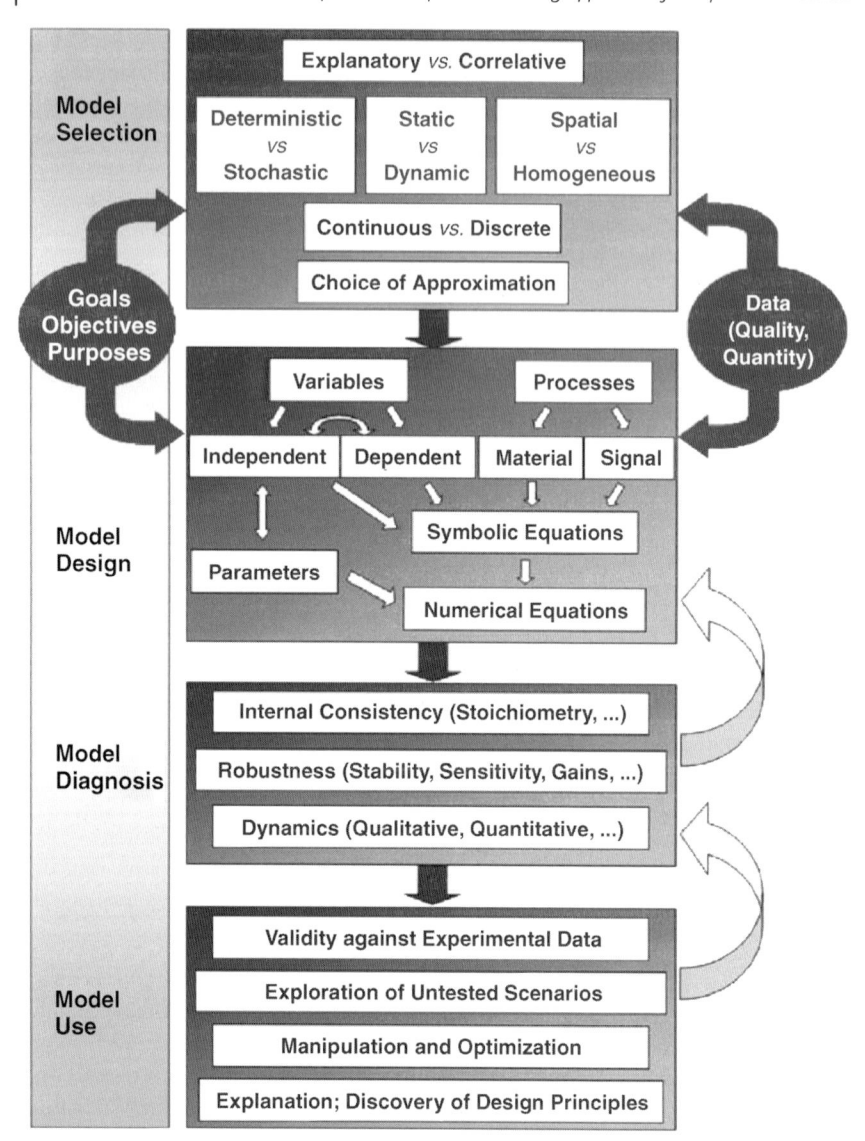

**Figure 11.2** Steps in modeling complex biological systems (Voit, Qi, and Miller, 2008; Reprinted with kind permission from Georg Thieme Verlag KG, Stuttgart).

iv) *Model use (and extension)*. Once the model has passed the diagnostic tests, we assume that it is ready for its envisioned purposes. A typical task is the analysis of "what-if" scenarios – these simulations could also include occasional or persistent changes. Another possibility is to use the model to explain why a certain stimulus leads to an certain observed behavior. It may also be possible to gain an understanding of "design principles" (i.e., why a natural system is composed in a particular fashion and not in an alternative way). Extensive use of models will lead to identification of shortcomings and trigger the iterative process of improvement and extension.

A wide variety of software tools support such processes of modeling biological systems, both in its entirety or in individual steps. The software packages that support established interchange standards such as SBML (Systems Biology Mark-Up Language) or emerging ones such as SBGN (Systems Biology Graphical Notation) are particularly useful – a current list and short descriptions of each tool's functionality can be found at http://sbml.org/Main_Page.

## 11.3
## From Diagrams to Qualitative Models through Petri Nets

A (discrete) Petri net is a graphical and mathematical formalism for describing concurrent processes. It consists of a bipartite, weighted, and directed graph, whose nodes are called "places" and edges are called "transitions," together with a marking (a function assigning a number of tokens to each place). The marking indicates the initial state of the system of concurrent processes while the weight is a measure of certain features of the transition. For example, in a Petri net model of a metabolic network, the weight is the stoichiometry of the substrate in the underlying chemical reaction. Petri nets as we understand them today were introduced by Carl Adam Petri in his PhD thesis in 1962. They have been applied in various technical and administrative systems since the mid-1970s, including engineering, computer science, and business management. The web site "Petri Nets World" (http://www.informatik.uni-hamburg.de/TGI/PetriNets/index.html) provides a wealth of information on application areas.

The idea to use the formalism to represent biochemical networks was mentioned by Petri himself in one his internal research reports on interpretation of net theory in the 1970s. A biochemical network was also used as an introductory example in Murata (1989). Reddy and collaborators are generally credited with the first applications of Petri nets in systems biology through their study of metabolic networks in 1993 (Reddy, Mavrovouniotis, and Liebman, 1993). Table 1 in Marin-Sanguino and Mendoza, 2008, illustrates a mapping of Petri net to biochemical terms. In our own joint work with experimentalists, discrete Petri nets have proven to be very effective in evolving qualitative computational models from the diagrams that they use. The rather intuitive graphical representation allows a consistent and sustainable dialogue between modelers and experimenters in the incremental modeling process.

The behavior of a Petri net is defined by a firing rule that consists of two parts – the precondition and the firing itself. A transition is enabled in a marking if every element of the transition's preset has at least as many tokens as the transition weight. An enabled transition may fire and when it does, a new marking is defined by the so-called "state equation" $m'(p) = m(p) + f(p,t) - f(t,p)$, where $m$ is the initial marking and $f$ the weight function of the Petri net. The firing is assumed to happen atomically and not to consume any time. The "dual character" firing rule accounts for the flexibility and usefulness of the Petri net formalism (see J. Wehler's lectures on Petri nets at Ludwig-Maximilians-Universität; available at http://www.pst.informatik.uni-muenchen.de/personen/wehler/): the firing itself consists of a linear structure as expressed in the "state equation," which is restricted by a nonlinear structure originating from the precondition.

Qualitative analysis of Petri nets are usually differentiated into structural and behavioral analysis. Structural analysis starts with basic topological properties such as connectivity and strong connectivity (i.e., the existence of a path from one place to another or a consistently directed path from one node to the other). The study of place invariants and transition invariants comprises a particularly important part of structural analysis, as these invariants have significant biochemical interpretation. The basis of invariant analysis is the concept of the incidence matrix $C$, whose rows and columns are indexed by the places and transitions, respectively, and whose elements are given by $C(p,t) = f(t,p) - f(p,t)$. A place invariant is then defined as a vector $x$ whose transpose $x^T$ is a nontrivial integer solution of the homogeneous system of linear equations $x^T C = 0$. In metabolic networks, place invariants represent compound conservation, while in signal transduction networks, they correspond to different states (active or inactive) of a given protein or protein complex. Analogously, a transition invariant is an integer-valued vector $y$ such that $Cy = 0$. The entries of transition invariants can be viewed as the relative firing rates of transitions, all of them occurring permanently and independently. This level of activity corresponds to the steady-state behavior of the system. Particularly important are the minimal place and transition invariants: an invariant is minimal if its support (defined as the set of nonzero entries) does not contain the support of any other invariant and the greatest common divisor of all its entries is 1. Minimal transition invariants stand for minimal self-contained subnetworks which are necessary for the network to function at steady state (e.g., in metabolic networks, they describe minimal sets of enzymes).

Graph representations of networks (such as Petri nets) become unwieldy when the numbers of nodes and edges become large. The problem of determining submodules representing particular functions (or vice versa, the composition of large networks from functional subsystems) becomes acute. For Petri net models, the work by Gilbert and collaborators (Gilbert, Heiner, and Lehrack, 2007) has introduced modularity analysis through the computation of structured sets of transitions, the maximal dependent transition (MDT) sets. The computation is based on the concept of dependency among transitions with respect to minimal transition invariants: two transitions $t_i$, $t_j$ are dependent on each other if they both belong to the support set of exactly the same minimal transition invariants, that is, $t_i \in supp(x)$ iff $t_j \in supp(x)$ for each minimal transition invariant $x$. Dependency introduces an equivalence relation on the set of all transitions $T$ and the equivalence classes partition $T$ into disjoint subsets, which define the MDTs. An application of the MDT approach is discussed in the apoptosis model in Section 11.4.

Basic behavioral properties include boundedness (the existence of a bound for the number of tokens at all places), reversibility (ability to revert to the initial marking), and liveness (existence of a transition that is enabled in every marking reachable from the initial marking). Behavioral analysis extends from testing for the above basic properties to sophisticated model checking with software tools such as PRISM (Gilbert, Heiner, and Lehrack, 2007).

## 11.4
### Petri Net Modeling of Apoptosis in Leukemic Cells and Neurons

Apoptosis or programmed cell death is a process by which a cell commits suicide when malfunctions arise due to cell stress, cell damage, or conflicting cell division signals. It is crucial to maintain a balance between cell death and survival because abnormalities in this mechanism can result in serious diseases. For example, failure to respond to apoptotic stimuli could cause cancer, while excessive apoptotic activity in certain types of neurons could cause neurodegenerative diseases such as Parkinson's, Alzheimer's, and Huntington's diseases.

Alberghina and Colangelo (2006) presented a modularization framework for investigating apoptosis of neurons in the context of Alzheimer's disease. They identified five major functional modules for apoptosis using what they call a "modular systems biology" approach. This approach consisted of the following steps: (i) analysis of the global function of an event, (ii) determination of the major functional modules involved in the process and their regulatory connections, (iii) determination of a basic modular blueprint or the process of validation of its dynamics, and (iv) analysis of the module, which includes the determination of its content, development and analysis of a mathematical model at the molecular level. Employing this approach based on a purely biological knowledge of apoptosis, they identified the following as the major functional modules: (i) signals, referring to the events and/or factors which trigger apoptosis; (ii) actuators, which are usually the initiator caspases activated by the trigger events and executioner caspases involved in the pathway; (iii) speeder, which involves factors or processes that integrate the initial apoptotic signaling that would otherwise proceed to either cell death or recover from stress events (with the main actor being the mitochondria); (iv) brakes, which are the potent regulators of caspase activity; and (v) response to apoptosis. Figure 11.3 shows

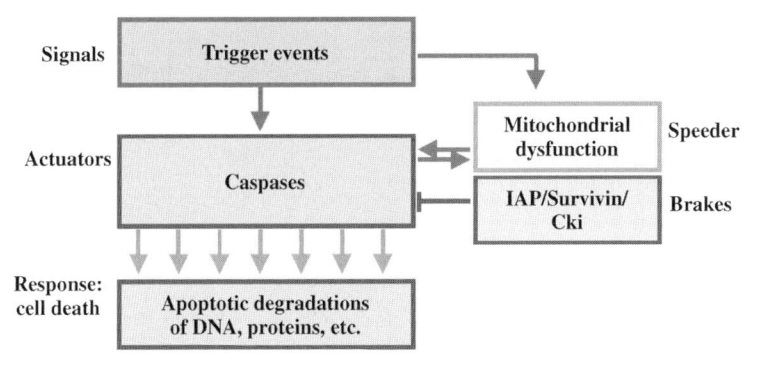

**Figure 11.3** Functional modules of neuronal apoptosis (Alberghina and Colangelo, 2006). IAP: inhibitor of apoptosis protein; Cki: cyclin kinase inhibitor.

these modules and how they are related to each other by positive and negative feedback loops. Alberghina and Colangelo used the model of apoptosis in Bentele et al. (2004) to illustrate the identification of these modules solely on the basis of the biological knowledge of mechanisms involved in such a model.

In her PhD thesis, Rodriguez (2008) studied whether the approach of Alberghina and Colangelo (2006) is applicable to a more complex system of apoptosis in leukemic cells induced by a novel marine compound, cephalostatin-1. In addition, she investigated whether, using the MDT techniques of Heiner et al. (Gilbert, Heiner, and Lehrack, 2007; Grafahrend-Belau and collaborators, 2008) on a Petri net model of cephalostatin-1 apoptosis, the modules of Alberghina and Colangelo (2006) could be recovered in an automated fashion. The essential step of building a more detailed, discrete Petri net model was accomplished in sustained and systematic dialogue with the experimental experts, resulting in a consensus on the underlying chemical reactions. Figure 11.4(a and b) shows the initial diagram and the resulting Petri net with 43 places and 59 transitions (representing 34 reactions, including seven reversible and 18 input transitions), which was constructed with the software tool SNOOPY. The invariants were computed with the companion tool Charlie. The 16 transition invariants determined represent compounds and processes which are the essential components of known apoptotic networks. The computation of MDT sets was confirmed as a reliable method for modularization of the model since the MDT sets partition the entire network into its smallest functional biological units.

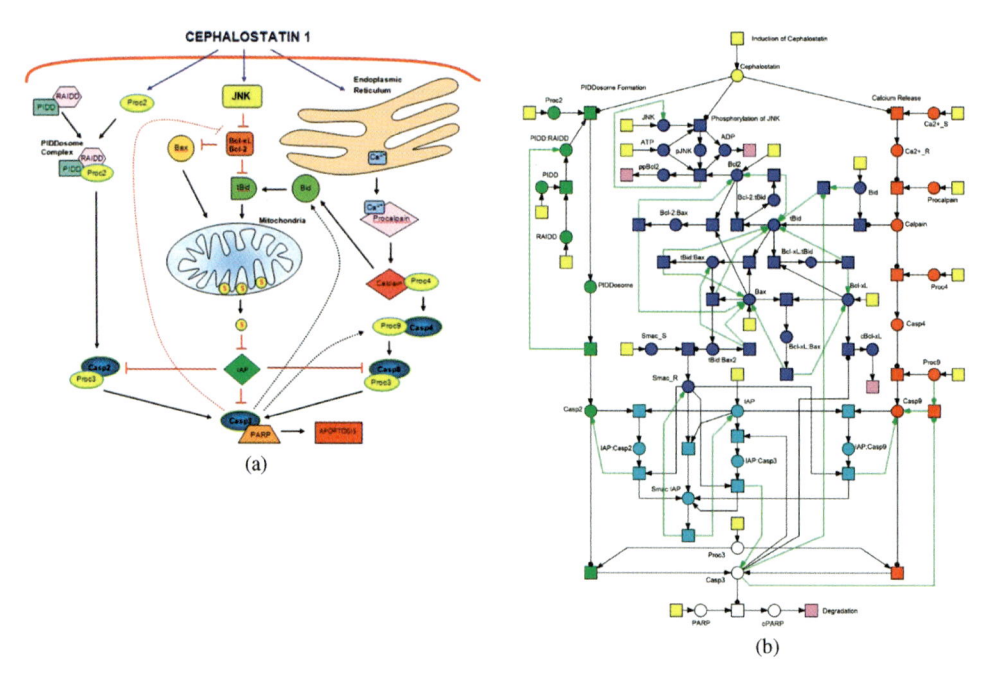

**Figure 11.4** (a) "Cartoon" and (b) discrete Petri net model of cephalostatin-1 apoptosis in leukemic cells (Rodriguez et al., 2009; Reprinted with kind permission from Springer Science + Business Media B.V.).

## 11.5
### From Stoichiometric (Qualitative) to Kinetic (Quantitative) Genome-Scale Models

Much progress has been made with genome-scale network reconstruction process, particularly in the field of metabolic networks (Palsson, 2006). The foundation of such genome-scale analysis is identification of the network's chemical components and the chemical transformations that they participate in, represented in a self-consistent and chemically accurate format in the stoichiometric matrix. Popular forms of "stoichiometric" modeling include flux balance analysis (FBA) and constraint-based analysis (CBA) – a detailed analysis of these approaches is provided by in Chapter 12 by Smith and Hütt.

With the growing availability of such genome-scale networks, the question of evolving them to full kinetic models has received much attention recently. In our work (Gonzalez et al., 2008) on a genome-scale model of the archaeon *Halobacterium salinarum*'s metabolism, we showed that a hybrid model combining the qualitative (modeled as a Petri net) network with nutrient consumption dynamics and applying a CBA allowed new insights into the organism's fluxome. Lee and Voit (unpublished) also recently introduced a novel approach which uses the information provided by the FBA model for careful parameter selection in the GMA model of monolignol biosynthesis in *Populus* to be constructed. More generally, Jamshidi and Palsson (2008) established a framework for transforming genome-scale stoichiometric to kinetic models. In the workflow that they propose based on the framework, they identified the following four mathematical challenges that need to be addressed in such transformations:

i) The fundamental structure of the Jacobian (i.e., its decomposition into the chemical, kinetic, and thermodynamic information).
ii) The structural similarity between the stoichiometric matrix and the transpose of the gradient matrix.
iii) The duality transformations enabling either fluxes or concentrations to serve as the independent variables.
iv) The timescale hierarchy in biological networks.

## 11.6
### From Diagrams Directly to Quantitative Canonical Models: The Concept Map Method

Qualitative models using discrete Petri nets can be extended to quantitative ones, both in the deterministic and stochastic forms. Gilbert, Heiner, and Lehrack (2007) provide a unifying framework for these transitions and transformations which are summarized in Figure 11.5. The same framework also easily accommodates the formulation of hybrid Petri net models (Marin-Sanguino and Mendoza, 2008). In the same paper, we pointed out that the use of canonical models, such those formulated in biochemical systems theory (BST) would be attractive, particularly because of the use of only two kinds of kinetic parameters, which are readily interpreted biologically.

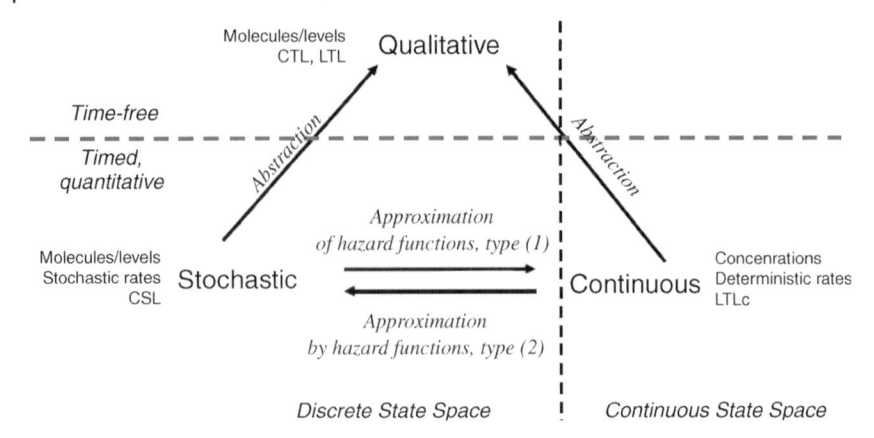

**Figure 11.5** Unifying conceptual framework for Petri net models (Gilbert, Heiner, and Lehrack, 2007). CTL: Computation Tree Logic; LTL: Linear Temporal Logic; CSL: Continuous Stochastic Logic.

However, it is also possible to directly construct such models (without the intermediary qualitative step) – an example of this is the concept map method (CMM) of Goel, Chou, and Voit (2006). The basic idea underlying the CMM is that bioscientists always know more than what is captured in the diagrams. Through an interactive dialogue between experimenter and modeler, the CMM aims to capture essential elements of this knowledge to transform the informal concept map to an initial quantitative model, which can be executed on a computer. In other words, the CMM's foremost goal is to integrate and formalize the knowledge of the experts into a preliminary mathematical model that can produce quantitative hypotheses through simulations. A comparison of simulations and experiment will refine such a model.

The CMM approach is structured into the following steps. The initial step of the dialogue consists in establishing lists of components and processes with relationships and rules that are visualized in the concept map. The modeler here should assume the role of a "sounding board," discussing, questioning, and revisiting in detail the biologists' concepts and perceived relationships in order to facilitate the construction of the initial coarse computational model. In a second step, after fixing a quantifiable scale at each node, the biologists' knowledge about stimulus–response dynamics of the particular component are extracted first by questions regarding its qualitative dynamics (i.e., goes up/down, fast, slow, linear, sigmoid, saturates, etc.). This is followed by queries about quantification (e.g., "at the end twice as high," "takes about 15 min," etc.). Figure 11.6 illustrates the expected output of this step. The figures are to be drawn by hand, and contain both experimental knowledge and intuition of the biologist. The experimenter–modeler team chooses all known components of the system and draws the expected or predicted dynamics of the concentration (the units of time and concentration are to be discussed by the experimenter–modeler team) under some chosen initial and boundary conditions and perturbations. A different set of curves should be drawn for systems with different initial and boundary conditions and perturbations. Note that not all of the components whose dynamics are drawn in this step could be included in the

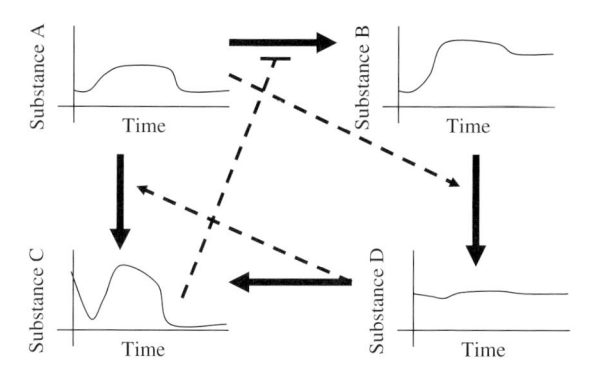

**Figure 11.6** Typical expected output of the second step of the CMM (Marin-Sanguino, Rosario, and Mendoza, 2009; Reprinted with kind permission from Georg Thieme Verlag KG, Stuttgart).

resulting quantitative (mathematical) model. The figures could just be estimates and hence the actual values of the concentrations and the time they reach steady state or peaks are to be taken by the modeling team as a guide. As shown in Figure 11.6, aside from the stimulus–response dynamics of each component, the network of the system is also implied in this map since mass flow (solid arrows in the figure), activation (dashed arrows), and inhibition (dashed lines) are also included.

In the third step, the CMM constructs a canonical model and employs an appropriate computational technique to estimate an initial set of kinetic parameters. In this so-called "inverse modeling" step, points are taken from the qualitatively drawn curves to form time series of "data" and these provide the basis for the initial model parameters. The big advantage of BST models (and canonical models, in general) lies in the vast literature of parameter estimation and structure identification algorithms that have been published in the last 5–7 years. In a recent review, Chou and Voit (2009) provide a comprehensive overview of the field and characterize various classes of the over 100 algorithms recently published. In the fourth (crucial) step, the model is tested and validated against further data and intuition in a joint biologist–modeler effort. This completes the first cycle and typically initiates the recursions needed to further improve the model.

In Marin-Sanguino, del Rosario, and Mendoza (2009), we elaborated how this method was used by Qi *et al.* (2008) in developing a preliminary but already quite useful model of dopamine metabolism in the presynapse.

## 11.7
## Summary and Outlook

In modeling complex biological systems, it is essential to follow a structured approach such as the method of Voit, Qi, and Miller (2008) that we have described. In addition, we have emphasized the need for methods which enable the evolution of informal diagrams produced by biologists (mainly for communicational purposes) into computational quantitative models. We discussed in detail two such approaches: an indirect way via qualitative models using discrete Petri nets (which in a second step are extended to continuous Petri nets) and direct way via the concept map method. Much progress has been made in the genome-scale reconstruction of metabolic networks and initial efforts in transforming these into kinetic models are underway. In the future, we expect more hybrid models, which are a combination of several modeling approaches, in order to cope with the growing complexity of the systems studied and the heterogeneous nature of available data.

## References

Alberghina, L. and Colangelo, A.M. (2006) The modular systems biology approach to investigate the control of apoptosis in Alzheimer's disease neurodegeneration. *BMC Neuroscience*, **7** (Suppl. 1), S2.

Bentele, M., Lavrik, I., Ulrich, M., Stösser, S., Heermann, D.W., Kalthoff, H., Krammer, P.H., and Eils, R. (2004) Mathematical modeling reveals threshold mechanism in CD95-induced apoptosis. *Journal of Cell Biology*, **166**, 839–851.

Chou, I. and Voit, E.O. (2009) Recent developments in parameter estimation and structure identification of biochemical and genomic systems. *Mathematical Biosciences*, **219**, 57–83.

Gilbert, D., Heiner, M., and Lehrack, S. (2007) A unifying framework for modelling and analysing biochemical pathways using Petri nets. Technical Report TR-2007-253, Department of Computer Science, University of Glasgow and Technical Report I-02/2007, Brandenburg University of Technology.

Goel, G., Chou, I., and Voit, E.O. (2006) Biological systems modeling and analysis: a biomolecular technique of the twenty-first century. *Journal of Biomolecular Techniques*, **17**, 252–269.

Gonzalez, O., Gronau, S., Falb, M., Pfeiffer, F., Mendoza, E., Zimmer, R., and Oesterhelt, D. (2008) Reconstruction, modeling and analysis of *Halobacterium salinarum* R-1 metabolism. *Molecular BioSystems*, **4**, 148–159.

Grafahrend-Belau, E., Schreiber, F., Heiner, M., Sackmann, A., Junker, B.H., Grunwald, S., Speer, A., Winder, K., Koch, I. (2008) Modularization of biochemical networks based on classification of Petri net T-invariants. *BMC Bioinformatics*, **9**, 90.

Jamshidi, N. and Palsson, B.O. (2008) Formulating genome-scale kinetic models in the post-genome era. *Molecular Systems Biology*, **4**, 171.

Marin-Sanguino, A. and Mendoza, E.R. (2008) Hybrid modeling in computational neuropsychiatry. *Pharmacopsychiatry*, **41** (Suppl. 1), S85–S88.

Marin-Sanguino, A., del Rosario, R.C., and Mendoza, E.R. (2009) Concept maps and canonical models in neuropsychiatry. *Pharmacopsychiatry*, **42** (Suppl. 1), S110–S117.

Murata, T. (1989) *Petri nets-Properties, analysis and applications.* Proceedings of the IEEE, vol 77.

Palsson, B.O. (2006) *Systems Biology – Properties of Reconstructed Networks*, Cambridge University Press, Cambridge.

Qi, Z., Miller, G.W., and Voit, E.O. (2008) Computational systems analysis of dopamine metabolism. *Plos One*, **3**(6): e2444.

Reddy, V.N., Mavrovouniotis, M.L., and Liebman, M.N. (1993) Petri net representations in metabolic pathways. *Proc. Int. Conf. Intell. Sys. Mol. Biol*, **1**, 328–336.

Rodriguez, E.M. (2008) A mathematical model for cepahlostatin-1 induced apoptosis in leukemic cells. PhD Thesis, University of the Philippines Diliman.

Rodriguez, E.M., Rudy, A., del Rosario, R.C.H., Vollmar, A., and Mendoza, E.R. (2009) A discrete Petri net model for cephalostatin 1-induced apoptosis in leukemic cells [Special issue on Petri Nets and Biosystems]. *Natural Computing*, doi:10.1007/s11047-009-9153-9.

Roenneberg, T., Chua, E.J., Bernardo, R., and Mendoza, E.R. (2008) Modelling biological rhythms. *Current Biology*, **18**, R826–R835.

Voit, E.O., Qi, Z., and Miller, G.W. (2008) Steps of modeling complex biological systems. *Pharmacopsychiatry*, **41** (Suppl. 1), S78–S84.

# 12
# Network Dynamics as an Interface between Modeling and Experiment in Systems Biology
*James Smith and Marc-Thorsten Hütt*

## 12.1
### Introduction

The hallmark signature of quantitative biology is the attempt of introducing quantitative methods into a discipline dominated historically by the qualitative interpretation of empirical observations and by reductionist approaches. The young and rapidly developing field of systems biology is becoming the most successful implementation of approaches taken in quantitative biology – the emerging new paradigm of modern biology in the postgenomic era. Systems biology is a melting pot of many disciplines, contributing to the understanding of the larger-scale of organization of living cells and their dynamic behavior in response to external and internal stimuli, including disease development (e.g., see Kitano, 2002, 2004; Klipp *et al.*, 2005; Alon, 2006; Kriete and Eils, 2006; Kholodenko, 2006).

Systems biology is situated at the intersection between the biological sciences, mathematics, statistical physics, chemical biophysics, and computer science. Topics include networks of gene regulation, metabolism, cellular signaling mechanisms, and the paths from the inventory of genome information to cellular and tissue function, from basic to the applied biosciences including the drug discovery process impacting on clinical practice. Serving as a powerful foundation for its success, systems biology is organized as a strongly iterative process between theory (establishing fundamental principles based on minimal models), experiment (particularly high-throughput measurements of cellular states), data analysis (the extraction of characteristic observables and statistical features from large data sets), and computer simulations (studying systems numerically using realistic mathematical models with predictive power).

To elucidate the mechanisms underlying cellular processes, the disciplines involved have to move closer together and bridge the gaps between them. Systems biology has the purpose of establishing the appropriate interdisciplinary links. In Joel Cohen's dictum (Cohen, 2004) that "Mathematics is biology's next microscope – only better," the importance of implementing modeling and quantitative approaches to experimental efforts in the biological sciences is summarized wonderfully. The other half of the title of Cohen's review, "Biology is mathematics' next physics – only better," goes even further, emphasizing an aspect that is frequently forgotten: mathematics and the other theory-oriented disciplines involved in this huge endeavor of systems biology receive immensely stimulating input from biology and, particularly from biological complexity, potentially leading to completely new theoretical research fields.

Recently, the development of systems biology, based on molecular and biochemical information, has been compared with the historical development of astronomy (Alter, 2006): from the design of instrumentation (by Galileo) and accumulation of data (by Brahe) to an understanding of natural laws (by Newton) progressing to an interpretation of patterns in the data (by Kepler). If this analogy holds, the biological sciences at present are at a most exciting point on this path involving the move from observing patterns towards an understanding of the natural governing laws and fundamental principles.

Neurological diseases, such as schizophrenia and Parkinson's disease, are often multifactorial and affect several layers of cellular organization simultaneously, from extra- and intracellular signaling to metabolism and gene regulation (Le Novere, 2008). Understanding how diseases differ from the normal or "healthy" states requires an integrated view at the system level, rather than a focus on separate components. Furthermore, diseases have temporal progressions, involving changes and complications over several timescales. Formulations built upon dynamical systems theory, for example as the concept of dynamical diseases (an der Heiden, 2006), may help to bridge between the observable and available data and any mathematical description of diseases.

In virtually all systems biology studies, statistical methods play an important role in identifying the intrinsic mechanisms behind the performance of a system. More precisely, the analysis of system-wide information with any statistical methods requires a clear concept of a null hypothesis; in other words, a random background, against which the observations can be compared. With appropriate null hypotheses, questions can be asked, such as whether the gene regulatory network (network of genes regulating other genes) activated under particular conditions is larger or smaller than expected at random? Is the network more or less densely connected? Is the amount of fragmentation unexpected? Is the number of essential reactions in a particular segment of metabolism high compared to (random) expectation or not?

As an example of the broader framework of "network biology" (Barabási and Oltvai, 2004), we will briefly discuss here how abstracting cellular processes into networks into a unifying language of small regulatory templates (Alon, 2007; Brandman et al., 2005; Brandman and Meyer, 2008) can help to identify deviations from randomness and contribute to an understanding of how such cellular systems function. The theory of networks is well established in mathematics in the field of "graph theory," where many applications of the mathematical concepts originated from studies of network architecture provided by (linear) systems theories in the early 1980s. In the biological sciences, the observation of networks is familiar from food webs in ecology and species relationships in population dynamics. In the biomedical sciences there are many familiar examples including biochemical reaction networks in metabolism (Jeong et al., 2000) and signaling (Papin et al., 2005) with pools and flows of metabolites and interacting molecules, and in the theory of neural networks developed in the 1960s and 1970s (Amit, 1989), in regulation networks (Shen-Orr et al., 2002; Xayaphoummine et al., 2007), and in genetic networks, particularly in relation to evolution (Bornholdt and Sneppen, 1998; Bornholdt, 2001; Cosentino-Lagomarsino et al., 2007; Evlampiev and Isambert, 2007, 2008). In a clinical context, networks are in use by epidemiologists to describe the social and sexual contacts of individuals in contagion and infection studies (Hethcote, 2000; Pastor-Satorras and Vespignani, 2001; Liljeros et al., 2001; Barthélemy et al., 2005; Hufnagel, Brockmann, and Geisel, 2004).

We summarize some of the most important concepts from graph theory in Section 12.2 and discuss their impact on systems biology in Section 12.3. Section 12.4 will introduce "network dynamics," which is an emerging field of study. The analysis of network dynamics is about investigating an ensemble of interacting nonlinear dynamic elements with a given architecture producing a "spatio"-temporal pattern (summarizing the time-courses of all nodes in the graph). In studying network dynamics, one attempts to understand how the network structures influence dynamic processes. Historically, some of the key ideas of this approach extend from the seminal work by Kauffman (1969) on the dynamics of networks consisting of binary elements (known as Boolean networks; see Section 12.4).

Learning to translate, for example, DNA microarray data into the language of graph theory (e.g., see Gebicke-Haerter and Tretter (2008) for an application to addiction), apart from revolutionizing our view of regulation, can also refine the categories of graph theory. In many ways such a description, when added to the repertoire of methods in bioinformatics, improves the interpretation of huge amounts of data and advances the field from just the administrative management of this data to the point of being able to ask insightful questions using conceptual commands.

## 12.2
### Aspects of Graph Theory

One of the most important developments in theoretical biology over the past few years has been certainly the application of innovations taken from graph theory to analyze biological processes. Watts and Strogatz (1998) proposed the construction of a graph with a short average path length and – at the same time – a high clustering coefficient that triggered a wave of research activity. It became apparent that such "small-world" graphs represented an adequate model for many technical, natural, and biological systems (Fell and Wagner, 2000; Newman, 2001; Strogatz, 2001). The construction of graphs with a scale-free degree distribution, as introduced by Barabási and Albert (1999), has been similarly successful for the modeling of real systems (Jeong et al., 2000; Wagner and Fell, 2001). Furthermore, it has been shown that especially biological networks, such as metabolic, protein, and gene networks, are based on scale-free topologies (Jeong et al., 2001; Farkas et al., 2003; Barabási and Oltvai, 2004).

Here in 12.2, we summarize some concepts from graph theory and outline the tools for analyzing graphs. Details and discussions about the implementation issues can be found in Hütt and Lüttge (2005), Müller-Linow, Hilgetag and Hütt (2008), and Hütt and Dehnert (2006). Further, in Sections 12.3 and 12.4, we apply this theoretical framework to a case study of simple forms of network dynamics. Our guiding questions then will be to what extent, and under what conditions, are the observed dynamics a reflection of internal topological properties of the underlying graph?

Networks consist of elements (nodes, vertices) and the interactions (connections, links) between them. The four classes of graphs, which up to now can still be considered most important for biological applications are (i) regular, (ii) random, (iii) "small-world," and (iv) scale-free graphs. We will summarize qualitatively the prescriptions for constructing such graphs, as well as their most important properties. For details and thorough mathematical discussions, see the excellent reviews by Albert and Barabási (2002), Newman (2003), and Strogatz (2001). The popular nontechnical book by Barabási (2002) gives a very readable account of the appearances of graphs in everyday life.

Regular graphs are used in the formulation of biological models, such as approximations of spatially organized elements or in extreme situations when network architecture is believed to be of minor importance for the behavior of the entire system. A fully connected graph underlies the classical investigation of spontaneous synchronization of coupled phase oscillators by Kuramoto (1984). A system of elements with coupling to spatial neighbors (e.g., on a two-dimensional spatial lattice with the nodes being placed at all lattice sites) is an adequate starting point for investigating spatio-temporal pattern formation. It is clear, however, that often a network is not related to real (physical) space. For example, the elements could be biochemically active substances in a solution, with the network being defined by the capacity of each substance to interact with (activate or inhibit) itself and the other substances in the solution. Such networks are called "relational networks" (as opposed to "spatial networks"). In such cases, the question of architecture often becomes a key issue when one wants to understand the dynamic properties of such a network.

The investigation of random graphs by Erdös and Rényi (1960) constituted a major breakthrough in graph theory. In the Erdös–Rényi (ER) model, there is a system of $N$ nodes and each possible link is given a probability $p$ of being present within the graph. From the ER model of a random graph, a variety of mathematical properties can be obtained analytically. However, what quantities are indeed useful for characterizing topological (architectural) features of a graph? On the level of individual nodes, the most important observable is the degree $k$ of a node, that is the number of nodes ("neighbors") linked to it. Averaging over all nodes gives the average degree of a graph. Often, particularly for scale-free graphs, instead of averaging, it is more informative to study the degree distribution $P(k)$, which is the (normalized) number of times the degree $k$ is found in the graph. A different property of graph architecture is addressed with the clustering coefficient, introduced by Watts and Strogatz (1998), to provide an efficient distinction between small-world graphs and other random graphs. As with the degree $k$ introduced above, the clustering coefficient is an observable at the level of an individual node. When considering a node with degree $k$, its clustering coefficient $C$ is determined by the number of links that the $k$ neighbors have between them. When this number $m$ is normalized by the number of possible links between these $k$ nodes, namely $k(k-1)/2$, one obtains the clustering coefficient $C$. Again, one can characterize the whole graph both by an average clustering coefficient and by a distribution $P(C)$ of clustering coefficients in this graph.

Examining ER graphs in more detail, one might expect that in random graphs generated by the ER model a wide range of degrees appears, each with a similar frequency. Counter-intuitively for ER graphs, the function $P(k)$ follows a Poisson distribution with a pronounced peak at some average degree $k$. Figure 12.1(a) gives an example of such a degree distribution for a graph consisting of 1000 nodes.

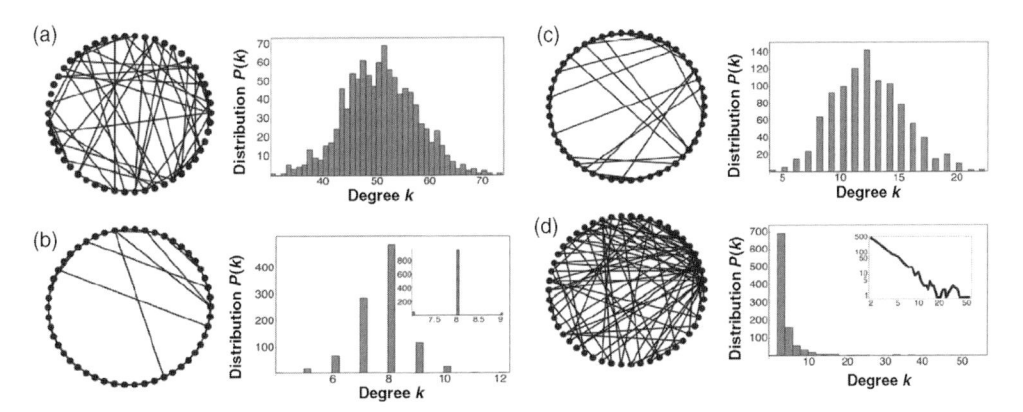

**Figure 12.1** Typical examples of graphs for four of the classes discussed in the main text, together with their degree distributions. The classes are (a) random graphs, small-world graphs obtained by (b) rewiring and (c) link-adding, and (d) scale-free graphs. The examples show graphs with 100 nodes, while the degree distributions have been computed for graphs consisting of 1000 nodes. For the random graph (a), the distribution $P(k)$ shows a pronounced peak, which approximates a Poisson distribution. Here, the connection probability has been set to $p=0.05$. The two types of small-world graphs are difficult to distinguish on the level of the graph itself. The degree distributions, however, show clear differences related to the respective prescriptions for generating these graphs. For the rewiring procedure (b), the degree distribution consists of isolated lines for low and intermediate rewiring probabilities. The inset shows an extreme case of very low rewiring probability, namely, $p=10^{-3}$. The signature of the regular backbone, a ring of nodes with links from each node to its eight nearest neighbors, from which the rewiring process is started, is clearly visible as an isolated line at $k=8$. Due to rewiring, two small lines appear at $k=7$ and $k=9$. The main part of the figure gives the degree distribution for $p=10^{-2}$. There, lines with a larger deviation from the $k=8$ line are found. At the same value of $p$, the other type of small-world graph (c) (obtained by the link-adding procedure) displays a degree distribution much more similar to the case of a random (ER) graph. The scale-free graph (d) has a completely different degree distribution, which can be described by a power law. The inset shows the same distribution in a double-logarithmic representation (a log-log plot). Already for 1000 nodes, the characteristic linear behavior in the log-log plot is the signature of a power law distribution. (Taken from Hütt and Lüttge, 2005.)

Furthermore, the average clustering coefficient is very low. This is the most striking difference to real (physical or natural) graphs, which usually, in addition to a certain amount of randomness, display a high clustering coefficient.

A remarkable topological (architectural) property of random graphs was discovered by Erdös and Rényi (1960) and concerns the connectedness of a graph. A graph G is connected if any node can be reached from any other node of G by following the links present in G. Clearly, at a very small connection probability $p$, the random graph will not be connected. Conversely, close to $p = 1$, connectedness is certain. However, an interesting question is, how does connectedness emerge when $p$ is increased from $p = 0$ to $p = 1$? Figure 12.2 shows the corresponding curve giving the percentage of connected graphs as a function of $p$, together with examples of disconnected graphs at low $p$ values with the largest connected component highlighted in grey.

Watts and Strogatz (1998) introduced a technique called "random rewiring" that allowed them to investigate small-world networks. Here, one starts from a regular

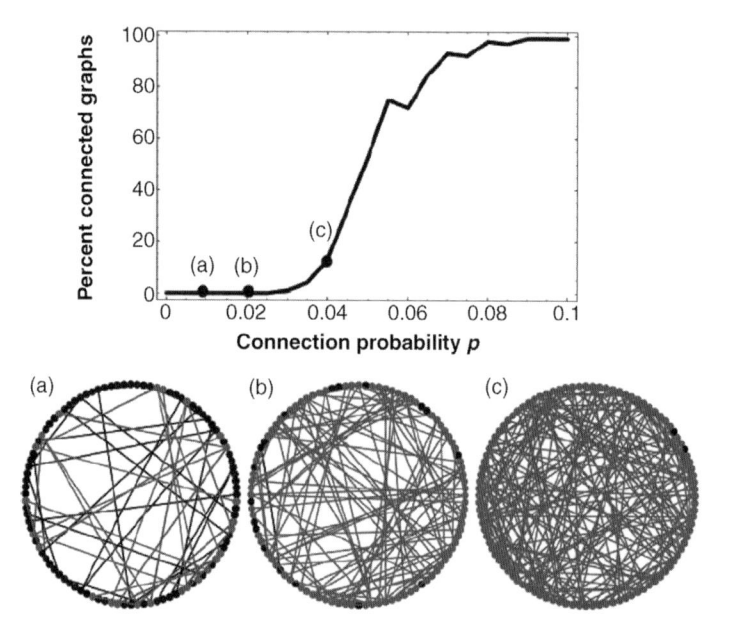

**Figure 12.2** Summary of the classical result of the ER random graph, namely that connectedness emerges spontaneously when the connection probability $p$ is increased and a critical value of $p$ is surpassed. The percentage of connected graphs as a function of the connection probability $p$ for the ER model is shown in the upper half of figure, while the lower part of the picture gives three examples of disconnected graphs that appear at small $p$. The curve has been obtained by generating 100 realizations of a graph consisting of 300 nodes for each value of $p$ and then evaluating the connectedness. Note that even for this comparatively small system size, the phase transition-like behavior of the curve is clearly seen. The three graphs given in the lower part of the figure have been obtained for $p = 0.01$ (a), 0.02 (b), and 0.04 (c), respectively. Each graph consists of 300 nodes. In all cases, the largest connected component is shown in gray. One sees that the size of this component increases rapidly with $p$. (Taken from Hütt and Lüttge, 2005.)

network structure (e.g., a chain of nodes, where each node is connected to its $k_0$ nearest neighbors) and then rewires the links in this graph randomly. Each link present in the regular "backbone" has a probability $p$ (the rewiring probability) of being disconnected from one of its nodes and randomly attached to another node. These structures were named small-world networks because typically the distance (the number of vertices on the path) between two elements is rather small in spite of the high clustering coefficient resulting from the regular backbone. This phenomenon is well known from social networks, as explored in the seminal study of social networks by Milgram (1967) and popularized, for example, in Guare's acclaimed theatre play *Six Degrees of Separation* (Guare, 1990). Furthermore, Watts (1999) argues that many social and, presumably also, many biological and biochemical networks are of this "small-world" type. An alternative method for obtaining small-world graphs is given by a "link-adding procedure," such as described by Strogatz (2001), where the links of the regular backbone are not rewired, but added randomly to the graph. The equivalence of this procedure with respect to graph properties is discussed by Albert and Barabási (2002).

Currently, an increasing number of networks is becoming publicly available at the genetic, protein or metabolic level. As pointed out in Section 12.1 and discussed in more detail below in Section 12.3, these networks are the striking hallmarks of current postgenomic research. One of the important results of the last decade of intense studies of biological networks, particularly at the level of genes, proteins, and metabolic components, is that a scale-free graph represents the topology of such networks more accurately than other types of random graphs (Jeong et al., 2000; Wagner and Fell, 2001). The defining property of a scale-free graph is a degree distribution obeying a power law, $P(k) \propto k^{-\gamma}$, with the power law exponent $\gamma$. A prescription for generating scale-free graphs with $\gamma = 3$ has been given by Barabási and Albert (1999). One starts with a small random graph consisting of $N_0$ nodes and adds nodes iteratively to this structure, where each added node is randomly linked to $m$ other nodes in the graph. The probability of such a link being formed, however, depends on the degree of the existing node, where the higher the degree of a node in the graph, the higher the probability of it linking to a newly added node. This rule is called "preferential attachment."

Figure 12.1 summarizes these introductory comments on the different types of graphs by showing typical examples of such graphs, together with the corresponding degree distributions. Maslov and Sneppen (2002) have been able to quantify the "disassortativity" of some biological networks: the hubs in protein networks have a higher probability of linking to nodes with very low degree. The investigations by Barabási and Oltvai (2004) showed that the class of scale-free graphs has to be subdivided according to the exponent $\gamma$ that characterizes the degree distribution. While the prescription based on preferential attachment leads to $\gamma = 3$, most biological networks in fact yield $2 < \gamma < 3$ where there is a pronounced "network-within-a-network" structure with hubs being present at different scales described as a "hierarchical network" by Ravasz et al. (2002).

Graph theory constitutes a promising framework for intense, biologically motivated theoretical studies as well as for experimental investigations guided by theoretical principles. Model systems and established data sets for a variety of biological networks are accessible (e.g., see Giot *et al.*, 2003; Li *et al.*, 2004; Tong *et al.*, 2004), and first ordering principles in the architecture of biological systems appear (especially via the notion of scale-free graphs).

As two last topological features, which have been discussed intensely over the last few years, we briefly discuss "modular networks" and "hierarchical networks." In particular in neuroscience, these graph types have received a large amount of attention. For more details and additional references, see Müller-Linow, Hilgetag, and Hütt (2008).

Biological systems were often found to be organized in network modules (Girvan and Newman, 2002; Newman and Girvan, 2004) or to contain characteristic circuits (motifs) that do not occur as frequently in other types of networks (Alon, 2007). "Hubs" can serve as central distributing elements or linkage points for many regions of a network (Barabási and Oltvai, 2004; Albert, Jeong, and Barabási, 1999; Jeong *et al.*, 2001) and this characteristic has been identified in several biological networks, such as protein–protein interaction networks or metabolic networks. Such hubs might also be present in neural systems and cortical networks (Kaiser, Görner, and Hilgetag, 2007; Kaiser *et al.*, 2007). A hub, for our purposes, is either a node with a high degree or a node with a high centrality, in other words with many shortest paths between nodes passing through it. The latter definition is more relevant to studying dynamics. Modules or network clusters, which are characterized by a higher frequency or density of connections within node clusters than between node clusters (Young *et al.*, 1995), have been identified in metabolic networks (Ravasz and Barabási, 2003; Guimerà and Amaral, 2005), and in neural networks at both the cellular level (Reigl, Alon, and Chklovskii, 2004) and the systems level (Hilgetag *et al.*, 2000). These modules often represent a specific function, such as a specific synthesis pathway in a metabolic reaction network (Ravasz *et al.*, 2002), and may shape the functional interactions within the networks at different scales (Stephan *et al.*, 2000; Zhou *et al.*, 2006; Honey *et al.*, 2007). Modules can be large or as small as a simple arrangement of a few interconnected nodes. Whatever their size, the recurrent patterns of interconnected nodes are referred to as "motifs." Interestingly, Milo *et al.* (2004), Kashtan *et al.* (2004), and Sporns and Kötter (2004) all argue that such "motifs" represent functionally relevant (micro) circuits.

In addition to the mentioned features, the organization of biological systems is often described as "hierarchical." However, there is no clear formal definition of how a "hierarchical" topology should appear. Typical descriptions of hierarchical organization use a modules-within-modules view (Ravasz and Barabási, 2003; Kaiser, Görner, and Hilgetag, 2007; Kaiser *et al.*, 2007), but other researchers have focused on the coexistence of modules and central (hub) nodes (Guimerà and Amaral, 2005; Han *et al.*, 2004) or have even related the concept of hierarchy to "fractality" (Sporns, 2006). The distinction between hubs that organize modules around them and hubs that connect modules on a higher topological level has been productive for understanding the functional categories of hubs in various observed (empirical) networks (Guimerà and Amaral, 2005; Han *et al.*, 2004; Sporns, Honey, and Kötter, 2007).

## 12.3
## A Network Perspective on Systems Biology

In the contribution to systems biology that we are describing here, the principal aim is not to have eventually a mathematical model representing a cell, but rather it is about understanding functions beyond the level of a few elements. In other words, how is robustness achieved? How can a system react rapidly to changes in the environment? Has a given pathway evolved to respond specifically to perturbations? How can both goals – robustness and sensitivity – be achieved within the same gene regulatory architecture? Which is the typical signal encoding a cell uses on a particular level of organization and why?

Many of these questions can be addressed by contrasting the observed structures with an appropriately selected random background (a null hypothesis). In essence, this task of formulating null models for observed complex patterns within a cell is one of the key challenges of systems biology. Evolution provides us (mostly) with "single runs" of complex cellular systems, but we can expect specific features of these systems to have evolved either from an economical construction or from an optimized function. Deviations of these complex systems from randomness may therefore indicate that properties of evolutionary processes have been shaping (and economically constructing) the system or reveal the need for efficient system performance.

Representing levels of cellular organization in terms of networks has turned out to be particularly helpful for the task of formulating the appropriate null models. The strength of graph theory is that it can represent a complex system in a unified formal language of nodes and links. In the case of gene regulatory networks the nodes are genes and the links denote the regulation of a gene by another. In the more specific case of transcriptional regulatory networks (TRNs), regulation is mediated by transcription factors. In a network representation of metabolism, the nodes are metabolites, whereas the edges are related to enzymatic reactions converting one metabolite into the other. Figure 12.3 illustrates this point for these two cases.

Within this chapter, we describe how large-scale systems can be abstracted as graphs and how the architecture can be related to (a dynamical) function. In the case of TRNs, we discuss the concept of network motifs – small frequent substructures within a given large network, which can be attributed a specific dynamical behavior.

For metabolism, we first summarize the typical approach of studying the dynamics of this system and then show how the network perspective helps in addressing fundamental questions about the organization of metabolism.

### 12.3.1
### Large-Scale Systems

A description of large-scale complex systems such as the complete intracellular metabolism or full TRN of a given organism is unavoidably an oversimplified representation the real system. However, the translation of the system into a graph representation will allow for the identification and comparison of a system's

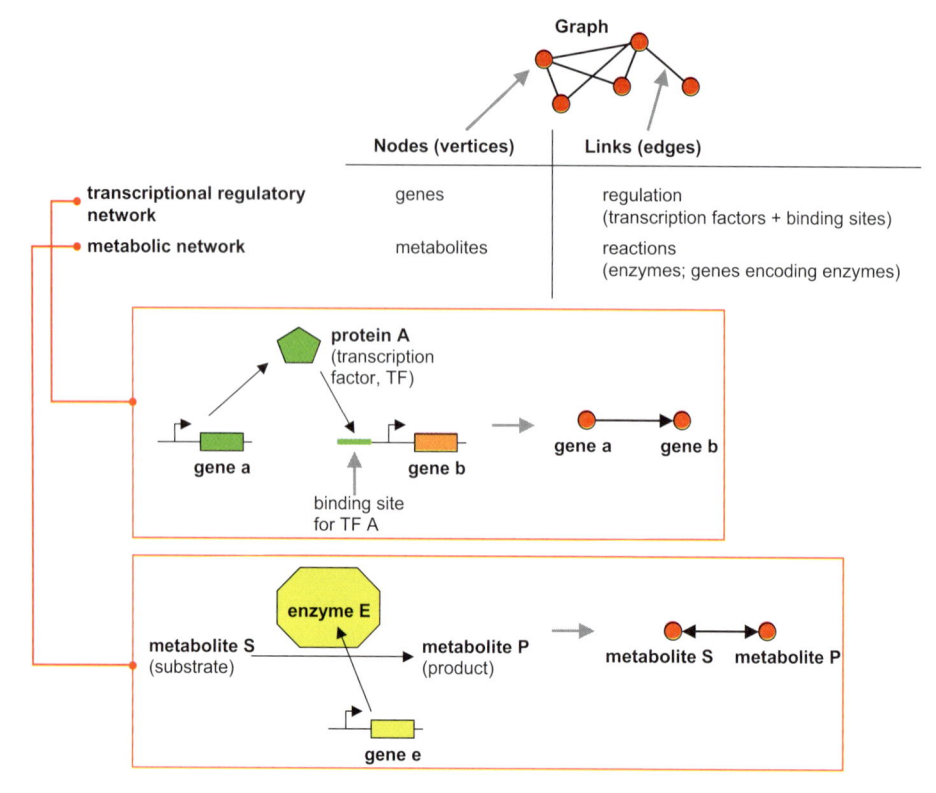

**Figure 12.3** Defining network representations for a TRN and a metabolic network, respectively. For the TRN (top half), the nodes are genes and the (directed) links describe the regulatory action of one gene onto another. More specifically, a link in the TRN represents a biological process, where gene a expresses protein A, which serves as a transcription factor (TF) binding in the regulatory region of gene b and thus controlling gene b. For metabolism (bottom half) the figure shows a metabolite-centric representation, where the nodes are metabolites, either substrates (S) or products (P), and links are either enzymes or reactions. In metabolic networks, links are often considered as bidirectional (e.g., to account for the reversibility of an enzyme reaction).

emergent properties. The observed properties of the system in a graph representation that relate to the dynamic behavior, can help answer some very basic but highly essential questions. When considering metabolism, for example, we can ask the following: why is metabolism robust with respect to small perturbations and, at the same time, highly flexible (enough) in its reaction to environmental changes? How can such a complicated yet organised system retain a steady-state as its dominant mode of operation? In this section, we will summarize a few recent findings on the organization of two large-scale intracellular systems – TRNs and metabolism.

## 12.3.2
### TRNs

The compilation of all known interactions between genes based on transcription factors for the bacterium *Escherichia coli* (Salgado *et al.*, 2004) into a database allowed for the construction of a TRN (Shen-Orr *et al.*, 2002). This view yielded remarkable topological insights into the hierarchical organization of TRNs (Ma *et al.*, 2004; Yu and Gerstein, 2006) and their composition from building blocks of specific network motifs (Shen-Orr *et al.*, 2002). Even a rather simple topological analysis of the system already reveals a fairly surprising design principle. Figure 12.4 shows the network representation of the gene–gene interactions in *E. coli* mediated by transcription factors, together with a plot of the number of regulating genes (the in-degree of a node) against the number of regulated genes (the out-degree of a node). One clearly sees that the range of in-degrees is far smaller than the range of out-degrees. In addition, nodes with a high out-degree tend to have a small in-degree. An interpretation is that this is a concept for information processing in the gene regulatory system, where a typical gene receives few inputs, but can affect a rather large number of other genes.

Beyond the analyses of network topology, TRNs have also been directly compared with expression profiles. Among the wide range of approaches (e.g., see Balazsi, Barabási, and Oltvai, 2005; Grondin, Raine, and Norris, 2007; Gutierrez-Rios *et al.*, 2003; Herrgard, Covert, and Palsson, 2003) some are particularly relevant for integrating experimental data and graph-theoretical considerations, because they formulate different variants of effective TRNs; Luscombe *et al.* (2004) showed that the topology of subnetwork structures in yeast is specific for cellular programs triggered by environmental conditions. Slow programs (e.g., in the cell cycle) employ a densely

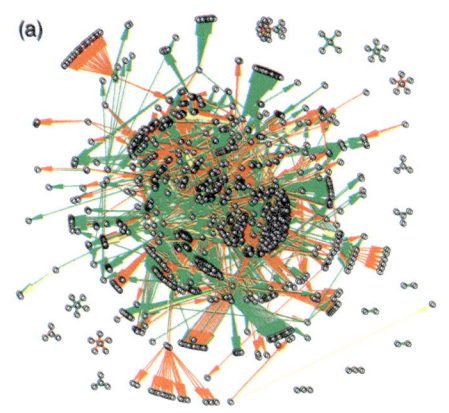

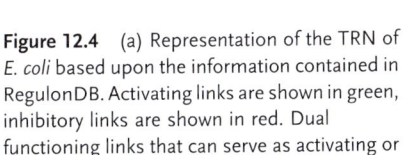

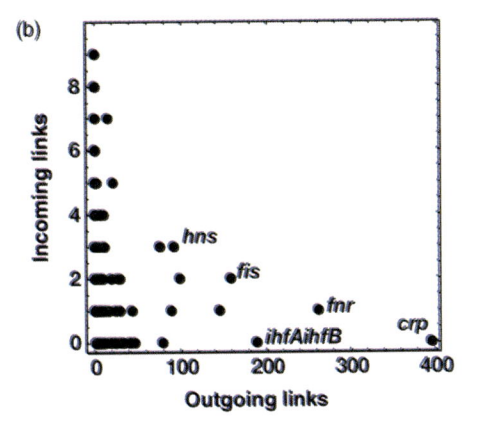

**Figure 12.4** (a) Representation of the TRN of *E. coli* based upon the information contained in RegulonDB. Activating links are shown in green, inhibitory links are shown in red. Dual functioning links that can serve as activating or inhibitory links according to the specific condition are depicted in yellow. (b) The in-degree is plotted against the out-degree for each gene. The names of the most prominent regulatory hubs (high out-degree) are given.

interconnected subnetwork structure, whereas programs required to act rapidly (e.g., in DNA repair) employ networks with shorter path lengths and, at the same time, less complex motif content. Such subnetwork structures have also been analyzed from an evolutionary perspective (Babu and Aravind, 2006) and by assessing the impact of gene duplication on TRN topology (Kim et al., 2006). Teichmann and Babu (2004) constructed condition-specific gene regulatory networks for yeast predicting new transcriptional regulatory links on this basis.

Recently, a systematic interplay between two types of control was discussed in gene expression profiles in *E. coli*, one network-mediated and the other mediated by DNA topology (Marr, Müller-Linow, and Hütt, 2008). Here, like in many situations in systems biology, the challenge is to formulate the suitable null model for contrasting the real-world observations. As in this study, an effective network is constructed (consisting of those links in the TRN in Figure 12.4 between genes with measured expression changes under the conditions at hand) and the null model is a network obtained from surrogate data on the expression level, such as randomly selected genes, to which we attribute an expression change.

An interesting topological feature of the TRN is that it has an almost tree-like structure, particularly for *E. coli* but also for yeast, *Saccharomyces cerevisiae* (see Alon, 2006, for a detailed discussion). This observation has several consequences. (i) Hierarchical levels in the network can be analyzed meaningfully (Yu and Gerstein, 2006). (ii) It leads to the question of how information can be circulated in the network when there is already a dominant directed flow in the network dictated by its architecture (topology). What becomes ever-more transparent is that additional interactions beyond the transcription factors, particularly regulation based on protein interactions and also based on small regulatory RNA, disrupt a general feed-forward structure and relay signals from the bottom layer again to top-level input nodes (Yu and Gerstein, 2006; Shimoni et al., 2007; Tsang, Zhu, and van Oudenaarden, 2007).

Alon and coworkers (Shen-Orr et al., 2002; Mangan and Alon, 2003; Milo et al., 2002, 2004; Alon, 2007) formulated the concept of network motifs, which are small few-node subgraphs with a specific link pattern occurring more often than expected at random. In a series of publications, they identified the most important motifs in the TRN of *E. coli* and studied their functional role using simple mathematical models. In order to recognize an over-representation of a certain subgraph in the network, they had to construct a suitable null model. Their technique was called network randomization, where one randomly selects two links in the network, detaches their endpoints and reassigns them to the previous two endpoint nodes, but in a crossed-over fashion. For an undirected graph this procedure is depicted in Figure 12.5. Iterating this procedure repeatedly (selecting two links from the network and swapping endpoints) yields a randomized version of the network, where each node, however, retains its degree. The rationale behind this randomization scheme is that the degree sequence (the list of degrees of all nodes in the network) as the network's most important topological feature is conserved, while all higher-order features are randomized. Performing a subgraph count on such randomized networks can identify the over-represented and under-represented subgraphs (motifs) in the real network.

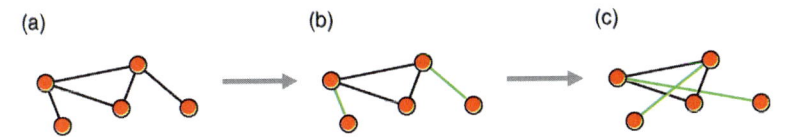

**Figure 12.5** Example of a network randomization scheme. In this small, undirected network (a), two links are selected at random (b) and then the endpoints are swapped (c).

## 12.3.3
### Metabolic Networks

The network of metabolic reactions in a cell is responsible for providing a wide range of substances at the right time in the right proportions. The logistic challenges are remarkable. Metabolism is at the same time a transport network, an assembly line, and a storage depot. Substances are taken up from the environment (by exchange reactions) and distributed in the cellular compartments (by transport reactions). Large parts of metabolism, however, are responsible for degrading complex substances into more elementary building blocks (catabolism) or for building up complexes from smaller units (anabolism) that are needed for use in disparate pathways, transported or even stored by the cell.

The organizational unit of metabolism is the individual biochemical reaction, often represented by the enzyme (or enzyme complex) serving as a catalyst for a given reaction. Qualitatively speaking, the exchange reactions can be regarded as an input layer, followed by a complex intracellular processing layer. In many modeling approaches the overall goal of metabolic function is abstracted as a (fictitious) biomass reaction, where each component known to contribute to cell growth enters this reaction. Figure 12.6 summarizes this situation and already provides the key idea of constraint-based analysis – a technique for predicting, for example, enzyme essentiality in metabolism.

Early studies on metabolic network topologies mostly focused on the broad degree distribution and the associated scale-free property (Jeong et al., 2000; Ma and Zeng, 2003; Arita, 2004). More recent work attempted to link topological properties with biological function, particularly with enzyme essentiality and flux organization (Samal et al., 2006; Almaas et al., 2004; Almaas, Oltvai, and Barabási, 2005; Wunderlich and Mirny, 2006). These studies profited from the predictive power of flux-balance analysis (FBA) (see below) where the steady-state fluxes of wild-type and mutants were simulated based on the stoichiometric matrix of the system and a biologically meaningful objective function (e.g., the biomass production or the ATP production, both of which can be observed empirically) under thermodynamic constraints and considering the reversibility of reactions and the maximal enzyme activities (Edwards and Palsson, 2000; Kauffman, Prakash, and Edwards, 2003).

Samal et al. (2006) analyzed clusters of "uniquely produced or uniquely consumed" metabolites and their associated reactions, and found a high essentiality of those genes and an overlap to the operon organization on the genome. Wunderlich and Mirny (2006) employed synthetic accessibility – a concept from organic chemistry

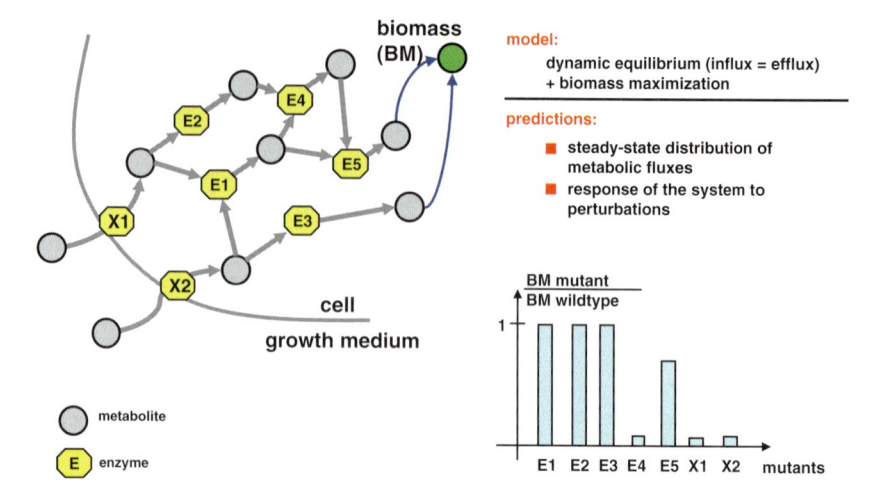

**Figure 12.6** Schematic representation of metabolism and the concept of FBA. With FBA, the fractional changes to the biomass can be predicted for a mutant cell (an entire metabolic network missing an enzyme) compared to the biomass calculated from the wild-type cell (an entire metabolic network without any mutations). Similarly, the dependence of biomass production on medium (extracellular) conditions can be studied using this framework.

that considers the number of steps necessary for synthesizing a compound from given predecessors – and analyzed how the path length towards the biomass changed under different mutations. They observed high agreement between their (topologically defined) set of significant mutations (in other words, those changing the synthetic accessibility of the biomass) and the set of known (or predicted) essential genes. Almaas *et al.* (2004) observed that the size distribution of metabolic fluxes is similar to the degree distribution and is also a power law. In a subsequent study, they identified a set of fluxes that would always be active under a wide range of medium (extracellular) conditions (Almaas, Oltvai, and Barabási, 2005). The existence of this "core component" can again be analyzed in terms of essentiality and topological properties. Currently, however, no unifying principle behind these different topological and dynamical attributes and their relation to biological function is visible.

Metabolic pathways can be classified into three types depending on how they use their exchanged metabolites and internal and external fluxes. The fluxes of exchanged metabolites can be internal, crossing boundaries of intracellular compartments (for example between the cytoplasm and organelles), or between independent metabolic pathways or processes. Alternatively they can be external crossing the boundary of the cell membrane with the surrounding extracellular environment. In the former, the exchanged metabolites are the individual requirements for an individual pathway, the output from that pathway, or alternatively the so-called currency metabolites, such as energy carriers, redox cofactors, coenzymes, and ions.

Metabolic pathways involve fluxes of exchange metabolites into and out of the extracellular environment and the exchange of the currency metabolites between intracellular compartments, and also incorporate cyclic reactions that result in a net flux of zero even across the boundaries of the compartments (Price, Papin, and Palsson, 2002). Considering the correct use of currency metabolites is important as they can either contribute directly to the exchange flux with the extracellular environment or alternatively are internalized within compartmentalized (and separated) pathways.

Strictly speaking, a graph representation of metabolism consists of two types of nodes – metabolites and enzymes (or reactions). Mathematically, this is called a bipartite graph. From this graph, two projections to single types of nodes are frequently used – the metabolite-centric graph and the enzyme-centric graph. In the first representation, metabolites are the nodes and a link denotes the involvement of any two (or more) metabolites used by a given enzyme reaction. Distinguishing between substrates and products of an enzyme reaction can provide directed links. In the second representation, enzymes are the nodes and edges indicate shared metabolites between their individual reactions.

All three representations (bipartite graph, metabolite-centric projection, and enzyme-centric projection) can have very highly connected nodes. A given enzyme can, of course, be involved with more than one substrate or more than one product metabolite or currency metabolite. The illustration in Figure 12.7 shows a bipartite representation of a (fictitious) metabolic subnetwork, together with its projections.

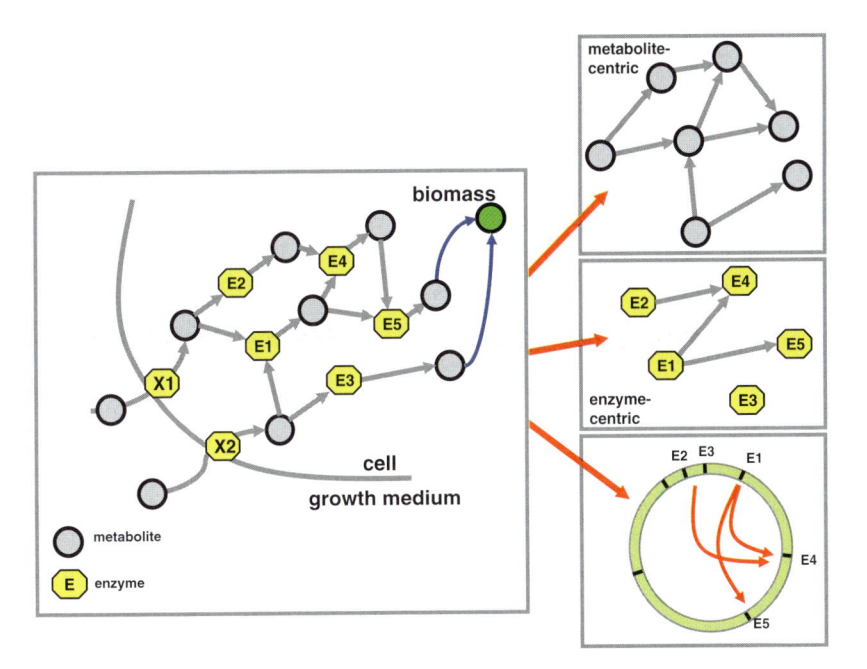

**Figure 12.7** Schematic view of the network representations of cellular metabolism (from Figure 12.6), with the projection of the bipartite representation of metabolism (left) to a metabolite-centric graph (right; top) and to an enzyme-centric graph (right; center). An advantage of the enzyme-centric representation is that it can be discussed in its genomic context. Here, the enzyme-centric metabolic network is shown inserted in the (circular) genome (right; bottom).

Using this formal arrangement, the functioning of the system can be studied from very different perspectives and with different questions. How does biomass production change under different media (extracellular and environmental) conditions? How does biomass production change under a somatic mutation (a gene knockout resulting in the loss or dysfunction of an enzyme)? How does biomass production change when there is an abnormal metabolite production or in the presence of a competitive inhibitor (faux metabolite)? Can the system be optimized for other objective functions instead of biomass production, such as a given ratio combination of metabolites?

There are many caveats regarding the metabolic network representations described above. One caveat is that any network topology of nodes and edges is independent of the real (intracellular) spatial dynamic distributions of the respective metabolites and enzymes and reactions. Even with the conventional metabolite-centric representation, there are a number of issues: many different irreversible enzyme reactions (directed edges) can contribute to one transformation between neighboring metabolites nodes $i$ and $j$. Furthermore, for a reversible reaction, many independent enzymes can contribute to the reaction in opposite directions between two neighboring metabolites. The choice of how loose or strict reversible enzyme reactions should be defined is important, especially when investigating how the mutual edge contributes to flux in small substructures or motifs.

Furthermore, in the second representation, the metabolite-centric projection, the convention is that an irreversible reaction is represented by a single directed edge and a reversible reaction represented by a bidirectional (mutual) edge. An edge could represent a general reaction or an enzyme reaction class (e.g., represented by an Enzyme Classification code with a well-defined reaction mechanism) or alternatively be one of a variety of enzymes that perform the same reaction (transformation) between two metabolites (nodes). This is true for pathways involved in both intermediate metabolism and endotoxic (waste product) metabolism, investigated in drug metabolism (Josephy, Guengerich, and Miners, 2005; Smith and Stein, 2009). The metabolite-centric representation is analogous to the representation of chemical reactions. Taking an example from phase I drug metabolism, by the mixed function oxygenases including the cytochrome P450 isoforms, a given substrate (represented by node $i$) could undergo enzyme hydroxylation from one enzyme (represented by edge $n$) into a cis-hydroxylated intermediate (node $j$) that then undergoes a second enzyme hydroxylation by another enzyme (edge $n + 1$) resulting in an epoxidated product (node $k$). This reaction $i$ to $k$ is a consequence of the labile intermediate ($j$) that is formed and often such short-lived metabolites are not easily identified or well-characterized enough to be included in the reaction network scheme.

As a final note, the currency metabolites in a metabolite-centric representation are often the most highly connected nodes. For an analysis of a given network, it may be more informative to remove some of, or indeed ignore all of, the currency metabolites leaving the so-called primary metabolites.

In summary, the use of constraint-based modeling and, in particular, FBA assumes a dynamic equilibrium of the system (where the influx at each node equals the efflux at all times) and requires a search for a solution (i.e., the flux distribution consistent with the equilibrium assumption) that maximizes some objective function (e.g., biomass production). Mathematically, the equilibrium assumption converts the system of ordinary differential equations governing the changes of metabolite concentrations into a homogeneous system of linear equations. Additional constraints come from maximal and minimal uptake rates of the exchange reactions. With this modeling approach, fascinating numerical experiments are possible. As an example, one can observe the reduction of predicted biomass production upon the removal of an enzyme from the system and use this as a predictor of the enzyme's essentiality (Figure 12.6; right-hand side).

Mathematical optimization methods formerly used mainly in engineering, economics, physics, and chemistry are becoming increasingly popular in the field of systems biology with applications ranging from model building and parameter estimation to optimal experimental design and synthetic biology (Banga, 2008). In particular, its capacity to predict gene essentiality with high accuracy for *E. coli* and *S. cerevisiae* has turned FBA into a widely accepted method for in silico studies of metabolic states (Varma and Palsson, 1994; Edwards and Palsson, 2000; Price, Reed, and Palsson, 2004).

Let us look at FBA in a little more detail. The conceptual starting point of FBA is the stoichiometric matrix **S** that relates the vector **v** of metabolic fluxes with changes $d\mathbf{m}/dt$ in the vector **m** of metabolite concentrations $d\mathbf{m}/dt = \mathbf{Sv}$. Assuming a steady state, where $d\mathbf{m}/dt = 0$, one obtains a homogeneous system of linear equations. Furthermore, assuming upper and lower bounds for each of the uptake rates of the compounds from the extracellular environment (medium), one obtains a well-defined subset of flux space (a "flux cone") as a potential solution space. The optimal solution can then be identified by maximizing the given objective function (e.g., the biomass production).

A variety of implementations of FBA are now available in the systems biology community (Urbanczik, 2006; Becker *et al.*, 2007), as well as for research in biochemical engineering, drug design, and other biotechnological contexts, helping to make FBA a very successful tool for studying this otherwise rather elusive level of cellular organization. Recent refinements of FBA focus on the distribution of fluxes upon mutations (Segre, Vitkup, and Church, 2002; Shlomi, Berkman, and Ruppin, 2005), the incorporation of temporal information beyond the steady state (Mahadevan, Edwards, and Doyle, 2002; Kauffman, Prakash, and Edwards, 2003), and gene regulatory information (Covert *et al.*, 2004; Shlomi *et al.*, 2007; Samal and Jain, 2008). The recent observation, however, that the majority of fluxes determined by metabolism alone is consistent with those obtained by evoking additional gene regulatory input (Shlomi *et al.*, 2007) strengthens the case for the use of FBA for large-scale system-wide studies of metabolism.

Of course, such a steady-state description has severe limitations. In particular, no pre-equilibrium and transient behaviors can be described. However, in order to understand the interplay between the network architecture and the dynamics, this steady-state description is quite helpful:

i) It contains few parameters and (as opposed to a fully nonlinear representations of the enzyme kinetics) can therefore been applied to whole metabolic networks.

ii) While transient behaviors and the changes from one condition to another are extremely interesting and informative, we may assume that the evolutionary shaping of the network architecture will have regulated the steady states (i.e., the features of the linear system).

Levels of cellular organization, like signal transduction, which often operate precisely in a transient regime (being switched on and off by external signals, etc.), cannot be suitably described by such methods (e.g., see Millat et al., 2007). For metabolism, these constraint-based modeling approaches work surprisingly well.

## 12.4
## Network Dynamics

As we have seen, network research employs the formal view of graph theory to understand the design principles of complex systems. In the last few years, the study of network dynamics has been investigated intensely in a wide range of disciplines – how the topology of a network shapes, regulates, or even enhances dynamic processes on the network.

### 12.4.1
### General Aspects

Despite the interesting findings concerning the relationship between topology and dynamics, such as for phase oscillators (Arenas, Díaz-Guilera, and Pérez-Vicente, 2006), synchronization in general (Nishikawa and Motter, 2006; Atay and Biyikoglu, 2005), epidemics (Moreno, Pastor-Satorras, and Vespignani, 2002), and excitable dynamics (Graham and Matthai, 2003; Roxin, Riecke, and Solla, 2004; Müller-Linow, Marr, and Hütt, 2006) or chaotic oscillators (Yook and Meyer-Ortmanns, 2006), very few universal laws governing the relationship between topology and dynamics have been identified.

Throughout systems biology, this relation is of importance. For gene regulatory networks this question is particularly interesting because an observed expression pattern may be viewed as the product of the on/off dynamics of the interacting genes in the gene regulatory networks (e.g., see Liebovitch et al., 2008, for more details). Obviously, many other biological mechanisms contribute to gene expression levels in similar proportions (e.g., DNA topology and the regulatory action of microRNA or protein–protein interactions not incorporated in the gene regulatory network based on transcription factors and their binding sites). The understanding of many biological phenomena is improved by applying the concept of coupled nonlinear oscillators. In such theoretical models the global behavior of many interacting dynamic elements can be discussed and then related to biological observations. Nonlinear dynamics and statistical physics provide the mathematical tools for this discussion. In the last few years, these tools have started to be combined with methods from graph theory, as the architecture of a network and its influence on pattern formation became the focus of scientific interest. Linked systems of coupled oscillators have been under investigation for several years now. The underlying graphs are regular arrangements like chains, grids, or fully connected systems (see also Section 12.2). With a statistical description of spontaneous synchronization in an ensemble of coupled phase oscillators beyond a critical coupling strength, Kuramoto (1984) formulated a model situation that is still today the basis for nearly every discussion covering synchronization (e.g., see Pikovsky, Rosenblum, and Kurths, 2001). An exemplary examination is the numeric simulation of a chain of biochemically motivated nonlinear oscillators describing the induction of complex spatio-temporal patterns via biological variability (Hütt, Busch, and Kaiser, 2003).

The relation of biological variability and connectivity in an ensemble of nonlinear oscillators is crucial for the synchronization in networks and has been studied by Hütt and Lüttge (2005). There, network variants of established spatio-temporal analysis tools have been formulated and extended into a procedure for reconstructing connectivity from the time series of the individual oscillators. For the sake of clarity and simplicity (and because we believe that this is particularly relevant for biology) we restrict the following discussion to discrete, few-state dynamics operating on a discretized timescale.

Discrete dynamics, and in particular binary dynamics, have proven helpful in the past for studying gene regulation. A simple and very successful mathematical model of gene regulation was formulated several decades ago by Kauffman (1969) describing the interaction of binary elements in a random graph. In a network consisting of nodes ("genes"), every single node is regulated by other randomly selected nodes via definite Boolean functions (random Boolean network). The pattern of transitions between system states – the attractor structure of the system – can be thought of as a highly simplified model of cell differentiation, where different attractors correspond to different cell types. A basin of attraction corresponds to the range of initial conditions (and, in a sense, of environmental cues) leading to this attractor (or cell type). Based on the general framework of random Boolean networks, and their extension to general network topologies and particularly the use of threshold dynamics, huge progress has been made in the last few years in linking observed properties of the dynamics with topological features of the graph (see, e.g., Bornholdt, 2005). However, the fundamental task of identifying the imprint of the gene regulatory network in a given gene expression pattern is still unsolved.

### 12.4.2
**Binary Dynamics and Cellular Automata on Graphs**

Let us look at Kauffman's model in a little more detail, which is remarkable both for its formalism and for its key results. The toy model system consisting of four genes displayed in Figure 12.8(a) can be (with some loss in generality, however) translated into a system of Boolean rules. The first feature discerned on this level is that gene $d$ has no input from the other genes. The state of gene $d$ ($0 = $ "off," $1 = $ "on") basically serves as an external switch for the other genes. It thus makes sense to discuss this system as a three-node network with an external input ($d = 0$ or $d = 1$).

(a)

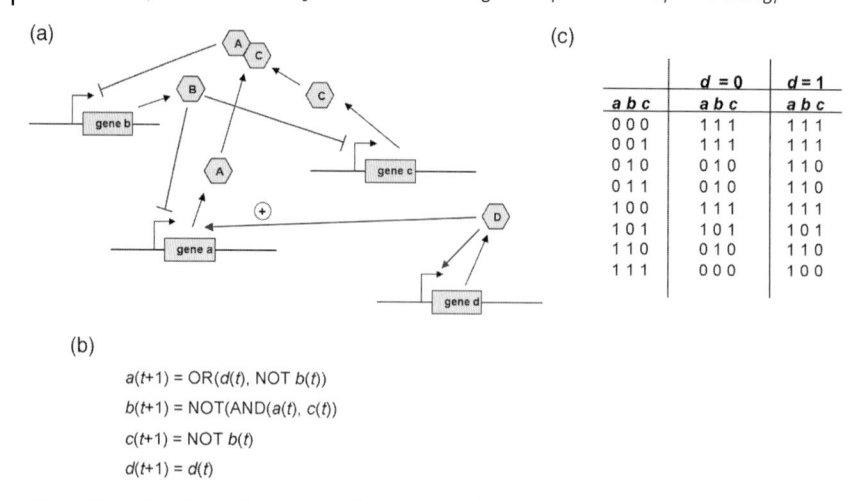

(c)

|   | $d=0$ | $d=1$ |
|---|---|---|
| a b c | a b c | a b c |
| 0 0 0 | 1 1 1 | 1 1 1 |
| 0 0 1 | 1 1 1 | 1 1 1 |
| 0 1 0 | 0 1 0 | 1 1 0 |
| 0 1 1 | 0 1 0 | 1 1 0 |
| 1 0 0 | 1 1 1 | 1 1 1 |
| 1 0 1 | 1 0 1 | 1 0 1 |
| 1 1 0 | 0 1 0 | 1 1 0 |
| 1 1 1 | 0 0 0 | 1 0 0 |

(b)

$a(t+1) = \text{OR}(d(t), \text{NOT } b(t))$

$b(t+1) = \text{NOT}(\text{AND}(a(t), c(t)))$

$c(t+1) = \text{NOT } b(t)$

$d(t+1) = d(t)$

**Figure 12.8** Simple regulatory network consisting of four genes (a) together with a mapping to Boolean rules (b) and the resulting update table for system states (c). (Partially adapted from Klipp et al., 2005.)

From the Boolean rules (Figure 12.8b) one can then extract the system update table (Figure 12.8c) for each of the two input states. The system state is given by the current states of the three genes $a$, $b$ and $c$. It is thus a three-digit binary number. For each of the possible system states (at time $t$), the update table gives the next system state (at time $t + 1$) according to the Boolean rules. The system state at time $t + 2$ is obtained by applying this update scheme to the system state at time $t + 1$. A full time-course of the system is obtained by iterating this procedure several times.

Kauffman discussed such systems from the perspective of Boolean attractors summarizing the network of system states linked by updates. For the network from Figure 12.8a) this is shown in Figure 12.9(a and b). For both possible input states $d$, the system behavior is fragmented between two fixed point attractors and a period-2 oscillation, each of which has a small (or even singular) basin of attraction.

The remarkable complexity of such systems is already discernible in the small network from Figure 12.8. When one flips the "sign" of action of the AC protein complex on gene $b$ (i.e., turning it from inhibitory to activating), the attractor structure changes substantially (Figure 12.9c and d). For both input states all system states now lead into a stable period-4 attractor.

Kauffman found that this model system undergoes a phase transition as soon as a critical value of the degree of linkage is exceeded. The behavior of the system changes from regular (ordered) to irregular (chaotic) at this point. Since experimental values are found to lie far above this critical value, it was for a long time an open question as to why deterministic chaos is almost absent in biological regulatory networks. More recent work shows ordered dynamics in a large parameter region if the random graph in Kauffman's model is replaced by a scale-free graph (Aldana, 2003). This is a

## 12.4 Network Dynamics

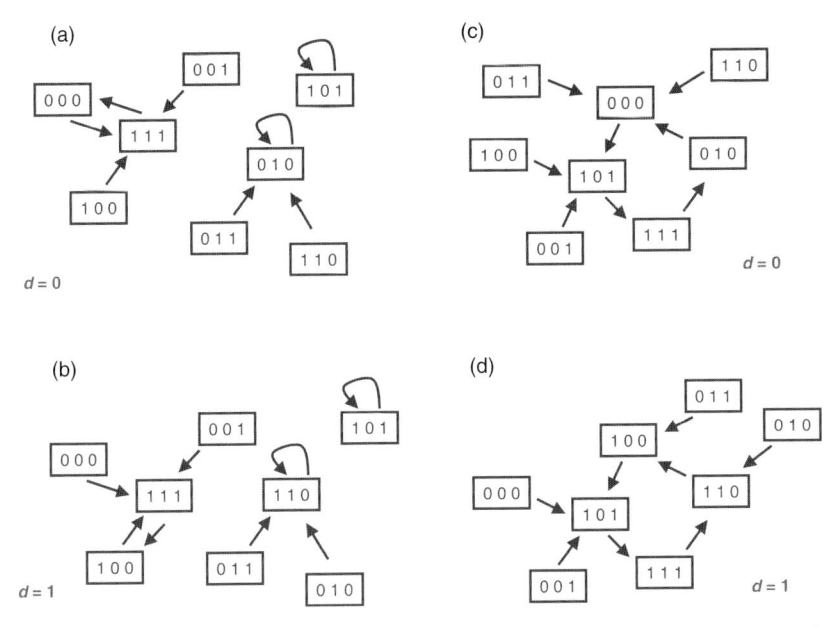

**(a)**

```
          0 0 1
0 0 0 ←         ← 1 0 1 ↺
      ↘   
       1 1 1
      ↗   ↖
            0 1 0 ↺
1 0 0
      ↗
       0 1 1
            1 1 0
```
$d = 0$

**(c)**

```
0 1 1 →              ← 1 1 0
          0 0 0
1 0 0 ↘        ↙ 0 1 0
          1 0 1
      ↗   ↖
0 0 1       1 1 1
```
$d = 0$

**(b)**

```
          0 0 1    1 0 1 ↺
0 0 0 ←
       1 1 1 ← 1 1 0 ↺
      ↗ ↖   ↖
1 0 0   0 1 1
                 0 1 0
```
$d = 1$

**(d)**

```
                      0 1 1
          1 0 0           ← 0 1 0
0 0 0 ↘        ↙ 1 1 0
          1 0 1
      ↗   ↖
0 0 1       1 1 1
```
$d = 1$

**Figure 12.9** Boolean attractors for the system from Figure 12.8, for the case $d = 0$ (a) and $d = 1$ (b). When the action of protein complex AC in Figure 12.8 is changed from "inhibitory" to "activating," the attractors change to those given in (c) and (d).

fundamental and novel example of the direct influence of architecture on possible forms of dynamics.

A huge success in understanding the intrinsic logic of (small-scale) gene regulatory networks has been the monitoring of binary flow patterns, the attractor landscape. The empirically observed dynamics of the yeast cell cycle network (11 genes passing through a specific sequence of active and inactive states during the cell cycle) has been shown to coincide with the dominant branch of the largest attractor basin under the framework of Boolean dynamics (Li et al., 2004). Interestingly, for the fission yeast *Schizosaccharomyces pombe* this protocol (and the associated features of the attractor basin) is implemented biologically in a completely different fashion, even though a Boolean dynamics analysis shows the same level of success in understanding the experimental sequence of states. Here, most of the links in the networks come from protein interactions rather than from transcription factors (Davidich and Bornholdt, 2008a).

The update rules are efficient (in terms of the parameters needed) representations of nonlinear regulations. A thorough comparison of a full ordinary differential equation-based description and a rule-based description in terms of Boolean networks for a single example – the yeast cell cycle – is given in Davidich and Bornholdt (2008b).

Another approach for modelling complex biological systems is that of cellular automata (CA), using Markovian dynamics in a finite state space. Proposed by von Neumann (see von Neumann, 2001) as a model system for biological self-reproduction, a surge of research activity from the 1980s onwards (Wolfram, 1983) established them as a standard tool of complex systems theory. CA on graphs allow an assessment of dynamical changes due to variation of graph topology (Marr and Hütt, 2005). CA have been studied in terms of complexity theory and computational universality and, moreover, they often serve as models of pattern formation.

Binary CA were recently reinvestigated by implementing them on graphs upon which the topological features were varied systematically (Marr and Hütt, 2005, 2006). How is the spatio-temporal pattern of a CA (e.g., the capacity to display oscillations or the complexity of an element's time evolution within the system) changed when, for example, some shortcuts are introduced in to the system or the regular neighborhood structure is completely substituted by a random graph topology? While it is clear that the change of a specific CA rule alters the dynamics and, consequently, the spatio-temporal pattern, it seems worthwhile to study the changes under topology variation at a fixed update scheme. Some approaches already deal with binary dynamics on complex topologies. The Ising model, for example, has been implemented on a small-world graph (Barrat and Weigt, 2000) and the scale-free topology has served as the backbone for Boolean dynamics (Aldana, 2003), as well as for epidemic models (Pastor-Satorras and Vespignani, 2001; Barthélemy et al., 2004). In Huang et al. (2003), the "game of Life" is studied on a small-world network and a phase transition at a critical network disorder is found.

Different attempts to classify the rule space of CA have been made. Wolfram (1984) divided CA qualitatively into four classes, according to the emerging spatio-temporal patterns and analogous to dynamical systems descriptions: I (homogeneous stationary state), II (heterogeneous stationary state or simple periodic structures), III (chaotic behavior), and IV (long-range correlations and propagating structures). The introduction of the Langton parameter $l$ (Langton, 1990) allowed a quantitative investigation of CA rules, even for large rule spaces. While Langton's scheme of generating a transition function for a given value of $l$ requires some statistical subsidiary conditions, for binary CA $l$ is simply the number of neighborhoods mapped on to state 1 divided by the number of all possible neighborhood states. For rules generated with the "random-table method," where $l$ is used as the probability for a neighborhood to be mapped on state 1, the Langton parameter defines trajectories through the CA rule space and, consequently, through the four different dynamical regimes. The order in which the corresponding Wolfram classes are passed as $l$ is increased is I to II to IV to III. Class IV automata, lying between periodic (II) and chaotic (III) behaviors, and exhibiting long-range correlations and propagating structures, are regarded as suitable scenarios for the study of complexity and self-organization.

Each node in the network was considered as a threshold device operating on a Boolean state space (that is on states 0 and 1). A node flips its state (from 0 to 1 or from 1 to 0), when the percentage of 1s in the neighborhood is larger than a threshold $K$ (for additional details, see Marr and Hütt, 2005, 2006).

Figures 12.10 and 12.11 present one investigation possible with this general framework. In Figure 12.10 two sets of spatio-temporal patterns on a ring graph are shown as typical examples for the impact of noise on such CA dynamics. In the first case, the regular graph topology is altered by rewiring the endpoints of randomly

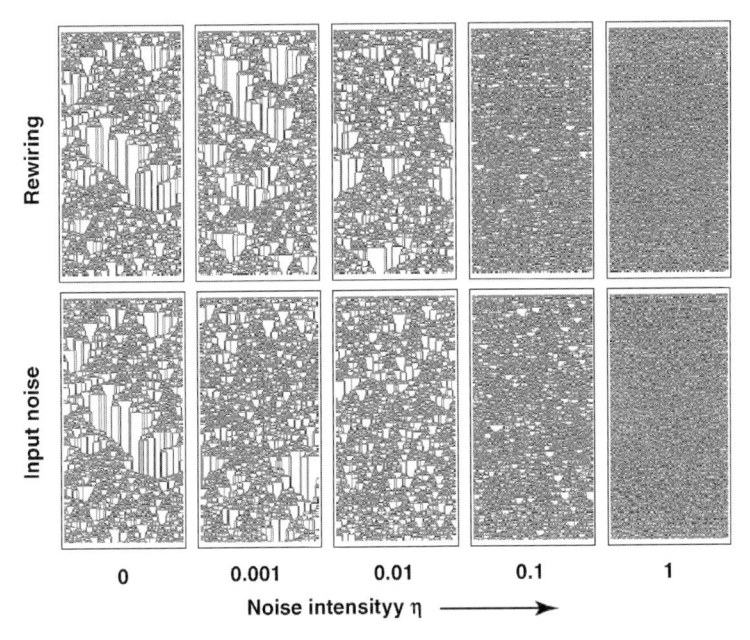

**Figure 12.10** Spatio-temporal patterns for different degrees of topological and dynamic noise. (Taken from Marr and Hütt, 2006.)

selected links. This procedure is similar to the one described in Watts and Strogatz (1998). Here, however, it is applied to directed graphs. In this context, the noise intensity $E$ specifies the fraction of rewired connections ("topological noise"). In the second case, it is possible that the communication between the elements is corrupted by noise, where the state $x_j$ of a linked element $j$ is substituted randomly with a random binary number with probability $E$ during the update of element $i$ ("input noise"). The visual impression of Figure 12.10 suggests that input noise, on the one hand, and rewiring the regular graph architecture, on the other, affect the complexity of the system dynamics in a similar way.

How can the similarity of two such different noise mechanisms be understood? Complex dynamics given, shortcuts insert information from distant regions of the network into local neighborhoods, which is – due to its lack of correlation to the neighborhood at hand and the overall dynamics of the network's elements – similar to random dynamic perturbations. Figure 12.11 depicts the two discussed noise mechanisms schematically. One prerequisite for the similar impact of both noise mechanisms is therefore clearly a globally running complex or chaotic dynamics that ensures an adequately random and uncorrelated signal from the linked distant node.

We believe that noise in high-throughput data will receive more attention in the future, as noise (beyond pure measurement noise) can be viewed as a sampling of the system's possibility space by small random perturbations. In addition, as noise avoidance is part of an efficient functioning system, so it is important to see the networks appearing on different levels of cellular organization also as noise processing devices.

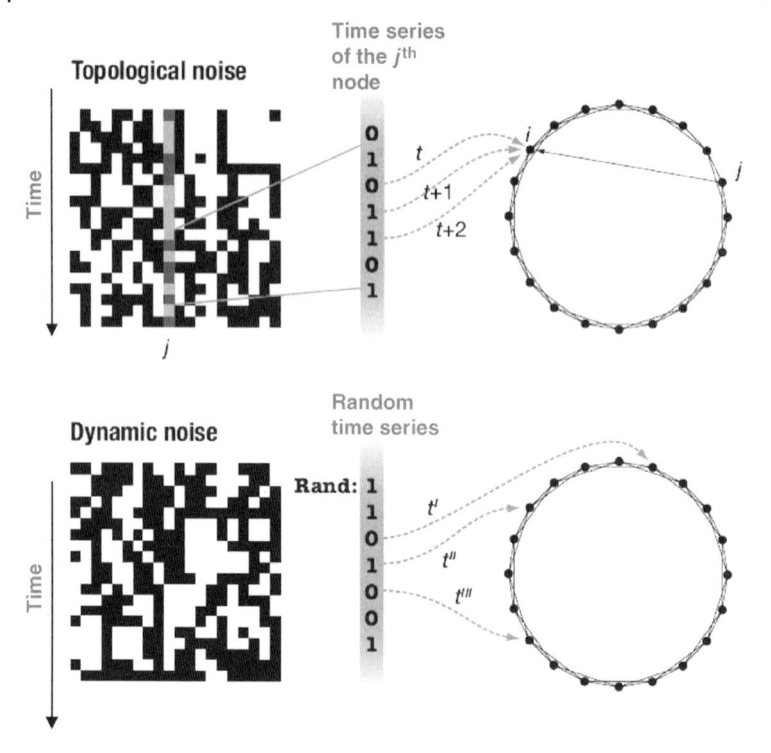

**Figure 12.11** Functional similarity of shortcuts (upper part) and noise (lower part) (adapted from Marr and Hütt, 2006). Qualitative visualization of the two noise mechanisms for an $N=50$, $d=4$ graph and a noise intensity of $\eta = 1/(dN)$. The schematic diagrams on the left show the result of a network dynamic, a spatio-temporal pattern of zeroes and ones where time runs downward and the 50 nodes are lined up horizontally according to their position in the unperturbed chain. In the upper picture, a simple directed link has been rewired and accordingly, the states of a distant node $j$ affect the update of node $i$ through this shortcut in every time step $t'$, $t'' + 1$, $t''' + 2$, ... The perturbing binary elements can be lined up as a noise vector, consisting of the time series of element $j$. In contrast, for dynamic perturbations this noise vector is made up of random binary numbers. These affect the update process of randomly chosen states at random times $t'$, $t''$, $t'''$, ..., substituting the state of a neighboring node with probability $\eta$.

The framework of CA on graphs, from Marr and Hütt (2005), reconciles elements of two well-established discrete models of complexity research – CA and Boolean networks. Together with an appropriate analysis scheme, it is well suited to measure, for example, the effect of different sources of noise on the pattern formation capacity of a network (Marr and Hütt, 2006). Using such dynamics (and the corresponding quantification framework), the question of how metabolic networks avoid complex dynamics and maintain a steady-state behavior was recently addressed by Marr, Müller-Linow, and Hütt (2007) (see also the subsequent debate in Holme and Huss (2008) and Marr, Müller-Linow, and Hütt (2008) for more information). Marr et al. found that the networks' responses are highly specific, where complex dynamics

Table 12.1 Summary of three case studies resorting to null models in order to extract nonrandom topological features of biological networks.

| System | Question | Null model | Result | Reference |
|---|---|---|---|---|
| Gene regulation | Which subnetwork structures appear more often than expected at random? | randomized graphs | importance of feed-forward loops in genetic circuits | Shen-Orr et al. (2002); Alon (2007) |
| | How well are observed changes in gene expression explained by the gene regulatory network? | randomly selected effective networks | interplay between different types of control | Luscombe et al. (2004); Marr, Müller-Linow, and Hütt (2008) |
| Metabolism | Why is metabolism operating essentially in a steady state? | variants of randomized graphs | systematic reduction of complex dynamics on metabolic networks | Marr et al. (2007) Marr, Müller-Linow, and Hütt (2007) |

are systematically reduced on metabolic networks compared to randomized networks with identical degree sequences. Already small topological modifications substantially enhance the capacity of a given network to host complex dynamic behavior and thus reduce its regularizing potential.

Table 12.1 summarizes the three cases studies of identifying nonrandom topological properties of biological networks that we have discussed so far in this chapter: network motifs, effective TRNs, and the complexity reduction in metabolic networks.

## 12.5
## Applicability in Psychiatry

One can thus discern the need of three general types of theoretical considerations in psychiatry: (i) system-oriented modeling approaches, (ii) a mathematical treatment of the nonlinear dynamical processes, and (iii) the theory of self-organization as one of the cornerstones of systems theory.

Providing these methods in a form suitable for psychiatry would pave the road for system-oriented theories of psychiatry (Bender et al., 2006), systemic modeling (Tretter and Scherer, 2006), and the new, emerging field of computational neuropsychiatry (Tretter and Albus, 2007).

The theory of self-organization has a particularly high potential as an interface between theory and experiment in psychiatry. Self-organization means that a set of system components or constituents under the influence of local interactions create long-range and frequently very complex structures that cannot be described any more by the degrees of freedom of the individual elements and must be assessed on the scale of the entire system. Qualitatively speaking, self-organization describes the formation of structures and patterns close to a phase transition.

Nevertheless, we must note a basic difference between theory and experiment. In simple models, complex structures (or patterns) result from the interaction of many identical elements. This step from simplicity to complexity is characteristic of theory. The units building up a biological system are in themselves already extremely complex. Each unit often has a multitude of different functions and rarely do two units resemble each other. Here, we encounter a relation between local complexity (i.e., the individual elements or constituents) and global complexity (i.e., the system). A good concrete example for such a relation between signal cascades and biological complexity is the interesting and important tumor suppressor protein p53, which is often termed the "guardian of the genome." It is centrally placed in various signal transduction pathways, and therefore can translate information on the state of cells into activations and deactivations of various signal transduction pathways. For instance, it can activate DNA repair mechanisms when DNA is damaged. It can stop the cell cycle and in the case of damage that cannot be repaired can trigger apoptosis (programmed cell death). The formation of p53 is regulated by more than 20 transcription factors. A large number of additional components modify the protein and therefore determine further functions. Due to this complexity it is not possible to reduce the property of a tumour repressor to the level of the single component p53. The system and its reactions on external effectors therefore becomes nearly unpredictable (Kitano, 2002).

Attempts have already been made to analyze psychotherapy data from the perspectives of self-organization and, in particular, the concept of attractors from nonlinear dynamics and dynamical systems theory (Schiepek, 1999; Strunk and Schiepek, 2002; see also Freund, Hütt, and Vec, 2004). By mapping protocols of psychotherapy consultation sessions onto a discretized scheme (with respect to possible states of certain key variables and time) Schiepek and coworkers could search for critical fluctuations – one of the hallmarks of a phase transition in self-organized processes. These approaches towards systemic psychotherapy could quantitatively make use of the concepts from synergetics, such as subthresholds and order/control parameters.

Applying systems biology modeling to psychiatry is helped considerably by a focus on the biochemical processes that include aspects of, and the responses from, transmitter systems. The theoretical approaches discussed throughout this chapter are still evolving. They are a long way from being able to predict reliably the activity of the complex and interacting systems involved in dysfunctional (and neurological) function. Many interesting first steps have been made, however, and initial models of psychiatric diseases are being developed. They include, for example, the function of dopamine receptors with regards to working memory and their relevance to schizophrenia (Durstewitz, Seamans, and Sejnowski, 2000; Durstewitz and Seamans, 2002; Wang et al., 2004).

The identification of the roles of local brain structures involved in psychiatric diseases has helped in the development of possible models for the healthy state and the diseased or dysfunctional state and different inhibited conditions under the action of different drugs. A good illustration is the modeling of the 30 interactions proposed by Cooper, Bloom, and Roth (2002) involving the activations and inhibitions between the six systems (dopamine $D_2$, serotonin 5-$HT_2$ system, hypoactivity of the excitatory glutamate N-methyl-D-aspartate transmission system, hypoactivity of the inhibitory $\gamma$-aminobutyric acid (GABA), transmission system, excitatory acetylcholine system, excitatory noradrenaline (norepinephrine) system) that make up the neurochemical interaction matrix – a network model first defined by Tretter and Albus (2004). The development of such interaction models into the more realistic functional models, reflecting the dynamics of the interacting elements, requires considerable empirical (kinetic and thermodynamic) detail of the biomolecular interactions involved.

An alternative approach to the neurochemical interaction matrix is the neurochemical mobile – a heuristic tool for representing the dynamics of the same system (Fritz, 1989). This tool was developed for clinical observation to help describe the neurochemical basis of several psychiatric disorders and the appropriate pharmacological interventions (Tretter and Albus, 2004). The "mobile," a metaphor for a hanging mobile ornament above a child's bed, consists of linked and oscillating scale pans each representing transmitter systems with mutually antagonistic effects. Each respective system component is weighted, and any activating and antagonizing systems substances are located in opposite positions on the respective scale pans. Each scale pan contains a transmitter and affects the other scale pans. With respect to schizophrenia, the hyperactivity of either of dopamine or serotonin, or the hypoactivity of glutamate or GABA, can be represented as an unbalanced configuration of the mobile. The effect of drug interventions and the subsystem of receptor subtypes (inhibitory $D_2$ and excitatory $D_1$ receptors) can also be explored. In summary, this use of the "mobile" is a simple but effective construct for a nonlinear neurochemical oscillator.

Disorders such as epilepsy have also been studied using multicellular models (Traub and Wong, 1982; Traub et al., 2005). Furthermore, in the literature there are many good illustrations of the modeling of signaling pathways involving glutamate and dopamine (e.g., see Fernandez et al., 2006; Lindskog et al., 2006; Gupta et al., 2007).

The example constructs above, in particular the neurochemical interaction matrix and the neurochemical mobile and specific models of signaling pathways, are useful starting points for more refined models. Even considering the complexity and diversity of many psychiatric diseases, the neurobiochemical networks of the functional and dysfunctional brain structures are becoming more amenable to quantitative descriptions. In line with the general goals of systems biology, quantitative microcircuit models of such neuronal systems incorporating details of gene expression, intracellular metabolic and signaling pathways, functional receptors, and intercellular interactions, will become the bedrock of future developments in this field.

### Acknowledgments

J.S. is supported by a Volkswagen Foundation grant to M.H. (program "Complex networks as a phenomenon across disciplines").

## References

Albert, R. and Barabási, A.-L. (2002) Statistical mechanics of complex networks. *Reviews of Modern Physics*, **74**, 47–97.

Albert, R., Jeong, H., and Barabási, A.-L. (1999) Diameter of the world-wide web. *Nature*, **401**, 130.

Aldana, M. (2003) Boolean dynamics of networks with scale-free topology. *Physica D*, **185**, 45–66.

Almaas, E., Kovács, B., Vicsek, T., Oltvai, Z.N., and Barabási, A.-L. (2004) Global organisation of metabolic fluxes in the bacterium *Escherichia coli*. *Nature*, **427**, 839–843.

Almaas, E., Oltvai, Z.N., and Barabási, A.-L. (2005) The activity reaction core and plasticity of metabolic networks. *PLoS Computational Biology*, **1**, e68.

Alon, U. (2006) *An Introduction to Systems Biology: Design Principles of Biological Circuits*, Chapman & Hall/CRC Press, London.

Alon, U. (2007) Network motifs: theory and experimental approaches. *Nature Reviews Genetics*, **8**, 450–461.

Alter, O. (2006) Discovery of principles of nature from mathematical modeling of DNA microarray data. *Proceedings of the National Academy of Sciences of the United States of America*, **103**, 16063–16064.

Amit, D.J. (1989) *Modelling Brain Function: The World of Attractor Neural Networks*, Cambridge University Press, Cambridge.

an der Heiden, U. (2006) Schizophrenia as a dynamical disease. *Pharmacopsychiatry*, **39** (Suppl. 1), 36–42.

Arenas, A., Díaz-Guilera, A., and Pérez-Vicente, C.J. (2006) Synchronization reveals topological scales in complex networks. *Physical Review Letters*, **96**, 114102.

Arita, M. (2004) The metabolic world of *Escherichia coli* is not small. *Proceedings of the National Academy of Sciences of the United States of America*, **101**, 1543–1547.

Atay, F.M. and Biyikoglu, T. (2005) Graph operations and synchronization of complex networks. *Physical Review E*, **72**, 016217.

Babu, M.M. and Aravind, L. (2006) Adaptive evolution by optimizing expression levels in different environments. *Trends in Microbiology*, **14**, 11–14.

Balazsi, G., Barabási, A.L., and Oltvai, Z.N. (2005) Topological units of environmental signal processing in the transcriptional regulatory network of *Escherichia coli*. *Proceedings of the National Academy of Sciences of the United States of America*, **102**, 7841–7846.

Banga, J. (2008) Optimization in computation systems biology. *BMC Systematic Biology*, **2**, 47.

Barabási, A.-L. (2002) *Linked: The New Science of Networks*, Perseus, London.

Barabási, A.-L. and Albert, R. (1999) Emergence of scaling in random networks. *Science*, **286**, 509–512.

Barabási, A.-L. and Oltvai, Z.N. (2004) Network biology: understanding the cell's functional organisation. *Nature Reviews Genetics*, **5**, 101–113.

Barrat, A. and Weigt, M. (2000) On the properties of small-world network models. *European Physical Journal B*, **13**, 547–560.

Barthélemy, M., Barrat, A., Pastor-Satorras, R., and Vespignani, A. (2004) Velocity and hierarchical spread of epidemic outbreaks in scale-free networks. *Physical Review Letters*, **92**, 178701–178704.

Barthélemy, M., Barrat, A., Pastor-Satorras, R., and Vespignani, A. (2005) Dynamical patterns of epidemic outbreaks in complex heterogeneous networks. *Journal of Theoretical Biology*, **235**, 275–288.

Becker, S.A., Feist, A.M., Mo, M.L., Hannum, G., Palsson, B.Ø., and Herrgard, M.J. (2007) Quantitative prediction of cellular metabolism with constraint-based models: the COBRA Toolbox. *Nature Protocols*, **2**, 727–738.

Bender, W., Albus, M., Möller, H.-J., and Tretter, F. (2006) Towards systemic theories in biological psychiatry. *Pharmacopsychiatry*, **39** (Suppl. 1), S4–S9.

Bornholdt, S. (2001) Modeling genetic networks and their evolution: a complex dynamical systems perspective. *Biological Chemistry*, **382**, 1289–1299.

Bornholdt, S. (2005) Less is more in modeling large genetic networks. *Science*, **310**, 449–451.

Bornholdt, S. and Sneppen, K. (1998) Neutral mutations and punctuated equilibrium in evolving genetic networks. *Physical Review Letters*, **81**, 236–239.

Brandman, O., Ferrell, J.E., Li, R., and Meyer, T. (2005) Interlinked fast and slow positive feedback loops drive reliable cell decisions. *Science*, **310**, 496–498.

Brandman, O. and Meyer, T. (2008) Feedback loops shape cellular signals in space and time. *Science*, **322**, 390–395.

Cohen, J.E. (2004) Mathematics is biology's next microscope, only better; biology is mathematics' next physics, only better. *PLoS Biology*, **2**, 2017.

Cooper, J.R., Bloom, F.E., and Roth, R.H. (2002) *The Biochemical Basis of Neuropharmacology*, Oxford University Press, New York.

Cosentino-Lagomarsino, M., Jona, P., Bassetti, B., and Isambert, H. (2007) Hierarchy and feedback in the evolution of the *E. coli* transcription network. *Proceedings of the National Academy of Sciences of the United States of America*, **104**, 5516–5520.

Covert, M., Knight, E., Reed, J.L., Herrgard, M.J., and Palsson, B.Ø. (2004) Integrating high-throughput and computational data elucidates bacterial networks. *Nature*, **429**, 92–96.

Davidich, M.I. and Bornholdt, S. (2008a) Boolean network model predicts cell cycle sequence of fission yeast. *PLoS ONE*, **3**, e1672.

Davidich, M.I. and Bornholdt, S. (2008b) The transition from differential equations to Boolean networks: a case study in simplifying a regulatory network model. *Journal of Theoretical Biology*, **255**, 269–277.

Durstewitz, D. and Seamans, J.K. (2002) The computational role of dopamine $D_1$ receptors in working memory. *Neural Networks*, **15**, 561–572.

Durstewitz, D., Seamans, J.K., and Sejnowski, T.J. (2000) Neurocomputational models of working memory. *Nature Neuroscience*, **3** (Suppl.), 1184–1191.

Edwards, J.S. and Palsson, B.Ø. (2000) The *Escherichia coli* MG1655 *in silico* metabolic genotype: its definition, characteristics, and capabilities. *Proceedings of the National Academy of Sciences of the United States of America*, **97**, 5528–5533.

Erdös, P. and Rényi, A. (1960) On the evolution of random graphs. *Publications of the Mathematical Institute of the Hungarian Academy of Sciences*, **5**, 17–61.

Evlampiev, K. and Isambert, H. (2007) Modelling protein network evolution under genome duplication and domain shuffling. *BMC Systems Biology*, **1**, 49.

Evlampiev, K. and Isambert, H. (2008) Conservation and topology of protein interaction networks under duplication-divergence evolution. *Proceedings of the National Academy of Sciences of the United States of America*, **105**, 9863–9868.

Farkas, I.J., Jeong, H., Vicsek, T., Barabási, A.-L., and Oltvai, Z.N. (2003) The topology of the transcription regulatory network in the yeast, *Saccharomyces cerevisiae*. *Physica A*, **381**, 601–612.

Fell, D.A. and Wagner, A. (2000) The small world of metabolism. *Nature Biotechnology*, **18**, 1121–1122.

Fernandez, E., Schiappa, R., Girault, J.A., and Le Novere, N. (2006) DARPP-32 is a robust integrator of dopamine and glutamate signals. *PLoS Computational Biology*, **2**, e176.

Freund, A., Hütt, M.-Th., and Vec, M. (2004) Selbstorganisation – Aspekte eines Begriffs- und Methodentransfers. *Systeme*, **18**, 1–21.

Fritz, J. (1989) *Einführung in die biologische Psychiatrie*, Fischer, Stuttgart.

Gebicke-Haerter, P.J. and Tretter, F. (2008) The systems view in addiction research. *Addiction Biology*, **13**, 449–454.

Giot, L., Bader, J.S., Brouwer, C., Chaudhuri, A., Kuang, B., Li, Y.L., Hao, Y.L., Ooi, C.E., Godwin, B., Vitols, E., Vijayadamodar, G., Pochart, P., Machineni, H., Welsh, M., Kong, Y., Zerhusen, B., Malcolm, R., Varrone, Z., Collis, A., Minto, M., Burgess, S., McDaniel, L., Stimpson, E., Spriggs, F., Williams, J., Neurath, K., Ioime, N., Agee, M., Voss, E., Furtak, K., Renzulli, R., Aanensen, N., Carrolla, S., Bickelhaupt, E., Lazovatsky, Y., DaSilva, A., Zhong, J., Stanyon, C.A., Finley, R.L. Jr., White, K.P., Braverman, M., Jarvie, T., Gold, S., Leach, M., Knight, J., Shimkets, R.A., McKenna, M.P., Chant, J., and Rothberg, J.M. (2003) A protein interaction map of *Drosophila melanogaster*. *Science*, **302**, 1727–1736.

Girvan, M. and Newman, M.E.J. (2002) Community structure in social and biological networks. *Proceedings of the National Academy of Sciences of the United States of America*, **99**, 7821–7826.

Graham, I. and Matthai, C.C. (2003) Investigation of the forest-fire model on a small world network. *Physical Review E*, **68** (3 Part 2), 036109.

Grondin, Y., Raine, D.J., and Norris, V. (2007) The correlation between architecture and mRNA abundance in the genetic regulatory network of *Escherichia coli*. *BMC Systematic Biology*, **1**, 30.

Guare, J. (1990) *Six Degrees of Separation: A Play*, Vintage Books, New York.

Guimerà, R. and Amaral, L.A.N. (2005) Functional cartography of complex metabolic networks. *Nature*, **433**, 895–900.

Gupta, S., Bisht, S.S., Kukreti, R., Jain, S., and Brahmachari, S.K. (2007) Boolean network analysis of a neurotransmitter signalling pathway. *Journal of Theoretical Biology*, **244**, 463–469.

Gutierrez-Rios, R.M., Rosenblueth, D.A., Loza, J.A., Huerta, A.M., Glasner, J.D., Blattner, F.R., and Collado-Vides, J. (2003) Regulatory network of *Escherichia coli*: consistency between literature knowledge and microarray profiles. *Genome Research*, **13**, 2435–2443.

Han, J.D.J., Bertin, N., Hao, T., Goldberg, D.S., Berriz, G.F., Zhang, L.V., Dupuy, D., Walhout, A.J.M., Cusick, M.E., Roth, F.P., and Vidal, M. (2004) Evidence for dynamically organised modularity in the yeast protein–protein interaction network. *Nature*, **430**, 88–93.

Herrgard, M.J., Covert, M.W., and Palsson, B.Ø. (2003) Regulatory network structures reconciling gene expression data with known genome-scale. *Genome Research*, **13**, 2423–2434.

Hethcote, H.W. (2000) Mathematics of infectious diseases. *SIAM Review*, **42**, 599–563.

Hilgetag, C.C., Burns, G., O'Neill, M., Scannell, J., and Young, M. (2000) Anatomical connectivity defines the organisation of clusters of cortical areas in macaque monkey and cat. *Philosophical Transactions of the Royal Society of London. B*, **355**, 91–110.

Holme, P. and Huss, M. (2008) Comment on "Regularizing capacity of metabolic networks". *Physical Review E*, **77**, 023901.

Honey, C.J., Kötter, R., Breakspear, M., and Sporns, O. (2007) Network structure of cerebral cortex shapes functional connectivity on multiple time scales. *Proceedings of the National Academy of Sciences of the United States of America*, **104**, 10240–10245.

Huang, S.Y., Zou, X.W., Tan, Z.-J., and Jin, Z.-Z. (2003) Network-induced nonequilibrium phase transition in the "game of Life". *Physical Review E*, **67**, 026107.

Hufnagel, L., Brockmann, D., and Geisel, T. (2004) Forecast and control of epidemics in a globalized world. *Proceedings of the National Academy of Sciences of the United States of America*, **101**, 15124–15129.

Hütt, M.-Th. and Dehnert, M. (2006) *Methoden der Bioinformatik. Eine Einführung*, Springer, Berlin.

Hütt, M.-Th. and Lüttge, U. (2005) The interplay of synchronisation and fluctuations reveals connectivity levels in networks of nonlinear oscillators. *Physica A*, **350**, 207–226.

Hütt, M.-Th., Busch, H., and Kaiser, F. (2003) The effect of biological variability on spatiotemporal patterns: model simulations for a network of biochemical oscillators. *Nova Acta Leopold*, **332**, 381–404.

Jeong, H., Tombor, B., Albert, R., Oltvai, Z.N., and Barabási, A.L. (2000) The large-scale organisation of metabolic networks. *Nature*, **407**, 651–654.

Jeong, H., Mason, S.P., Barabási, A.-L., and Ottavi, Z.N. (2001) Lethality and centrality in protein networks. *Nature*, **411**, 41–42.

Josephy, P.D., Guengerich, F.P., and Miners, J.O. (2005) Phase I and II drug metabolism: terminology that we should phase out? *Drug Metabolism Reviews*, **37**, 575–580.

Kaiser, M., Görner, M., and Hilgetag, C.C. (2007) Criticality of spreading dynamics in hierarchical cluster networks without inhibition. *New Journal of Physics*, **9**, 110.

Kaiser, M., Martin, R., Andras, P., and Young, M.P. (2007) Simulation of robustness against lesion of cortical networks. *European Journal of Neuroscience*, **25**, 3185–3192.

Kashtan, N., Itzkovitz, S., Milo, R., and Alon, U. (2004) Topological generalizations of network motifs. *Physical Review E*, **70** (3 Part 1), 031909.

Kauffman, S.A. (1969) Metabolic stability and epigenesis in randomly constructed genetic nets. *Journal of Theoretical Biology*, **22**, 437–467.

Kauffman, K.J., Prakash, P., and Edwards, J.S. (2003) Advances in flux balance analysis. *Current Opinion in Biotechnology*, **14**, 491–496.

Kholodenko, B.N. (2006) Cell-signalling dynamics in time and space. *Nature Reviews Molecular Cell Biology*, **7**, 165–176.

Klipp, E., Herwig, R., Kowald, A., Wierling, C., and Lehrach, H. (2005) *Systems Biology in Practice: Concepts, Implementation & Application*, Wiley-VCH Verlag GmbH, Weinheim.

Kim, P.M., Lu, L.J., Xia, Y., and Gerstein, M.B. (2006) Relating three-dimensional structures to protein networks provides evolutionary insights. *Science*, **314**, 1938–1941.

Kitano, H. (2002) Computational systems biology. *Nature*, **420**, 206–210.

Kitano, H. (2004) Biological robustness. *Nature Reviews Genetics*, **5**, 826–837.

Kriete, A. and Eils, R. (eds) (2006) *Computational Systems Biology*, Elsevier, London.

Kuramoto, Y. (1984) *Chemical Oscillations, Waves, and Turbulence*, Springer, Berlin.

Langton, C.G. (1990) Computation at the edge of chaos: phase transitions and emergent computation. *Physica D*, **42**, 12.

Le Novere, N. (2008) Neurological disease: Are systems approaches the way forward? *Pharmacopsychiatry*, **41** (Suppl. 1), S1–S4.

Li, S., Armstrong, C.M., Bertin, N., Ge, H., Milstein, S., Boxem, M., Vidalain, P.O., Han, J.D.J., Chesneau, A., Hao, T., Goldberg, D.S., Li, N., Martinez, M., Rual, J.F., Lamesch, P., Xu, L., Tewari, M., Wong, S.L., Zhang, L.V., Berriz, G.F., Jacotot, L., Vaglio, P., Reboul, J., Hirozane-Kishikawa, T., Li, Q., Gabel, H.W., and Elewa, A. (2004) A map of the interactome network of the metazoan *C. elegans. Science*, **303**, 540–543.

Liebovitch, L.S., Shehadeh, L.A., Viktor, K., Jirsa, V.K., Hütt, M.-Th., and Marr, C. (2008) Determining the properties of gene regulatory networks from expression data, in *Computational Methodologies in Gene Regulatory Networks* (eds. S. Das, D. Caragea, W.H. Hsu, and S.M. Welch), IGI Global Publishing, Hershey, PA, pp. 405–428.

Liljeros, F., Edling, C.R., Amaral, L.A.N., Stanley, H.E., and Åberg, Y. (2001) The web of human sexual contacts. *Nature*, **411**, 907–908.

Lindskog, M., Kim, M., Wikström, M.A., Blackwell, K.T., and Kotaleski, J.H. (2006) Transient calcium and dopamine increase $pK_a$ activity and DARPP-32 phosphorylation. *PLoS Computational Biology*, **2**, e119.

Luscombe, N.M., Babu, M.M., Yu, H., Snyder, M., Teichmann, S.A., and Gerstein, M. (2004) Genomic analysis of regulatory network dynamics reveals large topological changes. *Nature*, **431**, 308–312.

Ma, H.W. and Zeng, A.-P. (2003) The connectivity structure, giant strong component and centrality of metabolic networks. *Bioinformatics*, **19**, 1423–1430.

Ma, H.W., Kumar, B., Ditges, U., Gunzer, F., Buer, J., and Zeng, A.-P. (2004) An extended transcriptional regulatory network of *Escherichia coli* and analysis of its hierarchical structure and network motifs. *Nucleic Acids Research*, **32**, 6643–6649.

Mahadevan, R., Edwards, J.S., and Doyle, F. (2002) Dynamic flux balance analysis of diauxic growth in *Escherichia coli. Biophysical Journal*, **83**, 1331–1340.

Mangan, S. and Alon, U. (2003) Structure and function of the feed-forward loop network motif. *Proceedings of the National Academy of Sciences of the United States of America*, **100**, 11980–11985.

Marr, C. and Hütt, M.-Th. (2005) Topology regulates pattern formation capacity of binary cellular automata on graphs. *Physica A*, **354**, 641–662.

Marr, C. and Hütt, M.-Th. (2006) Similar impact of topological and dynamic noise on complex patterns. *Physics Letters A*, **349**, 302–305.

Marr, C., Müller-Linow, M., and Hütt, M.-Th. (2007) Regularizing capacity of metabolic networks. *Physical Review E*, **75**, 041917.

Marr, C., Müller-Linow, M., and Hütt, M.-Th. (2008) Reply to Comment on "Regularizing capacity of metabolic networks". *Physical Review E*, **77**, 023902.

Maslov, S. and Sneppen, K. (2002) Specificity and stability in topology of protein networks. *Science*, **296**, 910–913.

Milgram, S. (1967) The small world problem. *Psychology Today*, **1**, 61–67.

Millat, T., Bullinger, E., Rohwer, J., and Wolkenhauer, O. (2007) Approximations and their consequences in dynamic modelling of signal transduction pathways. *Mathematical Biosciences*, **47**, 40–57.

Milo, R., Shen-Orr, S., Itzkovitz, S., Kashtan, N., Chklovskii, D., and Alon, U. (2002) Network motifs: simple building blocks of complex networks. *Science*, **298**, 824–827.

Milo, R., Itzkovitz, S., Kashtan, N., Levitt, R., Shen-Orr, S., Ayzenshtat, I., Sheffer, M., and Alon, U. (2004) Superfamilies of evolved and designed networks. *Science*, **303**, 1538–1542.

Moreno, Y., Pastor-Satorras, R., and Vespignani, A. (2002) Epidemic outbreaks in complex heterogeneous networks. *European Physical Journal B*, **26**, 521–529.

Müller-Linow, M., Marr, C., and Hütt, M.-Th. (2006) Topology regulates synchronization patterns in excitable dynamics on graphs. *Physical Review E*, **74**, 016112.

Müller-Linow, M., Hilgetag, C., and Hütt, M.-Th. (2008) Organization of excitable dynamics in hierarchical biological networks. *PLoS Computational Biology*, **4**, e1000190.

Newman, M.E.J. (2001) The structure of scientific collaboration networks. *Proceedings of the National Academy of Sciences of the United States of America*, **98**, 404–409.

Newman, M.E.J. (2003) The structure and function of complex networks. *SIAM Review*, **45**, 167–256.

Newman, M.E.J. and Girvan, M. (2004) Finding and evaluating community structure in networks. *Physical Review E*, **69**, 026113.

Nishikawa, T. and Motter, A.E. (2006) Synchronization is optimal in non-diagonalizable networks. *Physical Review E*, **73**, 065106.

Papin, J.A., Hunter, T., Palsson, B.Ø., and Subramaniam, S. (2005) Reconstruction of cellular signalling networks and analysis of the their properties. *Nature Reviews Molecular Cell Biology*, **99**, 99–111.

Pastor-Satorras, R. and Vespignani, A. (2001) Epidemic spreading in scale-free networks. *Physical Review Letters*, **86**, 3200–3203.

Pikovsky, A., Rosenblum, M., and Kurths, J. (2001) *Synchronisation. A Universal Concept in Nonlinear Sciences, Cambridge Nonlinear Science Series*, Cambridge University Press, Cambridge.

Price, N.D., Papin, J.A., and Palsson, B.Ø. (2002) Determination of redundancy and systems properties of the metabolic network of *Helicobacter pylori* using genome-scale extreme pathway analysis. *Genome Research*, **12**, 760–769.

Price, N.D., Reed, J.L., and Palsson, B.Ø. (2004) Genome-scale models of microbial cells: evaluating the consequences of constraints. *Nature Reviews Microbiology*, **2**, 886–897.

Ravasz, E. and Barabási, A.-L. (2003) Hierarchical organisation in complex networks. *Physical Review E*, **67**, 026112.

Ravasz, E., Somera, A.L., Mongru, D.A., Oltvai, Z.N., and Barabási, A.-L. (2002) Hierarchical organization of modularity in metabolic networks. *Science*, **297**, 1551–1555.

Reigl, M., Alon, U., and Chklovskii, D.B. (2004) Search for computational modules in the *C. elegans* brain. *BMC Biology*, **2**, 25.

Roxin, A., Riecke, H., and Solla, S.A. (2004) Self-sustained activity in a small-world network of excitable neurons. *Physical Review Letters*, **92**, 198101.

Salgado, H., Gama-Castro, S., Martinez-Antonio, A., Diaz-Peredo, E., Sanchez-Solano, F., Peralta-Gil, M., Garcia-Alonso, D., Jiménez-Jacinto, V., Santos-Zavaleta, A., Bonavides-Martínez, C., and Collado-Vides, J. (2004) RegulonDB (version 4.0): transcriptional regulation, operon organization and growth conditions in *Escherichia coli* K-12. *Nucleic Acids Research*, **32**, D303–D306.

Samal, A., and Jain, S. (2008) The regulatory network of *E. coli* metabolism as a Boolean dynamical system exhibits both homeostasis and flexibility of response. *BMC Systematic Biology*, **2**, 21.

Samal, A., Singh, S., Giri, V., Krishna, S., Raghuram, N., and Jain, S. (2006) Low degree metabolites explain essential reactions and enhance modularity in biological networks. *BMC Bioinformatics*, 7, 118.

Schiepek, G. (1999) Selbstorganisation in psychischen und sozialen prozessen: Neue perspektiven der psychotherapie, in *Komplexe Systeme in Natur und Gesellschaft. Komplexitätsforschung in Deutschland auf dem Weg ins nächste Jahrhundert* (ed. K. Mainzer), Springer, Berlin, pp. 280–317.

Segre, D., Vitkup, D., and Church, G. (2002) Analysis of optimality in natural and perturbed metabolic networks. *Proceedings of the National Academy of Sciences of the United States of America*, 99, 15112–15117.

Shen-Orr, S., Milo, R., Mangan, S., and Alon, U. (2002) Network motifs in the transcriptional regulation network of *Escherichia coli*. *Nature Genetics*, 31, 64–68.

Shimoni, Y., Friedlander, G., Hetzroni, G., Niv, G., Altuvia, S., Biham, O., and Margalit, H. (2007) Regulation of gene expression by small non-coding RNAs: a quantitative view. *Molecular Systems Biology*, 3, 138.

Shlomi, T., Berkman, O., and Ruppin, E. (2005) Regulatory on/off minimization of metabolic flux changes after genetic perturbations. *Proceedings of the National Academy of Sciences of the United States of America*, 102, 7695–7700.

Shlomi, T., Eisenberg, Y., Sharan, R., and Ruppin, E. (2007) A genome-scale computational study of the interplay between transcriptional regulation and metabolism. *Molecular Systems Biology*, 3, 101.

Smith, J., and Stein, V. (2009) SPORCalc: a development of a database analysis that provides putative metabolic enzyme reactions for ligand-based drug design. *Computational Biology and Chemistry*, 33, 149–159.

Stephan, K.E., Hilgetag, C.C., Burns, G.A., O'Neill, M.A., Young, M.P., and Kötter, R. (2000) Computational analysis of functional connectivity between areas of primate cerebral cortex. *Philosophical Transactions of the Royal Society of London B*, 355, 111–126.

Sporns, O. and Kötter, R. (2004) Motifs in brain networks. *PLoS Biology*, 2, e396.

Sporns, O. (2006) Small-world connectivity, motif composition, and complexity of fractal neuronal connections. *Bio Systems*, 85, 55–64.

Sporns, O., Honey, C.J., and Kötter, R. (2007) Identification and classification of hubs in brain networks. *PLoS ONE*, 2, e1049.

Strogatz, S.H. (2001) Exploring complex networks. *Nature*, 410, 268–276.

Strunk, G., and Schiepek, G. (2002) Dynamische komplexität in der therapeut-klient-interaktion. *Psychotherapeut*, 47, 291–300.

Teichmann, S.A. and Babu, M.M. (2004) Gene regulatory network growth by duplication. *Nature Genetics*, 36, 492496.

Tong, A.H.Y., Lesage, G., Bader, G.D., Ding, H., Xu, H., Xin, X., Young, J., Berriz, G.F., Brost, R.L., Chang, M., Chen, Y.Q., Cheng, X., Chua, G., Friesen, H., Goldberg, D.S., Haynes, J., Humphries, C., He, G., Hussein, S., Ke, L., Krogan, N., Li, Z., Levinson, J.N., Lu, H., Ménard, P., Munyana, C., Parsons, A.B., Ryan, O., Tonikian, R., Roberts, T., Sdicu, A.M., Shapiro, J., Sheikh, B., Suter, B., Wong, S.L., Zhang, L.V., Zhu, H., Burd, C.G., Munro, S., Sander, C., Rine, J., Greenblatt, J., Peter, M., Bretscher, A., Bell, G., Roth, F.P., Brown, G.W., Andrews, B., Bussey, H., and Boone, C. (2004) Global mapping of the yeast genetic interaction network. *Science*, 303, 808–813.

Traub, R. and Wong, R.S.K. (1982) Cellular mechanisms or neuronal synchronisation in epilepsy. *Science*, 216, 745–747.

Traub, R., Contreras, D., Cunningham, M., Murray, H., LeBeau, F., Roopun, A., Bibbig, A., Wilent, W., Higley, M., and Whittington, M. (2005) Single-column thalamocortical network model exhibiting gamma oscillations, sleep spindles, and epileptogenic burst. *Journal of Neurophysiology*, 93, 2194–2232.

Tretter, F. and Albus, M. (2004) *Einführung in die Psychopharmakotherapie*, Thieme, Stuttgart.

Tretter, F. and Albus, M. (2007) "Computational neuropsychiatry" of working memory disorders in schizophrenia: the network connectivity in prefrontal cortex – data and models. *Pharmacopsychiatry*, 40 (Suppl. 1), S2–S16.

Tretter, F. and Scherer, J. (2006) Schizophrenia, neurobiology and the methodology of systemic modeling. *Pharmacopsychiatry*, **39** (Suppl. 1), S26–S35.

Tsang, J., Zhu, J., and van Oudenaarden, A. (2007) MicroRNA-mediated feedback and feed forward loops are recurrent network motifs in mammals. *Molecular Cell*, **26**, 753–767.

Urbanczik, R. (2006) SNA – a toolbox for the stoichiometric analysis of metabolic networks. *BMC Bioinformatics*, **7**, 129.

Varma, A. and Palsson, B.Ø. (1994) Stoichiometric flux balance models quantitatively predict growth and metabolic by-product secretion in wild-type *Escherichia coli* W3110. *Applied and Environmental Microbiology*, **60**, 3724–3731.

von Neumann, J. (2001) *J. von Neumann Collected Works*, vol. **5** (ed. A.H. Taub), Macmillan, New York.

Wagner, A. and Fell, D.A. (2001) The small world inside large metabolic networks. *Proceedings of the Royal Society of London B*, **268**, 1803–1810.

Wang, X.L., Tegner, J., Constandinidis, C., and Goldman-Rakic, P.S. (2004) Division of labor among distinct subtypes of inhibitory neurons in a cortical microcircuit of working memory. *Proceedings of the National Academy of Sciences of the United States of America*, **101**, 1368–1373.

Watts, D.J. (1999) *Small Worlds: The Dynamics of Networks Between Order and Randomness*, Princeton University Press, Princeton, NJ.

Watts, D.J. and Strogatz, S.H. (1998) Collective dynamics of 'small-world' networks. *Nature*, **393**, 440–442.

Wolfram, S. (1983) Statistical mechanics of cellular automata. *Reviews of Modern Physics*, **55**, 601–644.

Wolfram, S. (1984) Universality and complexity in cellular automata. *Physica D*, **10**, 1–35.

Wunderlich, Z. and Mirny, L.A. (2006) Using the topology of metabolic networks to predict viability of mutant strains. *Biophysical Journal*, **91**, 2304–2311.

Xayaphoummine, A., Viasnoff, V., Harlepp, S., and Isambert, H. (2007) Encoding folding paths of RNA switches. *Nucleic Acids Research*, **35**, 614–622.

Yook, S.H. and Meyer-Ortmanns, H. (2006) Synchronisation of Rössler oscillators on scale-free topologies. *Physica A*, **371**, 781–789.

Young, M.P., Scannell, J.W., O'Neill, M.A., Hilgetag, C.C., Burns, G., and Blakemore, C. (1995) Non-metric multidimensional scaling in the analysis of neuroanatomical connection data from the primate visual system. *Philosophical Transactions of the Royal Society of London B*, **348**, 281–308.

Yu, H. and Gerstein, M. (2006) Genomic analysis of the hierarchical structure of regulatory networks. *Proceedings of the National Academy of Sciences of the United States of America*, **103**, 14724–14731.

Zhou, C., Zemanová, L., Zamora, G., Hilgetag, C.C., and Kurths, J. (2006) Hierarchical organization unveiled by functional connectivity in complex brain networks. *Physical Review Letters*, **97**, 238103.

# 13
# Some Useful Mathematical Tools to Transform Microarray Data into Interactive Molecular Networks

*Franziska Matthäus, V. Anne Smith, and Peter J. Gebicke-Haerter*

## 13.1
## Introduction

With the advent of high-throughput technologies, the amount of data obtained from just a few experiments has skyrocketed. In particular, microarray technologies, where the expression of thousands of genes or proteins can be monitored simultaneously, have aggravated this situation significantly. Mathematicians and bioinformaticians have been quick to respond to these data-handling requirements and developed specific algorithms to process them. Despite many Jeremiah's prophecies and warnings that researchers keep collecting data without knowing what to do with them, it has turned out that statistical evaluations of them by various sophisticated tools is not the problem. The problem is the interpretation of the data in a biologically meaningful context. This problem evidently cannot be tackled only by computational scientists alone, but requires the help of experienced biologists. Other problems emerge during processing of the datasets, including the major issue of setting cut-offs at some level of significance.

Significantly regulated genes "stand out" above those levels, but are not necessarily the key players (Gass *et al.*, 2008). However, significantly regulated genes can be used as "indicators" to find the key players. They likely are part(s) of molecular networks, which include those key players, as well.

Hence, it is of value to identify the networks that those significantly regulated genes belong to. If there are many significantly regulated genes, they likely do not belong to only one network. However, is it very much possible that those subnetworks closely interact with each other and that the key players occupy pivotal nodes.

A variety of statistical approaches have been used to analyze data obtained from microarrays. In this chapter, we want to elaborate only on a few approaches pursued in our own laboratories: principal component analysis (PCA), clustering methods, statistical tests like analysis of variance (ANOVA), and naïve Bayes classifier that are often used to identify clusters of "indicator" molecules in two conditions as mentioned above. Typically, measurements are carried out at two different time-points – time 0, that is supposed to characterize the "healthy," unaffected condition, and time 1, the "disease" or "treated" condition. We are not going to touch here upon nonlinear differential equations, Petri nets, or cellular automata, which can be usefully applied in those more complex applications.

To visualize in more detail molecular interactions and unravel nodes of special interest, the use of Bayesian networks can be very helpful. They can be used for investigations on more dynamic processes, particularly when several timepoints are determined. Some drawbacks result from the fact that they can handle only limited numbers of nodes. Despite the construction of molecular networks by those approaches, the results remain completely unbiased and encompass no *a priori* biological meaning. At this point, it is up to experienced biologists to extract the biological messages.

## 13.2
## Microarray Data

A common way to access genetic influence on biological or medical conditions is through microarray data (Figure 13.1). The number of genes per microarray chip is thereby large, usually a few hundreds up to thousands. To quantify whether a certain gene is involved in the process or condition of interest, gene expression data is

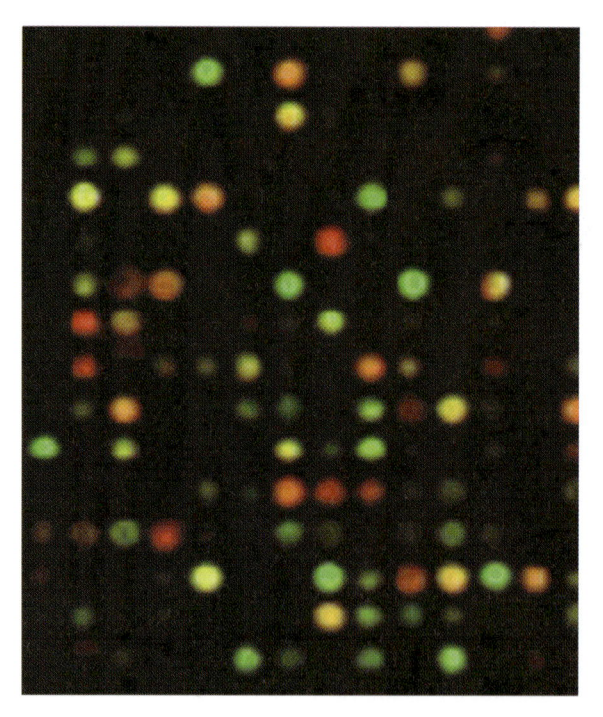

**Figure 13.1** Fluorescently labeled cDNAs hybridized to a microarray. Each spot represents one specific gene. Copy DNAs from control tissue are labeled green, from disease tissue red. When a gene is expressed in equal amounts in both tissues, the respective spot lights up yellow.

measured in a number of individuals and under different conditions (e.g., male and female, healthy and sick, treatment and control, or considering various treatment possibilities). In any case, the result consists of a very large number of genes and their expression values under various conditions or of a large number of individuals. The size of this dataset poses a challenge for its analysis, especially since the expression of a certain gene is not only determined by the condition under study, but usually also depends on the expression of other genes. It is now widely accepted that a change in the condition not only leads to differences in the expression of single genes, but rather to changes in the gene expression pattern. Even if a certain condition influences only a single gene (e.g., a gene knockout or knockdown through RNA interference), the modified activity of this gene again influences the activity of other genes and thus causes a shift of the expression pattern of the whole system. The experimental approach based on microarrays allows measurement of such system changes since a large number of genes can be included on a single chip. The vast amount of information contained in microarray data, however, requires subsequent computational processing to reveal which parts of the system or which interactions are most relevant to the posed question. A very nice overview addressing many issues concerning the analysis of microarray data can be found in *Systems Biology in Practice* (Klipp et al., 2005).

In the following sections we outline a few possible methods to process microarray data. First, we describe a few common methods of summarizing the data into smaller, more manageable units, dimensionality reduction, and clustering. This is necessary in order to collapse measurements of hundreds to thousands of genes down to structural or functional subunits of the system that might then be studied separately. The set of selected methods does not aim to be exhaustive, but rather exemplary. Especially when it comes to clustering, there exists a large variety of methods and we restrict ourselves to the most widely used approaches. In a further section (Section 13.3.2.3) we compare two classical approaches for the identification of key genes (i.e., genes that show significant changes in their expression levels under treatment): ANOVAs and naïve Bayes classifiers. Selection of a smaller number of genes is important if one wants to apply methods that yield useful information, but are computationally expensive and therefore cannot be applied to the entire dataset. One such method is to infer the interactions between genes using Bayesian networks, described in the previous section.

## 13.3
## Dimensionality Reduction

### 13.3.1
### Principal Components Analysis (PCA)

PCA is probably the most common approach to reduce the dimensionality of the data (Figure 13.2). It is based on the idea that multiple variables are likely to be highly inter-related (e.g., genes in the same pathway), even though they may each measure slightly different things. In order to tease out these "different things" from the

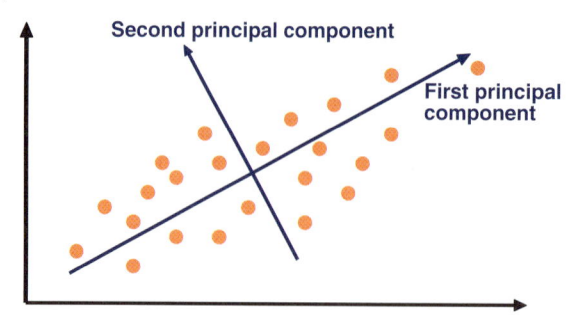

**Figure 13.2** PCA identifies the spatial dimensions in which the data show the largest variability.

measurements, without over-representing the features variables have in common, we look for a set of principal components that each represent independent characters of the measured data. These principle components are calculated from information on how each variable varies in relation to the others, as represented in a covariance matrix. PCA represents a linear transformation of the data that is based on computing the eigenvalues and eigenvectors of the covariance matrix (Jolliffe, 2002; Wall, Rechtsteiner, and Rocha, 2003). The eigenvectors are the principal components; they are ordered by the eigenvalues such that those with the largest eigenvalues have the feature that a projection of the data onto the subspace spanned by these vectors captures a very high percentage of the variation in the data. Thus, the so-obtained dimensionality reduction can be performed without loosing much information originally provided by the data. The relative amount of variation contained in the reduced dataset (after projection onto the first $k$ principal components) can be accessed through the ratio between the sum of the first $k$ largest eigenvalues and the sum of all eigenvalues of the covariance matrix. A projection onto a lower-dimensional subspace (usually two- or three-dimensional) only makes sense if the amount of explained variation is high.

PCA is often used to visualize high-dimensional data. If the data are given as a set of $n$ variables and $m$ observations, each variable can be represented as a point in an $m$-dimensional space (or, alternatively, each condition as a point in an $n$-dimensional space). Commonly, $m$ and $n$ are large, and visualization of the original dataset is not possible.

In the case of gene expression data, PCA can be performed using either the different genes or the conditions as variables. In the first case, the expression of every gene can be thought of as a point in a space where each of the $m$ conditions corresponds to another dimension. If two genes differ in their expression profiles over the different conditions, then their distance in this $m$-dimensional space will be large and vice versa. Groups of jointly regulated genes, or genes that strongly influence each other, will form clusters in this high-dimensional space. Data visualization, facilitated after dimensionality reduction through PCA, can be extremely helpful to recognize such inherent substructure.

Conversely, one can also consider every condition as a point in a space spanned by the expression values of $m$ different genes. PCA performed on this dataset will give an indication as to which conditions affect the gene expression levels most. If two conditions result in very similar expression profiles, then the two corresponding points will be close to each other in the high-dimensional (and also in the reduced) space and vice versa.

## 13.3.2
### Clustering Methods

The goal of any clustering algorithm is to partition the dataset into a number of groups such that the elements within one group are more similar to each other than elements belonging to other groups.

For an overview of clustering methods see, for instance, the book by Xu and Wunsch (2009), and for applications to microarray data the articles by Eisen et al. (1998), Hand and Heard (2005), or Shay (http://www.science.co.il/enuka/Essays/Microarray-Review.pdf).

The classification is subject to a chosen measure of similarity or dissimilarity of the elements and this measure depends crucially on the nature of the data. It is clear that, for instance, binary data needs a completely different similarity measure than symbolic features (say, colors or letters). Microarray data yields gene expression levels to which a numeric value can be assigned. An expression pattern (i.e., the change of gene expression over a set of different conditions) can be represented as an array of numerical values. Classical similarity measures for this type of data are based on the so-called $p$-norms. Here are some examples. First, we denote the expression level of the genes $v$ and $w$ under the $i$th condition as $v_i$ and $w_i$, with $1 \leq i \leq m$ and $m$ denoting the number of different conditions. The most widely used distance measure for this type of numerical data is the Euclidean distance $d_E$, relating to the case $p = 2$. It is defined as $d_2 = \sqrt{\sum_i (v_i - w_i)^2}$. Other often used similarity measures are the Manhattan distance, which is based on the 1-norm and defined as $d_1 = \sum_i |v_i - w_i|$, and the distance based on the maximum norm, $d_\infty = \max(|v_i - w_i|)$. Alternatively, one can use the correlation coefficient $\varrho_{vw}$ and define a distance as how far from perfectly correlated ($\varrho_{vw} = 1$) two elements are, $d_\varrho = 1 - [\text{cov}(v,w)/(\sigma_v \sigma_w)]$, where cov($v,w$) denotes the covariance between $v$ and $w$ and $\sigma_v$ and $\sigma_w$ the standard deviation of the expression levels of gene $v$ and $w$, respectively. A more exhaustive overview over the different similarity measures, including measures for non-numeric data, is given in the book on *Data Clustering* by Gan, Ma, and Wu (2007).

For a given similarity measure, clustering of a dataset is always possible. Even the number of groups into which the data are split can be arbitrary. The difficulty with every clustering approach is to choose the appropriate clustering approach and the parameters of the algorithm such that the resulting classification captures the true substructure of the data. As pointed out above, dimensionality reduction (e.g., through PCA) and subsequent data visualization can be very helpful to obtain a first estimate of the number of expected clusters. Clustering algorithms can be categorized into partitioning and hierarchical methods. The most widely used partitioning approach is $k$-means, which subdivides the data into $k$ groups, whereby the number of clusters k is a parameter of the algorithm. Hierarchical clustering methods join (bottom-up) or split (top-down) the data iteratively and yield a clustering on many scales. In this way they not only provide information on which elements are similar to each other, but also on how similar the clusters themselves are to each other. In the following, the two clustering methods are described in more detail.

### 13.3.2.1 Partitional Clustering: k-Means

The k-means clustering algorithm (Hartigan, 1975) is a partitioning method that splits the data into k groups. It is an efficient algorithm if the number of clusters is known and if the distribution of the cluster elements is well described by a Gaussian distribution. The algorithm starts with an arbitrary or heuristic partition into k groups and the calculation of the cluster means. Subsequently, the elements are successively moved from one cluster to another. For each move the cluster means are updated and a partition error $e_p$ is computed. If the partition error is reduced by the reassignment of an element, the element is classified into this cluster. The partition error is defined as the sum over all distances between the elements $v_{ij}$ (ith element of the jth cluster) and the corresponding means $\mu_j$. If the Euclidean distance is used, this leads to $e_p = \sum_i \sum_j (v_{ij} - \mu_j)^2$. The procedure is repeated until the partition error converges (i.e., until no assignment of any element to another group will lead to a reduction of $e_p$).

This method is computationally much less expensive (and therefore much faster) than the hierarchical methods described below. It has the disadvantage, though, that the number of clusters has to be known or guessed.

### 13.3.2.2 Hierarchical Clustering: Bottom-Up and Top-Down Approaches

Hierarchical clustering methods can be further categorized into bottom-up and top-down approaches. In the latter case, the data is initially considered to be one large cluster and is split iteratively into more and more groups. One possible top-down approach is to apply k-means partitioning to divide the data into two groups and then again on the two resulting clusters. This procedure can be iterated until every element forms its own cluster. In contrast to the top-down approach, a hierarchical bottom-up approach clusters elements in an agglomerative way, joining at each step the two elements or two clusters with the largest similarity.

In order to iteratively join the most similar clusters it is necessary to define a distance between two clusters, based on the distance of single elements. Therefore, there exist several possibilities, commonly referred to as single, average, or complete linkage (Figure 13.3). Let us briefly explain what these notions mean. For two clusters i and j we denote the elements they contain as $g_k^i$ and $g_l^j$ (so $g_k^i$ is the kth element in the ith cluster), and the distance between the two elements as $d(g_k^i, g_l^j)$. The distance of the clusters i and j, $d^{ij}$ is then given:

i) By single linkage as the minimum $d^{ij} = \min[d(g_k^i, g_l^j)]$.
ii) By average linkage as the average $d^{ij} = \langle d(g_k^i, g_l^j) \rangle$ and
iii) by complete linkage as the maximum $d^{ij} = \max[d(g_k^i, g_l^j)]$.

The result of a hierarchical clustering can be represented as a tree (dendrogram), which displays the substructure of the data (Figure 13.4). Branches of similar elements join first and branches relating groups whose elements are less similar

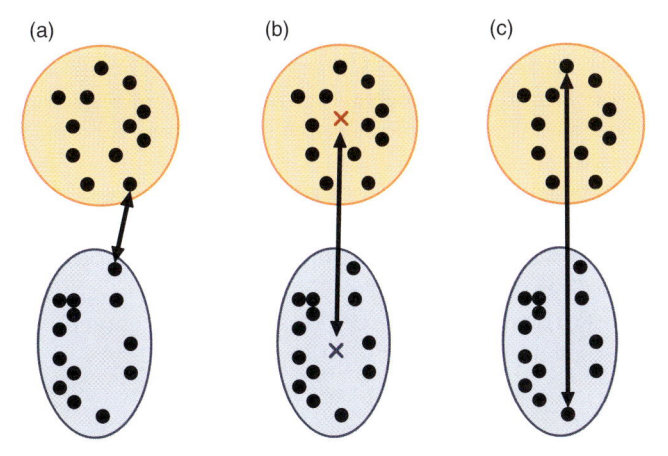

**Figure 13.3** Distance measures for clustering data points: (a) single linkage, (b) average linkage, and (c) complete linkage.

to each other join close to the root of the tree. The dendrograms contain information on the similarity structure of all elements and clusters on many scales. The disadvantage of hierarchical methods is that they take a lot of computation time, since the algorithms scale as $O(N^2)$, whereby $N$ denotes the number of elements of the dataset. Another problem is that the hierarchical structure of the data as given by the dendrogram still does not determine how to split the data into a number of clusters. For this, one still has to choose at what level of similarity the branches are cut. If elements of the same cluster should be very similar to each other, then the number of resulting clusters will be large and vice versa. There exist a number of methods to determine when to set the cutting level (e.g., Milligan and Cooper, 1985; Jung *et al.*, 2003). One often used method is to monitor the number of clusters for every joining (or splitting) distance. If some larger interval in the joining distances

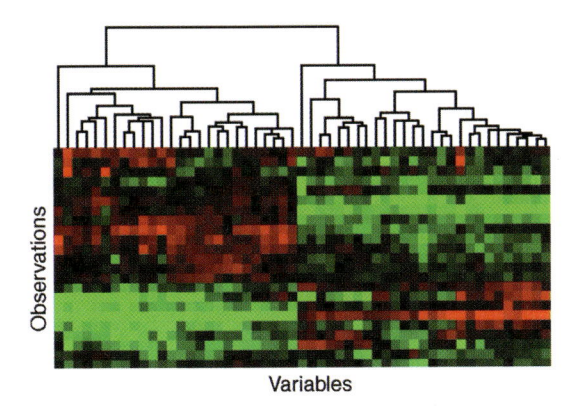

**Figure 13.4** Example of the result of a hierarchical clustering. The similarity structure of the variables and their clusters is contained in the dendrogram displayed above the data matrix. The picture was created using the open access software Cluster 3.0 by Eisen and Hoon (Eisen *et al.*, 1998), and Treeview (http://jtreeview.sourceforge.net).

exist over which the cluster number does not change, then this clustering is considered more robust and is therefore preferred to a clustering at a different scale.

To summarize, dimensionality reduction and clustering algorithms can be used to identify groups of genes with similar expression patterns and thus functional subunits or sets of coregulated genes.

However, coming back to the microarray data, the studies are usually carried out with some specific question in mind. Usually one wants to know which genes change their expression pattern under a certain condition. Clustering methods identify subunits of the genetic network, but do not help to identify which of these subunits play a role in the investigated process. For this purpose, statistical tests are applied, two of which are briefly presented below.

## 13.4
### Statistical Tests: ANOVA and the Naïve Bayes Classifier

The identification of subunits of the genetic network is one approach to gain information from large microarray datasets. Another objective of the study might be to determine a set of genes that is involved in a certain process (a disease, a regulatory process, etc.). A clustering approach can also be helpful in this case, because if some key genes of the process are already known, then the other genes clustered into the same group become interesting as well. If no prior knowledge is available, then statistical tests are used to identify genes that are characterized by a different expression level when two conditions are compared. A very naïve approach would be to simply compare the difference in the mean values of the two groups. The informative value of the difference in the average expression level depends, however, on how much the expression level scatters. If the variation in the expression level is large, then a big difference in the average expression level is less significant than when the variation is small. Hence, the variance of the data has to be taken into account. A standard method for this is called ANOVA (Ayroles and Gibson, 2006; Kerr, Martin, and Churchill, 2000; Kerr and Churchill, 2001). It quantifies significant differences in the mean of two sets of experiments by comparing the between-group variability to the within-group variability. The difference in the mean values is not significant if the variability within the groups does not differ from the variability of the whole set. A measure of the significance of the difference between two groups is the $p$-value, with $0 < p < 1$, which is given as the result of an $F$-test with the variance values as input. ANOVA can also be used when more than one condition is considered, like the comparison of treatment and control in a group containing males and females.

Another approach is to use a naïve Bayes classifier, which is related to Bayesian networks (see Section 13.5 on Bayesian network inference below). This Bayes classifier determines the statistical dependence between a given pattern in the gene expression level and the studied conditions. As in many Bayesian network approaches, a score is computed, called the Bayesian scoring metric, which quantifies how well a set of statistical dependencies explain the given dataset. To estimate the probability that a dependence between a variable and a condition exists, the scoring metric is computed for both cases (assuming statistical dependence and independence). The result is given in form of the so-called "log-ratio" (the logarithm of the ratio of the two values), which is large if a statistical dependence is much more probable than statistical independence.

A very common form is the discrete classifier, which requires the data to be transformed into a number of discrete states. For gene expression values this might be down- and upregulated or low, medium, and high. A discrete Bayesian classifier is then able to detect many kinds of statistical relationship, including linear, nonlinear, stochastic, and arbitrary combinatoric (Friedman, Murphy, and Russell, 1998; Heckerman, Geiger, and Chickering, 1995). The discrete states are coded as sequential integers starting from zero and enable the algorithm to return an "influence score" for each link produced (Yu *et al.*, 2004). This score indicates the magnitude and direction of influence between two variables. In case the score is zero, then the influence is nonmonotonic, for instance U-shaped or combinatoric.

The significance of a change in the expression level of a gene under some condition can thus be quantified by the *p*-values resulting from a statistical test like ANOVA or by the log-ratio given through the naïve Bayes classifier. It is interesting, that these two values are not correlated in many cases. Genes with a large log-ratio do not necessarily have to be characterized by low *p*-values. The reasons are the following. First, in the process of data discretization, needed for the discrete Bayesian classifier, information on the variability of the expression levels is lost. For the Bayesian classifier the variance of the two groups is thus of little importance, while for ANOVA it is the foundation on which the method is based. On the other hand, ANOVA returns low *p*-values only in case of monotonous changes in the expression levels. If a treatment leads to distinct upregulation of a gene in males, but to a spread between down- and upregulation in females, then the variance in the entire treatment group is large and the expression change will not be considered significant. In contrast, the naïve Bayes classifier is capable of dealing with such nonmonotonic relationship and can yield a large log-ratio in this case as well.

Both methods, ANOVA and the naïve Bayes classifier, return a list of genes sorted according to a numerical value that quantifies the significance of a change in the expression level of the gene and thus its importance for the considered process (Matthäus *et al.*, 2009). Each of the approaches has advantages and disadvantages, like the loss of information on the variance or the incapability of dealing with nonmonotonic statistical dependencies. The two methods complement each other and when combined they provide a great deal more information than any of them alone.

These analyses to reduce the huge numbers of genes in microarrays into more manageable units, either through dimension reduction, clustering, or selection of genes related to conditions, provide a good means to get a first handle on large datasets. They can be considered a form of summarizing the information in a microarray study to highlight the most relevant parts. However, once this is accomplished, it is often useful to further explore the data based on this summary. For example, one may wish to know how selected genes relate to each other – this can be done with model-based analysis (e.g., building a network from the genes). We will explore one such method in detail: Bayesian networks.

## 13.5
## Bayesian Networks

### 13.5.1
### Definition of a Bayesian Network

A Bayesian Network is a method to graphically display statistical dependence and conditional independence among a number of variables (Pearl, 1988; Heckerman, Geiger, and Chickering, 1995; Friedman, Murphy, and Russell, 1998). A BN is visualized as a graph where dependence relationships are represented as a link between two variables (or nodes), where the child is dependent on the parent (Figure 13.5). Dependence arises when knowledge of one variable provides information about another. For example, you might check the cloud cover to determine whether or not you plan to carry an umbrella: here the clouds (C) provide information about the umbrella (U); mathematically, your probability of carrying an umbrella would be represented $P(U|C)$, to indicate that carrying an umbrella is dependent on clouds. Conditional independence in a BN is read from the graph structure (Figure 13.5). Conditional independence involves at minimum three variables; it is the situation when knowledge of a third variable means that one variable no longer provides information about another. In the previous example, if you saw rain, you would not bother checking for clouds to determine whether to carry an umbrella:

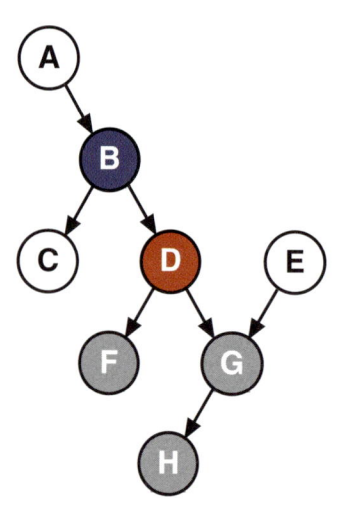

**Figure 13.5** A Bayesian Network. In this Bayesian Network, variable D (red) is the child of B (blue); B is the parent of D. Children are statistically dependent on their parents. The gray nodes (F, G, H) are known as the descendants of D. Conditional independence relationships are read from the network structure such that a variable is conditionally independent of its nondescendants given its parents. Here, D is conditionally independent of the white nodes given B. The joint probability distribution represented by this network is written as: $P(A, \ldots, H) = P(A)P(B|A)P(C|B)P(D|B)P(E)P(F|D)P(G|D,E)P(H|G)$.

knowledge of rain (R) creates the situation where knowledge of clouds provides no further useful information; mathematically, one would write P(U|C,R) = P(U|R); that is, when predicting umbrella, knowledge of both clouds and rain is no different from knowledge of rain alone. Here, clouds and umbrella are conditionally independent given rain. A BN representing these three variables would be C → R → U, showing that clouds does have an influence on umbrella, but only through a path mediated by rain. This encoding of conditional independence relationships is what enables a BN to distinguish direct influence among measured variables from indirect influence.

The statistical dependencies represented by the BNs graph are a factoring of the joint probability distribution across all the variables, such that given a BN we can reconstruct the joint probability distribution to be:

$$P(X_1, \ldots, X_n) = \prod_{i=1}^{n} P(X_i | Pa(X_i)),$$

where $X_1, \ldots, X_n$ are the variables in the network and $Pa(X_i)$ represents the parents of variable $X_i$ (Figure 13.5). The fact that a BN represents a factoring of a joint probability distribution leads to two features. First, in order for it to be a valid factoring, the BN must be acyclic (i.e., have no loop). (Later we will look at how dynamic Bayesian networks can represent loops in biological systems.) Second, since one can apply Bayes' rule to the factoring represented by one BN to produce a mathematically equivalent factoring, there can be multiple BNs representing the same distribution – known as an equivalence class – which differ only the direction of some of their links (Figure 13.6). Thus, the direction of links in a BN should not be interpreted in a causal manner; links are more usefully conceptualized as meaning that the parents are useful for predicting the child (again, however, dynamic Bayesian networks can represent causal information).

A fully described BN requires one more thing: precisely how the variable $X_i$ depends on its parents $Pa(X_i)$. This is known as the parameters of $X_i$. In discrete BNs, where all variables take on discrete states (e.g., on/off, high/medium/low), each parameter $\theta_{x_i | pa(X_i)}$ specifies the probability that variable $X_i$ has a particular discrete state $x_i$ when its parents $Pa(X_i)$ have the particular states $pa(X_i)$. As each of these

P(A,B,C) = **P(A|B)**P(B)P(C|A) = P(A)P(B|A)**P(C|A)** = P(A|C)P(B|A)P(C)

Bayes' Rule:
$$P(A|B) = \frac{P(A)P(B|A)}{P(B)}$$

Bayes' Rule:
$$P(C|A) = \frac{P(C)P(A|C)}{P(A)}$$

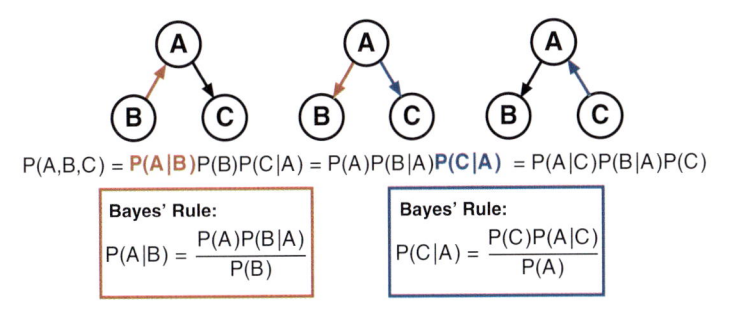

**Figure 13.6** Equivalence classes in Bayesian networks. Three BNs which form an equivalence class are shown, above their factorings of the joint probability distribution. Each application of Bayes' rule transforms one factoring into another mathematically equivalent factoring, whose BN has a link in the opposite direction.

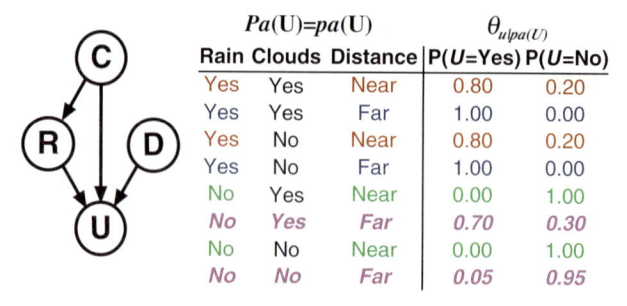

| $Pa(U)=pa(U)$ | | | $\theta_{u|pa(U)}$ | |
|---|---|---|---|---|
| Rain | Clouds | Distance | P(U=Yes) | P(U=No) |
| Yes | Yes | Near | 0.80 | 0.20 |
| Yes | Yes | Far | 1.00 | 0.00 |
| Yes | No | Near | 0.80 | 0.20 |
| Yes | No | Far | 1.00 | 0.00 |
| No | Yes | Near | 0.00 | 1.00 |
| No | Yes | Far | 0.70 | 0.30 |
| No | No | Near | 0.00 | 1.00 |
| No | No | Far | 0.05 | 0.95 |

**Figure 13.7** Parameters in Bayesian networks. On the left is a BN representing a slightly more complicated version of the umbrella example from the main text: A fourth factor, the distance to be travelled (D) is added. This changes the statistical dependence relationships such that umbrella (U) is now dependent on all three other variables. The table shows a potential set of parameters for the umbrella node. Each row represents a possible combination of umbrella's parents' values, shown on the left of the table. The right of each row shows the probability that an umbrella is either carried, P(U = Yes), or not, P(U = No), depending on the parent values to the left. Distance influences umbrella such that short distances are less likely to have an umbrella, even in rain. Distance also influences umbrella's relationship with clouds. It is only when distance is far that presence of clouds affect umbrella, leading to a higher probability of carrying an umbrella without rain. This is a combinatoric type of relationship. It can be seen in the highlighted purple rows of the table: only these rows have different values depending on clouds, all other combinations of rain and distance have identical probabilities regardless of clouds (pairs of rows in red, blue, and green).

parameters can be set independently, a discrete BN can represent many types of relationships between variables, including nonlinear, nonadditive, and even arbitrary combinatoric (Figure 13.7).

### 13.5.2
### Learning Bayesian Networks from Data

Early use of Bayesian networks was as expert systems, where a human expert encodes their knowledge of a system in a BN, which is then used to make predictions (Pearl, 1988). However, the last two decades has seen the growth of structure learning in BNs (i.e., using multiple measurements of variables to determine what BN structure represents the statistical dependencies present in the data). Historically, two approaches have been taken to this problem: (i) performing tests to identify all conditional independencies present in the data (e.g., Spirtes and Glymour, 1991; Cheng, Bell, and Liu, 1997; Bach and Jordan, 2002; Margaritis, 2005) and (ii) calculating a score for how well a network describes the data and searching for the best scoring network (e.g., Heckerman, Geiger, and Chickering, 1995). Cowell (2001) has shown that the two methods can be mathematically equivalent and the latter has generally been applied to biological problems (Friedman, 2004; Markowetz and Spang, 2007).

BN score calculation can be further subdivided into those which aim to provide a most concise description of both the model and the data (minimum description length, MDL (Suzuki, 1993; Lam and Bacchus, 1994) and minimum message length (Wallace, Korb, and Dai, 1996)), and those which calculate a Bayesian scoring metric

(BSM) that estimates the probability of the graph given the data, P(G|D) (Buntine, 1991; Cooper and Herskovits, 1992; Spiegelhalter et al., 1993; Heckerman, Geiger, and Chickering, 1995; Friedman, Murphy, and Russell, 1998; Bach and Jordan, 2002; Imoto et al., 2003). Generally, the BSM is represented as a log of this probability, and, expanded via Bayes' rule, can be written as:

$$\text{BSM}(G : D) = \log P(G|D) = \log P(D|G) + \log P(G) - \log P(D).$$

In practice, P(D) is left uncalculated as it is constant when comparing networks on the same data. P(G) is known as the "prior over graphs" and can be used to incorporate information about what networks are more likely, as we will see later. However, when there is no reason to prefer some structures over others, this, too, is left uncalculated. This leaves P(D|G), to be either calculated or approximated. As an exemplar, we describe the Bayesian Dirichlet equivalent (BDe) method (Buntine, 1991; Cooper and Herskovits, 1992; Spiegelhalter et al., 1993; Heckerman, Geiger, and Chickering, 1995), which is a direct calculation of P(D|G).

The BDe works on discrete data and provides a metric that determines how well a particular graph structure describes the data, given all possible settings of the graph's parameters. This provides the structure that is most robust to deviation in its parameters. Heckerman, Geiger, and Chickering (1995) provide a detailed derivation of the score; here, we will simply present the score in order to discuss its features. The BDe is calculated as:

$$\text{BDe}(G : D) = \log \left( \prod_{i=1}^{n} \prod_{j=i}^{q_i} \left\{ \frac{\Gamma(\alpha_{ij})}{\Gamma(\alpha_{ij} + N_{ij})} \prod_{k=1}^{r_i} \frac{\Gamma(\alpha_{ijk} + N_{ijk})}{\Gamma(\alpha_{ijk})} \right\} \right),$$

where $n$ is the number of variables, $r_i$ is the number of values of each variable, $q_i$ is the number of sets of parent values for each variable, $\Gamma(\cdot)$ is the gamma function (an extension of the factorial for noninteger values), $N_{ij}$ and $N_{ijk}$ are "sufficient statistics" (counts of the number of times the parents of a variable are in each of their possible states and of the number of times the variable takes on each of its values with the parents having particular states, respectively), and $\alpha_{ij}$ and $\alpha_{ijk}$ are calculated from an "equivalent sample size" (*ess*) parameter we provide (Heckerman, Geiger, and Chickering, 1995).

To provide an understanding of how this score relates to the data, consider the terms inside the final two products. The rightmost product term, $\Gamma(\alpha_{ijk} + N_{ijk})/\Gamma(\alpha_{ijk})$, is higher when the counts are more concentrated in a particular child value. This indicates that the scores are better when parent values are more useful for predicting the child, matching with the definition of statistical dependence. The other product term, $\Gamma(\alpha_{ij})/\Gamma(\alpha_{ij} + N_{ij})$, is higher with (i) fewer parent values and (ii) equal examples of each value. The first feature is an inherent penalty for complexity that only allows more complex networks (i.e., more parents) when the predictive value from an additional parent (rightmost term) can counterbalance the cost of adding an additional parent (leftmost term) (Yu, 2005). The second feature indicates that the score is higher when the data are distributed more evenly across their possible values.

In addition to the actual data, two other factors influence this score: the value used for *ess* and the number of discrete states of the variables. The *ess* can be thought of as representing how many data points worth of belief there is that "anything can happen." A higher *ess* means the score for any interaction is higher and thus it is easier for any link to be found; usually low values (e.g., 1–2) are used to put the burden on the data for revealing interactions. More discrete states of variables can lead to more detailed predictive power from parent to child, but also leads to more parent values and more sensitivity to noise. Since these features affect accuracy in opposing ways, when turning originally continuous biological data into discrete states, we need to balance using more discrete states to capture detail and fewer state to give statistical power (Yu *et al.*, 2004).

With a scoring method of evaluating BNs, the problem is then to find which network structure has the highest score. However, because the number of potential networks grows super-exponentially with the number of variables, finding the one with the highest score is computational intractable (Chickering, 1996; Chickering, Heckerman, and Meek, 2004). While efforts continue to find more efficient methods to determine the best BN for a dataset, they are in general too time consuming to be practical (a 20-variable network took over 2 days on a supercomputer; Ott, Imoto, and Miyano, 2004). Instead, heuristic search methods (Figure 13.8) are employed, such as

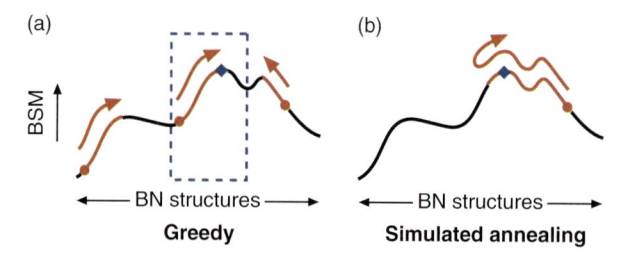

**Figure 13.8** Heuristic search methods. Heuristic search methods are a way of finding a good solution when it is impossible or impractical to directly calculate the one best solution. They consist of a set of rules ("heuristics") for moving from one potential solution to another, with the aim of finding successively better solutions. Two common heuristic search methods are illustrated – greedy search and simulated annealing. The plots of BN structures versus BSM are a conceptual visualization to show that BNs more similar in structure will have more similar scores (in reality, the solution space of potential networks is multidimensional, not linear). (a) A greedy search begins with a BN (red circle) and evaluates possible changes in structure (e.g., adding or removing links), accepting only those changes that increase the score. Thus, greedy search will climb hills in the solution space (red portions of plot show solutions visited; red arrows show path of search). Since the solution space is often bumpy, such that dips can occur before higher points (as on the right of our hypothetical solution space), greedy search must be run many times to increase chances of finding high scores. For example, the highest point on this plot (blue diamond) can only be found by a greedy search that starts on the slopes up to this peak (highlighted by dashed blue box). (b) In contrast, simulated annealing will occasionally move to a lower scoring solution, enabling crossing dips between neighboring hills. The probability of accepting lower scores decreases throughout the progress of a simulated annealing search, such that the search ranges widely early on, exploring many possibilities, before settling on a particular hill to climb. The example here illustrates a simulated annealing search progressing up and down in score, before climbing to the highest point visited.

greedy search (Yu *et al.*, 2004), simulated annealing (Hartemink *et al.*, 2002; Wang, Touchman, and Xue, 2004; Yu *et al.*, 2004), or genetic algorithms (Yu *et al.*, 2004; Auliac *et al.*, 2008). Additionally, model averaging approaches have been developed to either provide a summary of "best" solutions from a heuristic search (Hartemink *et al.*, 2002) or to provide a fully Bayesian picture of the probability of each link being in the solution (Friedman and Koller, 2003; Husmeier, 2003; Luna *et al.*, 2007).

### 13.5.3
### Bayesian Networks and Microarrays

Friedman *et al.* (2000) first applied Bayesian networks to microarray data with the intent to recover transcriptional regulatory networks. This was quickly followed by a number of similar applications (Hartemink *et al.*, 2001, 2002; Pe'er *et al.*, 2001; Imoto, Goto, and Miyano, 2002; Ong, Glasner, and Page, 2002; Pe'er, Regev, and Tanay, 2002). Since then, many researchers have developed and improved BNs specifically for application to gene expression data (Kim, Imoto, and Miyano, 2003; Friedman, 2004; Markowetz and Spang, 2007).

Bayesian networks indicate only the presence of a statistical dependence among variables and no information about the type of interaction (as was provided in the coefficients of previous equation-based methods of modeling gene regulatory networks; e.g., D'haeseleer *et al.*, 1999; Weaver, Workman, and Stromo, 1999). To provide such information relevant to gene regulation, Hartemink *et al.* (2001) developed a technique of annotating edges in a Bayesian network to indicate that the relationship was either upregulation ($+$), downregulation ($-$), or nonmonotonic (neither positive nor negative, e.g., U-shaped or combinatoric). Yu *et al.* (2004) further developed an *influence score* to provide both the sign ($+/-$) and magnitude (scaled 0–1) of the relationship between two genes.

Loops and causation are difficult to evaluate when applying BNs to biological data, due to the acyclicity and equivalence class issues addressed above. However, even before BNs had been applied to microarrays, Murphy and Mian (1999) suggested that dynamic Bayesian networks could provide both cycles and causal information – provided sufficient time series microarray data were available. A DBN models variables at two (or more) points in time, allowing a BN to connect variables in the past to those in the future, and thus models loops in time without any cycles in the underlying BN. Additionally, equivalence classes of BNs play no role, because the temporal sequence indicates direction of causality from past to future (Figure 13.9). Several simulation studies showed that DBNs could effectively model loops and causation in gene regulatory networks (Smith, Jarvis, and Hartemink, 2002, 2003; Husmeier, 2003; Yu *et al.*, 2004), and they have been applied to real data to recover gene regulatory networks (Ong, Glasner, and Page, 2002; Perrin *et al.*, 2003; Kim, Imoto, and Miyano, 2004; Zou and Conzen, 2005; Luna *et al.*, 2007; Sun and Hong, 2007). DBNs require time series data, which while currently increasing, are still the minority of microarray experiments (Simon *et al.*, 2005); thus, analysis of gene regulatory networks is likely to continue to make use of both DBNs and BNs.

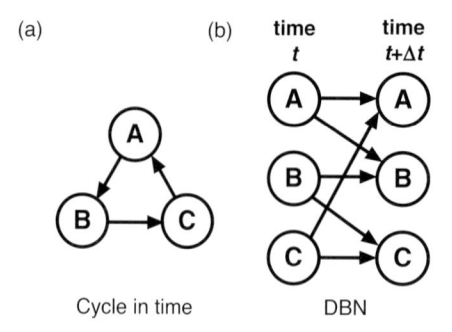

**Figure 13.9** Dynamic Bayesian network. A loop or cycle in time (a) can be represented by an acyclic BN (b) representing each variable at two points in time. Causation is inferred through the temporal relationship, such that links from variables at time $t$ to those at time $t + \Delta t$ are interpreted as causal influences.

Microarray data tend to have low data amounts relative to the demands of BNs (Heckerman, Geiger, and Chickering, 1995; Yu et al., 2004; Wang, Chen, and Cloutier, 2007). Thus, many techniques have been developed to enable more accurate recovery of BNs in the context of sparse data in biology. Using only provided microarray data, one can postprocess links based on magnitude of influence score (Yu et al., 2004), interpolate in time series (Yu et al., 2004), limit the number of parents per variable considered (Hansen, Ott, and Koentges, 2004; Yu, 2005), and use conditional independence tests to provide better starting networks for heuristic search (Wang, Chen, and Cloutier, 2007). Other methods take advantage of biological intervention experiments (e.g., mutations, knockouts, overexpression, drug interference, etc.), and develop BSMs and search techniques tailored for these types of experiments (Cooper and Yoo, 1999; Pe'er et al., 2001; Pe'er, Regev, and Tanay, 2002; Yoo, Thorsson, and Cooper, 2002; Tamada et al., 2005; Dojer et al., 2006).

The large body of genomics and bioinformatics data provides an additional avenue to improve BN representation of gene regulation. BNs are particularly well suited to make use of this additional information, via the prior over graphs P(G) component of the BSM. This is done by using different values of P(G) based on whether the graph matches with what would be expected from other biological information. At the most extreme, P(G) can be set to zero for structures that conflict with biological knowledge, removing those structures from consideration. For example, Hartemink et al. (2002) made use of DNA binding information from chromatin immunoprecipitation/ microarray ChIP-chip data to consider structures consisting only of links where the regulator gene was known to bind upstream of its target. Finer distinctions can be made by providing a probability value for P(G). Imoto et al. (2004) used higher values of P(G) for links supported by DNA binding data and lower values for those not; in this way, strong evidence from the gene expression data could overcome lack of prior knowledge. Bernard and Hartemink (2005) explicitly incorporated confidence in the DNA binding data, such that evidence from more trusted data sources provided higher values for P(G).

P(G) can incorporate other types of biological knowledge. Presence of known binding motifs can suggest potential regulators (Imoto et al., 2004); enrichment in motifs across a number of genes can suggest links to a common regulator (Tamada et al., 2003). Tamada et al. (2005) developed a method to adjust P(G) based on comparative analysis of phylogenetic relationships and Imoto et al. (2006) used results of gene underexpression to inform P(G). Werhli and Husmeier (2007) have developed an extensible framework to incorporate multiple sources of prior biological knowledge into a P(G) value simultaneously.

Biological knowledge can also be incorporated into BNs in other ways. For example, protein-protein interactions can suggest genes that could be combined into a single node to represent a protein complex in the network (Imoto et al., 2004; Nariai et al., 2004). Biological literature can be used to assist in the heuristic search: Djebbari and Quackenbush (2008) used known interactions to provide good starting BNs for a heuristic search; and Lee and Lee (2005) used gene annotations to subdivide genes into functional modules, building networks from the smaller number of genes in each module.

With all of these developments tailoring BNs to microarray data, BNs are now being employed to discover novel biological knowledge. BNs have been employed to reveal genome scale networks in *Plasmodium* for understanding parasite biology (Date and Stoeckert, 2006) and in adipose tissue for understanding disease traits (Sieberts and Schadt, 2007). BNs have also been used to explore networks related to specific biological functions, such as nitric oxide response in yeast (Zhu et al., 2006), development in *C. elegans* (Sun and Hong, 2007), and toxicity of acetaminophen (Toyoshiba et al., 2006). Rodriguez-Zas et al. (2008) used BNs to reveal multiple small functional networks from a genome-scale analysis of mammalian embryonic cells.

Networks recovered by BNs have been used for discovery of genes involved in biological processes or disease and to suggest potential drug targets. BNs have been used to identify candidate genes involved in adipose metabolism in pigs (Li et al., 2008), Alzheimer's disease (Liu et al., 2006), alcoholism (Matthäus et al., 2009), and immune response to herpesvirus (Takaku et al., 2005). BNs have been used to suggest therapeutic targets in leukemia (Dejori, Schuermann, and Stetter, 2004) and autoimmune diseases (Palacios et al., 2007). Savoie et al. (2003) applied BNs to yeast data to suggest alternate targets for antifungal drugs.

The structure of recovered BNs has also been used to identify cis-regualtory motifs specific to different human tissues (Chen and Blanchette, 2008) and related to breast cancer (Niida et al., 2008). Finally, BNs have been used to create predictive models: Toyoshiba et al. (2004) used BNs to model effects of carcinogens, and Gevaert et al. (2006) used BNs to help predict breast cancer based on both microarrays and clinical data.

Thus, as investigations into biological processes involve more genome-wide expression analyses and the role of networks of functional interactions becomes more apparent, BNs provide a promising method for discovery of relevant biological knowledge.

## 13.6
## Final Considerations

As mentioned above, there are presently only few expression profiling studies performed on a timescale. Some reasons for this strategy may reside in restricted financial support, because more experiments are required in this context. Additional, insurmountable problems exist when the human brain is to be investigated. In this case, animal studies may be extremely valuable when carried out at a number of different timepoints. Plus, it is not true that this strategy requires many more single experiments with many more $n$-numbers. From a statistical point of view, $n$-numbers can be reduced with increasing numbers of timepoints without loosing statistical power. When using only one timepoint (i.e., disease (treated) versus control) typically between six and 12 samples per group are needed. The decision to investigate only one timepoint encompasses the major drawback that all samples are in distinct states (of disease), and the need for more samples per group. This is particularly compounding in human studies, where patients who are of different age, and of course at different stages of the disease individually, are investigated. To obtain sufficient statistical power, samples of those people have to be pooled, which rather confounds than sharpens the real situation. The resultant standard deviations observed in all statistical evaluations are evidence of that problem. Even though individuals may be closely related, such as rats or mice from the same litter, and undergo the very same treatments, the individual measurements of any parameter will reveal statistical variation that is, to a large degree, caused by biology and not inaccurate (technological) data acquisition.

Introduction of two or three more timepoints would require approximately half $n$-numbers per timepoint. This measure increases the total number of experiments, but also provides more detailed knowledge about the progression of the biological processes. Exactly this point, as touched upon already above, is crucial for an understanding of dynamic changes of a system. A series of time-course investigations reflects individual changes over time that can be compared with each other to find corresponding changes in the other samples. In this way, it may be possible to identify the "true" disease-related "key players," that are often hidden in the "noise" of data and hence are statistically not significant.

Thinking along these lines, it turns out to be even more complicated. Disease-related "key players" may pass the baton during disease development and progression. Therefore, even though we are successful in finding the culprit(s) in the noise, we have not learned much about the dynamics of the disease.

If one, then, analyzes expression patterns at different timepoints, the significantly regulated genes ("peaks of the iceberg") are supposedly different at each time, but possibly partially overlapping, which shows in changing "peak" patterns. This indicates that the culprits in the noise likely change, too (Figure 13.10).

This inevitably leads us back to the above-mentioned nonlinearity of biological systems. Probability-based methods, such as BNs as described here, have the advantage to be used in both linear and nonlinear relationships, and they are noise-tolerant. Inherent limitations in these methods are data discretization,

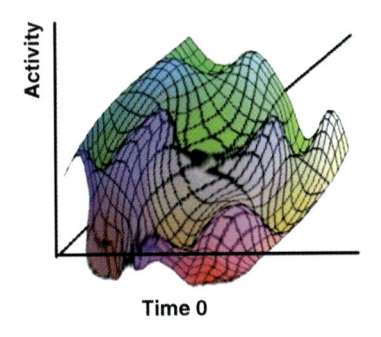

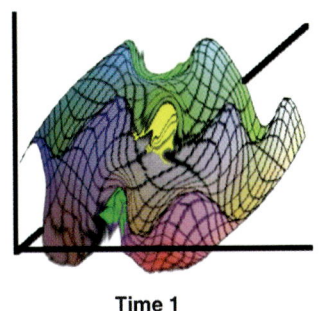

**Figure 13.10** Dynamic molecular networks. Consider activity of genes laid out for visualization in a plane, assuming similar acting genes are neighbors, and plotted at two time points. Because it is not independent genes that are changing over time, but a dynamic network connecting many processes–both those related to the questtion of interest and those unrelated– the entire surface shifts between time points.

pair-wise testing, and the focus on correlations rather than causality. An interesting approach to overcome these drawbacks has been elaborated recently (Luo, Hankenson, and Woolf, 2008). Other, additional promising computational strategies in this direction have also been developed (Ho et al., 2007; Hache et al., 2009).

Figure 13.10 shows that we also have to take into consideration up- and down-regulated genes – a feature that may also change over time.

This said, we can conclude that the development of a disease is individually distinct – it depends on both genetic predisposition and environmental impacts that emboss trillions of engrams. Subtle changes of key players influenced by environmental (epigenetic) cues may result in:

i) Dramatic acceleration of disease progression spiraling into catastrophe.
ii) Collapse of disease progression into a "healthy" state.
iii) Myriads of intermediary states between those two extremes.

## References

Auliac, C., Frouin, V., Gidrol, X., and d'Alché-Buc, F. (2008) Evolutionary approaches for the reverse-engineering of gene regulatory networks: a study on a biologically realistic dataset. *BMC Bioinformatics*, **9**, 91.

Ayroles, J.F. and Gibson, G. (2006) Analysis of variance of microarray data. *Methods in Enzymology*, **411**, 214–233.

Bach, F.R. and Jordan, M.I. (2002) Learning graphical models with Mercer kernels, in *Advances in Neural Information Processing Systems 15* (eds. S. Becker, S. Thrun, and K. Obermayer), MIT Press, Cambridge, MA, pp. 1033–1040.

Bernard, A. and Hartemink, A.J. (2005) Informative structure priors: joint learning of dynamic regulatory networks from multiple types of data, in *Proceedings of the Pacific Symposium on Biocomputing* (eds. R.B. Altman, A.K. Dunker, and L. Hunter), World Scientific, Singapore, pp. 459–470.

Buntine, W.L. (1991) Theory refinement of Bayesian networks, in *Proceedings of the 7th Annual Conference on Uncertainty in Artificial Intelligence* (eds. B. D'Ambrosio and P. Smets), Morgan Kaufmann, San Francisco, CA, pp. 52–60.

Chen, X. and Blanchette, M. (2008) Prediction of tissue-specific *cis*-regulatory modules using Bayesian networks and regression trees. *BMC Bioinformatics*, **8**, S2.

Cheng, J., Bell, D.A., and Liu, W. (1997) Learning belief networks from data: an information theory based approach, in *Proceedings of the Sixth International Conference on Information and Knowledge Management*, ACM Press, New York, pp. 325–331.

Chickering, D.M. (1996) Learning Bayesian networks is NP-complete, in *Learning from Data: AI and Statistics V* (eds. D. Fisher and H.-J. Lenz), Springer, New York, pp. 121–130.

Chickering, D.M., Heckerman, D., and Meek, C. (2004) Large-sample learning of Bayesian networks is NP-Hard. *Journal of Machine Learning Research*, **5**, 1287–1330.

Cooper, G.F. and Herskovits, E. (1992) A Bayesian method for the induction of probabilistic networks from data. *Machine Learning*, **9**, 309–347.

Cooper, G.F. and Yoo, C. (1999) Causal discovery from a mixture of experimental and observational data, in *Proceedings of the 15th Conference on Uncertainty in Artificial Intelligence* (eds. K. Laskey and H. Prade), Morgan Kaufman, San Francisco, CA, pp. 116–125.

Cowell, R.G. (2001) Conditions under which conditional independence and scoring methods lead to identical selection of Bayesian network models, in *Proceedings of the 17th Conference in Uncertainty in Artificial Intelligence*, Morgan Kaufmann, San Francisco, CA, pp. 91–97.

D'haeseleer, P., Wen, X., Fuhrman, S., and Somogyi, S.R. (1999) Linear modeling of mRNA expression levels during CNS development and injury, in *Proceedings of the Pacific Symposium on Biocomputing* (eds. R.B. Altman and K. Lauderdale), World Scientific, Singapore, pp. 41–52.

Date, S.V. and Stoeckert, C.J. Jr. (2006) Computational modeling of the *Plasmodium falciparum* interactome reveals protein function on a genome-wide scale. *Genome Research*, **16**, 542–549.

Dejori, M., Schuermann, B., and Stetter, M. (2004) Hunting drug targets by systems-level modeling of gene expression profiles. *IEEE Transactions on Nanobioscience*, **3**, 180–191.

Djebbari, A. and Quackenbush, J. (2008) Seeded Bayesian networks: Constructing genetic networks from microarray data. *BMC Systems Biology*, **2**, 57.

Dojer, N., Gambin, A., Mizera, A., Wilczynski, B., and Tiuryn, J. (2006) Applying dynamic Bayesian networks to perturbed gene expression data. *BMC Bioinformatics*, **7**, 249.

Eisen, M.B., Spellman, P.T., Brown, P.O., and Botstein, D. (1998) Cluster analysis and display of genome-wide expression patterns. *Proceedings of the National Academy of Sciences of the United States of America*, **95**, 14863–14868.

Friedman, N. (2004) Inferring cellular networks using probabilistic graphical models. *Science*, **303**, 799–805.

Friedman, N. and Koller, D. (2003) Being Bayesian about network structure: a Bayesian approach to structure discovery in Bayesian networks. *Machine Learning*, **50**, 95–126.

Friedman, N., Murphy, K., and Russell, S. (1998) Learning the structure of dynamic probabilistic networks, in *Proceedings of the 14th Annual Conference on Uncertainty in Artificial Intelligence* (eds. G.F. Cooper, and S. Moral), Morgan Kaufmann, San Francisco, CA, pp. 139–147.

Friedman, N., Linial, M., Nachman, I., and Pe'er, D. (2000) Using Bayesian networks to analyze expression data. *Journal of Computational Biology*, **7**, 601–620.

Gan, G., Ma, C., and Wu, J. (2007) *Data Clustering: Theory, Algorithms, and Applications, ASA/SIAM Series on Statistics and Applied Probability*, SIAM, Philadelphia, PA.

Gass, P., Leonardi-Essmann, F., Zueger, M., Spanagel, R., and Gebicke-Haerter, P.J. (2008) Transcriptional changes in insulin- and lipid metabolism-related genes in the hippocampus of olfactory bulbectomized mice. *Journal of Neuroscience Research*, **86**: 3184–3193.

Gevaert, O., De Smet, F., Timmerman, D., Moreau, Y., and De Moor, B. (2006) Predicting the prognosis of breast cancer by integrating clinical and microarray data with Bayesian networks. *Bioinformatics*, **22**, e184–e190.

Hache, H., Wierling, C., Lehrach, H., and Herwig, R. (2009) GeNGe: systematic generation of gene regulatory networks. *Bioinformatics*, **25**, 1205–1207.

Hand, D.J. and Heard, N.A. (2005) Finding groups in gene expression data. *Journal of Biomedicine and Biotechnology*, **2**, 215–225.

Hansen, A., Ott, S., and Koentges, G. (2004) Increasing feasibility of optimal gene network estimation. *Genome Informatics*, **15**, 141–150.

Hartemink, A.J., Gifford, D.K., Jaakkola, T.S., and Young, R.A. (2001) Using graphical models and genomic expression data to statistically validate models of genetic regulatory networks, in *Proceedings of the Pacific Symposium on Biocomputing* (eds. R.B. Altman, A.K. Dunker, L. Hunker, K. Lauderdale, and T.E.D. Klein), World Scientific, Singapore, pp. 422–433.

Hartemink, A.J., Gifford, D.K., Jaakkola, T.S., and Young, R.A. (2002) Combining location and expression data for principled discovery of genetic regulatory network models, in *Proceedings of Pacific Symposium on Biocomputing* (eds. R.B. Altman, A.K. Dunker, L. Hunter, and K. Lauderdale), World Scientific, Singapore, pp. 437–449.

Hartigan, J.A. (1975) *Clustering Algorithms*, John Wiley & Sons, Inc., New York.

Heckerman, D., Geiger, D., and Chickering, D.M. (1995) Learning Bayesian networks: the combination of knowledge and statistical data. *Machine Learning*, **20**, 197–243.

Ho, S.Y., Hsieh, C.H., Yu, F.C., and Huang, H.L. (2007) An intelligent two-stage evolutionary algorithm for dynamic pathway identification from gene expression profiles. *IEEE/ACM Transactions on Computational Biology and Bioinformatics*, **4**, 648–660.

Husmeier, D. (2003) Sensitivity and specificity of inferring genetic regulatory interactions from microarray experiments with dynamic Bayesian networks. *Bioinformatics*, **19**, 2271–2282.

Imoto, S., Goto, T., and Miyano, S. (2002) Estimation of genetic networks and functional structures between genes by using Bayesian networks and nonparametric regression, in *Proceedings of Pacific Symposium on Biocomputing* (eds. R.B. Altman, A.K. Dunker, L. Hunter, and K. Lauderdale), World Scientific, Singapore, pp. 175–186.

Imoto, S., Higuchi, T., Goto, T., Tashiro, K., Kuhara, S., and Miyano, S. (2004) Combining microarrays and biological knowledge for estimating gene networks via Bayesian networks. *Journal of Bioinformatics and Computational Biology*, **2**, 77–98.

Imoto, S., Kim, S., Goto, T., Miyano, S., Aburatani, S., Tashiro, K., and Kuhara, S. (2003) Bayesian network and nonparametric heteroscedastic regression for nonlinear modeling of genetic network. *Journal of Bioinformatics and Computational Biology*, **1**, 231–252.

Imoto, S., Tamada, Y., Araki, H., Yasuda, K., Print, C.G., Charnock-Jones, S.D., Sanders, D., Savoie, C.J., Tashiro, K., Kuhara, S., and Miyano, S. (2006) Computational strategy for discovering druggable gene networks from genome-wide RNA expression profiles, in *Proceedings of the Pacific Symposium on Biocomputing* (eds. R.B. Altman, A.K. Dunker, L. Hunter, T. Murray, and T.E. Klein), World Scientific, Singapore, pp. 559–571.

Jolliffe, I.T. (2002) *Principal Component Analysis*, 2nd edn, Springer Series in Statistics, Springer, New York.

Jung, Y., Park, H., Du, D.-Z., and Drake, B.L. (2003) A decision criterion for the optimal number of clusters in hierarchical clustering. *Journal of Global Optimization*, **25**, 91–111.

Kerr, M.K. and Churchill, G.A. (2001) Statistical design and the analysis of gene expression microarrays. *Genetical Research*, **77**, 123–128.

Kerr, M.K., Martin, M., and Churchill, G.A. (2000) Analysis of variance for gene expression microarray data. *Journal of Computational Biology*, **7**, 819–837.

Kim, S., Imoto, S., and Miyano, S. (2004) Dynamic Bayesian network and nonparametric regression for nonlinear modeling of gene networks from time series gene expression data. *BioSystems*, **75**, 57–65.

Kim, S.Y., Imoto, S., and Miyano, S. (2003) Inferring gene networks from time series microarray data using dynamic Bayesian networks. *Briefings in Bioinformatics*, **4**, 228–235.

Klipp, E., Herwig, R., Kowald, A., Wierling, C., and Lehrach, H. (2005) *Systems Biology in Practice*, Wiley-VCH Verlag GmbH, Weinheim.

Lam, W. and Bacchus, F. (1994) Learning Bayesian belief networks: an approach based on the MDL principle. *Computational Intelligence*, **10**, 269–293.

Lee, P.H. and Lee, D. (2005) Modularized learning of genetic interaction networks from biological annotations and mRNA expression data. *Bioinformatics*, **21**, 2739–2747.

Li, M., Zhu, L., Li, X., Shuai, S., Teng, X., Xiao, H., Li, Q., Chen, L., Guo, Y., and Wang, J. (2008) Expression profiling analysis for genes related to meat quality and carcass traits during postnatal development of backfat in two pig breeds. *Science in China C: Life Sciences*, **51**, 718–733.

Liu, B., Jiang, T., Ma, S., Zhao, H., Li, J., Jiang, X., and Zhang, J. (2006) Exploring candidate genes for human brain diseases from a brain-specific gene network. *Biochemical and Biophysical Research Communications*, **349**, 1308–1314.

Luna, T., Huang, Y., Yin, Y., Padillo, D.P.R., and Perez, M.C.C. (2007) Uncovering gene regulatory networks from time-series microarray data with variational Bayesian structural expectation maximization. *EURASIP Journal on Bioinformatics and Systems Biology*, 71312.

Luo, W., Hankenson, K.D., and Woolf, P.J. (2008) Learning transcriptional regulatory networks from high throughput gene expression data using continuous three-way mutual information. *BMC Bioinformatics*, **9**, 467.

Margaritis, D. (2005) Distribution-free learning of Bayesian network structure in continuous domains, in *Proceedings of the 20th National Conference on Artificial Intelligence*, AAAI, Menlo Park, CA, pp. 825–830.

Markowetz, F. and Spang, R. (2007) Inferring cellular networks – a review. *BMC Bioinformatics*, **8**, S5.

Matthäus, F., Smith, V.A., Fogtman, A., Sommer, W.H., Leonardi-Essmann, F., Lourdusamy, A., Reimers, M.A., Spanagel, R., and Gebicke-Haerter, P.J. (2009) Interactive molecular networks obtained by computer-aided conversion of microarray data from brains of alcohol-drinking rats. *Pharmacopsychiatry*, **42** (Suppl. 1), S118–S128.

Milligan, G.W. and Cooper, M.C. (1985) An examination of procedures for determining the number of clusters in a data set. *Psychometrika*, **50**, 159–179.

Murphy, K. and Mian, S. (1999) Modeling gene expression data using dynamic Bayesian networks. Technical Report, Computer Science Division, University of California, Berkeley, CA.

Nariai, N., Kim, S., Imoto, S., and Miyano, S. (2004) Using protein-protein interactions for refining gene networks estimated from microarray data by Bayesian networks, in *Proceedings of the Pacific Symposium on Biocomputing* (eds. R.B. Altman, A.K. Dunker, L. Hunter, and T.A. Jung), World Scientific, Singapore, pp. 336–347.

Niida, A., Smith, A.D., Imoto, S., Tsutsumi, S., Aburatani, H., Zhang, M.Q., and Akiyama, T. (2008) Integrative bioinformatics analysis of transcriptional regulatory programs in breast cancer cells. *BMC Bioinformatics*, **9**, 404.

Ong, I.M., Glasner, J.D., and Page, D. (2002) Modelling regulatory pathways in *E. coli* from time series expression profiles. *Bioinformatics*, **18**, S241–S248.

Ott, S., Imoto, S., and Miyano, S. (2004) Finding optimal models for small gene networks, in *Proceedings of the Pacific Symposium on Biocomputing* (eds. R.B. Altman, A.K. Dunker, L. Hunter, T.A. Jung, and T.E.D. Klein), World Scientific, Singapore, pp. 557–567.

Palacios, R., Goni, J., Martinez-Forero, I., Iranzo, J., Sepulcre, J., Melero, I., and Villoslada, P. (2007) A network analysis of the human T-cell activation gene network identifies Jagged1 as a therapeutic target for autoimmune diseases. *PLoS ONE*, **2**, e1222.

Pe'er, D., Regev, A., Elidan, G., and Friedman, N. (2001) Inferring subnetworks from perturbed expression profiles. *Bioinformatics*, **17**, S215–S224.

Pe'er, D., Regev, A., and Tanay, A. (2002) Minreg: inferring an active regulator set. *Bioinformatics*, **18**, S258–S267.

Pearl, J. (1988) *Probabilistic Reasoning in Intelligent Systems*, Morgan Kaufmann, San Francisco, CA.

Perrin, B.-E., Ralaivola, L., Mazurie, A., Bottani, S., Mallet, J., and d'Alche-Buc, F. (2003) Gene networks inference using dynamic Bayesian networks. *Bioinformatics*, **19**, ii138–ii148.

Rodriguez-Zas, S.L., Ko, Y., Adams, H.A., and Southey, B.R. (2008) Advancing the understanding of the embryo transcriptome co-regulation using meta-, functional, and gene network analysis tools. *Reproduction*, **135**, 213–224.

Savoie, C.J., Aburatani, S., Watanabe, S., Eguchi, Y., Muta, S., Imoto, S., Miyano, S., Kuhara, S., and Tashiro, K. (2003) Use of gene networks from full genome microarray libraries to identify functionally relevant drug-affected genes and gene regulation cascades. *DNA Research*, **10**, 19–25.

Sieberts, S.K., and Schadt, E.E. (2007) Moving toward a system genetics view of disease. *Mammalian Genome*, **18**, 389–401.

Simon, I., Siegfried, Z., Ernst, J., and Bar-Joseph, Z. (2005) Combined static and dynamic analysis for determining the quality of time-series expression profiles. *Nature Biotechnology*, **23**, 1503–1508.

Smith, V.A., Jarvis, E.D., and Hartemink, A.J. (2002) Evaluating functional network inference using simulations of complex biological systems. *Bioinformatics*, **18**, S216–S224.

Smith, V.A., Jarvis, E.D., and Hartemink, A.J. (2003) Influence of network topology and data collection on network inference, in *Proceedings of the Pacific Symposium on Biocomputing* (eds. R.B. Altman, A.K. Dunker, L. Hunter, and T.A. Jung), World Publishing, Singapore, pp. 164–175.

Spiegelhalter, D., Dawid, A.P., Lauritzen, S.L., and Cowell, R.G. (1993) Bayesian analysis in expert systems. *Statistical Science*, **8**, 219–283.

Spirtes, P., and Glymour, C. (1991) An algorithm for fast recovery of sparse causal graphs. *Social Science Computing Reviews*, **9**, 62–72.

Sun, X., and Hong, P. (2007) Computational modeling of *Caenorhabditis elegans* vulval induction. *Bioinformatics*, **23**, i499–i507.

Suzuki, J. (1993) A construction of Bayesian networks from databases based on an MDL principle, in *Proceedings of the 9$^{th}$ Annual Conference on Uncertainty in Artificial Intelligence* (eds. D. Heckerman and E.H. Mamdani), Morgan Kaufmann, San Francisco, CA, pp. 266–273.

Takaku, T., Ohyashiki, J.H., Zhang, Y., and Ohyashiki, K. (2005) Estimating immunoregulatory gene networks in human herpesvirus type 6-infected T cells. *Biochemical and Biophysical Research Communications*, **336**, 469–477.

Tamada, Y., Bannai, H., Imoto, S., Katayama, T., Kanehisa, M., and Miyano, S. (2005) Utilizing evolutionary information and gene expression data for estimating gene networks with Bayesian network models. *Journal of Bioinformatics and Computational Biology*, **3**, 1295–1313.

Tamada, Y., Kim, S., Bannai, H., Imoto, S., Tashiro, K., Kuhara, S., and Miyano, S. (2003) Estimating gene networks from gene expression data by combining Bayesian network model with promoter element detection. *Bioinformatics*, **19**, ii227–ii236.

Toyoshiba, H., Sone, H., Yamanaka, T., Parham, F.M., Irwin, R.D., Boorman, G.A., and Portier, C.J. (2006) Gene interaction network analysis suggests differences between high and low doses of acetaminophen. *Toxicology and Applied Pharmacology*, **215**, 306–316.

Toyoshiba, H., Yamanaka, T., Sone, H., Parham, F.M., Walker, N.J., Martinez, J., and Portier, C.J. (2004) Gene interaction network suggests dioxin induces a significant linkage between aryl hydrocarbon receptor and retinoic acid receptor beta. *Environmental Health Perspectives*, **112**, 1217–1224.

Wall, M.E., Rechtsteiner, A., and Rocha, L.M. (2003) Singular value decomposition and principal component analysis, in *A Practical Approach to Microarray Data Analysis* (eds. D.P. Berrar, W. Dubitzky, and M. Granzow), Kluwer, Norwell, MA, pp. 91–109.

Wallace, C., Korb, K.B., and Dai, H. (1996) Causal discovery via MML, *Proceedings of the 13th International Conference on Machine Learning* (ed. L. Saitta), Morgan Kauffman, San Francisco, CA, pp. 516–524.

Wang, M., Chen, Z., and Cloutier, S. (2007) A hybrid Bayesian network learning method for constructing gene networks. *Computational Biology and Chemistry*, **31**, 361–372.

Wang, T., Touchman, J.W., and Xue, G. (2004) Applying two-level simulated annealing on Bayesian structure learning to infer genetic networks, in *Proceedings of the IEEE*

Computational Systems Bioinformatics Conference, IEEE, New York, pp. 647–648.

Weaver, D.C., Workman, C.T., and Stromo, G.D. (1999) Modeling regulatory networks with weight matrices, in *Proceedings of the Pacific Symposium on Biocomputing* (eds. R.B. Altman and K. Lauderdale), World Scientific, Singapore, pp. 112–123.

Werhli, A.V. and Husmeier, D. (2007) Reconstructing gene regulatory networks with Bayesian networks by combining expression data with multiple sources of prior knowledge. *Statistical Applications in Genetics and Molecular Biology*, **6**, 15.

Xu, R. and Wunsch, D.C. (2009) *Clustering, IEEE Press Series on Computational Intelligence*, IEEE-Wiley, New York.

Yoo, C., Thorsson, V., and Cooper, G.F. (2002) Discovery of causal relationships in a gene-regulation pathway from a mixture of experimental and observational DNA microarray data, in *Proceedings of Pacific Symposium on Biocomputing* (eds. R.B. Altman, A.K. Dunker, L. Hunter, and K. Lauderdale), World Scientific, Singapore, pp 498–509.

Yu, J. (2005) Developing Bayesian network inference algorithms to predict causal functional pathways in biological systems. PhD Thesis. Duke University, Durham, NC.

Yu, J., Smith, V.A., Wang, P.P., Hartemink, A.J., and Jarvis, E.D. (2004) Advances to Bayesian network inference for generating causal networks from observational biological data. *Bioinformatics*, **20**, 3594–3603.

Zhu, J., Jambhekar, A., Sarver, A., and DeRis, J. (2006) A Bayesian network driven approach to model the transcriptional response to nitric oxide in Saccharomyces cerevisiae. *PLoS ONE*, **1**, e94.

Zou, M. and Conzen, S.D. (2005) A new dynamic Bayesian network (DBN) approach for identifying gene regulatory networks from time course microarray data. *Bioinformatics*, **21**, 71–79.

# 14
# Biochemical Networks in Psychiatric Disease
*Maria Lindskog, Geir Halnes, Rodrigo F. Oliveira, Jeanette Hellgren Kotaleski, and Kim T. Blackwell*

## 14.1
## Introduction

Effective treatment of psychiatric diseases is rare and even when it exists, it is not certain that we understand how the treatment works. A major cause of this is the unknown etiology of psychiatric diseases; another is the fact that psychiatric diagnosis is usually based on a cluster of symptoms and what is brought together as one disease may have many different biological origins. As an example of the complexity we can consider schizophrenia – a greatly disabling disease that affects almost 1% of the population (Perala *et al.*, 2007). The disease is composed of positive symptoms, such as psychosis and hallucinations, and negative symptoms, including poor attention and lack of motivation. In an attempt to try to understand the neurobiology of schizophrenia, the disease has been tackled at many different levels, including as a chemical imbalance, a genetic disturbance, or a malfunctioning at the network level.

## 14.2
## The Example of Schizophrenia

One of the first neurobiological attempts to explain the disease was the dopamine hypothesis, where schizophrenia is explained by an imbalance in the dopamine system. The rationale for this explanation is that dopamine $D_2$ receptor inhibitors are effective in treating schizophrenia. Although it has proven hard to find changes in the dopaminergic system in patients with schizophrenia compared to healthy individuals, modern imaging techniques have shown that there is an increased sensitivity in the dopaminergic system (Toda and Abi-Dargham, 2007). Although dopamine antagonists are still the most common treatment in schizophrenia, there is evidence for the involvement of other neurotransmitters as well, including glutamate through *N*-methyl-D-aspartate (NMDA) receptors, $\gamma$-aminobutyric acid (GABA)

(Dickinson and Harvey, 2008), and with serotonin as a strong modulatory candidate (Jones and McCreary, 2008).

The exact locus of the disease has also been elusive. From anatomical studies of patients we know that there are morphological abnormalities in patients with schizophrenia, including a smaller hippocampus and enlarged ventricles (Jindal and Keshavan, 2008). At the circuit level, there seems to be differences in connectivity in schizophrenic patients (Harrison and Eastwood, 2001) as well as in animal models of schizophrenia (Mukai et al., 2008). More recent studies involving new techniques looking at network organizations have shown a problem in the hierarchical network organization using imaging (Bassett et al., 2008) and it has also been suggested that γ oscillations, seen during cognition, are disturbed in schizophrenia (Ferrarelli et al., 2008).

Despite the fact that schizophrenia is one of the psychiatric diseases with the highest heritability, the search for susceptibility genes has not yielded the results hoped for. Targeted genes are diverse, including those for dystobrevin-binding protein, neuregulin-1 (*NRG1*), D-amino acid oxidase (*DOA*), regulator of G-protein signaling-4 (*RGS4*), and the metabotropic glutamate receptor (mGluR) 3 receptor (*GRM3*) (Kirov, O'Donovan, and Owen, 2005), and expressed in different systems of the brain, and their identification had not given the insight into the disease that was hoped for.

Although results from all levels give relevant information about what can be a biological problem in schizophrenia, no single level seems to give the whole answer. More recently, there have been attempts to consolidate the findings from different levels to find common systems that are affected. For example, Lisman et al. (2008) suggest that by studying brain circuits and their modulation, it is possible to reconcile a large part of what is known about schizophrenia, and they put the inhibition by hippocampal interneurons in focus.

## 14.3
### Looking for Nodes of Interaction

Schizophrenia is an extreme example in that it is one of the most complicated diseases that affect the brain. However, it is still valuable as an example of how difficult it is to understand the brain and to appreciate all the possible levels of interpretation. The same multilevel analysis as was done above for schizophrenia can be done for other psychiatric diseases. It is therefore not strange that systems analysis has been seen as the only way forward, to look at how genes, proteins, synapses, and cells work together to produce a function. A systems approach will also give us information about how the same malfunction can be the result of changes in many different components within the system.

Another inference that can be made from the multilevel analysis is that instead of looking for specific mechanisms of a disease, it may be more fruitful to try to find nodes of convergence. This approach has been put forward at a symposium for Advances in Neuroscience for Medical Innovation (Agid et al., 2007). One such node in psychiatric disease is the synapse – crucial in determining the network behavior by tuning the contacts between the cells and the locus of action of most of the susceptibility genes involved in schizophrenia. Moreover, by studying the synapse we can make sense of the studies that imply a role for fast neurotransmission in psychiatric disease, as well as for the importance of the modulatory neurotransmitter systems involved, including dopamine and serotonin: the role of the modulatory neurotransmitters is necessary to stabilize the network through modulation of synaptic strength.

Synaptic strength is regulated by several intracellular biochemical signaling pathways that are activated by electrical activity as well as modulatory transmitters. Thus, by understanding how intracellular signaling pathways are regulated we will be able to understand how synaptic strength will be modified. The intracellular signaling that regulates synaptic strength is relevant not only for learning and adaptation in the healthy nervous system. If we understand the signaling triggered by the involved neurotransmitters, we will understand the neurobiology of psychiatric disease such as schizophrenia and be able to suggest new potential therapies.

Intracellular signaling cascades, although small in the physical world, can be considered a system in themselves. When neurotransmitters bind to transmembrane receptors, intracellular biochemical events are triggered that have been described as signaling "pathways." Since these pathways have many overlapping steps, and are important points for convergence and interaction, intracellular signaling generates large information processing networks – rather then pathways – with nonintuitive, emerging properties. The interconnection of events activated by various receptors also makes this system very dynamic, because a change in one part of the network may affect how another part of the network responds to other stimuli. The study of intracellular signaling thus benefits from a systems biology approach, meaning that each component has to be studied in terms of its role within the containing whole. It is important to look at the known output and try to understand the contributions of various components. Experimental biochemical work by its nature is hard to use for this approach, since we cannot control all the components and the time resolution is very poor. A complementary approach therefore is to construct computer models, based on biochemical data, where each component can be followed with very high time resolution during various manipulations. In this chapter we give an example of a signaling network, the protein kinase A (PKA)/dopamine and cAMP-regulated phosphoprotein of 32 kDaDARPP-32/protein phosphatase-1 (PP1) cascade activated by dopamine, and show how this network can be modeled.

## 14.4
## Dopamine Signaling and DARPP-32

The effects of dopamine in the central nervous system have been known for 50 years now (Carlsson *et al.*, 1958) and this is still a hot topic of research. One of the reasons is that dopamine modulation of other systems makes dopamine a neurotransmitter involved in several behaviors and clinical conditions, including motivation, locomotion, reward, schizophrenia, attention deficit hyperactivity disorder, Parkinsons disease, and drugs of abuse. Intracellular signaling of dopamine has been thoroughly studied both in normal and pathophysiological conditions. Dopamine receptors are G-protein coupled and can either stimulate ($D_1$ and $D_5$ receptors) or inhibit ($D_{2-4}$ receptors) adenylate cyclase and cAMP production. Through cAMP activation of PKA they affect the phosphorylation state of the signaling cascade and modulate several intracellular processes, including gene transcription and synaptic transmission.

The phosphoprotein DARPP-32 has proven to be a key signaling molecule in dopaminergic transmission (Greengard, 2001). DARPP-32 is an acronym for dopamine and cAMP-regulated phosphoprotein of 32 kDa. As the name implies, it was first discovered as a substrate for cAMP-dependent phosphorylation in the caudate putamen (Hemmings et al., 1984; Ouimet et al., 1984; Walaas and Greengard, 1984) and was shown to be heavily expressed in areas with dense dopamine innervation. When phosphorylated at Thr34 in the rat sequence. DARPP-32 was shown to be a potent and specific inhibitor of PP1. This phosphatase is concentrated in spines and synapses, and thus puts the DARPP-32/PP1 complex in an important place to regulate synaptic strength and plasticity. This also means that DARPP-32/PP1, being regulated by dopamine, provides a mechanism for dopamine to modulate synaptic strength (Allen, 2004).

Lately it has been shown that DARPP-32 is regulated also by other first messengers. Not surprisingly, other transmitters acting on G-protein-coupled receptors can affect DARPP-32 regulation through the cAMP cascade, including adenosine (Lindskog et al., 2002; Svenningsson et al., 2000), serotonin (Svenningsson et al., 2002a) and opioids (Lindskog et al., 1999). Moreover, regulation of DARPP-32 phosphorylation at Thr34 is not limited to G-protein-coupled receptors, but neurotransmitters that act on ionotropic receptors can also regulate DARPP-32 phosphorylation through calcium signaling. Thus, the major excitatory and inhibitory transmitters glutamate (Halpain, Girault, and Greengard, 1990; Nishi et al., 2005) and GABA (Flores-Hernandez et al., 2000; Snyder, Fisone, and Greengard, 1994) can also regulate the phosphorylation of DARPP-32 at Thr34.

It was earlier shown that DARPP-32 had more phosphorylation sites than Thr34, namely at Ser102 and Ser137 in the rat sequence. Ser102 is phosphorylated by casein kinase-2 (Girault et al., 1989) and Ser137 is phosphorylated by casein kinase-1. (Desdouits et al., 1995). The first messengers regulating the phosphorylation of these sites are still not completely clarified, although it has been shown that serotonin can increase phosphorylation at Ser137 (Svenningsson et al., 2002a). The overall consequence of phosphorylation of DARPP-32 by casein kinase-1 or -2 first seems to be to increase the state of phosphorylation of Thr34 (Desdouits et al., 1995; Svenningsson et al., 2004) through intramolecular mechanisms. However, more recent data shows that phosphorylation at Ser102 (or Ser97 in mice) induces translocation of DARPP-32 to the nucleus, where it is important in regulating gene transcription (Stipanovich et al., 2008a).

A breakthrough in the study of DARPP-32 came with the discovery that DARPP-32 can be phosphorylated also at Thr75 (Bibb et al., 1999). When phosphorylated at this site DARPP-32 is an inhibitor of PKA, and thus has an opposite affect compared to when it is phosphorylated at Thr34. This means that DARPP-32 can have opposite effects depending on where it is phosphorylated: when phosphorylated on Thr34 DARPP-32 inhibits protein phosphorylation and thus enhances the effect of the cAMP/PKA cascade, whereas when it is phosphorylated at Thr75 it inhibits PKA activity and thus decreases the cAMP/PKA cascade (Lindskog et al., 2002; Nishi et al., 2000). Phosphorylation of DARPP-32 at Thr75 is catalyzed by the cyclin-dependent kinase Cdk5, a kinase that despite its name is not cyclin-dependent but activated by binding of the cofactor p35. The regulation of this pathway remains to be clarified, although recently glutamate has been shown to be involved as a first messenger. Activation of mGluRs increases Cdk5 activity (Liu et al., 2001), whereas activation of NMDA and kainate receptors decreases Cdk5 activity (Wei et al., 2005).

Due to its regulation of several of the major neurotransmitters, DARPP-32 has turned into something of a "model molecule" when it comes to the complex networks of intracellular signaling. The intracellular regulation and second messenger systems involved have been described in great detail and there is a large amount of data available of intracellular regulation. As the Thr34 site is phosphorylated by PKA, phosphorylation of this site is increased by different receptors increasing cAMP through adenylate cyclase. Likewise, DARPP-32 also can be affected by phosphodiesterases (Nishi et al., 2008), which decreases cAMP levels and thus reduces phosphorylation at Thr34. Thr34 is dephosphorylated by protein phosphatase 2B (PP2B), a calcium/calmodulin-dependent phosphatase (also known as calcineurin) (Halpain, Girault, and Greengard, 1990; Nishi, Snyder, and Greengard, 1997). The Thr75 site of DARPP-32 was discovered several years after the discovery of the Thr34 site and less is known of its regulation. This site is phosphorylated by Cdk5, an enzyme that is activated by binding of the cofactor p35. Despite the well-studied role of Cdk5 in development, much less is known about this enzyme in mature neurons and its role in modulating synaptic strength. Thr75 is dephosphorylated mainly by protein phosphatase 2A (PP2A) (Bibb et al., 1999 Nishi et al., 2000), an enzyme that can be regulated both by phosphorylation and calcium. The details of this regulation, however, still remain to be elucidated.

## 14.5
## Physiological Role of DARPP-32

Changes is synaptic strength depend on the balance between kinases and phosphatases in the intracellular signaling network, giving DARPP-32 an important role in regulation. Reinforcement learning theorizes that strengthening of synaptic connections in striatal medium spiny neurons occurs when glutamatergic input from cortex and dopaminergic input from substantia nigra are received simultaneously. Phosphorylation processes involving cAMP-dependent kinase (PKA) increases the insertion of α-amino-3-hydroxy-5-methyl-4-isoxazol-propionacid (AMPA) receptors in the membrane, whereas dephosphorylation processes involving PP1 has the opposite effect, and the balance between PKA and PP1 is heavily mediated by DARPP-32.

Experiments performed in DARPP-32 knockout mice have shed more light on the function of DARPP-32. The mice express no overt phenotype, but response to dopamine, especially $D_1$ receptor stimulation, is blunted in the basal ganglia at all levels, from intracellular signaling, to electrical properties and behavior (Fienberg and Greengard, 2000; Fienberg et al., 1998). The same effect is seen in response to other neurotransmitters, such as adenosine, where the response to the adenosine A2A receptor antagonist caffeine is decreased, and serotonin. By its dual role, with the possibility to both inhibit PKA and PP1, DARPP-32 plays an important role in regulating the state of phosphorylation of several effector proteins in the cell, including ion channels which affect synaptic strength and cell excitability. AMPA receptors (Snyder et al., 2000), NMDA receptors (Snyder et al., 1998), $Ca^{2+}$ channels (Surmeier et al., 1995), and $K^+$ channels (Flores-Hernandez et al., 2000) are all regulated through the PKA/DARPP-32/PP1 pathway, and the regulation of activity of these proteins is impaired in DARPP-32 knockout mice (Fienberg et al., 1998). The overall effect of DARPP-32 on neuronal excitability is complex since the modulation of different ion channels or ion pumps is affected in different ways, sometimes giving opposite effects on synaptic transmission and intrinsic excitability. Moreover, the result from stimulating one specific part of the pathway is not intuitive, since the different components in the signaling network are interconnected.

In addition to its effect on synaptic proteins, DARPP-32 also affects signaling to the nucleus. Phosphorylation of mitogen-activated protein kinase and cAMP response element-binding protein by dopamimetic drugs (Valjent et al., 2005) or $D_2$ receptor stimulation (Yan et al., 1999) are almost abolished in DARPP-32 knockout mice. This implies that signaling to the nucleus and gene transcription – events important for long term plasticity and adaptation to drugs – are controlled by DARPP-32. More recently, it was shown that DARPP-32 itself can translocate to the nucleus when it is phosphorylated at Ser97 (mice sequence, equivalent to Ser102 in rat sequence) (Stipanovich et al., 2008b), thus inhibiting nuclear PP1 and inhibiting histone H3 phosphorylation. The reduced gene transcription has effects on response to drugs of abuse.

## 14.6
### DARPP-32 in Psychiatric Disease

The fact that DARPP-32 is regulated by the two neurotransmitters that are the most often implied in schizophrenia (i.e., dopamine and glutamate) makes it an interesting candidate to study from this perspective. However, expression and genetic studies have given mixed signals about the role of DARPP-32 in the etiology of schizophrenia. Reduced levels of DARPP-32 have been shown in the prefrontal cortex of schizophrenic patients (Albert et al., 2002; Ishikawa et al., 2007) as well as animal models of the disease (Romero et al., 2007), whereas other studies show no change (Baracskay, Haroutunian, and Meador-Woodruff, 2006; Ishikawa et al., 2007). Likewise, polymorphism in the gene coding for DARPP-32, *PPP1R1B*, has been linked to schizophrenia in one study (Meyer-Lindenberg et al., 2007), but not in others (Akira et al., 2007; Hu et al., 2007). The role of DARPP-32 in schizophrenic disease is likely to be more subtle than revealed by expression analysis, since its function is regulated by phosphorylation and not levels of expression. Also, it has been shown that schizophrenic-like states can be induced in animals by drugs acting on different transmitter systems, but they all involve signaling through DARPP-32. In addition, the effect of these drugs is reduced in DARPP-32 knockout mice (Svenningsson et al., 2003). Thus, despite the inconclusive data on the role of DARPP-32 in the etiology of schizophrenia, the study of DARPP-32 regulation can be very useful to understand the intracellular signaling network in both pathological conditions as well as in response to potential therapeutic drugs. For example, antipsychotic drugs that block dopamine $D_2$ receptors can increase phosphorylation at DARPP-32 (Pozzi et al., 2003; Svenningsson et al., 2000).

The involvement of dopamine in other psychiatric disorders, as well as the key role of DARPP-32 as a hub integrating signals from several first messengers, make this phosphoprotein interesting for disorders other than schizophrenia. For example, it has been shown that the commonly used antidepressant fluoxetine regulates DARPP-32 phosphorylation and that the effect of this drug is diminished in DARPP-32 knockout mice (Svenningsson et al., 2002b). Not surprisingly, since dopamine is the target neurotransmitter system for drugs of abuse, DARPP-32 is also an important target in substance abuse (Svenningsson, Nairn, and Greengard, 2005).

The role of DARPP-32 as a hub in the signaling network, shaping the response to several modulatory neurotransmitters in the central nervous system makes it an interesting target for psychiatric medication (Reis et al., 2007). One striking feature in drug development is that less selective drugs are not always the least effective (e.g., clozapine is the best treatment for schizophrenia and is also the most "dirty" compound, with affinity for a wide range of G-protein-coupled receptors). An explanation may be that substances with several targets can fine-tune a cellular response. A better understanding of the complex reaction of these networks can be very beneficial in order to understand how activation of several receptors interact in the intracellular signaling and shapes the response of the neuron.

The regulation of DARPP-32 of all the major neurotransmitters in the basal ganglia, and its involvement in many different clinical conditions, make it hard to understand the specificity of the signaling pathway and to predict an outcome of a given stimulation. To look at a single pathway at a time becomes almost meaningless when we understand how well they are all interconnected. Instead, we need to take a systems approach, where we look at the effect of DARPP-32 phosphorylation within the entire signaling network. One tool to do this without loosing track of individual molecules is by computer modeling. With this aim we have constructed a computer model of the regulation of DARPP-32 based on biochemical data in order to investigate possible effects of stimulation of various pathways and make testable hypothesis (Lindskog et al., 2006). This model illustrates how DARPP-32 can mediate the balance between PP1 and PKA, and includes a subset of the reactions regulating DARPP-32 phosphorylation (Figure 14.1). When site Thr34 is phosphorylated by PKA, DARPP-32 becomes a potent inhibitor of PP1. When phosphorylated at Thr75, DARPP-32 inhibits PKA. An increased cAMP level (typically produced by a dopaminergic input) leads to an increase in free catalytic PKA (catalytic protein kinase A, PKAc), which reacts with PP2A and increases its rate of dephosporylation at Thr75. In turn, this decreases the

## BIOCHEMICAL NETWORK

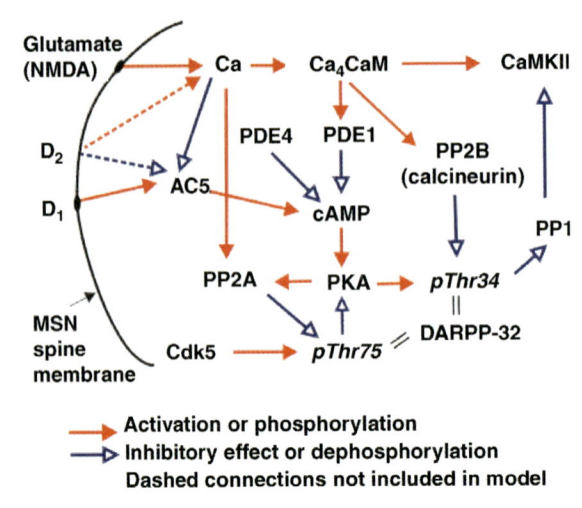

**Figure 14.1** Scheme of the biochemical pathways included in the model by Lindskog et al. (2006). When site Thr34 is phosphorylated by PKA, DARPP-32 becomes a potent inhibitor of PP1. When phosphorylated at Thr75, DARPP-32 inhibits PKA. An increased cAMP level (typically produced by a dopaminergic input) leads to an increase in free catalytic PKA (PKAc), which reacts with PP2A and increases its rate of dephosporylation at Thr75. In turn, this decreases the Thr75 inhibition of PKA, so that a positive feedback occurs. The initial effect of a glutamatergic input is that it (via an increased calcium level) increases further the PP2A dephosphorylation at Thr75, thus enhancing the dopamine effect. An increased calcium level will, however, also dephosphorylate DARPP-32 at Thr34 via the PP2B, counteracting the dopamine/PKA effect. AC: adenylate cyclase; CaM: calmodulin; CaMKII: $Ca^{2+}$/calmodulin-dependent kinase II; MSN: mesencephalic nucleus; PDE: phosphodiesterase; *pThr75*: phosphorylated Thr75.

Thr75 inhibition of PKA, so that a positive feedback occurs. The initial effect of a glutamatergic input is that it (via an increased calcium level) increases further the rate of PP2A phosphorylation at Thr75, thus enhancing the dopamine effect.

## 14.7
### Modeling Signaling Pathways with a Deterministic Model

The intracellular space contains a diversity of different signaling molecules. In biochemical models, these signaling pathways are described as chains of chemical reactions. A chemical reaction takes place when two substrate molecules (S1 and S2) interact to form a product (P):

$$S_1 + S_2 \leftrightarrow P \tag{14.1}$$

The forward rate constant ($k_f$) describes the frequency with which the product is formed, while the backward rate constant ($k_b$) describes the frequency with which the substrates ($S_1 + S_2$) are regenerated. In a deterministic model, the quantities of the species are typically described in terms of concentrations ($[S_1]$, $[S_2]$, and $[P]$). Time variations can be determined from differential equations:

$$\frac{d[S_1]}{dt} = \frac{d[S_2]}{dt} = \frac{-d[P]}{dt} = -k_f \cdot [S_1] \cdot [S_2] + k_b \cdot [P]. \tag{14.2}$$

Equations like these are commonly called "mass action kinetics" since they describe the behavior of large groups of molecules, assuming that the cell is "well stirred" and ignoring probabilistic interaction between individual molecules. If the system is in static steady state so that the time derivatives of the concentrations are zero, then:

$$K_d = \frac{k_b}{k_f} = \frac{[S_1^*] \cdot [S_2^*]}{[P^*]}, \tag{14.3}$$

where $K_d$ is called the dissociation constant and can be calculated from the steady-state concentrations ($[S_1^*]$, $[S_2^*]$, and $[P^*]$) of the involved species. Estimates of $k_b$ and $k_f$ require data on the system dynamics, which is often more difficult to obtain. In many experiments, therefore, only $K_d$ can be determined.

Most biochemical pathways are cascades of reactions, so that the product of one reaction may be substrate to another. A special example of a two step reaction is the enzyme reaction:

$$E + S \leftrightarrow ES \rightarrow E + P \tag{14.4}$$

In the first step, the enzyme and substrate combine to form a complex. The second and irreversible step (with rate constant $k_{cat}$) forms the product and regenerates the enzyme. The equations for this system become:

$$\frac{d[E]}{dt} = -k_f \cdot [E] \cdot [S] + (k_b + k_{cat}) \cdot ES \tag{14.5}$$

$$\frac{d[S]}{dt} = -k_f \cdot [E] \cdot [S] + k_b \cdot ES \tag{14.6}$$

$$\frac{d[ES]}{dt} = k_f \cdot [E] \cdot [S] - (k_b + k_{cat}) \cdot ES \tag{14.7}$$

$$\frac{d[P]}{dt} = k_{cat} \cdot ES \tag{14.8}$$

## 14.8
## Biological Conclusions from the DARPP-32 Model

When comparing the results on DARPP-32 phosphorylation from the model with what is known from biological experiments we confirmed known aspects of

DARPP-32 regulation, including the robustness of the phosphorylated state. This has also been shown by other similar (Fernandez et al., 2006) and more theoretical (Barbano et al., 2007) models. Upstream signaling can be transient and subject to rapid fluctuations, but the phosphorylation of DARPP-32, especially at the Thr34 site, seem to integrate and prolong the signal, by staying phosphorylated. This implies that DARPP-32 can work as a working memory molecule within the network, and can integrate subsequent signals over time. Being consistent findings in several models, the role of DARPP-32 as a stable integrator of several signaling pathways, and the stability of its phosphorylated state, may be useful for designing neuropharma. To fully exploit this aspect of the model, it would be useful to add receptors from additional neurotransmitters, to be able to do initial screens of drugs with various receptor binding profiles.

When it comes to the dynamic responses we obtained surprising results from our deterministic model. These fast reactions are an aspect that is hard to study using biochemical tools and our deterministic model gave us some surprising hypothesis about how the network behaves. In response to fast and high amplitude calcium signals, the free PKAc increases transiently despite the decreased cAMP because calcium activation of PP2A leads to Thr75 dephosphorylation, relieving PKAc of inhibition. The consequent increase in PKAc contributes to an increase in phosphorylated Thr34, with a higher peak value (though shorter half-life) than that caused by dopamine stimulation, and shows that a very brief calcium influx can actually lead to an increase in phosphorylation at Thr34. This response is potentiated by the PKA/PP2A/Thr75 loop acting as a sink, binding or releasing catalytic PKA. This increase in phosphorylation at Thr34 is in sharp contrast to the general understanding of DARPP-32, stemming from results observed with steady-state calcium signals, that calcium decreases phosphorylation of Thr34.

An important mechanism in the effect of Thr75 acting as a sink and affecting calcium regulation of Thr34 is the calcium activation of PP2A (Nishi et al., 2000; Ahn et al., 2007a, 2007b). This mechanism requires a moderate affinity, yet very steep (e.g., fourth-order) dependence on calcium. In contrast, recent experiments show that the amount of calcium needed to activate PP2A *in vitro* is unphysiologically high (50 mM). This seems unlikely to be reached in the cytosol under normal AMPA/NMDA signaling, indicating that some other, currently undetected interactions might be needed in order to explain the observations.

## 14.9
### Stochastic Models

Deterministic models (mass action kinetics) describe the behavior of a large number of molecules that are evenly distributed within a compartment. However, many important cellular signaling events take place in small substructures of the cell, such as in the spines, where the abundance of a molecular species may be much higher or lower than in the dendrites. In such cases, the number of binding events may fluctuate strongly and a stochastic modeling approach should be considered.

At a microscopic level, a chemical reaction is a probabilistic event that takes place when molecules collide and bind to each other. In stochastic models, reactions are described as probabilistic events. Essentially, this means that the (deterministic) reaction rates (in Eq. (14.2)) are reinterpreted as probabilities that the binding events occur within a small time interval. Several algorithms have been developed that treat bimolecular reactions as probabilistic events (e.g., see Gillespie, 1977; Gibson and Bruck, 2000; Cao, Gillespie, and Petzold, 2005).

In small substructures that are connected to larger structures (e.g., spines attached to dendrites), the assumption of homogeneous distribution of molecules is incorrect. The abundance of a molecular species is controlled by diffusion between the structures. To take into account diffusion as well as reactions, stochastic diffusion algorithms are available (e.g., see Bhalla, 2004; Blackwell, 2006).

## 14.10
## PKA Activation: A Case Study

One of the pathways contained in the Lindskog *et al.* (2006) model described how cAMP interacts with PKA to release active subunits of PKAc. We will use this as an illustrative example of how a model of a biochemical pathway is built and parameterized, and we will also highlight the typical obstacles met when building complex biochemical network models.

### 14.10.1
### Conceptual Model

First we need a conceptual model of what happens when cAMP binds to- and activates PKA. The cAMP binding process is known to happen in a sequential order to two tandem cAMP binding domains, B and A, where binding to site B precedes binding to site A (Zawadzki and Taylor, 2004). Experimental measurements have yet to reveal PKA forms with either one or three cAMP molecules bound; suggesting highly cooperative binding of two cAMP molecules to the B site, followed by highly cooperative binding of an additional two cAMP molecules to the A site. When the total of four cAMP molecules are bound, the complex dissociates and releases two catalytic subunits (PKAc) (Zawadzki and Taylor, 2004). This process can be described by three simplified reactions (Lindskog *et al.*, 2006):

$$PKA + 2(cAMP) \leftrightarrow PKAcAMP_2(k_{Bf}, k_{Bb}) \quad (14.9)$$

$$PKAcAMP_2 + 2(cAMP) \leftrightarrow PKAcAMP_4(k_{Af}, k_{Ab}) \quad (14.10)$$

$$PKAcAMP_4 \leftrightarrow PKAr + 2PKAc(k_{Cf}, k_{Cb}) \quad (14.11)$$

In reality, two cAMP molecules do not bind to PKA simultaneously. If we assume that the first cAMP binding is time-limiting and the second molecule binds almost instantaneously after the first has bound, the bindings (Eqs. (14.9) and (14.10)) can be treated as bimolecular reactions (only considering the first binding event), consistent with the observed linear dependence of each binding site on cAMP concentration. The differential equations for the reactions then become:

$$\frac{d(cAMP)}{dt} = 2 \cdot (-k_{Bf} \cdot PKA \cdot cAMP + k_{Bb} \cdot PKAcAMP_2) \\ - k_{Af} \cdot PKAcAMP_2 \cdot cAMP + k_{Ab} \cdot PKAcAMP_4) \quad (14.12)$$

$$\frac{d(PKA)}{dt} = -k_{Bf} \cdot PKA \cdot cAMP + k_{Bb} \cdot PKAcAMP_2 \quad (14.13)$$

$$\frac{d(PKAcAMP_2)}{dt} = k_{Bf} \cdot PKA \cdot cAMP - k_{Bb} \cdot PKAcAMP_2 \\ - k_{Af} \cdot PKAcAMP_2 \cdot cAMP + k_{Ab} \cdot PKAcAMP_4 \quad (14.14)$$

$$\frac{d(PKAcAMP_4)}{dt} = k_{Af} \cdot PKAcAMP_2 \cdot cAMP - (k_{Ab} + k_{Cf}) \cdot PKAcAMP_4 \\ + k_{Cb} \cdot PKAc \cdot PKAr \quad (14.15)$$

$$\frac{d(PKAc)}{dt} = 2 \cdot (k_{Cf} \cdot PKAcAMP_4 - k_{Cb} \cdot PKAc \cdot PKAr) \quad (14.16)$$

Note that the model (Eqs.(14.12)–(14.16)) is a simplification of a complex process, and that other interpretations are possible. Nonetheless, the level of detail in Eqs. (14.12)–(14.16) is consistent with experimental observations and is sufficient for addressing question about the PKA regulation of DARPP-32 in a large signaling network.

### 14.10.2
### Empirical Estimates of Rate Constants

A quantitative model is obtained when the forward and backward rate constants are specified for the reactions. As a first approach, we look for such estimates in the literature. Association ($k_{ass}$) and dissociation ($k_{diss}$) constants for cAMP binding have been estimated to $1.5 \times 10^{-5}\,nM^{-1}\,s^{-1}$ and $1.6 \times 10^{-3}\,s^{-1}$ at site B, and $4 \times 10^{-5}\,nM^{-1}\,s^{-1}$ and $13 \times 10^{-3}\,s^{-1}$ at site A (Øgreid and Døskeland, 1981). Association and dissociation of the catalytic subunit were estimated to $1.7 \times 10^{-4}\,nM^{-1}\,s^{-1}$ and $1.6 \times 10^{-3}\,s^{-1}$ (Zawadzki and Taylor, 2004). These estimates are a good starting point; however, there may be discrepancies between empirically derived rate constants and the rate at which these reactions occur in real cells under physiological conditions. For instance, the estimates of cAMP-binding rates (Øgreid and Døskeland, 1981) were based on concentration of bound versus unbound cAMP in buffer solutions. Furthermore, binding rates may differ between cell types and brain regions, and also may depend on the presence of substances (i.e., by $SO_4^{2-}$ and $Cl^-$ (Øgreid and Døskeland, 1981)) not included in the model. Therefore the model parameters may need to be adjusted to obtain agreement with empirical data collected under other conditions.

### 14.10.3
### Model Validation against Steady-State Data

The model should be able to reproduce empirically observed relationships between steady state concentrations of cAMP, PKA, and PKAc. In previous empirical studies, a solution of PKA (10 nM) was exposed to different concentrations of cAMP and the system was allowed to reach steady state (Zawadzki and Taylor, 2004). The steady-state concentration of PKAc was zero when no cAMP was applied and increased with cAMP concentration until saturation (100%) was reached (see Figure 14.2, "Empirical data").

The experimental conditions (initial PKA and cAMP concentrations) were replicated with our model (Eqs. (14.12)–(14.13)). When the parameters from the literature (Øgreid and Døskeland, 1981; Zawadzki and Taylor, 2004) were applied, PKAc saturated at far too low a cAMP concentration (see Figure 14.2, "Original parameters"). In line with the discussion above, this may indicate that the cAMP-binding rates (from Øgreid and Døskeland, 1981) are not directly applicable to the reactions in the model. These binding rates were therefore modified (through a series of trials and errors) until a parameter set was found that could produce a saturation curve that agreed with data (see Figure 14.2, "Modified parameters").

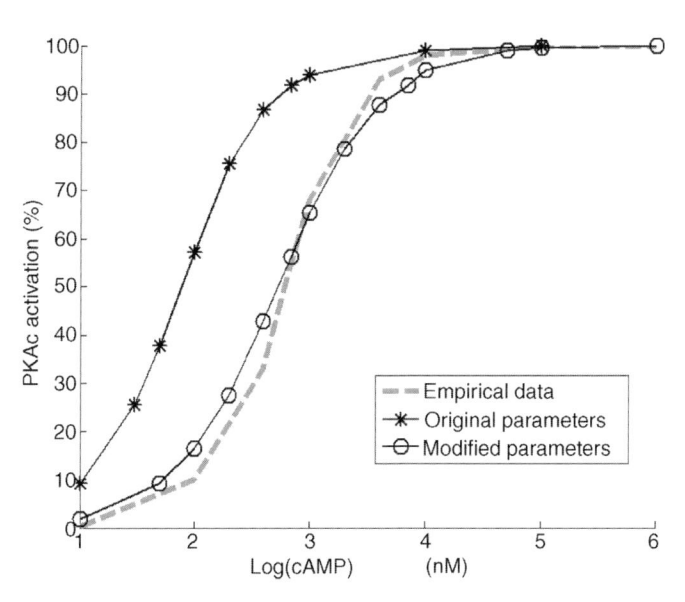

**Figure 14.2** Activation curved for PKAc. An original set of parameters was modified in order to fit the model to empirical steady-state data.

### 14.10.4
#### Model Validation against Dynamic Data

In dynamic experiments of PKA responses, it has been estimated that it takes about 20 s from cAMP stimulation until 90% of the steady-state PKAc concentration is reached (Gervasi et al., 2007). Response times will depend strongly on PKA and cAMP concentrations, which were not given for the experiments. Nevertheless, when using typical cAMP and PKAc concentrations from Lindskog et al. (2006), our model responded significantly slower than in the above experiments (see Figure 14.3, "×1"). Dynamic imaging techniques often reveal faster enzyme activation than standard static biochemical techniques. We therefore "sped up" our reactions by multiplying all rates by the same factor 3, which decreases response time from about 200 to about 50 s. Since all ratios $k_b/k_f$ are preserved, steady-state concentrations are unaffected by this increase in speed.

The final set of parameters (applied in Lindskog et al., 2006) give a good fit to steady-state data (Figure 14.2) and are a compromise between the order of magnitude reported in the literature (Table 14.1, "Original parameters") and the reaction speed reported in empirical dynamic studies.

### 14.10.5
#### Deterministic versus Stochastic Algorithms

The example above described a deterministic model for cAMP activation of PKA and illustrated how parameters were derived in the Lindskog et al. (2006) model. Here, we apply a recent stochastic algorithm that combines reactions and diffusion (Koh and

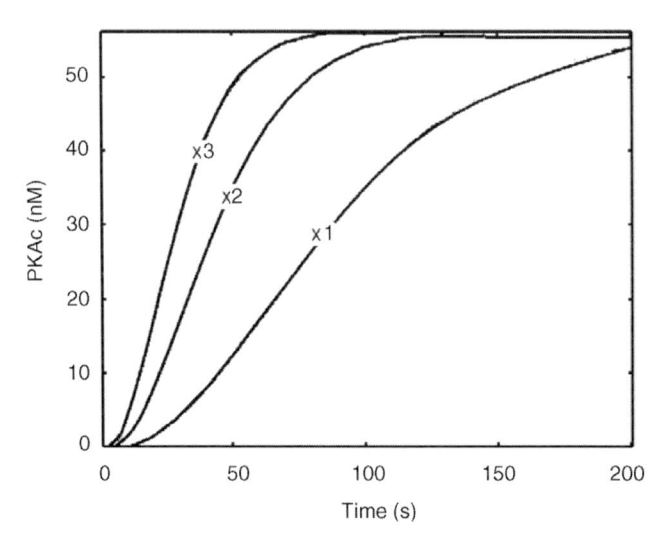

**Figure 14.3** The dynamic response (initial values: cAMP = 25 µM, PKA = 1200 nM, PKAc = 0) for the modified parameter set (×1), when their speed has been increased by a factor 2 (×2) and 3 (×3).

## 14.10 PKA Activation: A Case Study

**Table 14.1** Forward and backward reaction rates.

| Reaction | Original parameters | | Modified (steady-state data) | | Final (steady-state data + dynamic data) | |
|---|---|---|---|---|---|---|
| | $k_f$ (nM$^{-1}$s$^{-1}$) | $k_b$ (s$^{-1}$) | $k_f$ (nM$^{-1}$s$^{-1}$) | $k_b$ (s$^{-1}$) | $k_f$ (nM$^{-1}$s$^{-1}$) | $k_b$ (s$^{-1}$) |
| 1 | $1.50 \times 10^{-5}$ | $1.60 \times 10^{-3}$ | $0.86 \times 10^{-5}$ | $2.00 \times 10^{-3}$ | $2.60 \times 10^{-5}$ | $6.00 \times 10^{-3}$ |
| 2 | $4.00 \times 10^{-5}$ | $1.30 \times 10^{-2}$ | $1.16 \times 10^{-5}$ | $2.00 \times 10^{-2}$ | $3.50 \times 10^{-5}$ | $6.00 \times 10^{-2}$ |
| 3 | $1.70 \times 10^{-4}$ | $1.60 \times 10^{-3}$ | $1.70 \times 10^{-4}$ | $1.60 \times 10^{-3}$ | $5.10 \times 10^{-4}$ | $4.80 \times 10^{-3}$ |

Blackwell, 2008) in order to illustrate how the conclusions from deterministic simulations may not always be reproduced by models where stochastic fluctuations are taken into account.

Figure 14.4(a–d) shows the clear response patterns of cAMP, PKAcAMP$_2$, PKA-cAMP$_4$, and PKAc when a synchronized dopamine and calcium input was given to the model. The cAMP and PKA signal reaches amplitudes of several hundred nanomolar, while the PKAc concentration is as low as around 5 nM. A concentration of 5 nM corresponds to approximately 3 molecules/μm$^3$. If we assume that these reactions take place in a spine of volume 2 μm$^3$, this means that our system contains, at maximum, six PKAc molecules. This is not a high number, so in this case we would

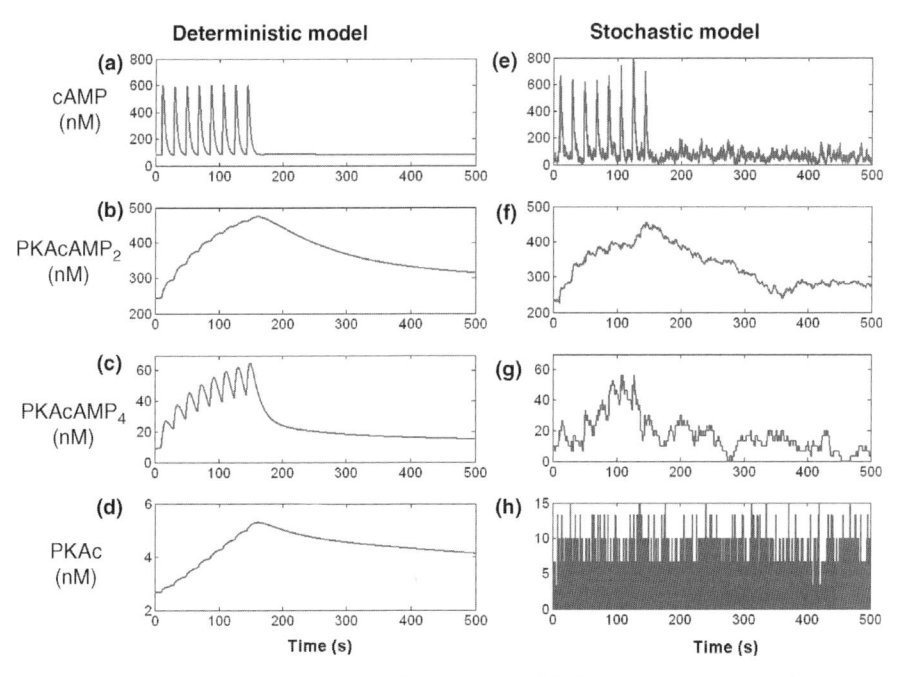

**Figure 14.4** Deterministic (a–d) versus stochastic (e–h) model, illustrating cAMP-activated PKA producing the active subunit PKAc.

expect the PKAc dynamics to be heavily affected by stochastic fluctuations. This suspicion was confirmed when we ran the corresponding simulations with a stochastic algorithm. The cAMP and PKAcAMP$_2$ signals were (roughly) reproduced in the stochastic simulations, because their concentrations are higher (Figure 14.4e–f), For PKAcAMP$_2$, the signal is highly distorted, while the PKAc signal is completely dominated by stochastic fluctuations (Figure 14.4h). This indicates that some of the model conclusions downstream of PKAc should be interpreted with some caution.

## 14.11
## Conclusions

For the dynamic responses in our deterministic DARPP-32 model (Figure 14.1), we found some discrepancy between model simulations and experiments, which highlights the usefulness of models based on biochemical data. If the reactions are all based on experimental data, but the outcome does not fit experimental results, we can conclude that the model is not complete. The discrepancies may stem from high-resolution effects that have not been incorporated in the model (see the PKAc example in Section 14.10, where stochastic fluctuations play a major role). In other cases, discrepancies between model and experiments may indicate that there are reactions taking place that experimentalists have overlooked (e.g., in the case of calcium-dependent PP2A activation discussed in Section 14.8,). Thus, computer models can help guide experimental work to identify missing parts in the puzzle and illustrate which experiments are critical. The best progress is achieved when experimentalists and modelers work together, to make sure the model is firmly based on relevant biology, that the most relevant way of modeling is chosen, and to identify new areas where more data is needed to fill-in the gaps. Iterating the model construction and data collection creates a powerful tool to test hypotheses, evaluate individual reactions with a systems perspective, and thus get a better, multilevel, understanding of how the brain works in health and disease.

## References

Agid, Y., Buzsaki, G., Diamond, D.M., Frackowiak, R., Giedd, J., Girault, J.-A., Grace, A., Lambert, J.J., Manji, H., Mayberg, H. et al. (2007) How can drug discovery for psychiatric disorders be improved? *Nature Reviews Drug Discovery*, **6**, 189–201.

Ahn, J.-H., McAvoy, T., Rakhilin, S.V., Nishi, A., Greengard, P., and Nairn, A.C. (2007a) Protein kinase A activates protein phosphatase 2A by phosphorylation of the B56δ subunit. *Proceedings of the National Academy of Sciences of the United States of America*, **104**, 2979–2984.

Ahn, J.-H., Sung, J.Y., McAvoy, T., Nishi, A., Janssens, V., Goris, J., Greengard, P., and Nairn, A.C. (2007b) The B″/PR72 subunit mediates Ca$^{2+}$-dependent dephosphorylation of DARPP-32 by protein phosphatase 2A. *Proceedings of the National Academy of Sciences of the United States of America*, **104**, 9876–9881.

Akira, Y., Nagahide, T., Shinichi, S., Yoshihito, I., Branko, A., Hinako, U., Yukiko, K., Yukari, W., Takeo, Y., Tadafumi, K. et al. (2007) Genetic analysis of the gene coding for DARPP-32 (*PPP1R1B*) in Japanese patients with schizophrenia or bipolar disorder. *Schizophrenia Research*, **100**, 334–341.

Albert, K.A., Hemmings, H.C. Jr., Adamo, A.I.B., Potkin, S.G., Akbarian, S., Sandman, C.A., Cotman, C.W., Bunney, W.E. Jr., and Greengard, P. (2002) Evidence for decreased DARPP-32 in the prefrontal cortex of patients with schizophrenia. *Archives of General Psychiatry*, **59**, 705–712.

Allen, P.B. (2004) Functional plasticity in the organization of signaling complexes in the striatum. *Parkinsonism and Related Disorders*, **10**, 287–292.

Baracskay, K.L., Haroutunian, V., and Meador-Woodruff, J.H. (2006) Dopamine receptor signaling molecules are altered in elderly schizophrenic cortex. *Synapse*, **60**, 271–279.

Barbano, P.E., Spivak, M., Flajolet, M., Nairn, A.C., Greengard, P., and Greengard, L. (2007) A mathematical tool for exploring the dynamics of biological networks. *Proceedings of the National Academy of Sciences of the United States of America*, **104**, 19169–19174.

Bassett, D.S., Bullmore, E., Verchinski, B.A., Mattay, V.S., Weinberger, D.R., and Meyer-Lindenberg, A. (2008) Hierarchical organization of human cortical networks in health and schizophrenia. *Journal of Neuroscience*, **28**, 9239–9248.

Bhalla, U. (2004) Signalling in small subcellular volumes. I. Stochastic and diffusion effects on individual pathways. *Biophysical Journal*, **87**, 733–744.

Bibb, J.A., Snyder, G.L., Nishi, A., Yan, Z., Meijer, L., Fienberg, A.A., Tsai, L.H., Kwon, Y.T., Girault, J.A., Czernik, A.J. et al. (1999) Phosphorylation of DARPP-32 by Cdk5 modulates dopamine signalling in neurons. *Nature*, **402**, 669–671.

Blackwell, K.T. (2006) An efficient stochastic diffusion algorithm for modeling second messengers in dendrites and spines. *Journal of Neuroscience Methods*, **157**, 142–153.

Cao, Y., Gillespie, D.T., and Petzold, L.R. (2005) Avoiding negative populations in explicit Poisson tau leaping. *Journal of Chemical Physics*, **123**, 054104.

Carlsson, A., Lindqvist, M., Magnusson, T., and Waldeck, B. (1958) On the presence of 3-hydroxytyramine in brain. *Science*, **127**, 471.

Desdouits, F., Siciliano, J.C., Greengard, P., and Girault, J.A. (1995) Dopamine- and cAMP-regulated phosphoprotein DARPP-32: phosphorylation of Ser-137 by casein kinase I inhibits dephosphorylation of Thr-34 by calcineurin. *Proceedings of the National Academy of Sciences of the United States of America*, **92**, 2682–2685.

Dickinson, D. and Harvey, P.D. (2008) Systemic hypotheses for generalized cognitive deficits in schizophrenia: a new take on an old problem. *Schizophrenia Bulletin*, **35**, 403–414.

Fernandez, R., Schiappa, R., Girault, J.-A., and LeNovere, N. (2006) DARPP-32 is a robust integrator of dopamine and glutamate signals. *PLoS Computational Biology*, **2**, e176.

Ferrarelli, F., Massimini, M., Peterson, M.J., Riedner, B.A., Lazar, M., Murphy, M.J., Huber, R., Rosanova, M., Alexander, A.L., Kalin, N., and Tononi, G. (2008) Reduced evoked gamma oscillations in the frontal cortex in schizophrenia patients: a TMS/EEG study. *American Journal of Psychiatry*, **165**, 996–1005.

Fienberg, A.A. and Greengard, P. (2000) The DARPP-32 knockout mouse. *Brain Research Reviews*, **31**, 313–319.

Fienberg, A.A., Hiroi, N., Mermelstein, P.G., Song, W., Snyder, G.L., Nishi, A., Cheramy, A., O'Callaghan, J.P., Miller, D.B., Cole, D.G. et al. (1998) DARPP-32: regulator of the efficacy of dopaminergic neurotransmission. *Science*, **281**, 838–842.

Flores-Hernandez, J., Hernandez, S., Snyder, G.L., Yan, Z., Fienberg, A.A., Moss, S.J., Greengard, P., and Surmeier, D.J. (2000) $D_1$ dopamine receptor activation reduces GABAA receptor currents in neostriatal neurons through a PKA/DARPP-32/PP1 signaling cascade. *Journal of Neurophysiology*, **83**, 2996–3004.

Gervasi, N., Hepp, R., Tricoire, L., Zhang, J., Lambolez, B., Paupardin-Tritsch, D., and Vincent, P. (2007) Dynamics of Protein Kinase A Signaling at the Membrane, in the

Cytosol, and in the Nucleus of Neurons in Mouse Brain Slices. *J. Neurosci.*, **27**, 2744–2750

Gibson, M.A. and Bruck, J. (2000) Efficient exact stochastic simulation of chemical systems with many species and many channels. *Journal of Physical Chemistry A*, **104**, 1876–1889.

Gillespie, D.T. (1977) Exact stochastic simulation of coupled chemical reactions. *Journal of Physical Chemistry*, **81**, 2340–2361.

Girault, J.A., Hemmings, H.C. Jr., Williams, K.R., Nairn, A.C., and Greengard, P. (1989) Phosphorylation of DARPP-32, a dopamine- and cAMP-regulated phosphoprotein, by casein kinase II. *Journal of Biological Chemistry*, **264**, 21748–21759.

Greengard, P. (2001) The neurobiology of slow synaptic transmission. *Science*, **294**, 1024–1030.

Halpain, S., Girault, J.A., and Greengard, P. (1990) Activation of NMDA receptors induces dephosphorylation of DARPP-32 in rat striatal slices. *Nature*, **343**, 369–372.

Harrison, P.J. and Eastwood, S.L. (2001) Neuropathological studies of synaptic connectivity in the hippocampal formation in schizophrenia. *Hippocampus*, **11**, 508–519.

Hemmings, H.C. Jr., Nairn, A.C., Aswad, D.W., and Greengard, P. (1984) DARPP-32, a dopamine- and adenosine 3′:5′-monophosphate-regulated phosphoprotein enriched in dopamine-innervated brain regions. II. Purification and characterization of the phosphoprotein from bovine caudate nucleus. *Journal of Neuroscience*, **4**, 99–110.

Hu, J.-X., Yu, L., Shi, Y.-Y., Zhao, X.-Z., Meng, J.-W., He, G., Xu, Y.-F., Feng, G.-Y., and He, L. (2007) An association study between PPP1R1B gene and schizophrenia in the Chinese population. *Progress in Neuro-Psychopharmacology and Biological Psychiatry*, **31**, 1303–1306.

Ishikawa, M., Mizukami, K., Iwakiri, M., and Asada, T. (2007) Immunohistochemical and immunoblot analysis of Dopamine and cyclic AMP-regulated phosphoprotein, relative molecular mass 32,000 (DARPP-32) in the prefrontal cortex of subjects with schizophrenia and bipolar disorder. *Progress in Neuro-Psychopharmacology and Biological Psychiatry*, **31**, 1177–1181.

Jindal, R.D. and Keshavan, M.S. (2008) Neurobiology of the early course of schizophrenia. *Expert Review of Neurotherapeutics*, **8**, 1093–1100.

Jones, C.A. and McCreary, A.C. (2008) Serotonergic approaches in the development of novel antipsychotics. *Neuropharmacology*, **55**, 1056–1065.

Koh, W. and Blackwell, K.T. (2008) An efficient multi-scale stochastic algorithm for simulating biochemical reactions and second messenger diffusion in dendrites and spines, presented at the Annual Meeting of the Society for Neuroscience, Washington, DC, poster 334.22/E17.

Kirov, G., O'Donovan, M.C., and Owen, M.J. (2005) Finding schizophrenia genes. *Journal of Clinical Investigation*, **115**, 1440–1448.

Lindskog, M., Svenningsson, P., Fredholm, B., Greengard, P., and Fisone, G. (1999) Mu- and delta-opioid receptor agonists inhibit DARPP-32 phosphorylation in distinct populations of striatal projection neurons. *European Journal of Neuroscience*, **11**, 2182–2186.

Lindskog, M., Svenningsson, P., Pozzi, L., Kim, Y., Fienberg, A.A., Bibb, J.A., Fredholm, B.B., Nairn, A.C., Greengard, P., and Fisone, G. (2002) Involvement of DARPP-32 phosphorylation in the stimulant action of caffeine. *Nature*, **418**, 774–778.

Lindskog, M., Kim, M., Wikström, M.A., Blackwell, K.T., and Kotaleski, J.H. (2006) Transient calcium and dopamine increase PKA activity and DARPP-32 phosphorylation. *PLoS Computational Biology*, **2**, e119.

Lisman, J.E., Coyle, J.T., Green, R.W., Javitt, D.C., Benes, F.M., Heckers, S., and Grace, A.A. (2008) Circuit-based framework for understanding neurotransmitter and risk gene interactions in schizophrenia. *Trends in Neurosciences*, **31**, 234–242.

Liu, F., Ma, X.H., Ule, J., Bibb, J.A., Nishi, A., DeMaggio, A.J., Yan, Z., Nairn, A.C., and Greengard, P. (2001) Regulation of cyclin-dependent kinase 5 and casein kinase 1 by metabotropic glutamate receptors. *Proceedings of the National Academy of Sciences of the United States of America*, **98**, 11062–11068.

Meyer-Lindenberg, A., Straub, R., Lipska, B., Verchinski, B., Goldberg, T., Callicott, J., Egan, M., Huffaker, S., Mattay, V., Kolachana, B. et al. (2007) Genetic evidence implicating DARPP-32 in human frontostriatal structure, function, and cognition. *Journal of Clinical Investigation*, **117**, 672–682.

Mukai, J., Dhilla, A., Drew, L.J., Stark, K.L., Cao, L., MacDermott, A.B., Karayiorgou, M., and Gogos, J.A. (2008) Palmitoylation-dependent neurodevelopmental deficits in a mouse model of 22q11 microdeletion. *Nature Neuroscience*, **11**, 1302–1310.

Nishi, A., Snyder, G.L., and Greengard, P. (1997) Bidirectional regulation of DARPP-32 phosphorylation by dopamine. *Journal of Neuroscience*, **17**, 8147–8155.

Nishi, A., Bibb, J.A., Snyder, G.L., Higashi, H., Nairn, A.C., and Greengard, P. (2000) Amplification of dopaminergic signaling by a positive feedback loop. *Proceedings of the National Academy of Sciences of the United States of America*, **97**, 12840–12845.

Nishi, A., Watanabe, Y., Higashi, H., Tanaka, M., Nairn, A.C., and Greengard, P. (2005) Glutamate regulation of DARPP-32 phosphorylation in neostriatal neurons involves activation of multiple signaling cascades. *Proceedings of the National Academy of Sciences of the United States of America*, **102**, 1199–1204.

Nishi, A., Kuroiwa, M., Miller, D.B., O'Callaghan, J.P., Bateup, H.S., Shuto, T., Sotogaku, N., Fukuda, T., Heintz, N., Greengard, P., and Snyder, G.L. (2008) Distinct roles of PDE4 and PDE10A in the regulation of cAMP/PKA signaling in the striatum. *Journal of Neuroscience*, **28**, 10460–10471.

Ouimet, C.C., Miller, P.E., Hemmings, H.C. Jr., Walaas, S.I., and Greengard, P. (1984) DARPP-32, a dopamine- and adenosine 3′:5′-monophosphate-regulated phosphoprotein enriched in dopamine-innervated brain regions. III. Immunocytochemical localization. *Journal of Neuroscience*, **4**, 111–124.

Perala, J., Suvisaari, J., Saarni, S.I., Kuoppasalmi, K., Isometsa, E., Pirkola, S., Partonen, T., Tuulio-Henriksson, A., Hintikka, J., Kieseppa, T. et al. (2007) Lifetime prevalence of psychotic and bipolar I disorders in a general population. *Archives of General Psychiatry*, **64**, 19–28.

Pozzi, L., Hakansson, K., Usiello, A., Borgkvist, A., Lindskog, M., Greengard, P., and Fisone, G. (2003) Opposite regulation by typical and atypical anti-psychotics of ERK1/2, CREB and Elk-1 phosphorylation in mouse dorsal striatum. *Journal of Neurochemistry*, **86**, 451–459.

Reis, H.J., Rosa, D.V.F., Guimaraes, M.M., Souza, B.R., Barros, A.G.A., Pimenta, F.J., Souza, R.P., Torres, K.C.L., and Romano-Silva, M.A. (2007) Is DARPP-32 a potential therapeutic target? *Expert Opinion on Therapeutic Targets*, **11**, 1649–1661.

Romero, E., Ali, C., Molina-Holgado, E., Castellano, B., Guaza, C., and Borrell, J. (2007) Neurobehavioral and immunological consequences of prenatal immune activation in rats. Influence of antipsychotics. *Neuropsychopharmacology*, **32**, 1791–1804.

Snyder, G.L., Fisone, G., and Greengard, P. (1994) Phosphorylation of DARPP-32 is regulated by GABA in rat striatum and substantia nigra. *Journal of Neurochemistry*, **63**, 1766–1771.

Snyder, G.L., Fienberg, A.A., Huganir, R.L., and Greengard, P. (1998) A dopamine/$D_1$ receptor/protein kinase A/dopamine- and cAMP-regulated phosphoprotein ($M_r$ 32 kDa)/protein phosphatase-1 pathway regulates dephosphorylation of the NMDA receptor. *Journal of Neuroscience*, **18**, 10297–10303.

Snyder, G.L., Allen, P.B., Fienberg, A.A., Valle, C.G., Huganir, R.L., Nairn, A.C., and Greengard, P. (2000) Regulation of phosphorylation of the GluR1 AMPA receptor in the neostriatum by dopamine and psychostimulants *in vivo*. *Journal of Neuroscience*, **20**, 4480–4488.

Stipanovich, A., Valjent, E., Matamales, M., Nishi, A., Ahn, J.-H., Maroteaux, M., Bertran-Gonzalez, J., Brami-Cherrier, K., Enslen, H., Corbille, A.-G. et al. (2008a) A phosphatase cascade by which rewarding stimuli control nucleosomal response. *Nature*, **453**, 879–884.

Stipanovich, A., Valjent, E., Matamales, M., Nishi, A., Ahn, J.H., Maroteaux, M., Bertran-Gonzalez, J., Brami-Cherrier, K., Enslen, H.,

Corbille, A.G. et al. (2008b) A phosphatase cascade by which rewarding stimuli control nucleosomal response. *Nature*, **453**, 879–884.

Surmeier, D.J., Bargas, J., Hemmings, H.C. Jr., Nairn, A.C., and Greengard, P. (1995) Modulation of calcium currents by a $D_1$ dopaminergic protein kinase/phosphatase cascade in rat neostriatal neurons. *Neuron*, **14**, 385–397.

Svenningsson, P., Lindskog, M., Ledent, C., Parmentier, M., Greengard, P., Fredholm, B.B., and Fisone, G. (2000) Regulation of the phosphorylation of the dopamine- and cAMP-regulated phosphoprotein of 32 kDa *in vivo* by dopamine $D_1$, dopamine $D_2$, and adenosine A2A receptors. *Proceedings of the National Academy of Sciences of the United States of America*, **97**, 1856–1860.

Svenningsson, P., Tzavara, E.T., Liu, F., Fienberg, A.A., Nomikos, G.G., and Greengard, P. (2002a) DARPP-32 mediates serotonergic neurotransmission in the forebrain. *Proceedings of the National Academy of Sciences of the United States of America*, **99**, 3188–3193.

Svenningsson, P., Tzavara, E.T., Witkin, J.M., Fienberg, A.A., Nomikos, G.G., and Greengard, P. (2002b) Involvement of striatal and extrastriatal DARPP-32 in biochemical and behavioral effects of fluoxetine (Prozac). *Proceedings of the National Academy of Sciences of the United States of America*, **99**, 3182–3187.

Svenningsson, P., Tzavara, E.T., Carruthers, R., Rachleff, I., Wattler, S., Nehls, M., McKinzie, D.L., Fienberg, A.A., Nomikos, G.G., and Greengard, P. (2003) Diverse psychotomimetics act through a common signaling pathway. *Science*, **302**, 1412–1415.

Svenningsson, P., Nishi, A., Fisone, G., Girault, J.-A., Nairn, A.C., and Greengard, P. (2004) DARPP-32: an integrator of neurotransmission. *Annual Review of Pharmacology and Toxicology*, **44**, 269–296.

Svenningsson, P., Nairn, A.C., and Greengard, P. (2005) DARPP-32 mediates the actions of multiple drugs of abuse. *AAPS Journal*, **7**, E353–E360.

Toda, M. and Abi-Dargham, A. (2007) Dopamine hypothesis of schizophrenia: making sense of it all. *Current Psychiatry Reports*, **9**, 329–336.

Valjent, E., Pascoli, V., Svenningsson, P., Paul, S., Enslen, H., Corvol, J.C., Stipanovich, A., Caboche, J., Lombroso, P.J., Nairn, A.C. et al. (2005) Regulation of a protein phosphatase cascade allows convergent dopamine and glutamate signals to activate ERK in the striatum. *Proceedings of the National Academy of Sciences of the United States of America*, **102**, 491–496.

Walaas, S.I. and Greengard, P. (1984) DARPP-32, a dopamine- and adenosine 3′:5′-monophosphate-regulated phosphoprotein enriched in dopamine-innervated brain regions. I. Regional and cellular distribution in the rat brain. *Journal of Neuroscience*, **4**, 84–98.

Wei, F.Y., Tomizawa, K., Ohshima, T., Asada, A., Saito, T., Nguyen, C., Bibb, J.A., Ishiguro, K., Kulkarni, A.B., Pant, H.C. et al. (2005) Control of cyclin-dependent kinase 5 (Cdk5) activity by glutamatergic regulation of p35 stability. *Journal of Neurochemistry*, **93**, 502–512.

Yan, Z., Feng, J., Fienberg, A.A., and Greengard, P. (1999) $D_2$ dopamine receptors induce mitogen-activated protein kinase and cAMP response element-binding protein phosphorylation in neurons. *Proceedings of the National Academy of Sciences of the United States of America*, **96**, 11607–11612.

Zawadzki, K.M. and Taylor, S.S. (2004) cAMP-dependent Protein Kinase Regulatory Subunit Type IIβ: ACTIVE SITE MUTATIONS DEFINE AN ISOFORM-SPECIFIC NETWORK FOR ALLOSTERIC SIGNALING BY cAMP. *J. Biol. Chem.*, **279**, 7029-7036.

Øgreid, D. and Døskeland, S.O. (1981) The kinetics of the interaction between cyclic AMP and the regulatory moiety of protein kinase II. *FEBS Letters*, **129**, 282–286.

# 15
# Local Cortical Dynamics Related to Mental Illnesses

*Marco Loh, Edmund T. Rolls, and Gustavo Deco*

## 15.1
## Introduction

The dynamics of small patches of the cortex is a fundamental property of cortical processing. The local dynamics is based on interactions between thousands of cortical neurons, which are typically connected locally in recurrent neural networks and yield interesting properties due to the interaction of the neurons. The local circuitry in the cerebral cortex is made up of two types of neurons. Pyramidal neurons have a high density of excitatory connections to each other within a local area (1–3 mm) (Braitenberg and Schütz, 1991; Abeles, 1991). These local recurrent collateral excitatory connections provide a positive-feedback mechanism. Long-range connections also exist and create connections to other brain regions. The recurrent excitation is kept under control by inhibitory interneurons, which act locally. This functional architecture enables neurons to maintain their activity for many seconds and to implement a short-term memory (Goldman-Rakic, 1995). Each memory is formed by a set of neurons in the local cortical network that are coactive during a learning phase and formed strengthened connections between themselves by associative synaptic modification. During recall, the whole set of neurons can be activated by a partial recall cue and can maintain their activity for a number of seconds because of the positive feedback within that set of neurons. We describe a model of such a network, and then go into the dynamic properties and effects that can arise. Thereby, we introduce the concepts of attractor networks, energy landscape, noise, stability, and distractibility. We conclude by relating these concepts to mental disorders.

## 15.2
## A Recurrent Neural Network

As an example of a cortical network model, we present the attractor neural network that has been used in many of our investigations of cortical dynamics (Brunel and

Wang, 2001; Deco and Rolls, 2003; Loh and Deco, 2005; Loh, Rolls, and Deco, 2007; Rolls, Loh, and Deco, 2008). Note that this is just one of several implementations (e.g., see Durstewitz, Kelc, and Gunturkun, 1999; Durstewitz, Seamans, and Sejnowski, 2000a) and that, depending on the questions addressed, a network may provide more or fewer details. More abstract connectionist models might model a whole set of neurons by one variable (a rate model), whereas more detailed models such as the Hodgkin–Huxley model include cell membrane and synaptic ion channels. As the neural basis for our model, we use a standard recurrent network model (Brunel and Wang, 2001), which is based on integrate-and-fire neurons and has synaptically activated ion channels. Apart from its relevance for working memory, the model also addresses processes involved in decision-making, attention and perceptual detection (Loh and Deco, 2005; Deco and Rolls, 2006; Deco et al., 2007). We first describe the neural level of the model and then present the network architecture.

The neural correlate of both excitatory and inhibitory neurons is represented by the leaky integrate-and-fire model (Tuckwell, 1988):

$$C_m \frac{dV(t)}{dt} = -g_m(V(t) - V_L) - I_{syn}(t), \quad (15.1)$$

where $V(t)$ is the membrane potential, $C_m$ is the membrane capacitance, $g_m$ is the leak conductance, and $V_L$ is the resting potential. The basic state variable of a single model neuron is the membrane potential $V(t)$ (see Figure 15.1). It decays in time when the neurons receive no synaptic inputs down to a resting potential $V_L$. When the synaptic input $I_{syn}$ increases the membrane potential to a threshold $V_{thr}$, a spike is emitted and the neuron is set to the reset potential $V_{reset}$ at which it is kept for the refractory period $t_{ref}$. Spikes generate a synaptic current in the receiving neurons.

The synaptic input $I_{syn}$ of a neuron is made up of two parts: an external excitatory input via α-amino-3-hydroxy-5-methyl-4-isoxazol-propionacid (AMPA)-type synapses and a recurrent input from the other neurons of the network. The latter is made up of AMPA and N-methyl-D-aspartate (NMDA) synaptic currents activated by glutamate from the excitatory neurons, and γ-aminobutyric acid (GABA)$_A$ synaptic currents activated by GABA from the inhibitory neurons. Thus, $I_{syn}$ reads:

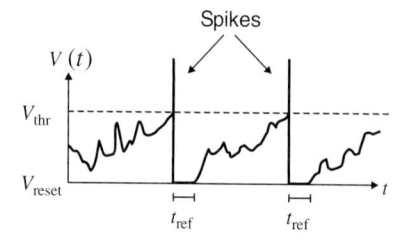

$$I_{syn}(t) = I_{AMPA,ext}(t) + I_{AMPA,rec}(t) + I_{NMDA,rec}(t) + I_{GABA}(t). \quad (15.2)$$

**Figure 15.1** Course of the membrane potential including spikes in the integrate-and-fire model.

The asynchronous external input has a rate of 2.4 kHz and can be viewed as originating from 800 external neurons firing at an average rate of $n_{ext} = 3$ Hz per neuron. The recurrent input currents are summed over all neurons with weights depending on the connectivity. The excitatory currents are mediated by the glutamatergic receptors AMPA and NMDA, which are both modeled by their effect in producing exponentially decaying currents in the postsynaptic neuron. The rise time of the AMPA current is neglected, because it is typically very short. The NMDA channel is modeled with an α function including both a rise and a decay term. In addition, the synaptic function of the NMDA current includes a voltage dependence controlled by the extracellular magnesium concentration (Jahr and Stevens, 1990). The recurrent excitation mediated by the AMPA and NMDA receptors is dominated by the NMDA current to avoid instabilities during the delay periods (Wang, 1999). The inhibitory neurons mediate their effects by GABA-A receptors described by a decay term. The parameters of the integrate-and-fire neurons and the synaptic channels for AMPA, NMDA, and GABA were chosen according to biological data. A full set of parameters and equations can be found in Brunel and Wang (2001) and Loh, Rolls, and Deco (2007).

Now we have described the neurons and the synapses that transmit the action potentials from neuron to neuron. The network set-up is next. Our standard network contains 400 excitatory and 100 inhibitory neurons, which is consistent with the observed proportions of pyramidal cells and interneurons in the cerebral cortex (Abeles, 1991; Braitenberg and Schütz, 1991).

The cortical network model consists of two selective pools S1 and S2 (Figure 15.2). The two selective pools represent memories that have been formed due to frequent simultaneous activations of these neurons. The connection weights between the neurons of each selective pool are called the intrapool connection strengths $w_+$. These are increased since we assume that a frequent simultaneous activation increases the connection strength. The increased strength of the intrapool connections is counterbalanced by the other excitatory connections ($w_-$) to keep the average input constant. The nonselective pool "NS" contains neurons that do not respond

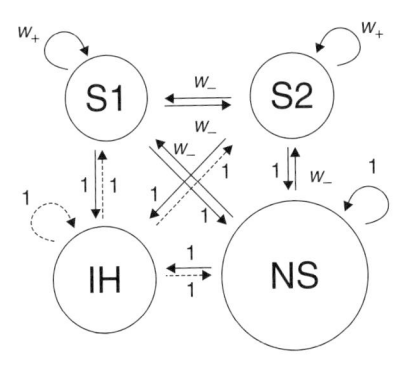

**Figure 15.2** Population and connection weights in a minimal neural network model to study stability and distractibility (see text).

selectively to the stimuli. These nonresponsive neurons are observed in many neurophysiological experiments. The nonselective neurons serve to generate an approximately Poisson spiking dynamics in the model, which is also observed in the cortex. The inhibitory pool "IH" contains the 100 inhibitory neurons. This architecture consisting of two selective populations is the minimal architecture to investigate stability and distractibility, which we envision are core characteristics affected by mental illnesses.

## 15.3
## Concept of an Attractor Network

The network described above has the properties of an attractor network. With no external influences, the activity of all neurons is low, around 3 Hz for excitatory neurons and 9 Hz for inhibitory neurons (see Figure 15.3b, "Spontaneous stable"). If the activity of the neurons of one pool (e.g., S1) is increased by an external stimulus, the activity of this pool increases. When the external stimulus is removed, the activity of S1 stays high (see Figure 15.3b, "Persistent stable"). The activity pattern of the network is self-sustained and has entered another state, which we call an attractor state. The network is an attractor network. In an attractor network, the state of the network is "attracted" towards the nearest possible attractor state in phase space, the space of all possible activity configurations. There are usually at least two distinct

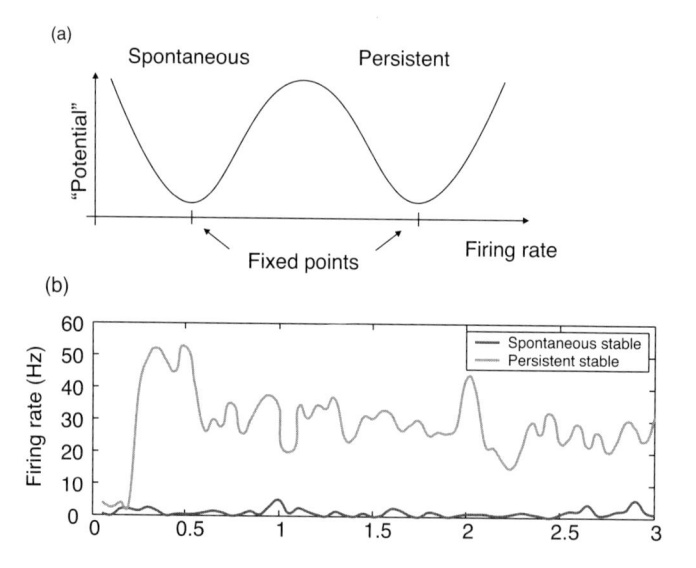

**Figure 15.3** Stable attractor configuration. (a) Hypothetical energy landscape showing two basins of attractions of the spontaneous and persistent state. (b) Two trials of a spiking simulation showing the activity of pool S1. In the persistent condition, a stimulus cue was applied to pool S1 in the period 0–0.5s.

states – a spontaneous state marked by low neuronal firing rates, and a persistent state with higher firing rates in which one of the selective populations has a high firing rate, here either S1 or S2. An attractor network can have many different attractor states, each consisting of a different subset of the neurons being active; any one subset of neurons can represent a short-term memory.

Attractor networks appear to operate in the prefrontal cortex – an area that is important in attention and short-term memory, as shown, for example, by firing in the delay period of a short-term memory task (Kubota and Niki, 1971; Fuster and Alexander, 1971; Funahashi, Bruce, and Goldman-Rakic, 1989; Fuster, 1995; Rolls, 2008) and is fundamental to top-down attention since everything that requires attention has to be maintained in a short-term memory (Desimone and Duncan, 1995; Rolls and Deco, 2002). The impairments of attention induced by prefrontal cortex damage may be accounted for in large part by an impairment in the ability to hold the object of attention stably and without distraction in the short-term memory systems in the prefrontal cortex (Goldman-Rakic, 1996).

The behavior of attractor networks can be illustrated by an energy landscape (Hopfield, 1982). Figure 15.3(a) shows such landscapes in which the attractor states (each one representing a memory) or fixed points are indicated by the valleys. One can imagine a ball moving in that landscape that is at rest at the bottom of a valley. A force in terms of input or noise is needed to move the ball from one valley to another. In general, the hypothetical landscape can be multidimensional with several distinct attractor states, each one representing a different stored memory. We envision that many parts of the brain operating as dynamical systems have characteristics of such attractor systems including statistical fluctuations, caused, for example, by the probabilistic firing of the neurons, in which the noise or fluctuations make the system jump over an energy barrier from one attractor state (including the spontaneous state) to another.

Although energy functions only apply to recurrent networks with symmetric connections between the neurons (Hopfield, 1982) and do not necessarily apply to more complicated networks with, for example, incomplete connectivity, the properties of these other recurrent networks are similar (Treves and Rolls, 1991; Rolls and Treves, 1998). The concept of an energy function and landscape is useful for discussion purposes.

## 15.4
## Noise and Stability

Up to now, we discussed how the neural network is constructed and how its behavior can be seen as an attractor system, which can be pictured in an energy landscape. External influences can change the activity of the system and thereby cortical processing takes place. However, we did not discuss an important factor that disturbs almost every biological system – noise. In the network, the information is passed from neuron to neuron via spikes – point-like events that induce currents in other neurons. There are two sources of noise in such spiking networks – the randomly arriving external Poissonian spike trains and the statistical fluctuations due to the spiking of the neurons within the network. These statistical fluctuations depend on the finite size of the network (Brunel and Hakim, 1999; Mattia and Del Giudice, 2002, 2004; Deco and Rolls, 2006). That is, the smaller the network, the higher the fluctuations due to finite size effects. In the brain, the noise arises not only from the probabilistic spiking of the neurons in finite size networks, but also from any other sources or the environment (Faisal, Selen, and Wolpert, 2008), including the effects of distracting stimuli. An important feature of brain dynamics is that it is sufficiently stable that processing is not normally disturbed, but also new inputs are able to affect the dynamics.

The stability of an attractor is characterized by the average time in which the system stays in the basin of attraction under the influence of noise. The noise provokes transitions to other attractor states. Note that the probability of moving from state A to state B might be different from the probability of moving from state B to state A, as the shape of the landscape is important. Two factors determine the stability. (i) If the depth of the attractors is shallow, then less force is needed to move a ball from one valley to the next. This is shown in Figure 15.4(a) in which the valleys are shallower compared to the normal condition (Figure 15.3a). (ii) Noise will also make it more likely that the system will jump over an energy boundary from one state to another. These two effects go together and it is hard to investigate them separately in network

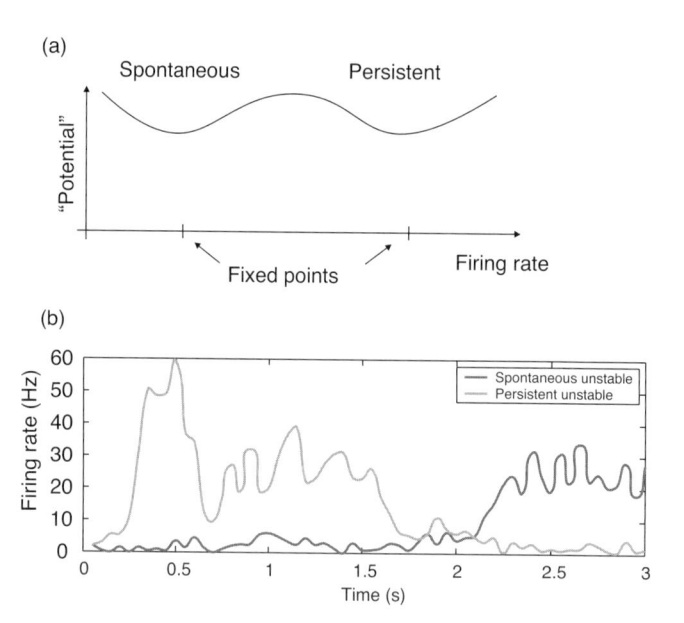

**Figure 15.4** Unstable attractor configuration. (a) Hypothetical energy landscape showing two shallow basins of attractions of the spontaneous and persistent state. (b) Two trials of a spiking simulation showing the activity of pool S1. In both trials, the activity is unstable either at the spontaneous state or the persistent state and switches to the other state. In these simulations, we reduced the conductances of the NMDA and GABA channels. In the persistent condition, a stimulus cue was applied to pool S1 in the period 0–0.5s.

simulations, since changing the depth of the attractors might also have an influence on the noise. Using the meanfield approach to neural networks (Brunel and Wang, 2001; Loh, Rolls, and Deco, 2007), it is possible to assess the changes in the attractor landscape and thereby the depth of the basin of attraction without the influence of noise. We showed clearly that changes in the synaptic conductances of NMDA and GABA changes the depth of the attractors (Loh, Rolls, and Deco, 2007). Another possibility is to change the size of the network and to see how the noise due to the finite size effects changes. This has, for example, an influence on stochastic resonance (Goldbach et al., 2007).

In general, the stability of the attractor states of the network can be measured by the probability of the system entering another attractor state in a certain amount of time. In our simulations, we address the stability of the spontaneous and the persistent attractor separately. In spontaneous simulations, we run spiking simulations for 3 s without any extra external input. The aim of this condition is to test whether the network is stable in maintaining a low average firing rate or whether it jumps into one of its persistent attractor states (those with high continuous firing rates) without any external input (as shown in Figure 15.4b, "Spontaneous unstable"). In persistent simulations, we shift the activity of the pool S1 with an additional external input to the persistent state and then run the simulation after removing this external cue for another 2.5 s. The aim of this condition is to investigate whether once in the persistent (high firing rate) attractor state, the network can maintain its activity stably, or whether it falls out of its attractor (Figure 15.4b, persistent unstable). We measure the stability of the network in many trials. We run the same simulation for typically 1000 trials each 3 s long with different random seeds for the external background input and then count how often the system has jumped, for example, from the spontaneous state to the persistent state. These statistics provide a measure for the stability of the spontaneous or persistent attractor state.

How can the stability of a system be changed? There are several factors, some of them already mentioned above. The influence of the synaptically activated ion channels is of especial interest, since changes in them are envisioned to play a major role in major mental illnesses such as schizophrenia and obsessive-compulsive disorder (OCD). Here, we address aspects of the NMDA and GABA synaptic channels.

The excitatory NMDA channels have a great influence on the shape of the attractor landscape and the firing rates, especially of the persistent attractor state. Reducing the conductances of the NMDA receptor activated ion channels makes the basin of attraction shallower and reduces the firing rate of the persistent attractor state. As a consequence the attractor becomes unstable (Durstewitz, Seamans, and Sejnowski, 2000b; Loh, Rolls, and Deco, 2007). However, there are also specific characteristics of the NMDA channel that contribute to the stability.

The NMDA receptor activated ion channel has a long time constant (around 100 ms as compared to the AMPA excitatory receptor (5–10 ms) or the GABA receptor (5–10 ms) activated ion channels). This smooths across time the statistical fluctuations that are caused by the random spiking of populations of neurons in the network. In addition, the relation of the excitatory and inhibitory time constants is crucial.

The excitatory component needs to be slower than the inhibitory one, since a system with fast-positive and slow-negative feedback is dynamically unstable (Wang, 1999; Compte et al., 2000). The NMDA channel also has voltage-dependent gating. If the cells are depolarized and fire strongly, the magnesium block is removed. This additionally promotes stability, since cells that are already firing at a high frequency can be excited more easily than cells which fire at resting frequencies. In addition, the slow unbinding of glutamate from the NMDA receptor can cause an upper bound on the firing rate, which together with the inhibitory feedback helps to keep high firing under control (Wang, 2001).

The inhibitory GABA channels also influence the attractor landscape. Reducing their conductance makes the basin of attraction of the spontaneous state shallower and increases the firing rate of the persistent attractor state. Thus, the system is more likely to jump from the spontaneous state to the persistent state.

## 15.5
## Mental Illnesses

We have now discussed the basic concepts of biophysical neural modeling, attractor dynamics, and stability in the context of a concrete implementation of a neural network. The underlying assumption of this modeling is that local dynamics and its rich properties are essential for cortical processing and normal mental function, and if operating incorrectly may contribute to mental illnesses. Changes in the synaptic conductances within the local neural networks can have drastic effects on the behavior that characterizes the networks, and can thereby have a wide range of effects on brain activity. Our hypotheses for mental disorders build on the concept of basins of attraction. We argue that an altered attractor landscape together with the noise causes altered transitions between the different attractors and the spontaneous firing state. We describe how we relate alterations in neural networks to mental illnesses.

### 15.5.1
### Schizophrenia

Schizophrenia is a major mental illness, which has a great impact on patients and their environment. The lifetime risk of a schizophrenic episode is about 1% and the onset of the symptoms typically is in young adulthood. The diagnosis is based on the self-reported experiences of the patients and observations of family members or friends. However, there is currently no biological marker available. Schizophrenia is characterized by three types of symptoms: positive symptoms, negative symptoms, and cognitive dysfunction. The positive symptoms of schizophrenia include bizarre (psychotic) trains of thoughts, hallucinations, and (paranoid) delusions. The illness is typically diagnosed with the onset of the first episode of positive symptoms. The positive symptoms come often in episodes, but medication (neuroleptics) can in most patients alleviate or eliminate those symptoms. The cognitive symptoms include distractibility, poor attention, and the dysexecutive syndrome, which are probably caused by a working memory deficit (i.e., a difficulty in maintaining items in short-term memory). The negative symptoms refer to the flattening of affect and a reduction in emotion. Behavioral indicators are blunted affect, emotional and passive withdrawal, poor rapport, lack of spontaneity, motor retardation, and disturbance of volition. The cognitive and negative symptoms typically appear together, and often exist before the first psychotic episode. They usually persist after the outbreak of the illness and the neuroleptics have a limited effect on these symptoms.

Several computational models have been developed to investigate the symptoms associated with schizophrenia. Of those we want to mention the work of Cohen and collaborators who developed high-level connectionist models for concrete tasks that show effects in schizophrenic patients (e.g., see Braver, Barch, and Cohen, 1999; Montague, Hyman, and Cohen, 2004). This type of model is important, since it allows a direct modeling of experimental evidence. Durstewitz, Seamans and colleagues developed detailed neurophysiological models based on neurophysiological evidences on dopamine (Durstewitz, Kelc, and Gunturkun, 1999; Durstewitz, Seamans, and Sejnowski, 2000a, 2000b; Durstewitz and Seamans, 2002). Their main finding is that dopamine might adapt the signal-to-noise ratio (Seamans *et al.*, 2001; Seamans and Yang, 2004). Based on this finding, they developed a dual-state theory of schizophrenia (Durstewitz and Seamans, 2008), which contains a dopamine $D_1$-dominated state and a dopamine $D_2$-dominated state. These are associated with a deep and shallow attractor landscape, respectively. Here, we develop a top-down approach based on the different types of symptom and relate them to instabilities in attractor neural networks in different brain regions (Loh, Rolls, and Deco, 2007). Note that we do not derive our hypothesis from neurophysiological evidence, which we discuss as a second step in the light of our findings and which may provide reasons why the stability has been reduced. For a more detailed review of the computational models on schizophrenia, see Rolls *et al.* (2008b).

We relate the positive symptoms to shallow basins of attraction of both the spontaneous and persistent states, which might occur in the temporal lobe semantic memory networks (Rolls, 2005; Loh, Rolls, and Deco, 2007). This could result in activations arising spontaneously and thoughts moving too freely round the energy landscape from thought to weakly associated thought. This might lead to bizarre thoughts and associations, which may eventually over time cause false beliefs and delusions. Consistently, neuroimaging studies suggest higher activation especially in areas of the temporal lobe (Scheuerecker *et al.*, 2007; Shergill *et al.*, 2000; Weiss and Heckers, 1999).

We propose that the cognitive symptoms may be related to instabilities of persistent states in attractor neural networks in the prefrontal cortex that implement short-term memory and attention (Rolls, 2005; Loh, Rolls, and Deco, 2007), consistent with the body of theoretical research on network models of working memory (Durstewitz, Seamans, and Sejnowski, 2000a, 2000b; Brunel and Wang, 2001; Durstewitz and Seamans, 2002). The basins of attraction of the persistent states are shallow and thus a stable short-term memory is difficult to maintain, normally the source of the bias in biased competition models of attention (Rolls and Deco, 2002; Deco and Rolls, 2005).

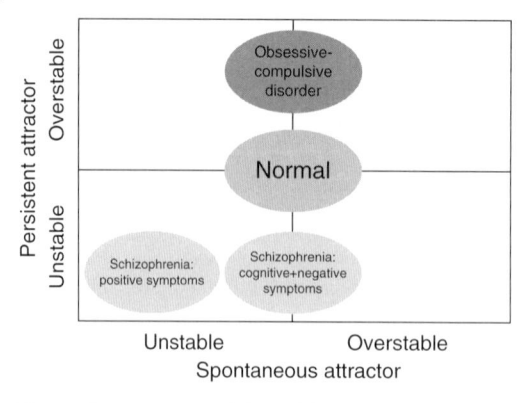

**Figure 15.5** Overview of the stabilities and their relationship to the symptoms of schizophrenia and OCD.

The negative symptoms may be related to decreases in firing rates in the orbitofrontal cortex and/or anterior cingulate cortex which are usually produced by manipulations that decrease the stability of attractor networks (Rolls, 2005; Loh, Rolls, and Deco, 2007). Functional magnetic resonance imaging (fMRI) investigations have shown that these regions code reward value and pleasure (Rolls, 2005; Rolls and Grabenhorst, 2008). Consistently, imaging studies have identified a relationship between negative symptoms and prefrontal hypofunction (i.e., a reduced activation of frontal areas) (Aleman and Kahn, 2005; Wolkin *et al.*, 1992).

These instabilities are illustrated in Figure 15.5, which suggests that the cognitive and negative symptoms are related to a decrease in the stability of persistent attractor states, whereas the positive symptoms feature additionally an instability of the spontaneous attractor. This might help to explain the relationship of the symptoms. The negative and cognitive symptoms typically precede the first psychotic episode of the positive symptoms, and may reflect decreased stability of attractors and decreased firing rates. The positive symptoms may appear when there is in addition an instability of the spontaneous attractor.

Our hypothesis as described so far merely describes the effects in a dynamical system and is not dependent on any particular biological cause. (These might be described at several levels of abstraction in the brain, ranging from the single neuron level up to systems-level neuroscience approaches (Carlsson, 2006).) As schizophrenia is a heterogeneous disease, various biological causes could lead to the same set of symptoms and thus the hypothesized alterations in the dynamical attractor system. The overall goal is to investigate the pathways and mechanisms that lead to the hypothesized alterations (Figure 15.6). Here, we have described an attractor network approach focusing on the contribution of NMDA and GABA receptor-activated synaptic currents. These synaptic receptors are important since there is evidence that hypofunctionality of the NMDA and GABA systems is related to the symptoms of schizophrenia (Coyle, Tsai, and Goff, 2003; Lewis, Hashimoto, and Volk, 2005).

As mentioned above, a reduction of the NMDA conductance reduces the stability of the persistent state drastically. We hypothesized that such a pattern might be related

## 15.5 Mental Illnesses | 331

**Changes in...**
synaptic conductances
dopamine signaling
genetic mechanisms
mesocortical influences
regional differences
neuropil
plasticity
....

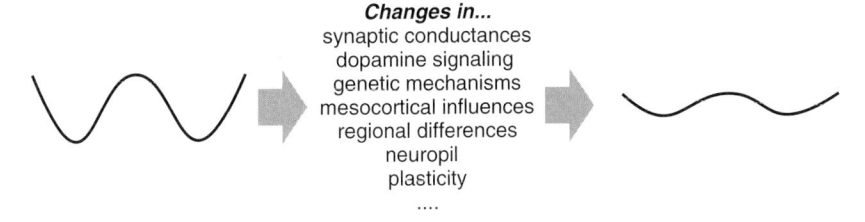

**Figure 15.6** Several biological mechanisms could cause the proposed alterations of the attractor landscape and probably there is no single pathway to heterogeneous disorders such as schizophrenia.

to the cognitive symptoms, since it shows a reduced stability of the working memory properties. We relate the negative symptoms to a reduction of the mean firing rate of the persistent state of networks in, for example, the orbitofrontal cortex. A reduction of the firing rate was produced when the NMDA current was reduced (see also Durstewitz, Seamans, and Sejnowski, 2000a; Brunel and Wang, 2001). Thus, the cognitive and negative symptoms can be related in our hypothesis to the same synaptic mechanism, namely a reduction of NMDA conductance.

When both NMDA and GABA are reduced one might think that these two counterbalancing effects (excitatory and inhibitory) would either cancel each other out or yield a tradeoff between the stability of the spontaneous and persistent state. However, this is not the case. The stability of both the spontaneous and the persistent state is reduced (Loh, Rolls, and Deco, 2007). We relate this pattern to the positive symptoms of schizophrenia, in which both the spontaneous and attractor states are shallow, and the system jumps due to the influence of statistical fluctuations between the different attractor states.

We further investigated the signal-to-noise ratio in relation to the changes in synaptic conductances. In an attractor network, a high signal-to-noise ratio could be used to refer to a scenario in which the network will maintain the attractor stably (which could be taken as the signal), as opposed to being disrupted by spiking-related statistical fluctuations which would make it enter the spontaneous state (which could be considered a noise outcome). We found that in all the cases in which the NMDA, the GABA conductance, or both, are reduced, the signal-to-noise ratio (e.g., measured by the proportion of trials on which the high firing rate persistent attractor state is correctly maintained) is also reduced (Loh, Rolls, and Deco, 2007). This relates to recent experimental observations that show a decreased signal-to-noise ratio in schizophrenic patients (Winterer *et al.*, 2000, 2004; Winterer *et al.*, 2006). We directly relate a decrease in the signal-to-noise ratio to changes (in this case decreases) in receptor-activated synaptic ion channel conductances (Loh, Rolls, and Deco, 2007; Rolls *et al.*, 2008b).

In our approach, we did not model dopamine specifically, and the modulatory effects of dopamine mediated through $D_1$ and $D_2$ receptors in cortical and subcortical circuits are complex. The dopamine hypothesis of schizophrenia originally focused on a hyperdopaminergic state in the striatum, partly due to the high density of $D_2$

receptors (Stevens, 1973). Schizophrenia is treated with neuroleptic drugs (typically dopamine receptor $D_2$ antagonists), which mainly alleviate the positive symptoms, whereas the cognitive and negative symptoms persist, especially for the typical neuroleptics (Mueser and McGurk, 2004). We found that a decrease in the GABA receptor-activated synaptic conductances makes the spontaneous firing state less stable so that the system jumps into an attractor even without a stimulus, and related this to the positive symptoms of schizophrenia (Loh, Rolls, and Deco, 2007; Rolls et al., 2008b). We reasoned that an effect of neuroleptics might be an increase in the GABA conductance. This is in fact consistent with experimental work – it has been found that activation of $D_2$ receptors decreases the GABA efficacy (Seamans et al., 2001; Trantham-Davidson et al., 2004). In a supersensitive $D_2$ receptor state (Seeman et al., 2005, 2006), $D_2$ antagonists would increase the GABA currents and thereby increase the inhibition in the network, and in particular reduce the instability of the spontaneous state.

Although the positive symptoms can be treated effectively in many patients, the negative and cognitive symptoms typically persist. To ameliorate the cognitive symptoms, the persistent state according to our hypothesis needs to be stabilized (see Figure 15.5). In the computational model investigated, the stability of the working memory high firing rate attractor state is increased by NMDA receptor-mediated currents, by increasing the firing rates making the attractors more stable, and by their long time constant smoothing effect on the fluctuations (Durstewitz, Seamans, and Sejnowski, 2000b; Wang, 2001; Deco, 2006). A possible way to increase the NMDA currents is by activation of $D_1$ receptors (Durstewitz and Seamans, 2002; Seamans and Yang, 2004). In agreement with our hypothesis, newer versions of the dopamine hypothesis propose that $D_1$ agonists may increase the signal-to-noise ratio by increasing the stability of high firing rate "persistent" attractor states (Winterer and Weinberger, 2004; Rolls et al., 2008b). The observation of "hypofrontality" in schizophrenia might be related to decreased stability in prefrontal cortical circuits that implement cognitive functions such as short-term memory and attention (Winterer et al., 2004; Winterer et al., 2006), which in turn has been indirectly linked to dopamine (Winterer et al., 2006).

### 15.5.2
### OCD

OCD is a chronically debilitating disorder with a lifetime prevalence of 2–3% (Robins et al., 1984; Karno et al., 1988; Weissman et al., 1994). It is characterized by obsessive, intrusive thoughts with accompanying compulsions. The symptoms take a considerable amount of time during the day and cause an impairment in everyday activities. OCD patients perform tasks (or compulsions) to neutralize obsession-related anxieties. Common themes of the obsessions are contamination and "germs," checking household items in case of fire or burglary, the order and symmetry of objects, and fears of harming oneself or others. Physical symptoms may be related to anxieties and unwanted thoughts, and may include rigidity, tremor, jerking arm movements, or involuntary movements of the limbs.

Analogously to the work on schizophrenia, we relate OCD and its symptoms to dynamical systems theory (Rolls, Loh, and Deco, 2008). In contrast to the instabilities in schizophrenia, we propose that overstability of attractor neuronal networks in cortical and related areas may be a core feature of OCD. Overstability could arise, for example, by overactivity in glutamatergic excitatory synapses, which increases the depth of the basins of attraction in part by increasing the firing rates of the high firing rate "persistent" attractor state and makes it more difficult to move the state of the system from one attractor to the next. This increased stability of cortical and related attractor networks, and the associated higher neuronal firing rates, could occur in different brain regions, and thereby produce different symptoms, as follows (Rolls, Loh, and Deco, 2008).

In high-order motor areas, the symptoms could include inability to move out of one motor pattern, resulting in repeated movements or actions, for example. In parts of the cingulate cortex and dorsal medial prefrontal cortex, this could result in difficulty in switching between actions or strategies (Rushworth *et al.*, 2007a, 2007b), as the system would be locked into one action or strategy.

The lateral prefrontal cortex is involved in short-term memory, attention, and executive function. The increased stability of prefrontal cortex attractor networks could produce symptoms such as difficulty in shifting attention and in cognitive set shifting (Menzies *et al.*, 2008). Planning may also be impaired in patients with OCD (Menzies *et al.*, 2008) and this could arise because there is too much stability of attractor networks in the dorsolateral prefrontal cortex, which could lead to a less flexible working memory and therefore a reduced capability of planning (Rolls, 2008).

In addition, emotional regions may be affected. An increased firing rate of neurons in the orbitofrontal cortex, and anterior cingulate cortex, would increase emotionality, which is frequently found in OCD. Part of the increased anxiety found in OCD could be related to an inability to complete tasks or actions in which one is locked. However, increased emotionality in OCD may also be directly related to increased firing rates produced by the increased glutamatergic activity in brain areas such as the orbitofrontal and anterior cingulate cortex (Rolls, 2008).

If the increased stability of attractor networks occurred in temporal lobe semantic memory networks, then this would result in a difficulty in moving from one thought to another and possibly in stereotyped thoughts, which again may be a symptom of OCD (Menzies *et al.*, 2008).

The obsessional states are thus proposed to arise because cortical areas concerned with cognitive functions have states that become too stable. The compulsive states are proposed to arise partly in response to the obsessional states, but also partly because cortical areas concerned with actions have states that become too stable. The theory provides a unifying computational account of both the obsessional and compulsive symptoms, in that both arise due to increased stability of cortical attractor networks, with the different symptoms related to overstability in different cortical areas. The theory is also unifying in that a similar increase in glutamatergic activity in the orbitofrontal and anterior cingulate cortex could increase emotionality, as described above.

Having proposed a generic hypothesis for OCD, we recognize of course that the exact symptoms that arise if stability in some systems is increased will be subject to the exact effects that these will have in an individual patient, who may react to these effects, and produce explanatory accounts for the effects, and ways to deal with them, that may be quite different from individual to individual.

Integrate-and-fire simulations show that an increase of NMDA or AMPA synaptic currents can increase the stability of attractor networks (Rolls, Loh, and Deco, 2008). The simulations also show that the stability of the spontaneous (quiescent) firing rate state is reduced by increasing NMDA or AMPA receptor-activated synaptic currents. We relate this to the symptoms of obsessive compulsive disorder in that the system is more likely than normally, under the influence of the spiking stochastic noise caused by the neuronal spiking in the system, to jump into one of the dominant attractor states, which might be a recurring idea or concern or action (see Figure 15.4b, "Spontaneous unstable").

This simulation evidence, that an increase of glutamatergic synaptic efficacy can increase the stability of attractor networks and thus potentially provide an account for some of the symptoms of OCD, is consistent with evidence that glutamatergic function may be increased in some brain systems in OCD (Rosenberg et al., 2000, 2004; Rosenberg, MacMillan, and Moore, 2001; Pittenger, Krystal, and Coric, 2006) and that cerebrospinal fluid glutamate levels are elevated (Chakrabarty et al., 2005). Consistent with this, agents with antiglutamatergic activity such as riluzole, which can decrease glutamate transmitter release, may be efficacious in OCD (Pittenger, Krystal, and Coric, 2006; Bhattacharyya and Chakraborty, 2007).

The theory about how the symptoms of OCD could arise in relation to the increased stability of cortical attractor networks has implications for possible pharmaceutical approaches to treatment. One is that treatments that reduce glutamatergic activity (e.g., by partially blocking NMDA receptors) might be useful. Another is that increasing the inhibition in the cortical system (e.g., by increasing GABA receptor-activated synaptic currents) might be useful, both by bringing the system from a state where it was locked into an attractor back to the normal level and by making the spontaneous state more stable, so that it would be less likely to jump to an attractor state (which might represent a dominant idea or concern or action).

## 15.6
## Outlook

Both models, for schizophrenia and for OCD, are based on the attractor framework and our simulations were done with the same neural network model. The model of OCD (Rolls, Loh, and Deco, 2008) is in a sense the opposite of the model of schizophrenia (Loh, Rolls, and Deco, 2007). In the schizophrenia model, we suggest that instabilities of the spontaneous and persistent attractors, and decreases in the firing rates of the high firing rate "persistent" attractor states are responsible for the diverse symptoms. In OCD, overstabilities and increases of firing rates might cause the symptoms. Our investigations and the underlying relationship to the biological data highlight the concept of attractor networks to cortical function (Rolls, 2008), and emphasize that any alteration in their stability, whether towards too little stability or too much, might have major consequences for the functioning of the cerebral cortex.

However, we emphasize that the way in which the network effects produce the symptoms in individual patients will be complex and there are various factors which can cause a predisposition for a disorder. Several factors could lead to the alterations of the attractor landscape. These include not only changes in the conductances of AMPA, NMDA, or GABA synaptic receptor-activated ion channels, but also of membrane ion channels of, for example, $Na^+$, $K^+$, $Ca^{2+}$, and $Cl^-$ ions, and several other receptor types, such as the metabotropic glutamate receptors. Beyond that, aberrant dopamine signaling, reduced neuropil, genetic mechanisms, and brain volume reduction might also be possible factors that influence the stability of neural networks (see Figure 15.6; Goldman-Rakic, 1999; Mueser and McGurk, 2004; Winterer and Weinberger, 2004). Differences between individuals in the symptoms may be related to differences in the stability of attractors in different cortical areas, and to regional differences in the functionality of the mesolimbic and mesocortical dopamine pathways (Capuano, Crosby, and Lloyd, 2002; Carlsson, 2006; Rolls, 2008). Our neurodynamical hypotheses might serve as a unifying framework for the different physiological causes that may all lead to similar alterations in the attractor landscape. We hope that our approach may help to deal with the heterogeneity and complexity of the neurobiology of schizophrenia by defining a common statistical dynamical framework based on the schizophrenic symptoms. Including aspects such as plasticity or genetic diversity might further stimulate the field on both sides.

An important contribution of our framework is that it shifts the interest to concepts like signal-to-noise ratio and stability, which might be more important than the absolute activities in the brain in many mental illnesses (Seamans and Yang, 2004; Winterer and Weinberger, 2004; Durstewitz and Seamans, 2008). Imaging studies using electroencephalography or fMRI have already targeted this issue (Winterer *et al.*, 2000; Winterer *et al.*, 2006), and our investigation of the neural network data show related effects (Rolls *et al.*, 2008b). A more detailed investigation of these concepts both in imaging studies and in neural network simulations might stimulate further thinking and research in this area.

## References

Abeles, M. (1991) *Corticonics*, Cambridge University Press, New York.

Aleman, A. and Kahn, R.S. (2005) Strange feelings: do amygdala abnormalities dysregulate the emotional brain in schizophrenia? *Progress in Neurobiology*, **77**, 283–298.

Bhattacharyya, S. and Chakraborty, K. (2007) Glutamatergic dysfunction–newer targets for anti-obsessional drugs. *Recent Patents CNS Drug Discovery*, **2**, 47–55.

Braitenberg, V. and Schütz, A. (1991) *Anatomy of the Cortex*, Springer, Berlin.

Braver, T.S., Barch, D.M., and Cohen, J.D. (1999) Cognition and control in schizophrenia: a computational model of dopamine and prefrontal function. *Biological Psychiatry*, **46**, 312–328.

Brunel, N. and Hakim, V. (1999) Fast global oscillations in networks of integrate-and-fire neurons with low firing rates. *Neural Computation*, **11**, 1621–1671.

Brunel, N. and Wang, X.J. (2001) Effects of neuromodulation in a cortical network model of object working memory dominated by recurrent inhibition. *Journal of Computational Neuroscience*, **11**, 63–85.

Capuano, B., Crosby, I.T., and Lloyd, E.J. (2002) Schizophrenia: genesis, receptorology and current therapeutics. *Current Medicinal Chemistry*, **9**, 521–548.

Carlsson, A. (2006) The neurochemical circuitry of schizophrenia. *Pharmacopsychiatry*, **39** (Suppl. 1), S10–S14.

Chakrabarty, K., Bhattacharyya, S., Christopher, R., and Khanna, S. (2005) Glutamatergic dysfunction in OCD. *Neuropsychopharmacology*, **30**, 1735–1740.

Compte, A., Brunel, N., Goldman-Rakic, P., and Wang, X.J. (2000) Synaptic mechanisms and network dynamics underlying spatial working memory in a cortical network model. *Cerebral Cortex*, **10**, 910–923.

Coyle, J.T., Tsai, G., and Goff, D. (2003) Converging evidence of NMDA receptor hypofunction in the pathophysiology of schizophrenia. *Annals of the New York Academy of Sciences*, **1003**, 318–327.

Deco, G. (2006) A dynamical model of event-related fMRI signals in prefrontal cortex: predictions for schizophrenia. *Pharmacopsychiatry*, **39** (Suppl. 1), S65–S67.

Deco, G. and Rolls, E.T. (2003) Attention and working memory: a dynamical model of neuronal activity in the prefrontal cortex. *European Journal of Neuroscience*, **18**, 2374–2390.

Deco, G. and Rolls, E. (2005) Attention, short-term memory, and action selection: a unifying theory. *Progress in Neurobiology*, **76**, 236–256.

Deco, G. and Rolls, E. (2006) Decision-making and Weber's law: a neurophysiological model. *European Journal of Neuroscience*, **24**, 901–916.

Deco, G., Pérez-Sanagustín, M., de Lafuente, V., and Romo, R. (2007) Perceptual detection as a dynamical bistability phenomenon: a neurocomputational correlate of sensation. *Proceedings of the National Academy of Sciences of the United States of America*, **104**, 20073–20077.

Desimone, R. and Duncan, J. (1995) Neural mechanisms of selective visual attention. *Annual Review of Neuroscience*, **18**, 193–222.

Durstewitz, D. and Seamans, J.K. (2002) The computational role of dopamine $D_1$ receptors in working memory. *Neural Networks*, **15**, 561–572.

Durstewitz, D. and Seamans, J. (2008) The dual-state theory of prefrontal cortex dopamine function with relevance to catechol-O-methyltransferase genotypes and schizophrenia. *Biological Psychiatry*, **64**, 739–749.

Durstewitz, D., Kelc, M., and Gunturkun, O. (1999) A neurocomputational theory of the dopaminergic modulation of working memory functions. *Journal of Neuroscience*, **19**, 2807–2822.

Durstewitz, D., Seamans, J.K., and Sejnowski, T.J. (2000a) Dopamine-mediated stabilization of delay-period activity in a network model of prefrontal cortex. *Journal of Neurophysiology*, **83**, 1733–1750.

Durstewitz, D., Seamans, J.K., and Sejnowski, T.J. (2000b) Neurocomputational models of working memory. *Nature Neuroscience*, **3** (Suppl.), 1184–1191.

Faisal, A., Selen, L., and Wolpert, D. (2008) Noise in the nervous system. *Nature Reviews Neuroscience*, **9**, 292–303.

Funahashi, S., Bruce, C., and Goldman-Rakic, P. (1989) Mnemonic coding of visual space in the monkey's dorsolateral prefrontal cortex. *Journal of Neurophysiology*, **61**, 331–349.

Fuster, J. and Alexander, G. (1971) Neuron activity related to short-term memory. *Science*, **173**, 652–654.

Fuster, J.M. (1995) *Memory in the Cerebral Cortex*, MIT Press, Cambridge, MA.

Goldbach, M., Loh, M., Deco, G., and Garcia-Ojalvo, J. (2007) Neurodynamical amplification of perceptual signals via system-size resonance. *Physica D*, **237**, 316–323.

Goldman-Rakic, P. (1995) Cellular basis of working memory. *Neuron*, **14**, 477–485.

Goldman-Rakic, P. (1996) Regional and cellular fractionation of working memory. *Proceedings of the National Academy of Sciences of the United States of America*, **93**, 13473–13480.

Goldman-Rakic, P.S. (1999) The physiological approach: functional architecture of working memory and disordered cognition in schizophrenia. *Biological Psychiatry*, **46**, 650–661.

Hopfield, J. (1982) Neural networks and physical systems with emergent collective computational abilities. *Proceedings of the National Academy of Sciences of the United States of America*, **79**, 2554–2558.

Jahr, C. and Stevens, C. (1990) Voltage dependence of NMDA-activated macroscopic conductances predicted by single-channel kinetics. *Journal of Neuroscience*, **10**, 3178–3182.

Karno, M., Golding, J., Sorenson, S., and Burnam, M. (1988) The epidemiology of obsessive-compulsive disorder in five US communities. *Archives of General Psychiatry*, **45**, 1094–1099.

Kubota, K. and Niki, H. (1971) Prefrontal cortical unit activity and delayed alternation performance in monkeys. *Journal of Neurophysiology*, **34**, 337–347.

Lewis, D.A., Hashimoto, T., and Volk, D.W. (2005) Cortical inhibitory neurons and schizophrenia. *Nature Reviews Neuroscience*, **6**, 312–324.

Loh, M. and Deco, G. (2005) Cognitive flexibility and decision making in a model of conditional visuomotor associations. *European Journal of Neuroscience*, **22**, 2927–2936.

Loh, M., Rolls, E.T., and Deco, G. (2007) A dynamical systems hypothesis of schizophrenia. *PLOS Computational Biology*, **3**, e228.

Mattia, M. and Del Giudice, P. (2002) Attention and working memory: a dynamical model of neuronal activity in the prefrontal cortex. *Physical Review E*, **66**, 51917–51919.

Mattia, M. and Del Giudice, P. (2004) Finite-size dynamics of inhibitory and excitatory interacting spiking neurons. *Physical Review E*, **70**, 052903.

Menzies, L., Chamberlain, S., Laird, A., Thelen, S., Sahakian, B., and Bullmore, E. (2008) Integrating evidence from neuroimaging and neuropsychological studies of obsessive-compulsive disorder: the orbitofronto-striatal model revisited. *Neuroscience and Biobehavioral Reviews*, **32**, 525–549.

Montague, P.R., Hyman, S.E., and Cohen, J.D. (2004) Computational roles for dopamine in behavioural control. *Nature*, **431**, 760–767.

Mueser, K.T. and McGurk, S.R. (2004) Schizophrenia. *Lancet*, **363**, 2063–2072.

Pittenger, C., Krystal, J., and Coric, V. (2006) Glutamate-modulating drugs as novel pharmacotherapeutic agents in the treatment of obsessive-compulsive disorder. *NeuroRx*, **3**, 69–81.

Robins, L., Helzer, J., Weissman, M., Orvaschel, H., Gruenberg, E., Burke, J., and Regier, D. (1984) Lifetime prevalence of specific psychiatric disorders in three sites. *Archives of General Psychiatry*, **41**, 949–958.

Rolls, E.T. (2005) *Emotion Explained*, Oxford University Press, Oxford.

Rolls, E.T. (2008) *Memory, Attention, and Decision-Making: A Unifying Computational Neuroscience Approach*, Oxford University Press, Oxford.

Rolls, E.T. and Deco, G. (2002) *Computational Neuroscience of Vision*, Oxford University Press, Oxford.

Rolls, E.T. and Grabenhorst, F. (2008) The orbitofrontal cortex and beyond: from affect to decision-making. *Progress in Neurobiology*, **86**, 216–244.

Rolls, E.T. and Treves, A. (1998) *Neural Networks and Brain Function*, Oxford University Press, Oxford.

Rolls, E., Loh, M., and Deco, G. (2008) An attractor hypothesis of obsessive-compulsive disorder. *European Journal of Neuroscience*, **28**, 782–793.

Rolls, E., Loh, M., Deco, G., and Winterer, G. (2008) Computational models of schizophrenia and dopamine modulation in the prefrontal cortex. *Nature Reviews Neuroscience*, **9**, 696–709.

Rosenberg, D., MacMaster, F., Keshavan, M., Fitzgerald, K., Stewart, C., and Moore, G. (2000) Decrease in caudate glutamatergic concentrations in pediatric obsessive-compulsive disorder patients taking paroxetine. *Journal of the American Academy of Child and Adolescent Psychiatry*, **39**, 1096–1103.

Rosenberg, D., MacMillan, S., and Moore, G. (2001) Brain anatomy and chemistry may predict treatment response in paediatric obsessive–compulsive disorder. *International Journal of Neuropsychopharmacology*, **4**, 179–190.

Rosenberg, D., Mirza, Y., Russell, A., Tang, J., Smith, J., Banerjee, S., Bhandari, R.,

Rose, M., Ivey, J., Boyd, C., and Moore, G. (2004) Reduced anterior cingulate glutamatergic concentrations in childhood OCD and major depression versus healthy controls. *Journal of the American Academy of Child and Adolescent Psychiatry*, **43**, 1146–1153.

Rushworth, M., Behrens, T., Rudebeck, P., and Walton, M. (2007a) Contrasting roles for cingulate and orbitofrontal cortex in decisions and social behaviour. *Trends in Cognitive Sciences*, **11**, 168–176.

Rushworth, M., Buckley, M., Behrens, T., Walton, M., and Bannerman, D. (2007b) Functional organization of the medial frontal cortex. *Current Opinion in Neurobiology*, **17**, 220–227.

Scheuerecker, J., Ufer, S., Zipse, M., Frodl, T., Koutsouleris, N., Zetzsche, T., Wiesmann, M., Albrecht, J., Bruckmann, H., Schmitt, G., Moller, H.J., and Meisenzahl, E.M. (2007) Cerebral changes and cognitive dysfunctions in medication-free schizophrenia: an fMRI study. *Journal of Psychiatric Research*, **42**, 469–476.

Seamans, J.K., Gorelova, N., Durstewitz, D., and Yang, C.R. (2001) Bidirectional dopamine modulation of GABAergic inhibition in prefrontal cortical pyramidal neurons. *Journal of Neuroscience*, **21**, 3628–3638.

Seamans, J.K., and Yang, C.R. (2004) The principal features and mechanisms of dopamine modulation in the prefrontal cortex. *Progress in Neurobiology*, **74**, 1–58.

Seeman, P., Weinshenker, D., Quirion, R., Srivastava, L.K., Bhardwaj, S.K., Grandy, D.K., Premont, R.T., Sotnikova, T.D., Boksa, P., El-Ghundi, M., O'Dowd, B.F., George, S.-R., Perreault, M.L., Mannisto, P.T., Robinson, S., Palmiter, R.D., and Tallerico, T. (2005) Dopamine supersensitivity correlates with $D_2$ High states, implying many paths to psychosis. *Proceedings of the National Academy of Sciences of the United States of America*, **102**, 3513–3518.

Seeman, P., Schwarz, J., Chen, J.F., Szechtman, H., Perreault, M., McKnight, G.S., Roder, J.C., Quirion, R., Boksa, P., Srivastava, L.K., Yanai, K., Weinshenker, D., and Sumiyoshi, T. (2006) Psychosis pathways converge via $D_2$ high dopamine receptors. *Synapse*, **60**, 319–346.

Shergill, S.S., Brammer, M.J., Williams, S.C., Murray, R.M., and McGuire, P.K. (2000) Mapping auditory hallucinations in schizophrenia using functional magnetic resonance imaging. *Archives of General Psychiatry*, **57**, 1033–1038.

Stevens, J.R. (1973) An anatomy of schizophrenia? *Archives of General Psychiatry*, **29**, 177–189.

Trantham-Davidson, H., Neely, L.C., Lavin, A., and Seamans, J.K. (2004) Mechanisms underlying differential $D_1$ versus $D_2$ dopamine receptor regulation of inhibition in prefrontal cortex. *Journal of Neuroscience*, **24**, 10652–10659.

Treves, A. and Rolls, E.T. (1991) What determines the capacity of autoassociative memories in the brain? *Network*, **2**, 371–397.

Tuckwell, H. (1988) *Introduction to Theoretical Neurobiology*, Cambridge University Press, Cambridge.

Wang, X.J. (1999) Synaptic basis of cortical persistent activity: the importance of NMDA receptors to working memory. *Journal of Neuroscience*, **19**, 9587–9603.

Wang, X.J. (2001) Synaptic reverberation underlying mnemonic persistent activity. *Trends in Neurosciences*, **24**, 455–463.

Weiss, A.P. and Heckers, S. (1999) Neuroimaging of hallucinations: a review of the literature. *Psychiatry Research*, **92**, 61–74.

Weissman, M., Bland, R., Canino, G., Greenwald, S., Hwu, H., Lee, C., Newman, S., Oakley-Browne, M., Rubio-Stipec, M., Wickramaratne, P., and The Cross National Collaborative Group (1994) The cross national epidemiology of obsessive compulsive disorder. *Journal of Clinical Psychiatry*, **55** (Suppl.), 5–10.

Winterer, G. and Weinberger, D.R. (2004) Genes, dopamine and cortical signal-to-noise ratio in schizophrenia. *Trends in Neurosciences*, **27**, 683–690.

Winterer, G., Ziller, M., Dorn, H., Frick, K., Mulert, C., Wuebben, Y., Herrmann, W.M., and Coppola, R. (2000) Schizophrenia: reduced signal-to-noise ratio and impaired phase-locking during information processing. *Clinical Neurophysiology*, **111**, 837–849.

Winterer, G., Coppola, R., Goldberg, T.E., Egan, M.F., Jones, D.W., Sanchez, C.E., and

Weinberger, D.R. (2004) Prefrontal broadband noise, working memory, and genetic risk for schizophrenia. *American Journal of Psychiatry*, **161**, 490–500.

Winterer, G., Egan, M.F., Kolachana, B.S., Goldberg, T.E., Coppola, R., and Weinberger, D.R. (2006a) Prefrontal electrophysiologic "noise" and catechol-*O*-methyltransferase genotype in schizophrenia. *Biological Psychiatry*, **60**, 578–584.

Winterer, G., Musso, F., Beckmann, C., Mattay, V., Egan, M.F., Jones, D.W., Callicott, J.H., Coppola, R., and Weinberger, D.R. (2006b) Instability of prefrontal signal processing in schizophrenia. *American Journal of Psychiatry*, **163**, 1960–1968.

Wolkin, A., Sanfilipo, M., Wolf, A.P., Angrist, B., Brodie, J.D., and Rotrosen, J. (1992) Negative symptoms and hypofrontality in chronic schizophrenia. *Archives of General Psychiatry*, **49**, 959–965.

# 16
# Epilogue

*Peter J. Gebicke-Haerter*

We discover a fascinating theory of the biosphere as one living system, where its numerous subsystems play diverse and mutually interdependent roles.

> (Lovelock, 1972)

General systems theory is a way of viewing the world as an interconnected hierarchy of matter and energy.

## 16.1
## Some General Remarks

At first sight, conceiving of living organisms as biological systems appears to be trivial, but this connotation has been revived only recently in the life sciences. Indeed, much of this thinking came out of the mid-1940s with the emergence of general systems theory laid out in the work of the biologists Ludwig von Bertalanffy and Paul Weiss (von Bertalanffy, 1976; Weiss, 1958). The number of publications on that issue has exploded within the last 5 years or so. I think it is safe to predict that within a short time no research approach in the life sciences can do without considering the systems' perspective.

The core of systems biology is to describe and understand biological systems on all levels, to dive into their complexities and to find mechanisms of reciprocal communication. There is no doubt that the brain is outstanding in its complexity; however, in psychiatric disorders, there is additional complexity through intricate, mutual influences between the brain and the periphery. Psychiatry is probably the best example where "systems" thinking is of the utmost importance. The sustained impacts of innumerable combinations of environmental factors affect brain functions on various levels, and continuously change its molecular and higher-order network structures. Those changes entail – on various timescales – short-term memories of optical, acoustical, and sensory inputs, as well as learning of huge amounts of skills including language, anticipation and planning for the future, consciousness, and emotions that may be consolidated in long-term memories ("engrams"). On the other hand, they entail aging and the development of chronic

*Systems Biology in Psychiatric Research.*
Edited by F. Tretter, P.J. Gebicke-Haerter, E.R. Mendoza, and G. Winterer
Copyright © 2010 WILEY-VCH Verlag GmbH & Co. KGaA, Weinheim
ISBN: 978-3-527-32503-0

diseases of the brain. This raises the question of the contribution that systems biology can lend to better pharmacotherapies of those disorders.

## 16.2
## Pharmaceutical Discourse

In the past, pharmaceutical companies have synthesized huge amounts of psychoactive drugs with the aim to increase specificities and to reduce unwanted side-effects. Typically, this strategy relies on substances with proven clinical effects or effects in animal experiments, where beneficial and toxic effects can be tested. Evidently, the large numbers of molecular targets in a biological system have impeded any systematic approach in drug discovery. Therefore, promising substances were often found by chance. The strategy of just modifying these drugs to obtain more specificity sometimes resulted in unexpected and more unwanted side-effects, which forced companies to withdraw those substances from the market. Hence, more and more, the pharmaceutical market has been suffering from a lack of discovery of new drugs. This dilemma has solicited a demand to find new ways of target identification. Systems biology may not be an immediate solution for the problem, but it may help in the longer term. It is a strategy that can also be used to study molecular networks on both the structural and the dynamic level. In combination with high-throughput technologies, in particular, it may become extremely valuable for the identification of better drug targets. In this way, it may help to reduce the number of compounds that need to be taken into consideration, which will save both time and money.

## 16.3
## From Molecular to Cellular Networks

After this short digression let us return to basics. It is hypothesized that memory is stored in widely distributed, interactive, and overlapping cellular networks of the cerebral cortex, from the sensory to the association cortex with numerous feedforward and feedback connections, and patterns of convergent, divergent, and lateral connectivities (Felleman and Van Essen, 1991) that undergo sustained changes through any new, incoming stimulus. Cognitive functions such as attention and perception, language, memory, and intelligence are based on neural transactions within and between those cortical memory networks. Maturation of these functions spreads from primary areas to progressively higher association areas. This "maturation" process depends greatly on the elaboration of neuronal connectivities. There seem to be extensive interactions between perceptual and executive memories that are pivotal in the integrative dynamics of the working memory. As pointed out by Sporns et al. (1989), integrative working memory operates by re-entrant cortical integration. Specialized "modules" of working memory in prefrontal cortex are nodes of heavy association in active executive-memory networks. Those nodes encode specific sensory-motor associations within the above-mentioned cortical networks, networks that encode the broad behavioral context in which those specific associations were formed and remain embedded. Working memory and long-term memory share most of the same cortical structures. A sensory stimulus to be saved in working memory activates a cortical network that not only encodes that stimulus, but also its former associative context in long-term memory.

This book tries to show a variety of means of how to achieve a better understanding of the wealth of interactions within this complex organ. Each aspect has merits on its own, but cannot be viewed separately from the others. Measuring electrical currents at the level of the skull does not reveal anything about molecular mechanisms, and altered molecular networks do not provide any clues as to what kind of behavior results. Nevertheless, it appears clear that changes on the molecular level can affect all levels and eventually cause a disease or improve well-being. A detailed understanding of those molecular changes will markedly help to predict consequences on other levels. Regardless of those levels, it has turned out that results from anatomy and neurochemistry do not reveal dynamics, and therefore are not very useful for expanding our view of the whole system and may even lead to misconceptions.

As pointed out in some contributions above, biological systems, and especially cellular and molecular systems, are highly dynamic systems that display self-similar, but not identical oscillatory behavior, because due to their open-system character they undergo subtle, but sustained changes over any given time interval. These oscillations may translate from the molecular to the cellular level and be correlated with low or high-frequency rhythmic or nonrhythmic activities. In dynamic, high-frequency re-entrant nonlinear oscillators, the situation may be described as such that (i) each neuron in a node of the network *does not* discharge with each iteration of the re-entrant oscillation, (ii) the probability of discharge function of re-entrant frequency and subthreshold membrane oscillations of individual neurons is determined by local neuronal properties, (iii) a single neuron can participate in multiple oscillators, (iv) there is weak coupling between different nodes within the same oscillator, and (v) a single oscillator can sustain multiple noncommensurate frequencies.

Those considerations may lead one to conclude that human beings are giant networks of chemical reactions between tens of thousands of biochemicals. These reactions occur in cells that are organized in tissues, organs, and whole organisms. All phenomena at higher levels of biological organization are emergent properties of the dynamics at this lowest, molecular level.

## 16.4
## Modeling Strategies

Mathematical models can aid in providing more insights into these processes by reducing them to their "essentials" – whatever that means – so that predictions can be based on mathematical analyses of the models. Hence, mathematical modeling is a process of selective ignorance, making simplifying assumptions. Decisions always have to be made as to what we can remain ignorant about and what should not be ignored. There are quite a number of mathematical tools already available that have been developed for analyses of chaotic systems, and that can be used to analyze and make predictions for those cellular and molecular networks (Auffray and Nottale, 2008; Nottale and Auffray, 2008). Dynamic networks such as artificial neuronal nets are used to determine changes in interneuron excitability as well as changes in coupling among oscillators due to changes in frequency and resonance. It turned out that neurons at nodes can become multipotential as a result of changing physiological structure. Weakly coupled nonlinear oscillators may be embedded in scale-free networks (Barabasi and Albert, 1999). These networks are very robust against random failure of a node, but highly vulnerable to deliberate attacks on hubs (critical nodes). They display small-world connectedness, which allows for rapid responses and resistance to lesions. Physiological functions are not represented in the nodes, but in the network. Further, they show parallel and distributed processing and holographic memory, and they are scalable. Other networks useful in this respect are networks with complex topology, such as random networks (Erdös and Rényi, 1959; Boccaletti *et al.*, 2006) and small-world networks (Watts and Strogatz, 1998). They reveal rich synchronization phenomena that result in self-organized structure formations. The approach extended thereof, of networks embedded in networks, appears to be promising for understanding some further aspects of structure formation. Presently, all these network approaches only encompass neuron–neuron interactions. Although this strategy is already extremely demanding, it has to be kept in mind that the brain is composed of approximately 10 times more glial cells that continuously and closely communicate with neurons, and substantially contribute to maintenance of their appropriate responses.

Systems biology of the brain is a formidable and extremely challenging task that can be tackled only by aid of ingenious computer programs and researchers of all required disciplines. Computer-aided applications to such a complex biological system, however, require a great deal of further developments, much better adaptations to the real biological world, and a qualified understanding of the workings of biological systems. As a matter of fact, we are only at the beginning of the grand venture to find more and more sophisticated means to explore and understand the impressive features of the most complex biological computer on Earth – the human brain.

## References

Auffray, C. and Nottale, L. (2008) Scale relativity theory and integrative systems biology: 1. Founding principles and scale laws. *Progress in Biophysics and Molecular Biology*, **97**, 79–114.

Barabasi, A.L. and Albert, R. (1999) Emergence of scaling in random networks. *Science*, **286**, 509–512.

Boccaletti, S., Latora, V., Moreno, Y., Chavez, M., and Hwang, D.U. (2006) Complex networks: structure and dynamics. *Physics Reports*, **424**, 175–308.

Erdös, P. and Rényi, A. (1959) On random graphs. *Publicationes Mathematicae*, **6**, 290–297.

Felleman, D.J. and Van Essen, D.C. (1991) Distributed hierarchical processing in the primate cerebral cortex. *Cerebral Cortex*, **1**, 1–47.

Lovelock, J.E. (1972) Gaia as seen through the atmosphere. *Atmospheric Environment*, **6**, 579–580.

Nottale, L. and Auffray, C. (2008) Scale relativity theory and integrative systems biology: 2. Macroscopic quantum-type mechanics. *Progress in Biophysics and Molecular Biology*, **97**, 115–157.

Sporns, O., Gally, J.A., Reeke, G.N. Jr., and Edelman, G.M. (1989) Reentrant signaling among simulated neuronal groups leads to coherency in their oscillatory activity. *Proceedings of the National Academy of Sciences of the United States of America*, **86**, 7265–7269.

von Bertalanffy, L. (1976) *General System Theory: Foundations, Development, Applications*, revised edition, George Braziller, New York.

Watts, D.J. and Strogatz, S.H. (1998) Collective dynamics of "small-world" networks. *Nature*, **393**, 440–442.

Weiss, P. (1958) *Modes of Being*, Southern Illinois University Press, Carbondale, IL.

# Index

## a

acetylcholine  162, 163
– classes of receptor  162
– physiological stimulation  166
active executive-memory networks  342
aggressive antismoking campaigns  163
Allen Brain Atlas  216, 220
Alzheimer's disease  4, 118, 293
α-amino-3-hydroxy-5-methyl-4-isoxazol-propionacid (AMPA)  146, 166
– dependent excitatory postsynaptic potentials (EPSPs)  150
– receptors  305, 327, 335
– type synapses  322
amplitude modulation (AM) patterns  136, 138
– γ AM patterns  138
– ECoG waveforms  136
– spatial patterns  136
analysis of variance (ANOVA)  277, 284
anatomical brain structures, mental functions  38
animal model systems  36
anterior cingulate cortex (ACC)  169
antipsychotics  61
– drugs  202, 307
– receptor binding profile  62
anxiety disorders  4, 27, 29, 38, 119, 333
apoptosis  237
– cephalostatin-1 apoptosis  238
– complex system  238
– functional modules, of neuronal apoptosis  237
– modular systems biology approach  237
– Petri net modeling  235
artificial networks
– cortical network models, structure  49
– display of states  49
– neural network models  19, 31, 47
attention deficit hyperactivity disorder (ADHD)  171
attractor network concept  324, 325
– behavior  325
– characteristics  325
– neurons of one pool  324
– prefrontal cortex, operating  325
– properties  324
– short-term memory task  325
– stability  330, 333
– stable attractor configuration  324
– unstable attractor configuration  326
atypical antipsychotics  61
autism, genetic liability, role  176

## b

bacterial endotoxin lipopolysaccharide (LPS)  152
Baye's rule  289
Bayes classifier  284
Bayesian networks (BNs)  212, 278, 286
– definition  286–288
– equivalence classes  287
– gene expression data  291
– microarrays  291–293
– parameters  288
– use  288–291
Bayesian picture, probability  291
Bayesian scoring metric (BSM)  284
benzodiazepine (BDZ)  74
biochemical systems theory (BST) models  232, 239
– advantage  241
– qualitative side, Petri nets  232
biological feedback system  68
– biological clock  69

biological models, formulation
- regular graphs  246
biological networks
- disassortativity  249
- null models  267
- regulatory networks  262
biological psychiatry  31. *see also* neuropsychiatry
biological system  325
bio-psychosocial model  31
bipartite graph  257
blood oxygen level-dependent (BOLD) signal  170
Boolean attractors  263
Boolean dynamics  264
Boolean networks  82, 231, 266
Boolean rules  261, 262
Box's law of modeling  233
brain  131, 161, 341
- automatic motor behavior  38
- behavioral examination  5
- cognitive functions  342
- comprehensive computer model  180
- default mode network  170
- disorders  33
- dopamine synthesis  198
- during excitation  131
- dynamics, feature  326
- electrical signaling  32
- function  161
- local networks, development of  44
- and mind
-- basic problem  6
-- behavioral examination  5
-- discourse, problem survey  21
-- inter-related mental functions  9
-- neurons, network  9
-- ontological quality  10
-- principle  6
-- relation  9
- neurochemical oscillator  70–73
- nicotinic cholinergic neurotransmission  162
- phenomena  4
- prevailing dogma  131
- psychiatric disorders  341
- signals, analog signals  9, 10
- structure  38–40, 180
-- identification of roles  268
-- mental functions  39
- system analysis  141
- systems biology  344
- waves, spatial structure  131–133
brainstem, modulatory pathways  39

*c*

calcineurin  305
calcium/calmodulin-dependent kinase (CaMK) pathway  167, 168
- CaMKII phosphorylates AMPA-type glutamate receptors  168
calcium/calmodulin-dependent phosphatase, *see* calcineurin
cAMP binding process  311
- association/dissociation constants  312
- rates  312
cAMP molecules  311, 315
- concentrations  313
- dependent kinase  305
- dependent phosphorylation  304
cAMP-response element binding protein (CREB)  168
cancer patients  94
candidate genes  218–220, 223
candidate pathways  218
- construction  219
canonical models  233
cardiac arrhythmia  102
cardiac cell models  98
cardiac pacemaker mechanism  99
cardiac rhythm  103
catalytic protein kinase A (PKAc), activation curved  313
catechol-*O*-methyltransferase (COMT) enzyme  180
- Val/Val allele  198
CellDesigner  93
cells
- global behavior  81
- signaling, downward causation  100
cellular/subcellular networks  61
cellular automata (CA)  263
- rule space  264
cellular metabolism
- network representations, schematic view  257
cellular organization  251
- levels  260
central nervous system  164, 303
- abundant receptors  164
- dopamine, effects  303
cerebral cortex, inhibitory neurons
- morphological appearance of  47
cholinergic transmission system  59
chronic neuropsychiatric disorders  176
circadian clocks  232
- taxonomy of models  232
circadian rhythms models  231
circadian system  231

circular causality mechanism  140
circular signaling pathways  67
CLOCK gene, in mice  104
clustering algorithms  281
clustering data points, distance measures for  283
clustering methods  85
CodeLink bioarrays  218
communicational models  233
complex biological systems  241, 344
– computer-aided applications  344
computational modeling  31
concept map method (CMM)  233, 239, 240
– constructs  241
– inverse modeling step  239
– steps  239
consciousness, secret  106
constraint-based analysis (CBA)  239
copy number variations (CNVs)  176, 180
cortex  130, 321
– excitatory/inhibitory neuron populations, negative feedback  136
– microscopic picture  45
– outnumber inhibitory neurons  134
– system properties  130
cortical network model  321
– intrapool connection strengths  323
– nonselective pool  323
– pools S1/S2  323
– population and connection weights  323
cortical system  140
– mesoscopic trajectory  141
– two levels of function, synthesis  140
cortical tissue, resistance  130
cortico-striato-cortical circuits  42
cytochrome P450 isoforms  258

d
DARPP-32 model  304, 305, 309
– biochemical pathways, schematic presentation  308
– biological conclusions  309
– biological experiments  309
– dynamic responses  316
– effect on synaptic proteins  306
– gene coding  306
– G-protein-coupled receptors, effect  304
– neurotransmitters, regulation  305
– phosphorylation  304, 310
– physiological role  305–307, 310
– PKA regulation  312
– PP1 inhibition  304
– in psychiatric disease  306–308
– regulation  307
– in signaling network  307
– study  304
– Thr75 site  305
– – effect  310
dementia praecox  4, 28, 114
3,4-dihydroxyphenylalanine (DOPA)  196
DNA  207
– binding data  292
– microarray data  245
– polymorphisms  207, 208, 219
– repair mechanisms  254, 268
– sequences  104, 213
– topology  260
dopamimetic drugs  306
– cAMP response element-binding protein  306
dopamine (DA)  145, 304
– actions, abnormal periadolescent maturation  151
– – in schizophrenia, developmental animal models  151–153
– antagonists  153
– based neurotransmission, wiring diagram  56
– based signal transmission  36
– dysfunction hypothesis  36
– electrophysiological actions in prefrontal cortical circuits  146–148
– [$^{18}$F]FDOPA study  198
– hyperfunction  72
– indirect agonists  197
– intracellular signaling  304
– intra-PFC administration  146
– maturation  151
– modulation, pyramidal neurons changes during adolescence  148
– periadolescent changes model  150
– prefrontal  146
– receptor  145, 201, 304
– – $D_1$ agonists  201
– – density  145
– – $D_{2/3}$ receptors  194, 197, 200–202
– role  145
– synaptic responses  146
– transmission system  48, 58
– transporter  196
dopamine $D_2$-like receptor  194, 196, 197
– antagonism  200
– competition model  197
– inhibitors  301
– PET/SPECT studies  195
drug therapies  27
dualism  7
dynamic molecular networks  295

## e

eigenvectors
– principal components   280
Einstein's equation   7
electrocorticograms (ECoG)   130, 132, 135, 140, 142
– band pass filtering   138
– β bursts   139
– mesoscopic order parameter   140
– spatial mapping   133
– spectrum   134, 140
– temporal structure   133–135
electroencephalograms (EEGs)   5, 32, 129, 132, 135, 142, 180
– activity   34
– band pass filtering   138
– electrophysiological methods   180
– patterns   35
– signals   8, 9
– spatial mapping   133
– spatio-temporal patterns, behavioral correlates   136–140
– temporal structure   133–135
electromyography (EMG) system   120
electrophysiological methods   120
– development   5
empirical research, cycle   13
encompassing directed network (EDN)   212
endophenotypes   34, 116, 214, 215
– genetic determination   117
– identification   118
– – genetic studies   214
– psychopathology, role   119
– requirements   118
endorphin/enkephalin receptors   60
enzyme
– apolipoprotein E (ApoE)   118
– centric projection   257
– classification code   258
– complex   255
– kinetics   259
– reactions   87
epigenetics   177
epilepsy, form   142
epistemic cycle   14, 15, 17
Erdös–Rényi (ER) model   246
– random graph, summary   248
*Escherichia coli*
– amino acid biosynthesis   91
– gene-gene interactions   253
ethanol   221
– putative pathways, acute functional tolerance   221

expression quantitative trait locus (eQTLs) analyses   209, 211, 218
– candidate gene   213
– characteristics   209
– classification   209
– confirmation   217
– data   215
– distal eQTLs   210
– local polymorphisms, statistical robustness   210
– overviews   209
– rodent studies   212
– whole-brain data   216
extracellular signal-regulated kinase (ERK)
– CREB pathway, dysregulation   168
– mitogen-activated protein kinase (MAPK) pathway   167
extrapyramidal side-effects (EPMSs)   200

## f

flux balance analysis (FBA)   239
– implementations   259
– predictive power   255
forebrain, illnesses   114
functional magnetic resonance imaging (fMRI)   115, 119, 330
– studies   170

## g

$\gamma$-aminobutyric acid (GABA)   41, 145, 269, 301, 331
– channels   328
– ergic inhibition   45
– ergic system   59
– function   46
– GABA-A antagonists   147–150
– hypofunctionality   73, 330
– markers   145
– receptor   327, 335
– – activated synaptic conductances   332
– synaptic currents   322
– – integrate-and-fire simulations   334
Gaussian distribution   282
gel electrophoreses   83
gene expression   209, 285
– control loops   68
– data   278
– expression, large-scale analyses   84
– genetic heritability   209
gene-gene interactions   176
gene ontology (GO)   104, 219, 220
– hierarchical set   219
gene-protein-lipid-cell network   105
gene regulatory networks   244, 260

genetical genomics  207, 213, 214
– candidate pathways, identification strategy  218–222
– complex/quantitative/physiologic/behavioral phenotypes  213
– eQTLs, characteristics  209
– extension, systems approach to phenomics  213
– gene expression studies  213
– generating networks  213
– genetical genomic phenomics  222
– – to systems biology and medicine, contributions  222
– history  207
– introduction  207
– marriage to  207
– potential pitfalls in phenotype selection/current technology  214–218
– recent developments  210
– traditional genetic analyses  213
genetic determinism, misconception  105
genetic/functional genomic/phenomic approaches  223
genetic maps  217
genetic marker datasets, sharing  210
genetic network  284
genetic program  98
genetic regulation  91
genome-scale models  239
genome-scale networks  239
– growing availability  239
– mathematical challenges  239
genome-wide expression analyses  293
genomic regions  208
– non-Mendelian manner  208
– quantitative trait locus (QTL) analysis  208
glial cells, important function  51
global brain circuits  75
– functional structure  44
global neuronal circuits
– connectivity, scheme  42
glutamate
– hypofunction  73
– neurotransmission  51
– receptors  54
glutamatergic receptors  323
– AMPA/NMDA  323
glutamatergic system  59
glycine receptors  60
G-proteins  57, 149, 304
– coupled receptors  304
graphs, typical examples  247
graph theory  249
– aspects  245–250

## h

hallmark signature  243
*Halobacterium salinarum*'s metabolism  239
– genome-scale model  239
Hamilton anxiety scale  28
Hanalyzer  220
– modules, system diagram  222
heart, potassium channel, types  98
Hebbian nerve cell  137, 140
Heidelberg psychopathological tradition  115
heuristic search methods  290
hidden Markov model  211
hierarchical clustering methods  85, 282, 283
hierarchical network  249
high-throughput analyzer, *see* Hanalyzer
high throughput cDNA sequencing technology  208
Hill constant  87
– reaction rates for  88
Hill kinetics  87
Hodgkin–Huxley model  322
human brain  161. *see also* brain
– cerebellum in  217
– $D_2$-like receptors  194
– nicotinic cholinergic signaling  161
– study  180
hybrid Petri net models  239
– unifying conceptual framework  240
hypothalamogram  131. *see also* local field potentials (LFPs)
hypothalamus  38
hypotheses, formation  13

## i

inbred long sleep (ILS)  211
inbred short sleep (ISS)  211
inhibitory neurons  47
– biological diversity  50
– classification  46
– interneurons, classification  46
inhibitory postsynaptic potential  52
*in silico* experiments  19
integrate-and-fire model, spikes in  322
interval mapping techniques  208
– development  208
intracellular molecular signaling network, components  70
intracellular signal processing  63, 64, 244
intracranial electric fields  130
– local field potentials (LFPs)  130

## k

Kauffman's model  261
– scale-free graph  262
knowledge cycle  15
Kraepelin's nosological dichotomy  114
Kyoto encyclopedia of genes and genomes (KEGG) system  92, 220
– PATHWAYdatabase  220

## l

large-scale complex systems  251
Libet's experiment  120, 121
likelihood-based causality model selection method  213
limbic system  130
local cortical network, neurons in  321
local field potentials (LFPs)  130, 132
loudness dependence of acoustical evoked potentials (LDAEPs)  119

## m

magnetic resonance imaging (MRI)  34, 130
magnetoencephalograms  130
magnetoencephalography (MEG)  180
– electrophysiological methods  180
mass action system  132
mass spectrometry (MS)  84
mathematical models, definition  343
mathematical optimization methods  259
maturation process  342
maximal dependent transition (MDT) sets  236, 238
– computation  238
medial prefrontal cortex (mPFC)  41
Mendelian approaches  214
mental disorders  3, 5, 28, 33, 35
– diagnoses  29
– prevention  27
mental dysfunctions  30
mental events
– identity  10
– – Psi\Phi level  10
– physical level, causality  11
mental illnesses  321, 328–335
– attractor network concept  324
– local cortical dynamics  321
– noise and stability  325–328
– OCD  332–334
– recurrent neural network  321–324
– schizophrenia  328–332
mentalism  6
mesocortical-PFC pathway  152
mesolimbic and mesocortical dopamine pathways  335

– functionality  335
metabolic control analysis (MCA)  89
metabolic fluxes, size distribution  256
metabolic network  235, 236
– Petri net model  235
– reaction network, synthesis pathway  250, 255
– topologies  255
metabolic pathways  256
– graph representation  257
metabolism, schematic presentation  256
metabolite-centric projection  257, 258
metabotropic receptors  56
3,4-methylenedioxymethamphetamine (MDMA)  195
Michaelis–Menten kinetics  194
microarray  217
– affymetrix  217, 218
– BXD RI panel  217
– cDNA sequencing technology  208
– data analysis  278, 279, 281, 284, 292
– – ChIP-chip data  292
– experiment  219
– fluorescently labeled cDNAs hybridized  278
– probe  217
– project  212
– spotted microarray
– – clustering methods  281
– – dimensionality reduction  279–284
– – hierarchical clustering  282–284
– – partitional clustering  282
– – statistical tests  284
mind  5
– and brain
– – computer simulations  19
– – dualistic concept  19
– – neurophilosophy, structure  20
– – systemic modeling, steps  20
– molecule meet  97
– – brain, programs in  105–107
– – genetic programs  104
– – higher-level simulation  101–104
– – modeling excitable cells  98–101
mind-body problem, mind philosophy  5–7, 105
– causation  11
– correlation  9
– dualism  8
– identity theory  9–11
– monism  8
– philosophers, scientific knowledge  11
– supervenience concept  11

modeling complex biological systems 233–235
– model design  233
– model diagnosis/validation  233
– model selection  233
– model use/extension  234
– steps  234
modeling excitable cells  98–101
modular networks  250
molecular biology  161, 207
– central dogma  207
– microarrays-techniques  33
– paradigm  161
molecular brain imaging, advantages  189
molecular genetic markers  208
molecular imaging techniques  189
– in living brain  189
– methodologic background  189–194
molecular network  343
– to cellular networks  342
– complex topology  344
– high-frequency re-entrant nonlinear oscillators  343
– *in vivo* behavior  64
– of signal transduction cascades  65
molecular phenotypes, genetic basis  209
movement disorders, Parkinsonism  215
mRNA synthesis  215–217
– DNA polymorphisms  208
– nAChR subunits  166
multilocus analysis  211

## n

naïve Bayes classifier  277, 279, 285
natural disease
– Emil Kraepelin's concept  124
neonatal ventral hippocampal lesion (NVHL)  151, 152
nervous network  106
network dynamics  245
– binary dynamics  261–267
– cellular automata  261–267
– general aspects  260
network randomization scheme  255
neural network model  325, 334
neural networks theory  245
neuregulin/Erb pathway, genes coding  169
neurobiological methods
– brain, global electrical signals  34
– data, complexity  18
– imaging methods  34
neurobiology  215
neurochemical mobile  71
– basic configuration  74
neurochemical transmission systems, interactions  72
neurodegenerative disease syndromes  223, 237
neuroimaging methods  119
neuroleptics  61
neurological diseases  244
neurological motor disorders  43
neuronal networks
– development  45
– discharge activity  54
– excitation, inhibition  45
– transmitter systems  179
– – computer models  179
– – nicotinic cholinergic/dopaminergic neurotransmission  179
neurons  52, 136
– $Ca^{2+}$-dependent signaling  167
– cell-cell contacts  40
– cellular network  63
– electrical signals  53
– interacting populations, properties  136, 139
– long-term depression (LTD)  167
– long-term potentiation (LTP)  167
– morphology, intracellular structures  52
– populations  129
– – conclusions/applications  141, 142
– – system properties in cerebral cortex  129
– terminals  216
– types  177
– *vs.* modulating excitability and firing rate  177
– working memory function, animal experiment  37
neurophilosophy  3, 5
– experimental research  12–17
– – bottom-up explanations  16
– – epistemic cycle  14
– – hypotheses  14
– – top-down analysis  15
– perspectives  19–22
– philosophy of
– – mind  3
– – science  3, 11
– speculation  12
neuropsychiatric disorders  176
– etiology  177
– genetics  176
neuropsychiatry  5, 33, 50
– computational  31
– computer-based modeling  32
– neurobiological fundamentals

－－ gene regulation   67–69
－－ genomics   66
－－ intracellular molecular processes   63
－－ intracellular signal cascades   63
－－ modeling signal transduction networks   64–66
− (neuro)biological psychiatry, basic findings   33
－－ brain, structure and function   38–40
－－ experimental set-ups   35–38
－－ global circuits connectivities   40–43
－－ neurobiological methods   34
－－ neurons, local networks   43–46
－－ neuropsychopathology   33
－－ schizophrenia, prefrontal network   46–50
− neuron   50–52
－－ electrical signaling   52–54
－－ synapse   51
－－ systems biology   69
− proteomics   67
− psychiatric fundamentals   27–33
－－ general psychiatry   27
－－ psychiatric diagnoses, criteria   30
− psychopathology   28, 33
－－ quantitative   28
－－ theoretical   29
− reductionistic research strategy   16
− synapse   54–62
－－ neurotransmitter receptors   56–62
− theoretical psychiatry   30
－－ computational neuropsychiatry   31
－－ systems neuropsychiatry   32
neuroscience, goal   161
neurotransmitters   55
− dopamine, postsynaptic site   55
− receptors   56, 179, 202
－－ types   202
− systems   194, 200, 202
－－ characterization with molecular imaging   194
－－ PET/SPECT studies   194
nicotine   162, 169, 171
− addiction   162, 163, 176
－－ effects   162
－－ genetic liability   163
－－ pathogenesis/symptomatology   178
− applications   175
− beneficial effects   169, 171
− CaMKs activation   168
− cellular effects   163–169
− and cognition   169–171
− high-affinity receptor   174
− metabolism   175
− nAChrRs   163–169
− naïve probands   169
− pentameric receptors   163
− prefrontal microcircuits, modulation   170
− and reward   171–173
− and stress response   173
nicotinic acetylcholine receptors (nAChRs)   162
− administration   169
− $Ca^{2+}$ influx   164
− $\beta_2$-containing   166
− functional states   164
− genes and smoking, variation   174–178
− intracellular signal transduction cascades downstream   177, 179
− nicotine effect   167
− numerous neuronal systems, modulation   164
− pharmacological manipulation   162
− physiological ligand   163
− post-translational mechanism   166
− role   174
− stimulation   164
－－ intracellular signaling downstream   165
− subtypes, localization/stimulation effect   164
− subunits   175
－－ genes encoding   174
－－ mRNA levels   166
nicotinic cholinergic system   162
− epidemiological relevance   162
− neurotransmission   162, 178
－－ physiological effects   164
nigrostriatal dopamine system   195
− pharmacological challenges   195
N-methyl-D-aspartate (NMDA)   54, 146, 166
− channels   327, 328, 335
− intra-VTA injection
－－ antagonists   151
－－ induced increased excitability   148
− receptors   147, 199, 301, 327, 328, 331
－－ antagonism   199
－－ hypofunctionality   330
− synaptic currents   322, 323, 332
－－ integrate-and-fire simulations   334
noise
− black noise   135
− brown noise   134
− functional similarity   266
− Rayleigh noise   139
− white noise   134
norepinephrinergic/noradrenergic transmission system   57
− $\alpha_1$ receptors   61

nosology 28
nucleus accumbens (NA) neurons 151
null hypothesis 211

## o

obsessive-compulsive disorder (OCD) 327, 332–334. *see also* mental illness
– anxiety 333
– brain systems 334
– core feature 333
– dynamical systems theory 333
– generic hypothesis 334
– lifetime prevalence 332
– patients 332
– prefrontal cortex attractor networks 333
– symptoms 334
– – physical symptoms 332
– – stability 330
olfactory bulb, system properties 137
olfactory cortex 131
– pyramidal cells, high packing density 131
olfactory system 135
– excitatory/inhibitory neurons 135
– oscillation cycle 135
ordinary differential equation (ODE) models 82, 231

## p

pacemaker rhythm, computer model 99
Parkinsonism syndrome, *see* extrapyramidal side-effects (EPMSs)
Parkinson's disease 36, 44, 196, 304
– symptoms 196
– treatment 200
parvalbumin (PV)
– containing fibers 145
– containing interneurons 46
Petri net 235, 242
– applications 235
– behavior 235
– edges 235
– formalism 236
– graph representations 236
– marking 235
– nodes 235
– qualitative analysis 236
– theory 233
pharmaceutical market 342
phenomics 207, 214
– candidate pathways, identification strategy 218–222
– eQTLs, characteristics 209
– genetical genomic phenomics 222

– – to systems biology and medicine, contributions 222
– marriage to 207
– potential pitfalls in phenotype selection/ current technology 214–218
– recent developments 210
phenotype/trait 214
– robustness, definition 214
phospholipase C (PLC) 57, 146
– postsynaptic activation 146
PI3K/Akt/GSK-3β pathway 168, 169
protein kinase A (PKA) 304
– activation 311–316
– – conceptual model 311
– cAMP activation 311
– catalytic subunits 314
– deterministic model 314
– – *vs.* stochastic model 314, 315
– dynamic experiments 314
– forward and backward reaction rates 315
– model validation against dynamic data 314
– model validation against steady-state data 313
– rate constants, empirical estimates 312
– signaling pathways, inhibition 146
*Populus*, monolignol biosynthesis 239
– GMA model 239
positron emission tomography (PET) 34, 180, 189, 190
– advantages/limitations 190
– annihilation/LOR/temporal coincidence 192
– – basic principal 192
– data analysis 193
– – time-activity curve (TAC) 193
– imaging 191
– isotopes 190
– line of response (LOR) 191
– photoelectron detector crystals/ photoelectron multipliers 191
– – polygonal array 191
positron-emitting isotopes 190
postsynaptic receptors
– activation 61, 63
– blockade 60
postsynaptic signaling cascade
– receptor activation 59
post-traumatic stress disorder 173
prefrontal cortex (PFC) 145, 198
– DA actions, complexity 145
– – primary cellular mechanisms 147
– dependent behavioral measures 151
– long-term potentiation (LTP) 147

– ventral hippocampus impairing development 152
prefrontal cortical circuits 153
– periadolescent maturation 153
prefrontal cortical networks 145
– dopamine (DA) modulation, role 145
– electrophysiology 145
presynaptic receptors, blockade 61
principal component analysis (PCA) 277, 279–281
– principal components 280
probability-based methods 294
productive psychosis, psychopharmaceutical treatment 74
programmed cell death, see apoptosis
promiscuous pharmacology, benefits 223
protein
– channels 99
– phosphatase 2A 305
protein-protein interactions 293
– databases 221
– networks 250
psychiatric disease 17, 162, 178, 214, 299
– biochemical networks 301
– clinical examination, symptoms, checklist 28
– cluster of symptoms 301
– DARPP-32 model 303–305
– – biological conclusions 309
– – physiological role 305
– – in psychiatric disease 306–308
– – role 307
– deterministic model, modeling signaling pathways 308
– diagnosis, problems 113
– dopamine signaling 303–305
– etiology 301
– multiple genes, impact 178
– neurobiology 303
– neurotransmission 303
– nodes of interaction, looking for 302
– schizophrenia, example 301
– stochastic models 310
– treatment 301
– pathogenesis/therapy 162
psychiatric research 231
– modeling process, steps 233–235
– Petri net modeling of apoptosis in leukemic cells/neurons 235
– from qualitative models through Petri nets 235–237
– from quantitative canonical models 239–241
– from stoichiometric to kinetic genome-scale models 239

– systems modeling approaches 231
psychiatrists 4, 28, 31, 33, 76, 114, 115, 141
psychiatry 341
– applicability 267–269
– central and controversial issue 113
– computational modeling 29
– conceptual and historical introduction 113–116
– diagnostic process 113
– research and development 4, 5
– system-oriented theories 267
– systems biology modeling 268
psychoactive drugs 342
– beneficial/toxic effects 342
psychopathology
– atomic unit 116
– – susceptibility genes 117
– future role 123–125
psychopharmacotherapy 189
– focus schizophrenia 196–199
– neurotransmitter systems, characterization 194–196
– progress via molecular imaging 189
– – methodologic background 189–194
– – techniques in living brain 189
– psychopharmaceuticals action in schizophrenia 199–202
pyramidal cells 43, 135
– re-excitation 135

*q*

qualitative models 231
– canonical models 239
– theoretical models 15
– usefulness 231
quantitative trait locus (QTL) analysis 208, 211, 218
– behavioral, identification, advantage 218
– physiologic/behavioral 218
quantitative traits, genetics 208

*r*

radiotracers 192
– molecular imaging agent 193
– synthesis 192
random Boolean network 261
random graph, ER model 246, 247
readiness potential (RP) 120
recombinant inbred (RI) rodent panels 210, 211
recurrent neural network 321–324
reductionistic paradigm 16
relational networks 246
reuptake mechanisms 54

REV-ERBα proteins  68
reward system  172, 173
– dopaminergic stimulation  173
– hypodopaminergic state  172
– *in vivo* nicotine effects  172
Rice's conversion factor  139
RNA interference  279
3R system  220, 221
– classes of algorithms  220
– goal  220
Rubinstein–Taybi syndrome  168

**s**

*Saccharomyces cerevisiae*  208
– genetical genomics study  208
– strains  210
scale-free degree distribution  245
scale-free graph  249
schizophrenia  5, 17, 153, 162, 163, 196, 301, 328–332. *see also* mental illness
– anatomical studies of patients  302
– animal models  302
– biological mechanisms  331
– biological problem  302
– brain areas  40
– diagnosis  30, 31, 328
– dopamine hypothesis  197, 198, 301
– dual-state theory  329
– etiology  306, 307
– example  301
– functional magnetic resonance imaging (fMRI) investigations  330
– global circuitry relevant for  43
– hierarchical network organization  302
– hypofrontality  332
– molecular imaging studies  202, 301
– multilevel analysis  302
– neurobiology  335
– neuroleptic drugs  332
– nicotinic cholinergic neurotransmission, role  163
– paranoid subtype  197
– pathogenesis  169, 180
– pathophysiology  153, 169, 197
– – implications  153
– – treatments  153
– patients, prefrontal cortex  306
– PI3K/Akt/GSK-β pathway, role  169
– PKA activation, case study  311–316
– prodromal symptoms  198
– psychopharmaceuticals, action  199–202
– psychopharmacology research  202
– susceptibility genes  67, 303
– symptoms  30, 31, 328–332

– – computational models  47
– systems-level neuroscience approaches  330
– targeted genes  302
– treatment  60
– working memory functions  37
*Schizosaccharomyces* pombe  263
scientific knowledge, cycle  15
Sen–Churchill framework  212
serotonergic transmission system  58
serotonin, hyperfunction  72
signaling pathways  92, 308
– chemical reaction, rate constant  308
– DARPP-32, role  310
– dissociation constant  309
– mass action kinetics  309
signal-to-noise ratio  146, 170, 331, 335
signal transduction networks  236
– cascades, molecular network  65
– *in silico* test  66
– pathways, intracellular loops  65
signal transmission  51, 54
simple regulatory network
– consisting of genes  262
single-cell organisms, molecular systems  75
single nucleotide polymorphisms (SNPs)  174, 176, 180
– behavioral correlates  174
– cellular effects  174
– dataset  210
– localization  218
– 5′-and 3′-untranslated regions (UTRs)  174
single photon emission computed tomography (SPECT)  34, 189, 191
– advantages  190, 192
– data analysis  193
– – time-activity curve (TAC)  193
– limitations  190
single-photon-emitting isotopes  191
smoking behavior, aspects  174
smoking-related causes  162
spatio-temporal patterns, dynamic noise  265
spiking networks  325
– Poissonian spike trains  326
standard recurrent network model  322
– decision-making  322
state variables, definitions  141
stochastic modeling approach  310
– stochastic diffusion algorithms  311
stochastic models  83
stress response system  173
– hypothalamic-pituitary-adrenal (HPA) axis  173

– structures   173
subcortical structures
– neurobiologically based diagram   48
substantia nigra (SN)   43, 44
supervenience concept, scheme   12
synapse   52, 136, 303, 323
– dendrito-dendritic synapses   54
– intracellular biochemical signaling pathways   303
– strengths   136
– systems biology   55
synaptic conductances   327, 328
synaptic plasticity   167
– LTP/LTD   167
– type   171
synaptic receptors   330
systems biology   179, 223, 243, 303, 341
– benefits   303
– contribution   342
– core   341
– data analysis
– – clustering approaches   85
– – components identification   84
– – detection   83
– – large-scale analyses   84
– – principal component analysis   85
– – purification   83
– definition   33, 179
– genetic background   179
– modeling
– – levels   82
– – purpose   81
– molecular and biochemical information   244
– network perspective
– – large-scale systems   251
– – metabolic networks   255–260
– – TRNS   253–255
– ODE modeling
– – differential equations   86
– – metabolic control analysis   88
– – parameter estimation   90
– – reaction kinetics   87
– – simulating models   89, 90
– – steady states   88
– – stoichiometric matrix   86
– principles   98
– research and development   93, 94
– results, just-in-time transcription   91
– scientific advancements   223
– standardized formats   91
– – databases   91

– – ODE models, XML-based formats   91, 92
– – tools   93
– statistical methods   244
systems biology graphical notation (SBGN)   235
systems biology mark-up language (SBML) model   235
– extensible markup language (XML)   91, 92
systems neuropsychiatry   32
systems neuroscience   179
– goal   179
– role   179

*t*

taxonomy tree   231
– feature   232
temporal synchrony   133
theoretical (neuro)psychiatry   17,18
theoretical neuroscience   31
theory-derived hypotheses   14
time of flight (TOF)   84
topographic mapping   133
transcriptional regulatory networks (TRNs)   251
– *E. coli*   254
– gene duplication   254
– hierarchical organization   253
– network representations, definition   252
– representation   253
– topological feature   254
transcription factors   168, 209
– coding sequence   209
transmitter systems
– drugs acting on   307
– substances   41

*v*

ventral tegmental area (VTA)   43, 146, 171
– stimulation   151

*w*

whole-brain transcriptome analysis   216
Wisconsin card sorting test   29
working memory   30
– conceptualization   30
– deficiency   47
– – computational modeling   47
– – in functionalist language   29
– experiments   36
– function   29
– neurobiological basis of   36–38
– simulation   50